Ticks of Europe and North Africa

Agustín Estrada-Peña
Andrei Daniel Mihalca
Trevor N. Petney
Editors

Ticks of Europe and North Africa

A Guide to Species Identification

With maps by Cristian Domşa and illustrations by Jacob Gragera

Springer

Editors
Agustín Estrada-Peña
Faculty of Veterinary Medicine
University of Zaragoza
Zaragoza
Spain

Trevor N. Petney
State Museum of Natural History
Karlsruhe
Germany

Andrei Daniel Mihalca
Faculty of Veterinary Medicine
University of Agricultural Sciences and Veterinary Medicine Cluj-Napoca
Cluj-Napoca
Romania

ISBN 978-3-030-09704-2 ISBN 978-3-319-63760-0 (eBook)
https://doi.org/10.1007/978-3-319-63760-0

Softcover re-print of the Hardcover 1st edition 2017

Printed on acid-free paper

This Springer imprint is published by Springer Nature
The registered company is Springer International Publishing AG
The registered company address is: Gewerbestrasse 11, 6330 Cham, Switzerland

This book is based upon work from COST Action TD1303 European Network for Neglected Vectors and Vector-Borne Infections (EURNEGVEC), supported by COST (European Cooperation in Science and Technology).

EUROPEAN COOPERATION
IN SCIENCE & TECHNOLOGY

COST (European Cooperation in Science and Technology) is a funding agency for research and innovation networks. Our Actions help connect research initiatives across Europe and enable scientists to grow their ideas by sharing them with their peers. This boosts their research, career and innovation.

Funded by the Horizon 2020 Framework Programme of the European Union

http://www.cost.eu

Foreword

Agustín Estrada-Peña, Trevor Neil Petney, and Andrei Daniel Mihalca have undertaken a huge task, bringing together much of the scattered knowledge on the ticks of Europe and the Mediterranean Basin and examining it critically. Animal owners have realized the importance of ticks and the diseases that they transmit, both to agricultural and companion animals, for a long time. In addition, the general public, and more particularly public health authorities, are increasingly recognizing the importance of zoonotic tick-borne diseases to human health, particularly since the discovery of Lyme borreliosis and the realization of how common this infection in fact is.

These three scientists have not only edited and illustrated this book "Ticks of Europe and North Africa", published by Springer, but also figured prominently among the contributors to its text. Other contributors are also among the best specialists of their subject.

The result of this undertaking is admirable and the book will certainly become and remain for a long time an essential tool to all scientists working on ticks and tick-borne agents, particularly in the regions covered.

The editors and contributors also discuss areas in which information is still incomplete or missing and present contradictory opinions about classification and the many problems which remain to be solved. Some of the problems are tenacious and for the moment unsolvable, especially when type specimens have been lost or never existed: using the names under which gene sequences have been deposited in GenBank does not help, as these names are just as unreliable.

One of the vexing problems in the genus *Rhipicephalus* is the *R. sanguineus* complex; many authors distinguish *R. sanguineus* from *R. turanicus* on morphological characters, but gene sequences of some of the ticks identified morphologically as *R. turanicus* are identical with sequences in *R. sanguineus.* It even appears to be impossible to define *R. sanguineus* sensu stricto! Ticks of the genus *Hyalomma* are also difficult to identify and present great morphological variability with some specimens being impossible to identify on their morphology to the species level. Nymphs are particularly difficult, and the recommendation is to allow engorged specimens to molt to the adult stage and then try to identify them.

I would advise scientists in this part of the world, when in doubt, to contact specialists among the editors and contributors of the book for help in obtaining identifications that are as reliable as possible.

Two of the three families of ticks are known to occur in the area covered: The Argasidae, with 2 genera, *Argas* and *Ornithodoros*, and the Ixodidae with 5 genera, *Dermacentor*, *Haemaphysalis*, *Hyalomma*, *Ixodes*, and *Rhipicephalus*. Those occurring in Europe and northern Africa are all treated exhaustively.

There are general parts on each genus, keys for the determination of the species in the area covered by the book, as well as distribution maps. Each species is then dealt with in terms of its life cycle and hosts, its ecology and distribution, and its medical and veterinary relevance. The beautiful, newly prepared drawings of all adults and most immatures will be of great help in using the identification keys. All authors have agreed on these keys and they have been tested independently by other researchers in the field.

References are given for almost all statements, which should make this book a basic one for research on ticks and tick-borne agents and eliminate many of the errors that are usually repeated from publication to publication, without going to the sources of the data.

In conclusion, this book will prove to be essential for all researchers and laboratory personnel involved in identification of European and Northern African ticks and tick-borne infections.

Gerrit Uilenberg
Retired Professor of Tropical Veterinary Medicine
Utrecht, The Netherlands;
Retired Director Animal Health
CIRAD, France

Acknowledgements

The following people contributed with georeferenced data for tick species distribution (listed alphabetically according to the surname): Manuela Amorim, Manuel António, Fábia Azevedo, Niina Bang, Susana Baptista, Birgit U. Baumgarten, Silvia Bino, Jovan Bojkovski, Antra Bormane, María Casero, Cristina Cazan, Miguel Cerqueira, Jan Chirico, Dimosthenis Chochlakis, Alexandra Corduneanu, António Cotão, Gianluca D'Amico, Isabel L. de Carvalho, Daniele De Meneghi, Rita de Sousa, Georgiana Deak, Daan Dekeukeleire, Enkelejda Dikolli, Mihael Dinev, Niamh Dolan, Rajkovic Dragana, Justine Dutraive, Agustin Estrada-Peña, Cláudia Fernandes, Leonel Fernandes, Carlos Fonseca, Frederico Freitas, Nickals Fritzen, Helena Garcia, Adriana Geraldes, Diogo S. Gomes, Salomé Gonçalves, Estelle Gonzalez, Mari Haapavaara, Jari Hänninen, Nina Helajärvi, Petr Heneberg, Sara Heponiemi, Václav Hönig, Tapani Hopkins, Sandor Hornok, Lena Hulden, Larry Hulden, Petteri Ilmoinen, Mariana Ioniţă, Ivana Ivanovic, Thomas Jaenson, Aleksandar Jurišić, Perparim Kadriaj, Marita Kallinen, Kari Kaunisto, Matej Kautman, Kieran Kenny, Dorota Kiewra, Tero Klemola, Seppo Koponen, Anna Koskela, Päivi Koskinen, Emma Kosonen, František Krause, Aleksandra Krawczyk, Milos Krist, Anton Kristin, Pipsa Kuusisto, Stina Laine, Jaakko Laitonen, Laetitia Lempereur, Monique L'Hostis, Ivan Literák, V. Lizana, Isis V. Llopis, Carla Maia, Igor Majláth, Viktória Majláthová, Juha Mäki, Jussi Mäki, Alessandro Mannelli, Alexander Mathis, Barry J. McMahon, Pedro Melo, Helena Metiäinen, Andrei D. Mihalca, Caroline Millins, Alexandr Morozov, Fergal Morris, Sara Moutailler, Vicenzo Musella, Marion Nieder, Mikko Niemi, Minna Niemi, Okko Niemi, Marketa Novakova, Maria S. Núncio, Mónica Nunes, Nina Nupponen, Anna Obiegala, Jack O'Connor, Tanja Ojanperä, Manuel Oliveira, Jasmin Omeragic, Ika Österblad, Jose A. Oteo, Domenico Otranto, Evelyn Overzier, Ana M. Palomar Urbina, Elias Papadopoulos, Ricardo Parreira, Anna Pauna, Ritva Penttinen, André Pereira, Aleksandra Petrovic, Martin Pfeffer, Carlos Pinto, Helena Pinto, Manuela Pinto, Olivier Plantard, Sabine Poljak, Arantza Portillo, Aleksandar Potkonjak, Oli Pyysalo, Carla Quintas, Markus Rahikainen, Kalle Rainio, Laura Rinaldi, Veikko Rinne, Monica Robinson, Luca Rossi, Kai Rruohomäki, Cédric Samorinha, Attila D. Sándor, S. Santibéñez, Ana S. Santos, Maria J. Santos, Maria M. Santos-Silva, Karoliina Savolainen, Dino Scaravelli, Henrique Sebastião, Vesa Selänne, Cornelia Silaghi, Domingos Silva, Sílvia Silva, Cristina Soares, Martti Soikkeli, Sanna Soini, Jani Sormunen, Jouni Sorvari, Hein Sprong, Peter Stuart, Ivica Sucic, Snežana Tomanović, Laura Tomassone, Tomi Trilar, Friederike v. Loewenich, Varpu Vahtera, Ana Vasić, Zati Vatansever, Muriel Vayssier-Taussat, Sónia Veloso, Hélène Verheyden, Outi Vesakoski, Cristina Vide, Vracar Vuk, Gernot Walder, Lisa Weis, Nemanja Zdravković, Annetta Zintl. We wish to thank them all sincerely for their help without which this book would not have been complete!

Contents

Presentation

A. Estrada-Peña

Department of Animal Health, Faculty of Veterinary Medicine, Zaragoza, Spain

After years of work on the systematics of ticks, I realized that most of my workplace was occupied by many reprints about the morphology, illustrations, keys, and, in summary, the "tools" to make an identification of a tick. Since it is assumed I am an expert on the topic (but very different opinions exist about this), I have always considered the serious difficulties the topic may present for researchers initiating their work in this field. For example, we can consider experts on microbiology who want to approach some aspect of tick-transmitted pathogens, or specialists working on ecological associations (parasitism is one of these) without an adequate knowledge or the necessary background to approach a project. Furthermore, in the recent years, we have been subjected to an explosion of online material. All of this material has something in common: it is recent. I realized this after contemplating my pile of old reprints dating back from the 1970s and 1980s: if something is not in Internet, it simply does not exist. It is not necessary to state that many sequences already available in GenBank are the product of an unreliable identification of tick specimens. Therefore, they generate a very dangerous background noise that obscures our abilities to identify a tick. If we cannot identify a tick by morphological means, and the molecular data available are "dangerously biased", then every further result coming from that tick is simply unreliable, including reports on pathogens, hosts, or distribution. In my mind, the idea of preparing a book compiling and summarizing our knowledge on ticks took shape. However, I could not make all that effort by myself. Not only to write, but also to prepare illustrations, to compile the known distributions, to separate the facts from the artifacts, or to evaluate what is a good report from others which are unreliable. This should be a task carried out by several specialists, with at least some funds to support the basics of the work. I never abandoned the idea but postponed time after time.

Sometimes dreams come true. One day I had a telephone talk with Trevor Petney, and other than a long discussion on our research projects, he suggested that we collaborate on preparing exactly what I wanted along the lines of a kind of "Atlas" on the ticks in the Western Palaearctic: compiling basic texts on all of the species present, as well as keys and illustrations in only one book. Of course, some other texts already existed, namely the one headed by Alan Walker, in which I also participated. But we anticipated including every stage of every species reported in the target territory. A monumental work! Let me repeat: sometimes dreams come true. A few days later I met Andrei Mihalca for the first time. He was leading an action funded by the European Union under the umbrella of the COST (European Cooperation in Science and Technology) Office. After the kick-off meeting of the COST Action TD1303 in Brussels, we had another meeting in Cluj-Napoca (Romania) and immediately it was obvious that both Andrei Mihalca and I were completely "tuned-on" about how to carry out this immense project. Therefore, around a good glass of Transylvanian wine in the city of Cluj-Napoca, Andrei and I realized all these caveats. And the idea quickly developed involving the three coeditors who originally planned this work. The keys were developed after an intensive 2-year work and validated using a blind test (Estrada-Peña et al. 2017a, b). In parallel, for each species and developmental stage, professional technical illustrations were prepared to accompany the keys. The result of all this work is the most comprehensive set of identification keys published so far for Western Palearctic ticks.

Thanks to the policy of "putting researchers in contact" carried out by the COST Action, we identified the potential contributors, who are obviously the coauthors of this book. It is not necessary to explain the gigantic task of compilation of records of every species of tick, the preparation of adequate keys, and of the illustrations, which should be academic, clear, and informative. All of this could not be possible without the voluntary contributions of many researchers (listed in the acknowledgments section) and the labor of the coauthors. A special person must be introduced here: Gerrit Uilenberg, although already retired he had never left his real work. Gerrit "Oele" Uilenberg, who has been a fundamental pillar in tick studies for decades, agreed to prepare the foreword for this book. As expected, he immediately became involved in the revision of potentially unreliable statements in the book, and he finished making an almost complete revision of the text. My personal special thanks, for many years of friendship, correspondence, support, and collaborations, are my tribute to his contributions to this book.

After a brief section on collecting ticks, we discuss first the Family Argasidae and then the Family Ixodidae. For each family, there is a key to the genera for all life history stages. For each genus, there is a relatively long revision of the basic references (commonly not available in Internet), which should be studied before considering the identity of each species, together with an explicit mention of the keys found in each reference, and that were drawn from the publication by Estrada-Peña et al. (2017a). All of these relatively old references are adequately included in the main text for each genus. After the discussion of the genus, we present newly developed keys for all life history stages of the species therein. The book was planned with a "morphological design" in mind, and therefore the keys and illustrations are parts of the basic "recipe" for identifying a tick. After the keys, each species is individually treated: we provide basic data about the ecology, hosts, life cycle, distribution, and potential involvement in the transmission of agents with medical and/or veterinary significance. At the end of each species chapter, the illustrations for each stage are included together with maps that show the recorded species distribution. The maps contained in this book are based on a GIS database, compiled from the information sent by the contributors (the final database contains more than 65,000 records). The species distribution maps were compiled using a 10 × 10 km grid (the ETRS89 LAEA European official grid, expanded to cover North Africa). For some countries, we found records of certain tick species without georeferenced data, therefore we also produced country distribution maps so this information could be included.

The keys are far from complete, since some species lack adequate features for the separation of some stages. This is particularly the case of immatures of the genus *Hyalomma*, and, to some extent, for genus *Ixodes*. Some species may be relatively easy to separate, while others share the most prominent morphological features. We therefore preferred not to include those species for which we were unable to find reliable diagnostic characteristics. We do provided adequate coverage of previous references which may help the readers. The same applies to the illustrations. It is probably paradoxical that we still do not know many of the most basic features of some species of ticks. Probably the best example are the species of *Ixodes* that are parasites of pelagic birds. Information about the morphology of these species is scarce, specimens are unavailable, and therefore adequate material to prepare adequate illustrations is simply missing. These few species lack illustrations.

References

Estrada-Peña A, Pfäffle M, Baneth G, Kleinerman G, Petney TN (2017a) Ixodoidea of the Western Palaearctic: a review of available literature for identification of species. Ticks Tick Borne Dis 8:512–525

Estrada-Peña A, D'Amico G, Palomar AM, Dupraz M, Fonville M, Heylen D, Habela MA, Hornok S, Lempereur L, Madder M, Núncio MS, Otranto D, Pfaffle M, Plantard O, Santos-Silva MM, Sprong H, Vatansever Z, Vial L, Mihalca AD (2017b) The results of a comparative test of ixodid tick identification by a network of European researchers. Ticks Tick Borne Dis 8:540–546

Contributors

G. Baneth Koret School of Veterinary Medicine, Hebrew University, Rehovot, Israel

G. D'Amico Department of Parasitology and Parasitic Diseases, University of Agricultural Sciences and Veterinary Medicine Cluj-Napoca, Cluj-Napoca, Romania

F. Dantas-Torres Centro de Pesquisas Aggeu Magalhães, Fundação Oswaldo Cruz, Recife, Brazil

C. Domşa Department of Parasitology and Parasitic Diseases, University of Agricultural Sciences and Veterinary Medicine, Cluj-Napoca, Romania

M. O. Dumitrache Department of Parasitology and Parasitic Diseases, University of Agricultural Sciences and Veterinary Medicine Cluj-Napoca, Cluj-Napoca, Romania

A. Estrada-Peña Department of Animal Health, Faculty of Veterinary Medicine, Zaragoza, Spain

S. Hornok Department of Parasitology and Zoology, University of Veterinary Medicine, Budapest, Hungary

T. G. T. Jaenson Medical Entomology Unit, Department of Organismal Biology, Evolutionary Biology Centre, Uppsala University, Uppsala, Sweden

G. Kleinerman Koret School of Veterinary Medicine, Hebrew University, Rehovot, Israel

M. Madder DG06/07 Ruisseau Creole, La Mivoie, Black River, Mauritius

A. D. Mihalca Department of Parasitology and Parasitic Diseases, University of Agricultural Sciences and Veterinary Medicine Cluj-Napoca, Cluj-Napoca, Romania

D. Otranto Dipartimento di Medicina Veterinaria, Università degli Studi di Bari, Bari, Italy

T. N. Petney State Museum of Natural History, Karlsruhe, Germany

M. P. Pfäffle Department of Ecology and Parasitology, Institute of Zoology, Karlsruhe Institute of Technology, Karlsruhe, Germany

A. D. Sándor Department of Parasitology and Parasitic Diseases, University of Agricultural Sciences and Veterinary Medicine, Cluj-Napoca, Romania

M. M. Santos-Silva Centro de Estudos de Vectores e Doenças Infecciosas Dr. Francisco Cambournac (CEVDI), Instituto Nacional de Saúde Doutor Ricardo Jorge, Águas de Moura, Portugal

H. Sprong Centre for Zoonoses & Environmental Microbiology, Centre for Infectious Disease Control, National Institute for Public Health and the Environment, Bilthoven, The Netherlands

Z. Vatansever Department of Parasitology, Faculty of Veterinary Medicine, Kafkas University, Kars, Turkey

Definitions and Abbreviations

COST	European Cooperation in Science and Technology
SME	Small and medium enterprise

Introduction

A. D. Mihalca, A. Estrada-Peña, and T. N. Petney

A simple search query of the generalist biomedical database PubMed shows that about 5000 papers were published on ticks and tick-borne diseases between 1996 and 2006; this number has almost doubled in the last 10 years. The increasing interest in the research of ticks and tick-borne diseases has various causes, the main ones being climate and land-use change with the consequent change in vector distribution, the recognition of new medically important pathogens and the increasing economic, social and public health impact of vector-borne diseases. The medical importance of ticks resides mostly in their capacity to transmit a wide range of microbial (viruses, bacteria) and parasitic (protozoans, helminths) agents to their vertebrate hosts, including humans.

Unlike most blood-sucking vectors (i.e. insects such as dipterans, fleas, lice and bugs, etc.), ticks have a rather unique life history involving multiple hosts and resulting in complex eco-epidemiological patterns of pathogen transmission. In regions with a temperate climate, ticks are estimated to be responsible for more human and animal infections than any other hematophagous vector, transmitting important pathogens like *Borrelia burgdorferi* sensu lato (s.l.), tick-borne encephalitis virus, spotted fever rickettsiae, piroplasms among others, some of them with a high clinical or economic importance. The recent dynamic environmental and landscape changes, coupled with an ever-increasing human population, animal movement, tourism and social habits, have induced changes in the epidemiological patterns of tick-borne diseases, including tick distribution. Moreover, due to development of novel molecular tools, new pathogen species, strains or genetic variants are constantly being recognized, even in relatively well-surveyed territories like Europe.

Ticks (Ixodida) are blood-feeding arthropods, with around 900 described species in 19 genera in three families (Barker and Murrell 2008), having a worldwide distribution and infesting virtually all terrestrial vertebrates. The Western Palearctic includes all of Europe, North Africa, northern and central parts of the Arabian Peninsula, and part of temperate Asia, roughly to the Ural Mountains, with a mainly temperate climate. According to Guglielmone et al. (2014), around 13.2% of the hard tick species (Ixodidae) of the world (93 species) are known from the Palaearctic region. The most well-represented genus is *Ixodes* (44 species), followed by *Haemaphysalis* (19 species), *Dermacentor* (13 species), *Rhipicephalus* and *Hyalomma* (each with 7 species) (Table 1). Of these, in the target territory covered by the current book, we have included only those occurring in Europe and North Africa, plus some Afrotropical species which have occasionally been reported, mainly from northern Africa (Table 1).

In the Western Palaearctic, the vast majority of the studies focus on the ecology and epidemiology of ticks and tick-borne diseases. However, they rely on a crucial step: the correct identification of the tick species. Although the diversity of ticks in Europe and North Africa is not as big as for other vectors (i.e. mosquitoes, biting midges, etc.), their identification is often not a routine procedure. The causes are multiple. Many specimens are collected from hosts and are hence at some stage of engorgement and with distorted morphological features. In the case of immature stages, the features of engorged specimens are even more distorted. Moreover, the uncertain taxonomic status of several tick groups, with poor descriptions or ambiguous morphological characters, creates real difficulties in species differentiation. The validity of some poorly known species (i.e. *Ixodes*

A. D. Mihalca
Department of Parasitology and Parasitic Diseases, University of Agricultural Sciences and Veterinary Medicine Cluj-Napoca, Cluj-Napoca, Romania
e-mail: amihalca@usamvcluj.ro

A. Estrada-Peña (✉)
Department of Animal Health, Faculty of Veterinary Medicine, Zaragoza, Spain
e-mail: aestrada@unizar.es

T. N. Petney
State Museum of Natural History, Erbprinzenstr. 13, 76133 Karlsruhe, Germany
e-mail: trevor.petney@kit.edu

A. Estrada-Peña et al. (eds.), *Ticks of Europe and North Africa*,
https://doi.org/10.1007/978-3-319-63760-0_1

Table 1 Diversity of ticks reported in the Palaearctic region and in the target territory of the present book

Genus	Species reported in the Palaearctic region[a]	Species in Europe and North Africa[b]
Ixodes	44	25
Dermacentor	14	2
Haemaphysalis	19	8
Hyalomma	7	11
Rhipicephalus	7	8
Argas	Not reliably estimated	6
Ornithodoros	Not reliably estimated	7
Total	91 (for Ixodidae)	67

[a]According to Guglielmone et al. (2014)
[b]Included in the present book. Some Afrotropical species have been included, as they are sporadically reported from Northern African countries

festai, *I. eldaricus* and *I. kaiseri*) or the synonymy of others (i.e. *I. redikorzevi* and *I. acuminatus*) is yet to be confirmed, even in well-explored geographical territories like Europe. Additionally, for certain species of ticks, some stages remain undescribed (i.e. the males of *I. berlesei*, the larvae and nymphs of several others). Some other species (i.e. *Rhipicephalus sanguineus* sensu lato) form complexes of cryptic or sibling species, while some others show a large range of intraspecific morphologic variability (i.e. *Hyalomma*). To complicate things even more, closely related species (which are anyway difficult to differentiate) may naturally hybridize, yielding offspring with confusing phenotypes.

So far, several morphological keys have been published for the identification of ticks, and certainly, most of them have had and still play an essential role in the advancement of tick research. They were compiled by prominent authorities in tick taxonomy, some of whom were the direct or indirect academic mentors of the authors of this book. All of the available literature for the identification of ticks in the Western Palaearctic has recently been revised by Estrada-Peña et al. (2017a). However, the world is changing, as are the times. Some of the past keys include only species of medical and/or veterinary importance, omitting many others, mainly those parasitic primarily on wildlife. Some other keys are limited only to adult ticks or to certain geographic regions (i.e. there are many keys at country level). Last but not least, most of the keys are outdated and do not incorporate the most recent tick taxonomic concepts. No key so far includes all of the tick species and stages reported in Europe and North Africa (as part of the Mediterranean Basin). Moreover, many lack reliable illustrations.

A recent blind-test (Estrada-Peña et al. 2017b) evaluated the comparative performance of 14 teams of researchers involved in the study of ticks and tick-borne pathogens in Europe on the morphological identification of 11 species of ticks reported as established in Europe and North Africa. The general aim was not only to identify the challenges when dealing with the identification of ticks, but also the causes for misidentifications, and the best procedures for consistent results. The results say a great deal: species in the different genera had variable misidentification rates of 7% (*Dermacentor*), 14% (*Ixodes*), 19% (*Haemaphysalis*), 36% (*Hyalomma*) and 54% (*Rhipicephalus*), showing the need for comprehensive and up-to-date keys as well as accurate and detailed illustrations of all species.

As most pathogens are specifically associated with certain tick species or groups of species, the correct taxonomic identification of these arthropod vectors is crucial for understanding the epidemiology of tick-borne diseases. So should we expect help from molecular genetics? Although the recent development of molecular taxonomy is becoming more and more common for tick identification, certain limitations remain. Although these technologies are useful, they will always rely on reference specimens for which morphological identification needs to be correctly conducted. 'Garbage sequences' obtained from unreliably identified specimens that accumulate in databanks are commonly a source of molecular misidentification and inaccurate phylogenetics. Moreover, as typically large sample sets are needed to obtain statistical significance especially in field studies, molecular approaches in these cases are costly and time-consuming.

The present book includes identification keys for 67 species of hard and soft ticks from Europe and North Africa together with a concise and updated overview of each species, including summarized data on the life cycle, ecology and medical importance. An essential component are the illustrations, prepared by Jacob Gragera and based on representative specimens which were either provided by the co-authors and completed with available illustrations. For each ticks species, we have prepared an updated grid-based distribution, compiled from the available literature references and completed by the personal records of participants. We thus produced detailed maps at the grid and country level.

We consider this book to be an important aid for researchers around Europe for the correct morphological identification of ticks, providing a basis for determining the DNA sequence-specific identity of specimens used for molecular taxonomical purposes. It will also increase the probability of detecting species invasive to new geographic territories.

References

Barker SC, Murrell A (2008) Systematics and evolution of ticks with a list of valid genus and species names. In: Bowman AS, Nuttall P (eds) Ticks: biology, disease and control. Cambridge University Press, Cambridge, pp 1–39

Estrada-Peña A, Pfäffle M, Baneth G, Kleinerman G, Petney TN (2017a) Ixodoidea of the Western Palaearctic: a review of available literature for identification of species. Ticks Tick Borne Dis 8:512–525

Estrada-Peña A, D'Amico G, Palomar AM, Dupraz M, Fonville M, Heylen D, Habela MA, Hornok S, Lempereur L, Madder M, Núncio MS, Otranto D, Pfaffle M, Plantard O, Santos-Silva MM, Sprong H, Vatansever Z, Vial L, Mihalca AD (2017b) The results of a comparative test of ixodid tick identification by a network of European researchers. Ticks Tick Borne Dis 8:540–546

Guglielmone AA, Robbins RG, Apanaskevich DA, Petney TN, Estrada-Peña A, Horak IG (2014) The hard ticks of the world (Acari: Ixodida: Ixodidae). Springer, Dordrecht, p 738

How to Collect Ticks and Interpret These Collections

T. N. Petney, M. P. Pfäffle, H. Sprong, A. D. Mihalca, and A. Estrada-Peña

Introduction

To date, almost 900 species of ticks have been described. The three existing families consist of the Argasidae (soft ticks; ca. 191 species), the Ixodidae (hard ticks; ca. 702 species) and the Nuttalliellidae with only one species (*Nuttalliella namaqua*) (Guglielmone et al. 2010; Estrada-Peña et al. 2014). They transmit more pathogen species to humans, livestock, companion animals and wildlife than any other blood-sucking arthropod species (Durden 2006). However, research focuses predominantly on the most abundant tick species, such as *Ixodes ricinus* in Europe, *Ixodes scapularis* in the USA or *Rhipicephalus sanguineus* s.l. worldwide, and their role in the transmission of diseases. For most other less abundant tick species, information on the biology, ecology and vector capability is rare, as described in the present book. However, tick species that are not a direct threat to humans or animals can still be important for the maintenance of enzootic cycles for different pathogens (Skuballa et al. 2010). Therefore, to understand the eco-epidemiology of various diseases, it is essential to study the interactions between ticks, their hosts, the pathogens and their biotic and abiotic environment. In order to achieve such an understanding, collecting ticks is necessary. The type of collection, as well as the quality of results, depends on the host-seeking behaviour, the life history stages and the natural habitat of the ticks.

Collection

Most tick species spend up to 97% of their lives off-host (Needham and Teel 1991). Depending on whether the species, or a certain life stage, is exophilic (on the vegetation or actively host seeking) or endophilic (in or close to burrows or nests), tick collecting methods vary. Therefore, due to incorrect sampling (in the wrong season, in the wrong habitat, using wrong techniques, etc.), some tick species are assumed to be absent from certain areas, leading to 'no presence' data with impact on distribution models. Estrada-Peña et al. (2013) provide a methodological template for the study of host-tick-pathogen cycles. They emphasise the need for suitable times between collections in the specific target areas in order to provide quantitative seasonal patterns of tick abundance for all life history stages.

When collecting, take personal measures to prevent tick bites. Light coloured clothing is useful as the ticks can be seen relatively easily. After collecting, it is necessary to thoroughly check yourself for attached ticks. These are best removed with a fine forceps applied as close as possible to the attachment point. Simply pull the tick off. Seek medical advice when disease symptoms develop after a tick bite.

Tick Dragging and Flagging

A standard method to collect ticks from the vegetation is flagging or dragging. Here a cloth or blanket, preferably

T. N. Petney (✉)
State Museum of Natural History, Erbprinzenstr. 13, 76133 Karlsruhe, Germany
e-mail: trevor.petney@kit.edu

M. P. Pfäffle
Department of Ecology and Parasitology, Institute of Zoology, Karlsruhe Institute of Technology, Karlsruhe, Germany
e-mail: miripfaeffle@web.de

H. Sprong
Centre for Zoonoses & Environmental Microbiology, Centre for Infectious Disease Control, National Institute for Public Health and the Environment, Bilthoven, The Netherlands
e-mail: hein.sprong@rivm.nl

A. D. Mihalca
Department of Parasitology and Parasitic Diseases, University of Agricultural Sciences and Veterinary Medicine Cluj-Napoca, Cluj-Napoca, Romania
e-mail: amihalca@usamvcluj.ro

A. Estrada-Peña
Department of Animal Health, Faculty of Veterinary Medicine, Zaragoza, Spain
e-mail: aestrada@unizar.es

A. Estrada-Peña et al. (eds.), *Ticks of Europe and North Africa*,
https://doi.org/10.1007/978-3-319-63760-0_2

white, is dragged over the vegetation for a certain amount of time or over a certain area. The use of an objective reference frame, such as area, is preferable to time, since some people may devote more time to cover the same surface. The cloth simulates a passing host, and the ticks transfer from the vegetation to the cloth. For dragging, a cloth of about 1 × 1 m in size is usually used sometimes with weights attached to the trailing edge to hold it down, while the leading edge is fixed to a wooden stick. The cloth is then dragged after the operator. When flagging, the cloth is also normally about 1 × 1 m in size and is swept over the vegetation next to the operator while walking along a certain distance. These methods are very simple, but can only cover a certain section of a study site, and they also only collect active host-seeking ticks. In addition, these methods are strongly influenced by vegetation type (the cloth becomes stuck on thorny vegetation), the temperature, the weather (the method is not useful on wet vegetation or in rainy conditions) and the operator (speed of walking, method of dragging/flagging). The strip-flag method, where the flag is divided into several strips, seems to be especially useful for collecting larvae on the vegetation of varying height (Gray and Lohan 1982). A magnifying glass is a useful tool with which to detect the larval stages.

Sampling can either be carried out over a specific time period, or, more commonly, over a certain distance. A distance of 100 m is commonly used, with the drag or flag being checked every 10 m to reduce the chances that ticks that had climbed onto the cloth would be brushed off again before collection (Estrada-Peña et al. 2013; Pfäffle et al. 2015). The 10 × 10 m also provides replicate data for potential statistical analysis. Here, two points should be taken into consideration: (1) Ticks are not randomly distributed even in suitable habitats but tend to be aggregated, especially the larvae which hatch from individual egg batches; (2) the distance dragged per replicate will depend on the tick species and the vegetation type so that some trial and error may be necessary to find the optimal sampling strategy.

Search by Eye and Walking Method

The walking and the search by eye method is one of the easiest methods to collect host-seeking, exophilic ticks from the vegetation. The sampler either walks through the vegetation and collects the ticks from his clothes or searches the vegetation, mostly the tips of grass stems, etc., for ticks and collects them by hand. Depending on the tick species, adults and nymphs can commonly be found using this method. In Central Europe, *I. ricinus* can be collected this way, as can the adults of *D. marginatus* and *D. reticulatus*. Walking seems to give a more precise insight on the infestation risk for humans in certain areas, larvae that stay in the leaf litter or in nest and burrows of hosts can, however, not be collected in this way (Ginsberg and Ewing 1989). Terassini et al. (2010) compared dragging with a visual collection of ticks on animal trails in the Amazonian rainforest in Brazil. It was easier to find adults of various *Amblyomma* species with the visual method than by dragging. Immature stages, however, could only be collected using the dragging method. The visual method is particularly suited for tick species that dwell in houses or shelters of domestic animals (e.g. *Argas reflexus, Ornithodoros moubata* and *R. sanguineus* s.l.).

CO_2 Traps

This method can be used to attract ticks that are in search of a host, with dry ice being the most convenient source for CO_2. There are CO_2 traps designed for both the Ixodidae and Argasidae (Miles 1968; Wilson et al. 1972). When dry ice is not available, a combination of baking soda and diluted lactic acid can be used to produce CO_2. CO_2 traps can also be combined with pheromones. There are a variety of different trap types. For example, the dry ice can be placed in a perforated container in the middle of a 1 m^2 white cloth. Ticks will crawl onto the cloth and can be collected there. The collection can be made after a specific time (e.g. after 30 min) providing a quantitative estimate of tick density. The counted numbers can be expressed as ticks per time/dry ice sample. Most active ticks in an area should be caught by this method within 3 h. It is also possible to put a sticky barrier around the dry ice in order to capture the ticks. This was done by Wilson et al. (1972), who studied the response of CO_2 traps to different life history stages of the lone star tick, *Amblyomma americanum*. The average number of ticks caught with traps was significantly higher and was also less variable than when using the flagging method. The effective sampling area of one trap was approximately 25 m^2. With a certain number of these traps, it is possible to detect in which part of a habitat the tick activity is the highest. CO_2 trapping can also be used for endophilic ticks in the burrows of their hosts (Miles 1968). The advantage of CO_2 trapping is that it can be used in different habitat types (Bryson et al. 2000), including those in which other collecting methods are not suitable due to the type of vegetation present. Ginsberg and Ewing (1989) compared different methods of collecting the American ticks *A. americanum* and *Ixodes scapularis* (as

Ixodes dammini), using flagging, walking, CO_2 trapping and collecting from mammal hosts. CO_2 traps worked better on *A. americanum* than on *I. scapularis*, even in areas with higher *I. scapularis* density. This can be explained by the different mobility of the two species: *A. variegatum* is a hunting tick moving freely in the environment, *I. scapularis* a questing tick found stationary on the vegetation (see also Falco and Fish 1991; Schulze et al. 2011). Petry et al. (2010) found that significantly more nymphs could be captured by CO_2 than by dragging, whereas more larvae could be collected by dragging, probably due to the potential distances that could be moved. Importantly, these two methods showed different success rates depending on whether forest or field was sampled.

CO_2 sampling can also be used successfully for certain species of argasid tick (Garcia 1962; Caiado et al. 1990).

Nests and Burrows

It is more complicated to collect endophilic ticks from the nests and burrows of their hosts than collecting exophilic species. The nest material can be collected and placed on a mesh over a Berlese-Tullgren funnel. Most of these funnels have a 26–60 W light bulb placed over the nest material, which slowly dries it out (Clayton and Walther 1997). This 'forces' the living ticks, and also other arthropods, to move downwards to the bottom of the funnel where they drop into a jar with alcohol. This way Mumcuoglu et al. (2005) were able to collect over 10,000 specimens of *Argas arboreus* from 210 nests of cattle egrets (*Bubulcus ibis*). However, in some countries, many avian and mammalian nests are protected by law, which makes it more complicated to collect ticks this way (e.g. Pfäffle et al. 2011). While collecting ticks from avian nests is more practicable, extracting ticks from burrows can be difficult due to difficulty of access, and the ticks may escape into cracks and holes, making the recovery rate low. Vacuuming seems to be better than burrow excavation or other survey methods that are sometimes not even able to determine the presence of ticks (Butler and Gibbs 1984). Apart from CO_2 trapping, Niebuhr et al. (2013) used a debris-filtering method with screens of varying gauges to collect unfed *Otobius megnini* from both animal shelters and more natural habitats.

Ticks on Hosts

Another way to collect both endophilic and exophilic ticks is the collection of specimens directly from their host. This can raise certain issues and sources of sampling error, such as the host type and skin cover (fur, feather and reptile skin), the time available for inspection (sedated, alert or dead animals), tick size (larvae, in particular, may be difficult to see), the infestation rate and the experience of the investigator. However, this method is especially practicable with small mammals, which can be easily caught in live traps (Pfäffle et al. 2015). Engorged ticks are easy to find on the animals and with some practice flat specimens, especially larvae, can be found and collected. The fur of the host should be carefully combed and ticks should be collected with fine forceps. On small mammals, most ticks can be found on the head, the tail and between the toes. Falco and Fish (1992) compared small mammal trapping, dragging and CO_2-traps as collecting methods for *I. scapularis* (as *I. dammini*). Especially early in their activity period, the larvae of *I. scapularis* could be better collected from hosts than from the vegetation. This is probably also the case for *I. ricinus*. Additionally, when searching for a certain tick species, it is necessary to be aware of the phenology of the varying life stages. For example, the immatures of *D. reticulatus* can only be found in the summer months on small mammals, while they are absent throughout the rest of the year (Pfäffle et al. 2015). Sampling in the wrong season will give the collector a false idea about the absence or prevalence of a tick in an area. Another possibility, and a chance to recover all ticks feeding on a host, is to keep the animal caged for few days and let the ticks drop into water trays (see e.g. Pfäffle et al. 2011). This is, however, not possible for animals of all sizes and also raises some logistical issues (location of the field site, space in the laboratory), as well as negative influence on the local host population by accidentally killing animals, by removing mothers from their young or disrupting territories. Additionally, the number of males or immatures of one- or two-host ticks might be underestimated with this technique.

Collecting ticks from larger animals is more difficult, even when the animals are sedated, leading more likely to underestimations of tick burden. Visual inspection of stock and game animals, both those held in cages and those recently killed, can provide a wealth of ticks. Here, certain species may have preferred attachment sites to their hosts. Various additional methods have been used in an attempt to obtain accurate estimates of tick burden; however, these are usually time-consuming and potentially expensive. For dead hosts, scrubbing the hide with a wire brush provided a better estimate of tick burden than collecting by eye (MacIvor et al. 1987), although this was inferior to alkali digestion (Van Dyk and McKenzie 1992).

It is important to remember that tick life history stages are likely to be aggregated within their host population, with a few hosts harbouring most ticks. This should be tested (for example, using the variance to mean ratio, Petney et al.

1990) and is critical for the choice of the correct statistical analysis.

Road kills can be an important source of ticks, particularly from protected species which are not available for examination by hunting (Skerratt 1998; Oakwood and Spratt 2000), as well as from game animals (Nelder and Reeves 2005). However, ticks frequently detach from and leave dead hosts (Degenhardt and Degenhardt 1965) so that quantitative estimates of tick density are related to the time post-mortem of the host.

Removal to the Laboratory and Preservation

As high but not extreme humidity is essential for tick survival, it is useful to add a few blades of green grass to and collecting vial if the ticks are to be transferred alive to the laboratory. Too much moisture, however, may cause fungal growth and the destruction of the tick.

The correct preservation of ticks after collecting is essential for their later identification, as well as pathogen analysis. Moreover, when collecting ticks from the same batch, their division into several preservation media has to be done in a way that the collector is sure that all specimens belong to the same species. Tick specimens used for DNA work have been preserved in various ways including absolute ethanol, storage at –20 °C, ultracold freezing, TE buffer at −70 °C, 70% ethanol/5% glycerol and 70% ethanol (for references see Mtambo et al. 2006). Corriveau et al. (2010) compared the effect of eight storage methods on the DNA preservation of *Balaustium* sp. (Acariformes: Erythraeidae), including 70–75% ETOH at 18 °C which is the standard storage regime for most fluid-preserved specimens in most acarology collections, 70–75% ETOH in the freezer at −20 °C, 95–100% ETOH in the freezer, 95–100% ETOH at room temperature, dry storage in a freezer, storage in Koenikes's fluid at room temperature (usually used for water mites), ETOH after clearing with lactic acid and storage in lysis buffer (Tris-HCL, EDTA, NaCl, SDS). All specimens were stored for 7–8 months. In general, all ETOH-based treatments showed very good DNA recovery abilities; dry storage was less successful than storage in alcohol. Koenike's fluid, lactic acid and the lysis buffer showed very poor performances. Storage in a mixture of alcohol and glycerine as suggested by Boardman (1944) is not recommended as the ticks become sticky and difficult to study.

Freezing

For arthropods, one of the most effective methods of preserving DNA is ultracold freezing at −80 °C (Cruickshank 2002; Ardia 2005). RNA viruses can also be detected if the samples are preserved in RNA later at −20 °C, for example, tick-borne encephalitis virus (Melik et al. 2007; Carpi et al. 2009).

Identification

Cleaning: Flat ticks found on vegetation or the ground are usually clean and require little or no extra work before placing them under the microscope. This is rarely the case with ticks collected from host animals, the morphological features of which are often obscured by dirt, host skin secretions or faecal matter. The African *Rhipicephalus evertsi evertsi* and Afrotropical/Palearctic *Hyalomma rufipes*, for example, prefer to attach around the host's perianal area (MacLeod 1975), while *Cosmiomma hippopotamensis* will even crawl a short distance into the anus of its rhinoceros hosts (Apanaskevich et al. 2013).

The hypostomes of ticks removed from their hosts often carry a sleeve of attachment cement that should be carefully removed with a fine, jeweller's forceps so as not to damage the denticles.

Several methods have been used to clean ticks. The simplest is to use a fine brush dipped in alcohol in an attempt to remove the dirt. This may be more or less successful depending on how tenaciously the dirt particles adhere to the tick. Various methods based on ultrasound have also been developed (Keirans et al. 1976; Corwin et al. 1979; Rupp 1990; Cruz and Estrada-Peña 1992; Dixon et al. 2000). This book provides the most up-to-date keys to all of the species found in the Western Palearctic. Additional information can be found in Estrada-Peña et al. (2017).

Interpretation

Be careful when using GenBank or other molecular reference systems: the quality of the molecular data is only as good as the initial morphological identification of the tick.

With the advent of molecular identification methods, it has become evident that certain tick 'species' are in fact groups of morphologically similar cryptic species (Beati et al. 2013 *Amblyomma cajennense* spp. group). However, genetic data are not always adequate for defining separate taxa as shown by the comprehensive morphological, ecological and breeding study of *Amblyomma parvum* in South America (Nava et al. 2016).

In the past, the association of immature stages with specific adults was best carried out by rearing a series, i.e. an engorged female was allowed to lay eggs that, once hatched to larvae, were then fed through the different immature stages to new adults. Thus, samples of larvae and nymphs could definitely be assigned to specific adults. Although this is still a recommended method as it provides comparative material at all life history stages, it is also time-consuming and requires suitable animal facilities for feeding the ticks. Molecular methods now make the identification of immatures possible without the need for rearing them (Andrews et al. 1992).

References

Apanaskevich DA, Walker JB, Heyne H, Bezuidenhout JD, Horak IG (2013) First description of the immature stages and redescription of the adults of *Cosmiomma hippopotamensis* (Acari: Ixodidae) with notes on its bionomics. J Med Entomol 50:709–722

Ardia DR (2005) Cross-fostering reveals an effect of spleen size and nest temperatures on immune responses in nestling European starlings. Oecologia 327–334. https://dx.doi.org/10.1007/s00442-005-0120-6

Beati L, Nava S, Burkman EJ, Barros-Battesti DM, Labruna MB, Guglielmone AA, Cáceres AG, Guzmán-Cornejo CM, León R, Durden LA, Faccini JL (2013) *Amblyomma cajennense* (Fabricius, 1787) (Acari: Ixodidae), the Cayenne tick: phylogeography and evidence for allopatric speciation. BMC Evol Biol 13:267

Boardman ET (1944) Methods for collecting ticks for study and delineation. J Parasitol 30:57–59

Bryson NR, Horak IG, Venter EH, Yunker CE (2000) Collection of free-living nymphs and adults of *Amblyomma hebraeum* (Acari: Ixodidae) with pheromone/carbon dioxide traps at 5 different ecological sites in heartwater endemic regions of South Africa. Exp Appl Acarol 24:971–982. https://dx.doi.org/10.1023/A:1010639113793

Butler JF, Gibbs EPJ (1984) Distribution of potential soft tick vectors of African swine fever in the Caribbean region (Acari: Argasidae). Prev Vet Med 2:63–70. https://dx.doi.org/10.1016/0167-5877(84)90049-7

Caiado JM, Boinas FS, Melo MA, Louza AC (1990) The use of carbon dioxide insect traps for the collection of *Ornithodoros erraticus* on African swine fever-infected farms. Prev Vet Med 8:55–59

Carpi G, Bertolotti L, Rosati S, Rizzoli A (2009) Prevalence and genetic variability of tick-borne encephalitis virus in host-seeking *Ixodes ricinus* in northern Italy. J Gen Virol 90:2877–2883

Clayton DH, Walther BA (1997) Appendix C: Collection and quantification of arthropod parasites of birds. In: Clayton DH, Moore J (eds) Host-parasite evolution: general principles and avian models. Oxford University Press, Oxford, pp 419–440

Corriveau M, Uppstrom K, Klompen H (2010) Effect of eight storage modes on DNA preservation. In: Trends in Acarology, Proceedings of the 12th International Congress. pp 553–556

Corwin D, Clifford CM, Keirans JE (1979) An improved method for cleaning and preparing ticks for examination with the scanning electron microscope. J Med Entomol 16:352–353

Cruickshank RH (2002) Molecular markers for the phylogenetics of mites and ticks. Syst Appl Acarol 7:3–14

Degenhardt WG, Degenhardt PB (1965) The host-parasite relationship between *Elaphe subocularis* (Reptilia: Colubridae) and *Aponomma elaphensis* (Acarina: Ixodidae). Southwest Nat 1:167–178

De La Cruz J, Estrada-Peña A (1992) A simple, new improved method for preparing ticks for examination by scanning electron microscopy. Acarologia 33:321–323

Dixon B, Petney TN, Andrews RH (2000) A simplified method of cleaning ixodid ticks for microscopy. J Microsc 197:317–319

Durden LA (2006) Taxonomy, host associations, life cycles and vectorial importance of ticks parasitizing small mammals. In: Morand S, Krasnov BR, Poulin R (eds) Micromammals and macroparasites. From evolutionary ecology to management. Springer Japan, Tokyo, pp 91–102. https://dx.doi.org/10.1007/978-4-431-36025-4

Estrada-Peña A, Gray JS, Kahl O, Lane RS, Nijhoff AM (2013) Research on the ecology of ticks and tick-borne pathogens—methodological principles and caveats. Front Cell Infect Microbiol 3:29

Estrada-Peña A, Nava S, Petney T, (2014) Description of all the stages of *Ixodes inopinatus* n. sp. (Acari: Ixodidae). Ticks Tick Borne Dis 5:734–743. https://dx.doi.org/10.1016/j.ttbdis.2014.05.003

Estrada-Peña A, Pfäffle M, Baneth G, Kleinerman G, Petney TN (2017) Ixodoidea of the Western Palearctic: a review of available literature for identification of species. Ticks Tick Borne Dis 8(4):512–525

Falco RC, Fish D (1991) Horizontal movement of adult *Ixodes dammini* (Acari: Ixodidae) attracted to CO_2-baited traps. J Med Entomol 28:726–729. https://dx.doi.org/10.1093/jmedent/28.5.726

Falco RC, Fish D (1992) A comparison of methods for sampling the deer tick, *Ixodes dammini*, in a Lyme disease endemic area. Exp Appl Acarol 14:165–173

Garcia R (1962) Carbon dioxide as an attachment for certain ticks (Acarina: Argasidae and Ixodidae). Ann Entomol Soc Am 55:605–606

Ginsberg HS, Ewing CP (1989) Comparison of flagging, walking, trapping, and collecting from hosts as sampling methods for northern deer ticks, *Ixodes dammini*, and lone-star ticks, *Amblyomma americanum* (Acari: Ixodidae). Exp Appl Acarol 7:313–322. https://dx.doi.org/10.1007/BF01197925

Gray JS, Lohan G (1982) The development of a sampling method for the tick *Ixodes ricinus* and its use in a redwater fever area. Ann Appl Biol 101:421–427. https://dx.doi.org/10.1111/j.1744-7348.1982.tb00842.x

Guglielmone AA, Robbins RG, Apanaskevich DA, Petney TN, Estrada-Peña A, Horak IG, Shao R, Barker SC (2010) The Argasidae, Ixodidae and Nuttalliellidae (Acari: Ixodida) of the world: a list of valid species names. Zootaxa 2528:1–28

Keirans JE, Clifford CM, Corwin D (1976) *Ixodes sigelos*, n. sp. (Acarina: Ixodidae), a parasite of rodents in Chile, with a method for preparing ticks for examination by scanning electron microscopy. Acarologia 18:217–225

MacLeod J (1975) Apparent host selection by some African tick species. Oecologia 19:359–370

MacIvor KM, Horak IG, Holton KC, Petney TN (1987) A comparison of live and destructive sampling methods of determining the size of parasitic tick populations. Expt Appl Acarol 3:131–143

Melik W, Nilsson AS, Johansson M (2007) Detection strategies of tick-borne encephalitis virus in Swedish *Ixodes ricinus* reveal evolutionary characteristics of emerging tick-borne flaviviruses. Adv Virol 152:1027–1034

Miles VI (1968) A carbon dioxide bait trap for collecting ticks and fleas from animal burrows. J Med Entomol 5:491–495. https://dx.doi.org/10.1093/jmedent/5.4.491

Mtambo J, Van Bortel W, Madder M, Roelants P, Backeljau T (2006) Comparison of preservation methods of *Rhipicephalus appendiculatus* (Acari: Ixodidae) for reliable DNA amplification by PCR. Exp Appl Acarol 38:189–199. https://dx.doi.org/10.1007/s10493-006-0004-4

Mumcuoglu KY, Banet-Noach C, Malkinson M, Shalom U, Galun R (2005) Argasid ticks as possible vectors of West Nile virus in Israel. Vector Borne Zoonotic Dis 5:65–71

Nava S, Gerardi M, Szabó MP, Mastropaolo M, Martins TF, Labruna MB, Beati L, Estrada-Peña A, Guglielmone AA (2016)

Different lines of evidence used to delimit species in ticks: a study of the South American populations of *Amblyomma parvum* (Acari: Ixodidae). Ticks Tick Borne Dis. 7:1168–1179

Needham GR, Teel PD (1991) Off-host physiological ecology of ixodid ticks. Annu Rev Entomol 36:659–681. https://dx.doi.org/10.1146/annurev.ento.36.1.659

Nelder MP, Reeves WK (2005) Ectoparasites of road-killed vertebrates in northwestern South Carolina, USA. Vet Parasitol Parasitol 129:313–322

Niebuhr CN, Breeden JB, Lambert BD, Eyres AI, Haefele HJ, Kattes DH (2013) Off-host collection methods of the *Otobius megnini* (Acari: Argasidae). J Med Entomol 50:994–998. https://dx.doi.org/10.1603/ME13020

Oakwood M, Spratt DM (2000) Parasites of the northern quoll, *Dasyurus hallucatus* (Marsupialia: Dasyuridae) in tropical savanna, Northern Territory. Aus J Zool 48:79–90

Petney TN, van Ark H, Spickett AM (1990) On sampling tick populations: the problem of overdispersion. Onderstepoort J Vet Res 57:123–127

Petry WK, Foré SA, Fielden LJ, Kim HJ (2010) A quantitative comparison of two sample methods for collecting *Amblyomma americanum* and *Dermacentor variabilis* (Acari: Ixodidae) in Missouri. Exp Appl Acarol 52:427–438

Pfäffle M, Littwin N, Petney T (2015) Host preferences of immature *Dermacentor reticulatus* (Acari: Ixodidae) in a forest habitat in Germany. Ticks Tick Borne Dis 6:508–515. https://dx.doi.org/10.1016/j.ttbdis.2015.04.003

Pfäffle M, Petney T, Skuballa J, Taraschewski H (2011) Comparative population dynamics of a generalist (*Ixodes ricinus*) and specialist tick (*I. hexagonus*) species from European hedgehogs. Exp Appl Acarol 54:151–164. https://dx.doi.org/10.1007/s10493-011-9432-x

Rupp MB (1990) An abbreviated method for preparing *Ixodes damini* ticks for scanning electron microscopy observation. Microsc Res Tech 15:99–100

Schulze TL, Jordan RA, Dolan MC (2011) Experimental use of two standard tick collection methods to evaluate the relative effectiveness of several plant-derived and synthetic repellents against *Ixodes scapularis* and *Amblyomma americanum* (Acari: Ixodidae). J Econ Entomol 104:2062–2067. https://dx.doi.org/10.1603/EC10421

Skerratt LF (1998) Diseases and parasites of the common wombat *Vombatus ursinus* in the Healesville area of Victoria. In: Wells RT, Pridmore PA (eds) Wombats. Surrey Beatty & Sons, Chipping Norton, NSW, Australia, pp 317–328

Skuballa J, Petney T, Pfäffle M, Taraschewski H (2010) Molecular detection of *Anaplasma phagocytophilum* in the European hedgehog (*Erinaceus europaeus*) and its ticks. Vector Borne Zoonotic Dis 10:1055–1057. https://dx.doi.org/10.1089/vbz.2009.0150

Terassini FA, Barbieri FS, Albuquerque S, Szabó MPJ, Camargo LMA, Labruna MB (2010) Comparison of two methods for collecting free-living ticks in the Amazonian forest. Ticks Tick Borne Dis 1:194–196. https://dx.doi.org/10.1016/j.ttbdis.2010.08.002

Van Dyk PJ, McKenzie AA (1992) An evaluation of the effectivity of the scrub technique in quantitative ectoparasite ecology. Expt Appl Acarol 15:271–283

Wilson JG, Kinzer DR, Sauer JR, Hair JA (1972) Chemo-attraction in the lone star tick (Acarina: Ixodidae): I. Response of different developmental stages to carbon dioxide administered via traps. J Med Entomol 9:245–252. doi:10.1093/jmedent/9.3.245

Part I

Family Argasidae Koch, 1844

The genus and species-level taxonomy of the family Argasidae is less clearly established than that of the Ixodidae. The lack of reliable guidelines based on stable morphological characteristics probably has several reasons. One involves the differences found among schools of taxonomists regarding which features are considered most relevant for providing coherent systematic clustering among genera and/or species. Another reason is the high level of diversity among the family Argasidae, which is mostly not well supported by reliable genetic data. The disagreement on the generic composition is such that 137 out of the 198 argasid species described have been assigned to more than one genus by different scientists. A basic illustration of terms of the body parts of the ticks of the family Argasidae is included in Fig. 1.

Key to the Genera of Family Argasidae in Europe and Northern Africa

Adults and Nymphs

1a. Body flattened dorsoventrally. Dorsal side clearly separated from ventral side by the presence of a differentiated tegumental tissue, or a suture, along the posterolateral margins consisting of folds in the cuticle or sclerites. Anterior margin of the body without extensions of the cuticle. Camerostome without cheeks. Tegument granular, with many discs usually in radial rows. Eyes absent... **Genus *Argas*** (Fig. 1A, B).

1b. Body thick. Dorsal side merging into ventral side due to the absence of a suture. Anterior margin of the body with extensions of the cuticle as rounded hood-like cones. Camerostome usually with cheeks. Tegument squamous, with discs usually not arranged in rows. Eyes either present or absent...**Genus *Ornithodoros*** (Fig. 1F, G).

Larvae

1a. Dorsal plate present or absent. Tarsus I without trumpet-shaped sensillum extending posteriorly from the capsule of Haller's organ. Dorsal surface of the body contains 13–50 pairs of setae and ventral body surface 7–9 pairs (except in the subgenus *Ornithodoros* which has only 7–9 pairs dorsally and 5 pairs ventrally)...**Genus *Ornithodoros***.

1b. Dorsal plate present. Tarsus I with a trumpet-shaped sensillum extending posteriorly from the capsule of Haller's organ (except in *A. persicus*)...**Genus *Argas***.

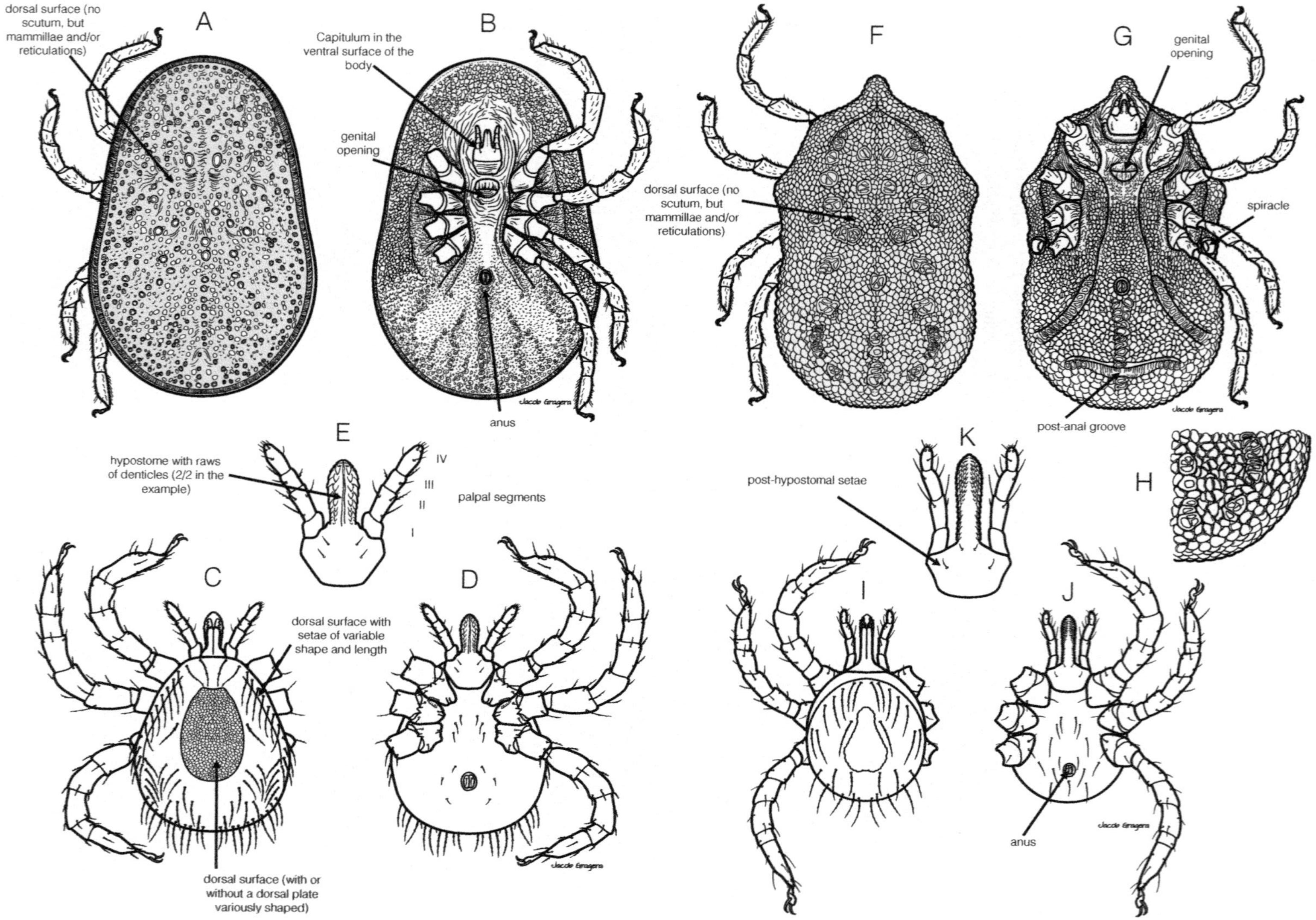

Fig. 1 The main morphological features of Argasid ticks. The genus *Argas* (**A–E**) and the genus *Ornithodoros* (**F–K**). The adults of the genus *Argas* (**A**: dorsal, **B**: ventral) and the larvae of the genus *Argas* (**C**: dorsal view, **D**: ventral view; **E**: capitulum, ventral). The adults of the genus *Ornithodoros* (**F**: dorsal, **G**: ventral), with general details of its dorsal surface reticulated pattern (**H**). The larvae of the genus *Ornithodoros* (**I**, dorsal view, **J**: ventral view, **K**: hypostome, ventral view, to show the differences in the pattern of the post-hypostomal setae with the genus *Argas*)

Genus *Argas* Latreille, 1795

A. Estrada-Peña, G. Kleinerman, and G. Baneth

There are six species of the genus *Argas* reported from the Western Palaearctic (Guglielmone et al. 2010). The two species commonly reported on bats are *A. transgariepinus* White, 1846 and *A. vespertilionis* (Latreille, 1796). *Argas transgariepinus* has been reported to be commonly associated with bats in Spain, Italy, Switzerland, Greece, Israel, Afghanistan, and Madagascar. Its wide distribution is probably associated with migrating bats that transport feeding larvae (Hoogstraal 1985). A morphological description and illustrations of adults and immature stages are available in Hoogstraal (1957). *Argas vespertilionis* (Latreille, 1796) parasitizes cave-dwelling insectivorous bats. This species is widely distributed in the Palaearctic region, and occurs in Asia and Africa. Morphological characterization and illustrations of adults and larvae are available in Hoogstraal (1958). A detailed redescription and illustrations were provided by Filippova (1966). An English translation of a short description of this tick species is also available (Filippova 1964).

A group of four other species, *A. reflexus* (Fabricius, 1794), *A. persicus* (Oken, 1818), *A. macrostigmatus* Filippova, 1961, and *A. polonicus* Siuda, Hoogstraal, Clifford and Wassef, 1979, are parasites of birds. *Argas reflexus* is widely distributed in the Western Palaearctic. It is a parasite of domestic pigeons and wild birds. The nymphal stages were described by Filippova (1964) and a more complete description of the species, with keys and illustrations, is available in Filippova (1966). A redescription of all life stages with illustrations and keys was provided by Hoogstraal and Kohls (1960). *Argas persicus* is a parasite of arboreal nesting birds and is common on domestic poultry. Keys and a redescription with illustrations of all the life stages can be found in Kohls et al. (1970). A short description of the adults is provided by Filippova (1964), with a more detailed description and illustrations published later (Filippova 1966). *Argas macrostigmatus* was first described from nymphs. Filippova (1966) later published a redescription of the nymphs and described the adults. Keys and illustrations can be found in both references. The larvae were first described by Hoogstraal et al. (1984). This species can be found on islands in the Black Sea and the Mediterranean where it parasitizes crested cormorants (Hoogstraal et al. 1984). *Argas polonicus* was adequately described, illustrated and separated from closely related species in the original publication (Siuda et al. 1979).

Key to the Species of Genus *Argas* in Europe and Northern Africa

Adults and Nymphs

Note: *Argas polonicus* and *A. macrostigmatus* are not illustrated because the lack of adequate material.

1a. Body subcircular, almost as wide as long or even wider than long…**2**.
1b. Body oval, longer than wide…**3**.
2a. Discs on the tegument not clearly visible and not arranged in clear radial rows. Coxa I clearly separated from coxa II. Transverse post-anal groove absent…***Argas transgariepinus*** (Fig. 11).
2b. Discs on the tegument well visible and arranged in clear radial rows on both dorsal and ventral surfaces. Coxae I and II contiguous. Transverse post-anal groove present…***Argas vespertilionis*** (Fig. 9).
3a. Dorsal side clearly separated from the ventral side by a peripheral suture or fold with large, rectangular or sub-rectangular cells. Peripheral cells of the dorsal portion

A. Estrada-Peña (✉)
Department of Animal Health, Faculty of Veterinary Medicine, Zaragoza, Spain
e-mail: aestrada@unizar.es

G. Kleinerman · G. Baneth
Koret School of Veterinary Medicine, Hebrew University, Rehovot, Israel
e-mail: gabriela.kleinerman@mail.huji.ac.il

G. Baneth
e-mail: gad.baneth@mail.huji.ac.il

A. Estrada-Peña et al. (eds.), *Ticks of Europe and North Africa*,
https://doi.org/10.1007/978-3-319-63760-0_3

with a single large setiferous pit occupying most of surface area...***Argas persicus*** (Fig. 3).

3b. Dorsal side separated from the ventral side by a peripheral suture or fold with striped cells. Peripheral cells of the dorsal portion with several setiferous pits...**4**.

4a. Capsule of Haller's organ open. Spiracular plate very large...***Argas macrostigmatus***.

4b. Capsule of Haller's organ closed. Spiracular plate small...**5**.

5a. With one pair of long post-hypostomal setae and about 12 pairs of small setae distributed over the ventral surface of the basis capituli...***Argas polonicus***.

5b. With one pair of long post-hypostomal setae and only 1 pair of small setae on the ventral surface of the basis capituli. Other small setae present on the lateral margins of the absis capituli...***Argas reflexus*** (Fig. 5).

Larvae

Note: *Argas polonicus* and *A. macrostigmatus* are not illustrated because the lack of adequate material.

1a. Dorsum with 25 or more pairs of setae...**2**.

1b. Dorsum with 13–15 pairs of dorsal setae, 11–12 being marginal...***Argas vespertilionis*** (Fig. 9).

2a. Dorsal setae long...**3**.

2b. Dorsal setae short...**4**.

3a. About 30 pairs of dorsal setae, of which 20–21 pairs are on the dorsal margin. Hypostome with a blunt rounded apex, with 6 rows of 2/2 basal denticles, 2 rows of 3/3 and a corona of 4/4 denticles...***Argas macrostigmatus***.

3b. About 46 pairs of dorsal setae, Hypostome with 8–12 rows of 2/2 basal denticles, and a corona of 3/3 or 4/4 denticles...**5**.

4a. Basis capituli trapezoidal, with an almost 90° angle between lateral and posterior margins. Palpal articles 1 and 2 fused. Dorsal plate absent or only slightly visible. With 18–20 pairs of setae on the dorsal margin...***Argas transgariepinus*** (Fig. 11).

4b. Basis capituli triangular with sharp and acute angles between the lateral and posterior margins. Palpal articles 1 and 2 not fused. Dorsal plate present. With 14–16 pairs of setae on the dorsal margin...***Argas persicus*** (Fig. 3).

5a. With about 26-27 pairs of setae on the dorsal margin. Dorsal setae barbed... ***Argas reflexus*** (Fig. 5).

5b. With about 8–10 pairs of setae on the dorsal margin. Dorsal setae not barbed... ***Argas polonicus***.

References

Filippova NA (1964) Data on ticks of the subfamily Argasinae (Ixodoidea, Argasidae). Report II—Taxonomy of Palaearctic Argasinae and diagnoses of the species of the USSR Fauna for all active phases of the life cycle. Parazitol Shorn 22:7–27 (In English by Division of Research Services National Institute of Health)

Filippova NA (1966) Argasid ticks (Argasidae). Fauna SSSR 4:1–255 (In Russian)

Guglielmone AA, Robbins RG, Apanaskevich DA, Petney TN, Estrada-Peña A, Horak IG, Shao R, Barker SC (2010) The Argasidae, Ixodidae and Nuttalliellidae (Acari: Ixodida) of the world: a list of valid species names. Zootaxa 2528:1–28

Hoogstraal H (1957) Bat ticks of the genus *Argas* (Ixodoidea, Argasidae), 2. *Secretargas* new subgenus and *A. transgariepinus* White, 1846, its adult and immature stages; with a definition of the subgenus *Argas*. Ann Entomol Soc Am 50:544–549

Hoogstraal H (1958) Bat ticks of the genus *Argas* (Ixodoidea, Argasidae), 3. The subgenus *Carios*, a description of *A.* (*C.*) *vespertilionis* (Latreille, 1802), and variation within an Egyptian population. Ann Entomol Soc Am 51:19–26

Hoogstraal H (1985) Argasid and nuttalliellid ticks as parasites and vectors. Adv Parasitol 24:135–238

Hoogstraal H, Kohls GM (1960) Observations on the subgenus *Argas* (Ixodoidea, Argasidae, *Argas*) 1. Study of *A. reflexus reflexus* (Fabricius 1794), the European bird argasid. Ann Entomol Soc Am 53:611–618

Hoogstraal H, Wassef HY, Guiguen C, Beaucournu JC (1984) Observations on the subgenus *Argas* (Ixodoidea: Argasidae: *Argas*). 19. Description of the larva of the Palaearctic cormorant parasite *A.* (*A.*) *macrostigmatus*. J Parasitol 70:292–294

Kohls GM, Hoogstraal H, Clifford CM, Kaiser MN (1970) The Subgenus *Persicargas* (Ixodoidea, Argasidae, *Argas*). 9. Redescription and New World Records of *A.* (*P.*) *persicus* (Oken), and Resurrection, Redescription, and Records of *A.* (*P.*) *radiatus* Railliet, *A.* (*P.*) *sanchezi* Duges, and *A.* (*P.*) *miniatus* Koch, New World ticks misidentified as *A.* (*P.*) *persicus*. Ann Entomol Soc Am 63:590–606

Siuda K, Hoogstraal H, Clifford CM, Wassef HY (1979) Observations on the subgenus *Argas* (Ixodoidea: Argasidae: *Argas*). 17. *Argas* (*A.*) *polonicus* sp. n. parasitizing domestic pigeons in Kraków, Poland. J Parasitol 65:170–181

Argas persicus (Oken, 1818) (Figs. 2 and 3)

M. P. Pfäffle and T. N. Petney

Before the work of Hoogstraal and colleagues in the 1960s and 1970s, there was considerable confusion as to the taxonomic definition, distribution, ecology and biology of *A. persicus* due to confusion with a variety of other species (Kohls et al. 1970). Records prior to 1970, particularly those from outside of the currently known range of this species, should therefore be treated sceptically.

Life Cycle and Host Preferences

The life cycle is typical of argasid ticks with up to four nymphal stages (Walker et al. 2003) and is greatly influenced by the environmental conditions. For example, Nosek et al. (1980) report the life cycle to take about 17 months under favourable laboratory conditions (21 °C, 75–95% relative humidity), whereas in a study from Egypt, the life cycle took only 63–178 days in the laboratory (30–32 °C, 75% relative humidity) and 111–260 days outdoors under natural conditions (El-Kammah and Abdel-Wahab 1980). Domestic fowls, turkeys, ducks and geese are the typical hosts of this tick. It also feeds on pigeons and a variety of wild birds (Walker et al. 2003), but only rarely on humans (Estrada-Peña and Jongejan 1999).

Ecology

Argas persicus can be found under wood, in crevices, under the loose bark of trees used as roosts by chickens or in debris under such roosts (Nosek et al. 1980). It is a domestic and endophilic tick, found in the fabric of poultry houses, bird nests and roosting sites (Walker et al. 2003). In a study by Abdussalam and Sarwar (2009) in Pakistan, *A. persicus* was found under the bark of 16 different tree species that were not used as roosts by domestic birds, but were used by vultures (*Pseudogyps bengalensis*) and herons (*Ardea cinerea*) for nesting or roosting. The tick occurs in regions with climates ranging from desert to Mediterranean temperate to rainforest (Walker et al. 2003). In Slovakia, for example, it is mostly found in semi-arid regions and is restricted by the 650–700 mm annual average isohyet and the +8 °C annual average isotherm (Nosek et al. 1980). In climates with a winter season, the larvae and first nymphal stage are most active in early summer, nymphal stages 2–4 in midsummer and adults in late summer and autumn (Walker et al. 2003). In Slovakia, the tick develops and reproduces during the warm season in the wild, larvae hatch from spring to autumn and nymphal and adult moulting occur chiefly in July and August (Nosek et al. 1980). In Egypt, females undergo ovarian diapause from November to April (El-Kammah and Abdel-Wahab 1980). Overwintering takes place in the adult stages or as eggs (Walker et al. 2003). The tick is nocturnal, with nymphs and adults preferably feeding at night (Nosek et al. 1980). The larvae can survive without feeding for 2 months or more, the nymphs for 1 year and the adults for up to 3 years without a blood meal (Walker et al. 2003). Moulting takes place in cracks and crevices, and the moult to the adult can occur from the second nymphal stage onwards (Walker et al. 2003).

Distribution

This species originally came from central Asia, however; because of its association with domestic birds (primarily domestic chickens), it has been introduced to almost all parts of the world (Gothe 1999). It has been reported from Armenia, Australia, Bangladesh, Bulgaria, China, Corsica, Cyprus, Egypt, Ethiopia, France, Republic of Macedonia, Ghana, India, Iran, Israel, Libya, Morocco, Pakistan,

M. P. Pfäffle (✉)
Department of Ecology and Parasitology, Institute of Zoology, Karlsruhe Institute of Technology, Karlsruhe, Germany
e-mail: miripfaeffle@web.de

T. N. Petney
State Museum of Natural History, Erbprinzenstr. 13, 76133 Karlsruhe, Germany
e-mail: trevor.petney@kit.edu

A. Estrada-Peña et al. (eds.), *Ticks of Europe and North Africa*,
https://doi.org/10.1007/978-3-319-63760-0_4

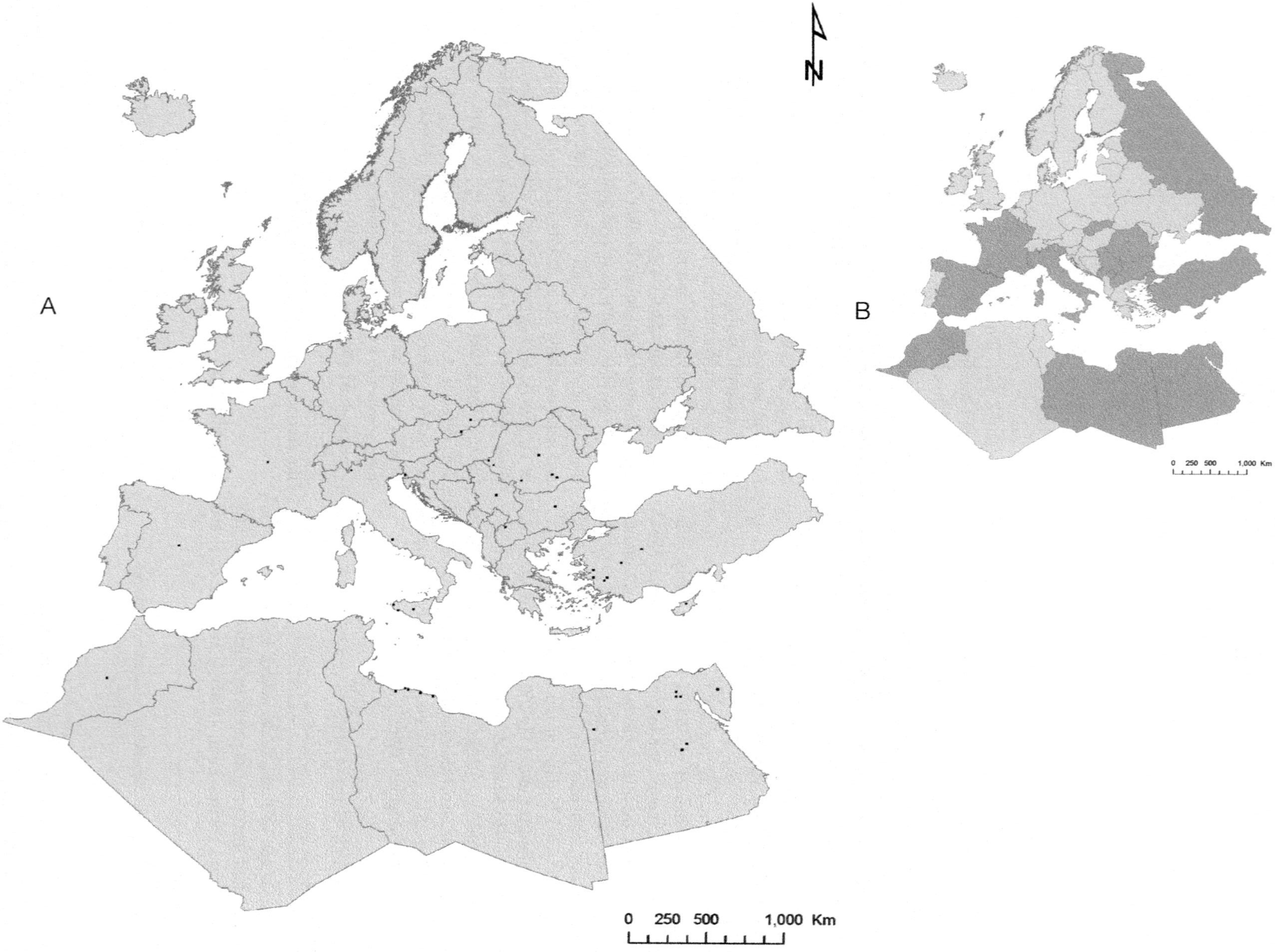

Fig. 2 **A** Distribution of *A. persicus* in Europe and Northern Africa (10 × 10 km grid presence with black dots). **B** Countries where the species has been reported are marked in dark grey

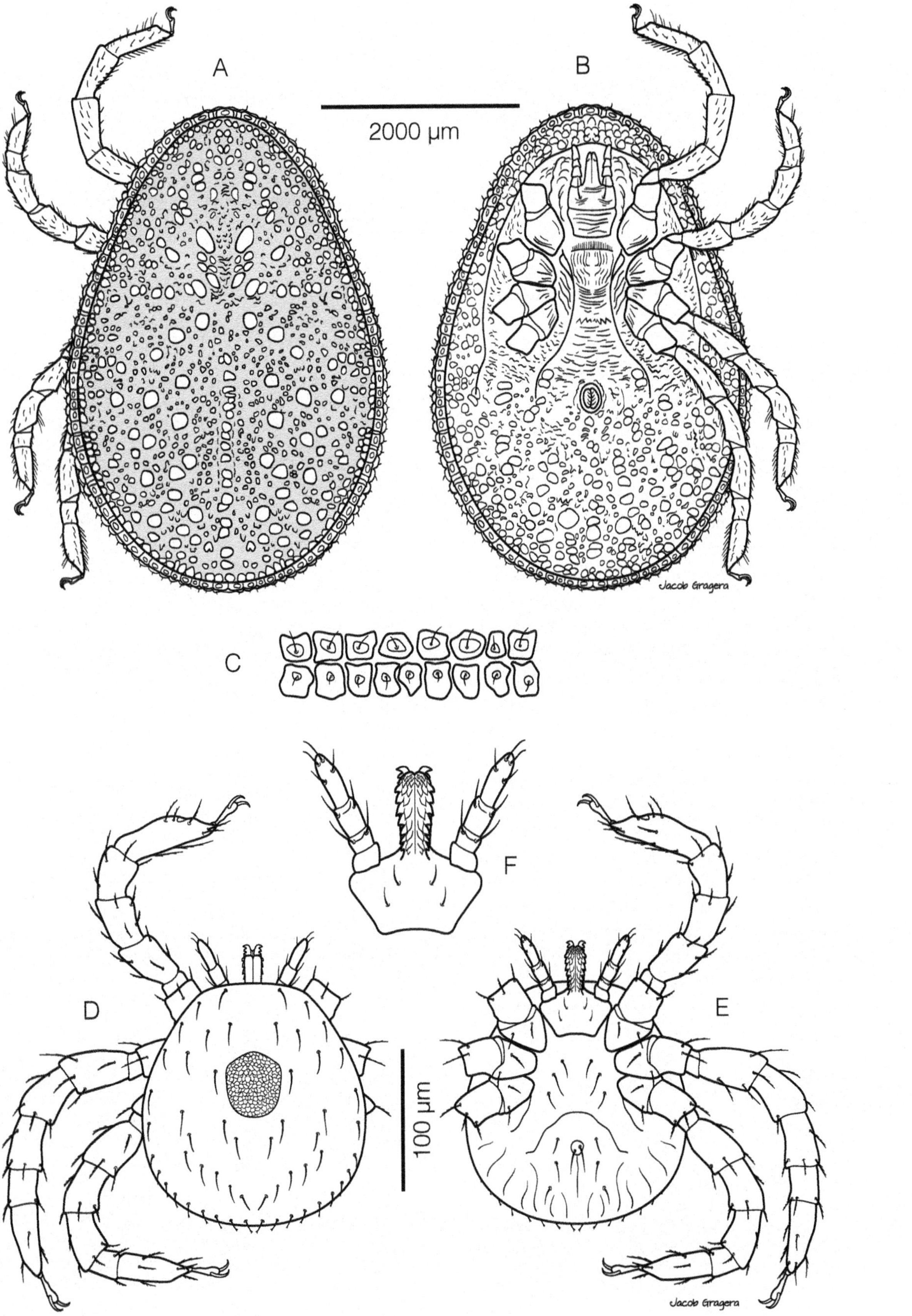

Fig. 3 **A–C** The adult of *A. persicus*. **A** Dorsal, **B** ventral, **C** details of the structure of the cells between the dorsal and ventral surfaces. **D–F** The larva of *A. persicus*. **D** Dorsal, **E** ventral, **F** details of the capitulum. Illustrations from specimens collected in Spain and the Czech Republic

Paraguay, Niger, Romania, Russia, Serbia, Slovakia, Spain and the USA (Hoffmann and Lindau 1971; Rageau 1972; Le Riche et al. 1974; Rosenstein 1976; Rehácek et al. 1977; El-Kammah and Abdel-Wahab 1980; Nosek et al. 1980; Vatandoost et al. 2003; Petney et al. 2004; Mumcuoglu et al. 2005; Abdussalam and Sarwar 2009; Coipan et al. 2012; Manzano-Román et al. 2012; Pader et al. 2012). It has been reported from Italy on migrating birds (Walter and Massa 1987).

Vectorial Capacity and Pathogen Burden

Argas persicus has been reported to induce tick paralysis in chickens by Kunze and Gothe (1971). However, this study was conducted with a population of ticks originating from Onderstepoort, South Africa, where this species does not exist and has been confused with *A. walkerae* (Gothe 1999). Later, Gothe (1999) discusses different studies which provide evidence that the larvae, but not the nymphs or the adults, are able to induce tick paralysis in fowl.

Crimean–Congo haemorrhagic fever virus as well as Slovakia virus and Kyasanur forest disease virus have been isolated from this tick (Hoogstraal 1985; Nuttall et al. 1994; Labuda and Nuttall 2004; Manzano-Román et al. 2012). It has, however, been demonstrated that the Crimean–Congo haemorrhagic fever virus cannot be transmitted by argasids under laboratory conditions. *Rickettsia* sp. of the spotted fever group, closely related to *Rickettsia slovaca*, has been detected, as has *Candidatus* 'Rickettsia hoogstraalii' (Rehácek et al. 1977; Pader et al. 2012). It is the main vector of *Borrelia anserina* and *Aegyptianella pullorum* (Gothe 1967; Walker et al. 2003; Petney et al. 2004). It has also been implicated in the transmission of *Mycobacterium avium* and *Pasteurella avicida* to poultry, may be involved in the circulation of West Nile virus and is known to harbour persistent infections of *Salmonella gallinarum/pullorum*, which it then excretes in the faeces (Stefanov et al. 1975; Petney et al. 2004). *Rahnella aquatilis*, *Enterobacter cloacae* and *Chryseobacterium meningosepticum* have been found in this tick (Montasser 2005; Manzano-Román et al. 2012).

References

Abdussalam M, Sarwar MM (2009) Trees as habitats of the fowl tick, *Argas persicus* (Oken). Bull Entomol Res 44:419–420. https://dx.doi.org/10.1017/S0007485300025529

Coipan EC, Vladimirescu AF, Ciolpan O, Teodorescu I (2012) Tick species (Acari: Ixodoidea) distribution, seasonality and host associations in Romania. Trav. du Muséum Natl. d'Histoire Nat. "Grigore Antipa" 54:301–317. https://dx.doi.org/10.2478/v10191-011-0018-y

El-Kammah KM, Abdel-Wahab KS (1980) *Argas* (*Persicargas*) *persicus* life-cycle under controlled and outdoor conditions. Acarologia 21:163–172

Estrada-Peña A, Jongejan F (1999) Ticks feeding on humans: a review of records on human-biting Ixodoidea with special reference to pathogen transmission. Exp Appl Acarol 23:685–715. https://dx.doi.org/10.1023/A:1006241108739

Gothe R (1967) Zur Entwicklung von *Aegyptianella pullorum* Carpano, 1928, in der Lederzecke *Argas* (*Persicargas*) *persicus* (Oken, 1818) und Übertragung. Zeitschrift Parasitenkd 29:103–118. https://dx.doi.org/10.1007/BF00260198

Gothe R (1999) Zeckentoxikosen. Hieronymus, München

Hoffmann G, Lindau M (1971) Zecken an Nutz- und Wildtieren in Niger. Zeitschrift Angew Entomol 69:72–82. https://dx.doi.org/10.1111/j.1439-0418.1971.tb03185.x

Hoogstraal H (1985) Argasid and nuttalliellid ticks as parasites and vectors. Adv Parasitol 24:135–238

Kohls GM, Hoogstraal H, Clifford CM, Kaiser MN (1970) The subgenus *Persicargas* (Ixodoidea, Argasidae, *Argas*). 9. Redescription and new world records of *Argas* (*P.*) *persicus* (Oken), and resurrection, redescription, and records of *A.* (*P.*) *radiatus* Railliet, *A.* (*P.*) *sanchezi* Duges, and *A.* (*P.*) *miniatus* Koch. Ann Entomol Soc Am 63:590–606. https://dx.doi.org/10.1093/aesa/63.2.590

Kunze K, Gothe R (1971) Die durch *Argas* (*Persicargas*) *persicus*-Larven bedingte Paralyse der Hühner. Zeitschrift Parasitenkd 36:251–264. https://dx.doi.org/10.1007/BF00348563

Labuda M, Nuttall PA (2004) Tick-borne viruses. Parasitology 129: S221–S245. https://dx.doi.org/10.1017/S0031182004005220

Le Riche PD, Altan Y, Campbell JB, Efstathiou GC (1974) Ticks (Ixodoidea) of domestic animals in Cyprus. Bull Entomol Res 64:53–63. https://dx.doi.org/10.1017/S0007485300026973

Manzano-Román R, Díaz-Martín V, de la Fuente J, Pérez-Sánchez R (2012) Soft ticks as pathogen vectors: distribution. INTECH Open Access Publisher, Surveillance and Control

Montasser AA (2005) Gram-negative bacteria from the camel tick *Hyalomma dromedarii* (Ixodidae) and the chicken tick *Argas persicus* (Argasidae) and their antibiotic sensitivities. J Egypt Soc Parasitol 35:95–106

Mumcuoglu KY, Banet-Noach C, Malkinson M, Shalom U, Galun R (2005) Argasid ticks as possible vectors of West Nile virus in Israel. Vector-Borne Zoonotic Dis 5:65–71

Nosek J, Hoogstraal H, Labuda M, Cyprich D (1980) Bionomics and health importance of fowl tick *Argas* (*Persicargas*) *persicus* (Oken, 1818) (Ixodoidea: Argasidae). Zeitschrift Parasitenkd 63:209–212. https://dx.doi.org/10.1007/BF00931982

Nuttall PA, Jones LD, Labuda M, Kaufman WR (1994) Adaptations of arboviruses to ticks. J Med Entomol 31:1–9. https://dx.doi.org/10.1093/jmedent/31.1.1

Pader V, Nikitorowicz Buniak J, Abdissa A, Adamu H, Tolosa T, Gashaw A, Cutler RR, Cutler SJ (2012) *Candidatus* Rickettsia hoogstraalii in Ethiopian *Argas* persicus ticks. Ticks Tick Borne Dis 3:45–338. https://dx.doi.org/10.1016/j.ttbdis.2012.10.021

Petney TN, Andrews RH, McDiarmid LA, Dixon BR (2004) *Argas persicus* sensu stricto does occur in Australia. Parasitol Res 93:9–296. https://dx.doi.org/10.1007/s00436-004-1141-5

Rageau J (1972) Repartation geographique et role pathogene des tiques (Acariens: Argasidae et Ixodidae) en France. Wiad Parazytol 18:707–719

Rehácek J, Urvölgyi J, Kovácová E (1977) Massive occurrence of rickettsiae of the spotted fever group in fowl tampan, *Argas persicus*, in the Armenian S.S.R. Acta Virol 21:8–431

Rosenstein M (1976) Paralysis in chickens caused by larvae of the poultry tick. *Argas persicus* Avian Dis 20:407–409

Stefanov V, Matev I, Balimezov I (1975) Role of ticks of the species *Argas persicus* Oken, 1818, in the epizootology of pullorum disease in birds. Vet Med Nauki 12:45–50

Vatandoost H, Ghaderi A, Javadian E, Rassi Y, Piazak N, Shaeghi M, Telmodarreiy Z, Abolhasani M (2003) Distribution of soft ticks and their infection with *Borrelia* in Hamadan. Iran J Public Health 32:22–24

Walker AR, Bouattour A, Camicas J-L, Estrada-Peña A, Horak IG, Latif AA, Pegram RG, Preston PM (2003) Ticks of domestic animal in Africa: a guide to identification of species. Bioscience Reports, Edinburgh

Walter G, Massa R (1987) Ein Beitrag zur Ektoparasitenfauna der Zugvögel in Norditalien. J Appl Entomol 103:523–527. https://dx.doi.org/10.1111/j.1439-0418.1987.tb01018.x

Argas reflexus (Fabricius, 1794) (Figs. 4 and 5)

M. P. Pfäffle and T. N. Petney

Life Cycle and Host Preferences

The life cycle of *A. reflexus* comprises eggs, larvae, two to four nymphal stages and the adults. Laboratory studies showed that the number of nymphal instars seems to depend on the outside temperature, with cooler temperatures leading to fewer instars (Dautel and Knülle 1997). The life cycle might be as long as 3–11 years in Central Europe (Dautel and Knülle 1997). Hosts are predominantly domestic pigeons (*Columba livia domestica*). Other hosts include rock pigeons (*C. livia*), turtle doves (*Streptopelia turtur*), swifts, swallows, owls, crows, several passerine birds and chickens (Hillyard 1996). Humans can be bitten, especially those living close to pigeons or old pigeon roosts or nests (Hoogstraal and Kohls 1960; Dautel et al. 1999).

Ecology

Argas reflexus is a nidicolous tick, and it is found in cracks and slits in walls or woodwork in the vicinity of pigeon nests (Dautel et al. 1991). In Central Europe, it is restricted to buildings where pigeons breed (Dusbabek and Rosicky 1976; Dautel et al. 1991). Dautel et al. (1994) report an activity peak for all life history stages from March to early June and for immatures another peak in late summer and autumn. Oviposition seems to be limited to the summer months (Dautel and Knülle 1998).

Distribution

Argas reflexus is found widely in the Palaearctic, with its presence being related to the distribution of domestic pigeons. In Europe, it has been reported from Italy, Spain, France, Switzerland, Netherlands, Austria, Belgium, Germany, Denmark, Sweden, Poland, Portugal, Czech Republic, Slovakia, Hungary, Romania, Bulgaria and United Kingdom (Hoogstraal and Kohls 1960; Dusbabek and Rosicky 1976; Hueli and Garcia Fernandez 1983; Hillyard 1996; Rosa et al. 2004). Other reports come from Egypt (as larvae), Libya and Tunisia (Hoogstraal and Kohls 1960), Israel (Theodor and Costa 2009), Turkey (Nemenz 1962; Erman et al. 2007), Iran (Manzano-Román et al. 2012), Pakistan (Manzano-Román et al. 2012), Afghanistan (Arsen'eva and Neronov 1980), Kazakhstan (Hoogstraal and Kohls 1960), the Crimea (Filippova 1966) and the North Caucasus (Dikaev 1980). In the north, the distribution is restricted to about 55°N, as the average temperature in the summer affects embryonic development in the summer and autumn (Dautel and Knülle 1998).

Vectorial Capacity and Pathogen Burden

Argas reflexus bites can have a substantial pathological effect on humans. They are known to cause skin reactions of varying intensity (Dautel and Kahl 1999) and in some individuals may lead to a severe allergic response, including anaphylactic shock (Laubstein et al. 1993; Kleintebbe et al. 2006; Śpiewak et al. 2006; Weckesser et al. 2010).

Argas reflexus is associated with a variety of disease agents affecting birds. It is the vector of *Aegyptianella pullorum* causing aegyptianellosis in birds (Jongejan and Uilenberg 2005). There is no direct evidence that *A. reflexus* is a competent vector of human pathogens. However, a variety of different agents have been isolated from this tick (Uspensky et al. 2008). These include Crimean–Congo

M. P. Pfäffle (✉)
Department of Ecology and Parasitology, Institute of Zoology, Karlsruhe Institute of Technology, Karlsruhe, Germany
e-mail: miripfaeffle@web.de

T. N. Petney
State Museum of Natural History, Erbprinzenstr. 13, 76133 Karlsruhe, Germany
e-mail: trevor.petney@kit.edu

A. Estrada-Peña et al. (eds.), *Ticks of Europe and North Africa*,
https://doi.org/10.1007/978-3-319-63760-0_5

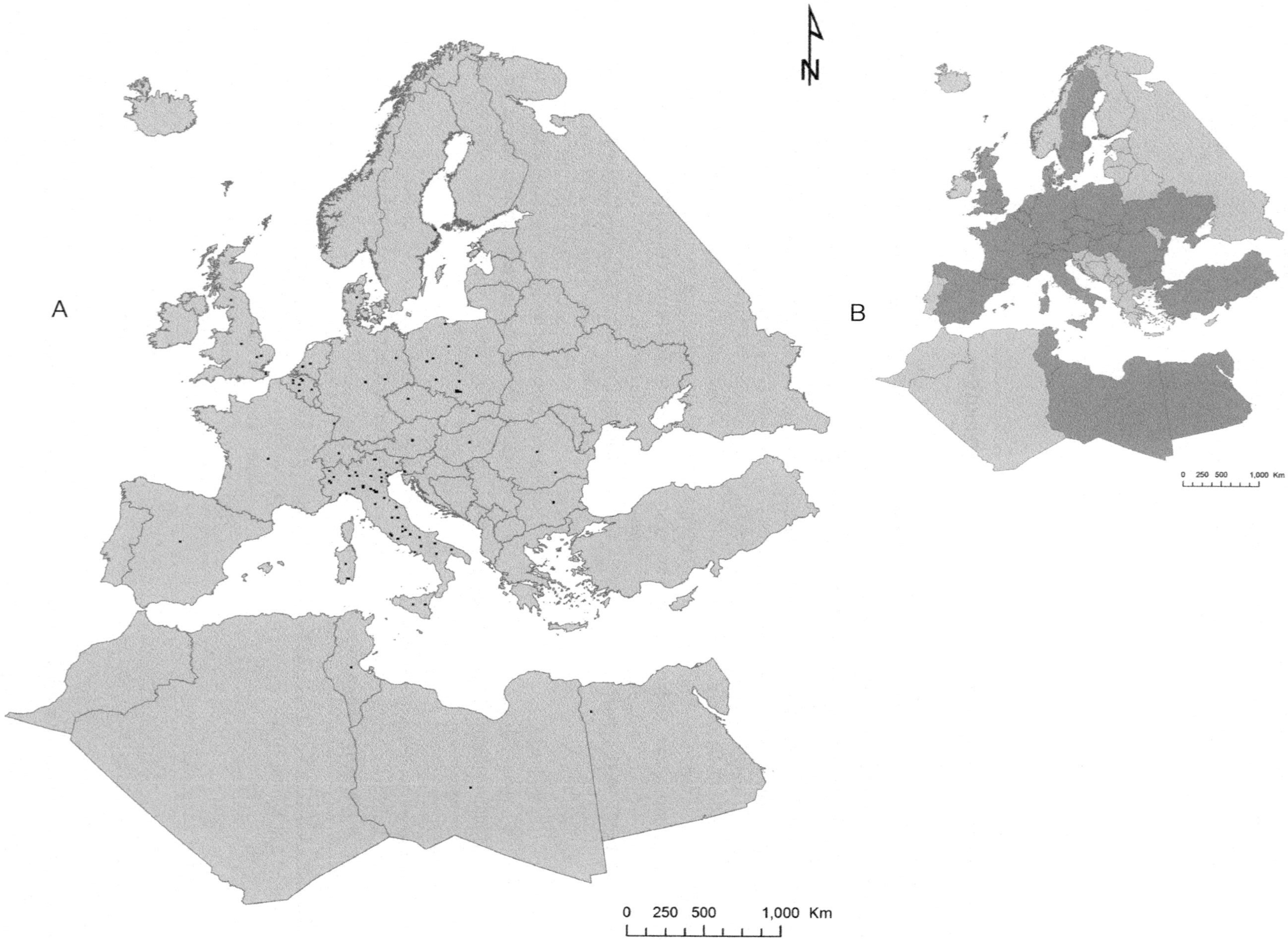

Fig. 4 **A** Distribution of *A. reflexus* in Europe and Northern Africa (10 × 10 km grid presence with black dots). **B** Countries where the species has been reported are marked in dark grey

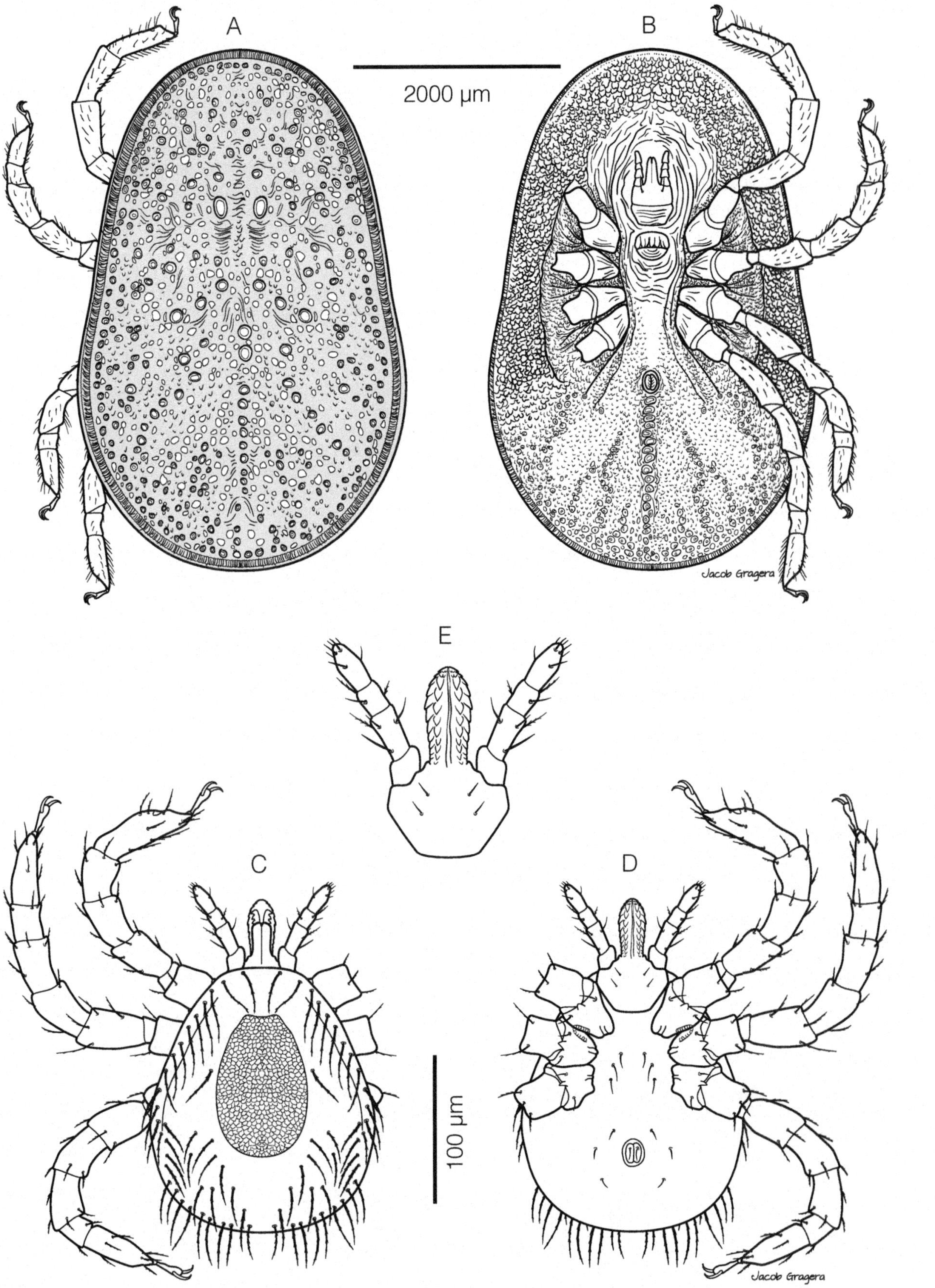

Fig. 5 **A–B**: The adult of *A. reflexus*. **A** Dorsal, **B** ventral. **C–E** The larva of *A. reflexus*. **C** Dorsal, **D** ventral, **E** details of the capitulum. Illustrations from specimens collected in Setubal and Viana do Castelo (North Portugal) and Italy (Rosa et al. 2004)

haemorrhagic fever virus, Uukuniemi virus, Grand Arbaud virus, Ponteves virus, Tunis virus, West Nile virus, Chenuda virus, Nyamanini virus and Quaranfil virus (Nuttall et al., 1994; Vermeil et al. 1996; Labuda and Nuttall 2004; Tahmasebi et al. 2010; Petney et al. 2012). This tick has been proven to be an inefficient vector of some of these agents under laboratory conditions. *A. reflexus* has been associated with *Borrelia burgdorferi* s.l. (see, e.g. Genchi et al. 1989); however, of 800 ticks screened for *Borrelia* by IFA monoclonal and polyclonal antibody, only one was positive and this proved to be negative for OspA (Dautel et al. 1999).

References

Arsen'eva LP, Neronov VM (1980) Ticks and mites, ectoparasites of wild and domestic animals of Afghanistan. Med Parazitol (Mosk) 49:37–42

Dautel H, Kahl O (1999) Ticks (Acari: Ixodoidea) and their medical importance in the urban environment. In: Proceedings of 3rd International Conference on Urban Pests, Czech Repub, 73–82

Dautel H, Kahl O, Knülle W (1991) The soft tick *Argas reflexus* (F.) (Acari, Argasidae) in urban environments and its medical significance in Berlin (West). J Appl Entomol 111:380–390. https://dx.doi.org/10.1111/j.1439-0418.1991.tb00338.x

Dautel H, Kahl O, Scheurer S, Knulle W (1994) Seasonal activities of the pigeon tick *Argas reflexus* (Acari: Argasidae) in Berlin. Germany Folia Parasitol (Praha) 41:155–160

Dautel H, Knülle W (1998) Seasonal oviposition and temperature requirements of eggs may limit Northern distribution of European *Argas reflexus* (Acari: Argasidae). J Med Entomol 35:26–37. https://dx.doi.org/10.1093/jmedent/35.1.26

Dautel H, Knülle W (1997) Life cycle and seasonal development of postembryonic *Argas reflexus* (Acari: Argasidae) at two thermally different locations in Central Europe. Exp Appl Acarol 21:697–712. https://dx.doi.org/10.1007/BF02803512

Dautel H, Scheurer S, Kahl O (1999) The pigeon tick (*Argas reflexus*): its biology, ecology, and epidemiological aspects. Zentralblatt für Bakteriol 289:745–753. https://dx.doi.org/10.1016/S0934-8840(99)80049-8

Dikaev BY (1980) Contribution to the argasid tick fauna in Checheno-Ingush ASSR (Argasidae). Parazitol Leningr 15:75–78

Dusbabek F, Rosicky B (1976) Argasid ticks (Argasidae, Ixodoidea) of Czechoslovakia. Prirodoved. Pr. Ust. Ceskoslov. Akad, Ved v Brne, p 10

Erman O, Özkan M, Ayyildiz N, Doğan S (2007) Checklist of the mites (arachnide: Acari) of Turkey. Second supplement. Zootaxa 21:1–21

Filippova NA (1966) Argasid ticks (Argasidae). Fauna SSSR 4:1–255 (In Russian)

Genchi C, Magnino S, Pacetti A, Simeoni J, Sacco BD, Falagiani P (1989) *Argas reflexus*, a possible vector, of *Borrelia burgdorferi*. G Mal Infett Parassit 41:403–408

Hillyard PD (1996) Ticks of North-West Europe. Field Studies Council, Shrewsbury

Hoogstraal H, Kohls GM (1960) Observations on the Subgenus *Argas* (Ixodoidea, Argasidae, *Argas*) 1. Study of *A. reflexus reflexus* (Fabricius, 1794), the European Bird Argasid. Ann Entomol Soc Am 53:611–618. https://dx.doi.org/10.1093/aesa/53.5.611

Hueli LE, Garcia Fernandez P (1983) *Apus apus* L. nuevo hospedador para *Argas* (*Argas*) *reflexus* (Fabricius, 1794) (Ixodoidea, Argasidae) en Espana. Rev Iber Parasitol 43:203–204

Jongejan F, Uilenberg G (2005) The global importance of ticks. Parasitology 129:S3. https://dx.doi.org/10.1017/S0031182004005967

Kleintebbe J, Heinatz A, Graser I, Dautel H, Hansen G, Kespohl S, Rihs H, Raulfheimsoth M, Vater G, Rytter M (2006) Bites of the European pigeon tick (*Argas reflexus*): risk of IgE-mediated sensitizations and anaphylactic reactions. J Allergy Clin Immunol 117:190–195. https://dx.doi.org/10.1016/j.jaci.2005.08.056

Labuda M, Nuttall PA (2004) Tick-borne viruses. Parasitology 129: S221–S245. https://dx.doi.org/10.1017/S0031182004005220

Laubstein B, Herold D, Audring H, Buchholtz I (1993) Nächtliche Anaphylaxie durch *Argas reflexus*. Allergologie 16:370–373

Manzano-Román R, Díaz-Martín V, de la Fuente J, Pérez-Sánchez R (2012) Soft ticks as pathogen vectors: distribution. INTECH Open Access Publisher, Surveillance and Control

Nemenz H (1962) Zecken aus der Türkei und dem Karakorum (Acari, Ixodidae). Zeitschrift Parasitenkd 22:111–113. https://dx.doi.org/10.1007/BF00329208

Nuttall PA, Jones LD, Labuda M, Kaufman WR (1994) Adaptations of arboviruses to ticks. J Med Entomol 31:1–9. https://dx.doi.org/10.1093/jmedent/31.1.1

Petney TN, Pfäffle MP, Skuballa JD (2012) An annotated checklist of the ticks (Acari: Ixodida) of Germany. Syst Appl Acarol 17:115–170. https://dx.doi.org/10.11158/saa.17.2.2

Rosa F, Crespo MV, Esteves L, Mendes L (2004) Distribuição de *Argas* (*Argas*) *reflexus* (Fabricius 1794) no Norte e Centro de Portugal. Acta Parasitologica Portuguesa 11:33–35

Śpiewak R, Lundberg M, Johansson SGO, Buczek A (2006) Allergy to pigeon tick (*Argas reflexus*) in Upper Silesia. Poland Ann Agric Environ Med 13:107–112

Tahmasebi F, Ghiasi SM, Mostafavi E, Moradi M, Piazak N, Mozafari A, Haeri A, Fooks AR, Chinikar S (2010) Molecular epidemiology of Crimean-Congo hemorrhagic fever virus genome isolated from ticks of Hamadan province of Iran. J Vector Borne Dis 47:211–216

Theodor O, Costa M (2009) New species and new records of Argasidae from Israel. Observations on the rudimentary scutum and the respiratory system of the larvae of the Argasidae. Parasitology 50:365. https://dx.doi.org/10.1017/S0031182000025464

Uspensky I, Robinson WH, Bajomi D (2008) Ticks (Acari: Ixodoidea) as urban pests and vectors with special emphasis on ticks outside their geographical range. In: 6th International Conference on Urban Pests, Budapest, Hungary, 13–16 July 2008. International Conference on Urban Pests (ICUP), pp. 333–347

Vermeil C, Marjolet M, Chastel C (1996) *Argas* et arbovirus actualités. Bull la Société Pathol Exot 89:363–365

Weckesser S, Hilger C, Lentz D, Jakob T (2010) Anaphylactic reactions to bites of the pigeon tick *Argas reflexus*. Eur J Dermatol 20:244–245. https://dx.doi.org/10.1684/ejd.2010.0886

Argas polonicus Siuda, Hoogstraal, Clifford and Wassef, 1979 (Fig. 6)

T. N. Petney, M. P. Pfäffle, and A. Estrada-Peña

Life Cycle and Host Preferences

Argas polonicus is a parasite of rock pigeons (*Columba livia*). It was first found in Krakow, Poland, where trumpeters of St. Mary's Church were frequently bitten by this tick when their hosts were driven from their resting places (Siuda 1982). Molecular (Bruin et al. 1999) and biochemical tests based on cuticular hydrocarbon composition (Estrada-Peña and Dusbábek 1993) supported the early morphological observations reporting this tick as a new species (Siuda et al. 1979).

Ecology

This tick is nocturnal and lives close to its hosts' nests, preferably in shallow cracks in walls (Siuda et al. 1979; Siuda 1982). Ticks seem not to be active in winter (Siuda et al. 1979). First activity has been reported in April continuing to late October/early November. Unless the ticks are disturbed, no daytime activity occurs (Siuda et al. 1979). An activity peak is reached in the summer months (July, August), in warm years also in September (Siuda et al. 1979). Larvae are mostly active in spring and late summer, while the other life stages are most abundant during summer (Siuda et al. 1979). Overall, this species shows a wide range of thermal preferences. For larvae, this is around 21 °C at night and 16 °C during the day (Siuda 1979). Adult thermal preferences vary within the range 14–17 °C at night and 15–23 °C during the day, depending on the season (Siuda 1979).

Distribution

This species has a very small reported distribution. It replaces or occurs with *A. reflexus* in some parts of Eastern Europe. It has been recorded from Poland and Slovakia (Siuda et al. 1979; Dusbábek 1985).

Vectorial Capacity and Pathogen Burden

Nothing is known about the role of *A. polonicus* in the transmission of disease pathogens, but itchy, erythematous lesions were recorded on legs and arms of humans bitten by this tick (Estrada-Peña and Jongejan 1999).

T. N. Petney (✉)
State Museum of Natural History, Erbprinzenstr. 13, 76133 Karlsruhe, Germany
e-mail: trevor.petney@kit.edu

M. P. Pfäffle
Department of Ecology and Parasitology, Institute of Zoology, Karlsruhe Institute of Technology, Karlsruhe, Germany
e-mail: miripfaeffle@web.de

A. Estrada-Peña
Department of Animal Health, Faculty of Veterinary Medicine, Zaragoza, Spain
e-mail: aestrada@unizar.es

A. Estrada-Peña et al. (eds.), *Ticks of Europe and North Africa*,
https://doi.org/10.1007/978-3-319-63760-0_6

Fig. 6 **A** Distribution of *A. polonicus* in Europe and Northern Africa (10 × 10 km grid presence with black dots). **B** Countries where the species has been reported are marked in dark grey

References

Bruin J, van der Geest LPS, Sabelis MW (eds) (1999) Ecology and evolution of the Acari. Springer Netherlands, Dordrecht. https://dx.doi.org/10.1007/978-94-017-1343-6

Dusbábek F (1985) Identity of *Argas* (*Argas*) *polonicus* populations in Czechoslovakia and Poland. Folia Parasitol (Praha) 32:163–171

Estrada-Peña A, Dusbábek F (1993) Cuticular hydrocarbon gas chromatography analysis of *Argas vulgaris*, *A. polonicus*, and their hybrids. Exp Appl Acarol 17:365–376. https://dx.doi.org/10.1007/BF00058599

Estrada-Peña A, Jongejan F (1999) Ticks feeding on humans: a review of records on human-biting Ixodoidea with special reference to pathogen transmission. Exp Appl Acarol 23:685–715. https://dx.doi.org/10.1023/A:1006241108739

Siuda K (1982) The differentiation of nymphal *Argas* (*Argas*) *polonicus* Siuda, Hoogstraal, Clifford and Wassef, 1979 (Acarina, Ixodides, Argasidae). Wiad Parazytol 28:51–55

Siuda K (1979) Investigations on the biology of the tick *Argas* (*Argas*) *polonicus* Siuda, Hoogstraal, Clifford et Wassef, 1979 (Acarina: Ixodides: Argasidae). 2. Thermal preferendum of larval and male and female adult stages. Folia Biol (Praha) 27:209–217

Siuda K, Hoogstraal H, Clifford CM, Wassef HY (1979) Observations on the subgenus *Argas* (Ixodoidea: Argasidae: *Argas*). 17. *Argas* (*A.*) *polonicus* sp. n. parasitizing domestic pigeons in Krakow, Poland. J Parasitol 65:170–181

Argas macrostigmatus Filippova, 1961 (Fig. 7)

T. N. Petney, M. P. Pfäffle, and A. Estrada-Peña

Life Cycle and Host Preferences

Argas macrostigmatus is a parasite of the crested cormorant *Phalacrocorax aristotelis* (Hoogstraal et al. 1984).

Ecology

Unlike most *Argas* spp., which prefer dry and/or hot regions, *A. macrostigmatus* lives in damp habitats (Oliver 1989). It infests wet nests of *P. aristotelis* as does *Ornithodoros maritimus*, which makes this the only case where *Argas* and *Ornithodoros* spp. coexist (Oliver 1989; Estrada-Peña et al. 2010).

Distribution

This tick is known from islands of the Black Sea and the Mediterranean (Corsica) (Filippova 1966; Beaucournu et al. 1982; Hoogstraal et al. 1984).

Vectorial Capacity and Pathogen Burden

Nothing is known about the vector capacity of this tick.

T. N. Petney (✉)
State Museum of Natural History, Erbprinzenstr. 13, 76133 Karlsruhe, Germany
e-mail: trevor.petney@kit.edu

M. P. Pfäffle
Department of Ecology and Parasitology, Institute of Zoology, Karlsruhe Institute of Technology, Karlsruhe, Germany
e-mail: miripfaeffle@web.de

A. Estrada-Peña
Department of Animal Health, Faculty of Veterinary Medicine, Zaragoza, Spain
e-mail: aestrada@unizar.es

A. Estrada-Peña et al. (eds.), *Ticks of Europe and North Africa*,
https://doi.org/10.1007/978-3-319-63760-0_7

Fig. 7 **A** Distribution of *A. macrostigmatus* in Europe and Northern Africa (10 × 10 km grid presence with black dots). **B** Countries where the species has been reported are marked in dark grey

References

Beaucournu JC, Monnat JY, Launay H (1982) *Ceratophyllus* (*Ceratophyllus*) *chasteli* n. sp. parasit du cormoran huppé *Phalacrocorax aristotelis* (L.) en Corse (Siphon. Ceratophyllidae). Bull Soc Entomol Fr 87:38–43

Estrada-Peña A, Mangold AJ, Nava S et al (2010) A review of the systematics of the tick family Argasidae (Ixodida). Acarologia 50:317–333. https://dx.doi.org/10.1051/acarologia/20101975

Filippova NA (1966) Argasid ticks (Argasidae). Fauna SSSR 4:1–255 (In Russian)

Hoogstraal H, Wassef HY, Guiguen C, Beaucournu JC (1984) Observations on the subgenus *Argas* (Ixodoidea: Argasidae: *Argas*). 19. Description of the larva of the Palaearctic cormorant parasite *A.* *(A.) macrostigmatus*. J Parasitol 70:292–294

Oliver JH (1989) Biology and systematics of ticks (Acari: Ixodida). Annu Rev Ecol Syst 20:397–430. https://dx.doi.org/10.1146/annurev.es.20.110189.002145

Argas vespertilionis (Latreille, 1796) (Figs. 8 and 9)

T. N. Petney, T. G. T. Jaenson, and M. P. Pfäffle

Hoogstraal (1958) and Filippova (1966) provide detailed descriptions of the morphology of the different stages of *A. vespertilionis*.

Life Cycle and Host Preferences

The short-legged bat tick, *A. vespertilionis*, is a homoxenous parasite of bats, including the northern bat (*Eptesicus nilssonii*), the common pipistrelle (*Pipistrellus pipistrellus*), Kuhl's pipistrelle (*Pipistrellus kuhlii*) and the lesser horseshoe bat (*Rhinolophus hipposideros*) (Imaz et al. 1999; Walter and Rackow 2007; Obsomer et al. 2013). Like *A. reflexus*, *A. vespertilionis* is known to attack humans and has been reported to be 'highly aggressive' (Hoogstraal 1985; Jaenson et al. 1994; Estrada-Peña and Jongejan 1999). It can occasionally be found on dogs (Jaenson et al. 1994) and birds (Shah et al. 2006).

The larvae stay on the bat host and feed for about three weeks. In contrast, the nymphs and the adults feed only briefly (20–50 min) so that they are rarely found on their hosts. Individuals that moult to adult males usually pass through only one nymphal stage. The females, on the other hand, usually pass through two nymphal stages before they reach the adult stage (Hoogstraal 1956, 1958).

Ecology

Argas vespertilionis is an endophilic and nocturnal species that occurs in the habitats of both migrating and nesting populations of many different bat species, e.g. attics, burrows, tree holes and caves (Arthur 1963; Siuda et al. 2009).

Distribution

Argas vespertilionis has a wide geographical distribution, which is typical for ticks associated with bats. It is found in Europe, Asia and Africa (Hoogstraal 1985). Countries where *A. vespertilionis* has been recorded include Austria (Siuda et al. 2009), Belgium (Obsomer et al. 2013), Czech Republic (Siuda et al. 2009), France (Socolovschi et al. 2012), Egypt (Hoogstraal 1956), Germany (Petney et al. 2012), Ghana (Manzano-Román et al. 2012), United Kingdom (Jameson and Medlock 2011), Hungary (Siuda et al. 2009), India (Manzano-Román et al. 2012), Iran (Hosseini-Chegeni and Tavakoli 2013), Iraq (Estrada-Peña and Jongejan 1999), Japan (Yamaguti et al. 1971), Korea (Lee et al. 1998), Kyrgyzstan (L'vov et al. 1973), Netherlands (Obsomer et al. 2013), Norway (Mehl 1983), Pakistan (Shah et al. 2006), Poland (Siuda et al. 2009), Portugal (Caeiro 1999), Republic of Kazakhstan (Zhmaeva et al. 1966), Russia (Estrada-Peña and Jongejan 1999), Slovakia (Siuda et al. 2009), Spain (Estrada-Peña and Jongejan 1999; Imaz et al. 1999), Sweden (Jaenson et al. 1994), Tadzhikistan (Manzano-Román et al. 2012), Turkey (Bursali et al. 2012), Turkmenistan (Manzano-Román et al. 2012) and Ukraine (Siuda et al. 2009). It has been recorded also in Ireland, Italy,

T. N. Petney (✉)
State Museum of Natural History, Erbprinzenstr. 13, 76133 Karlsruhe, Germany
e-mail: trevor.petney@kit.edu

T. G. T. Jaenson
Medical Entomology Unit, Department of Organismal Biology, Evolutionary Biology Centre, Uppsala University, Uppsala, Sweden
e-mail: thomas.jaenson@ebc.uu.se

M. P. Pfäffle
Department of Ecology and Parasitology, Institute of Zoology, Karlsruhe Institute of Technology, Karlsruhe, Germany
e-mail: miripfaeffle@web.de

A. Estrada-Peña et al. (eds.), *Ticks of Europe and North Africa*,
https://doi.org/10.1007/978-3-319-63760-0_8

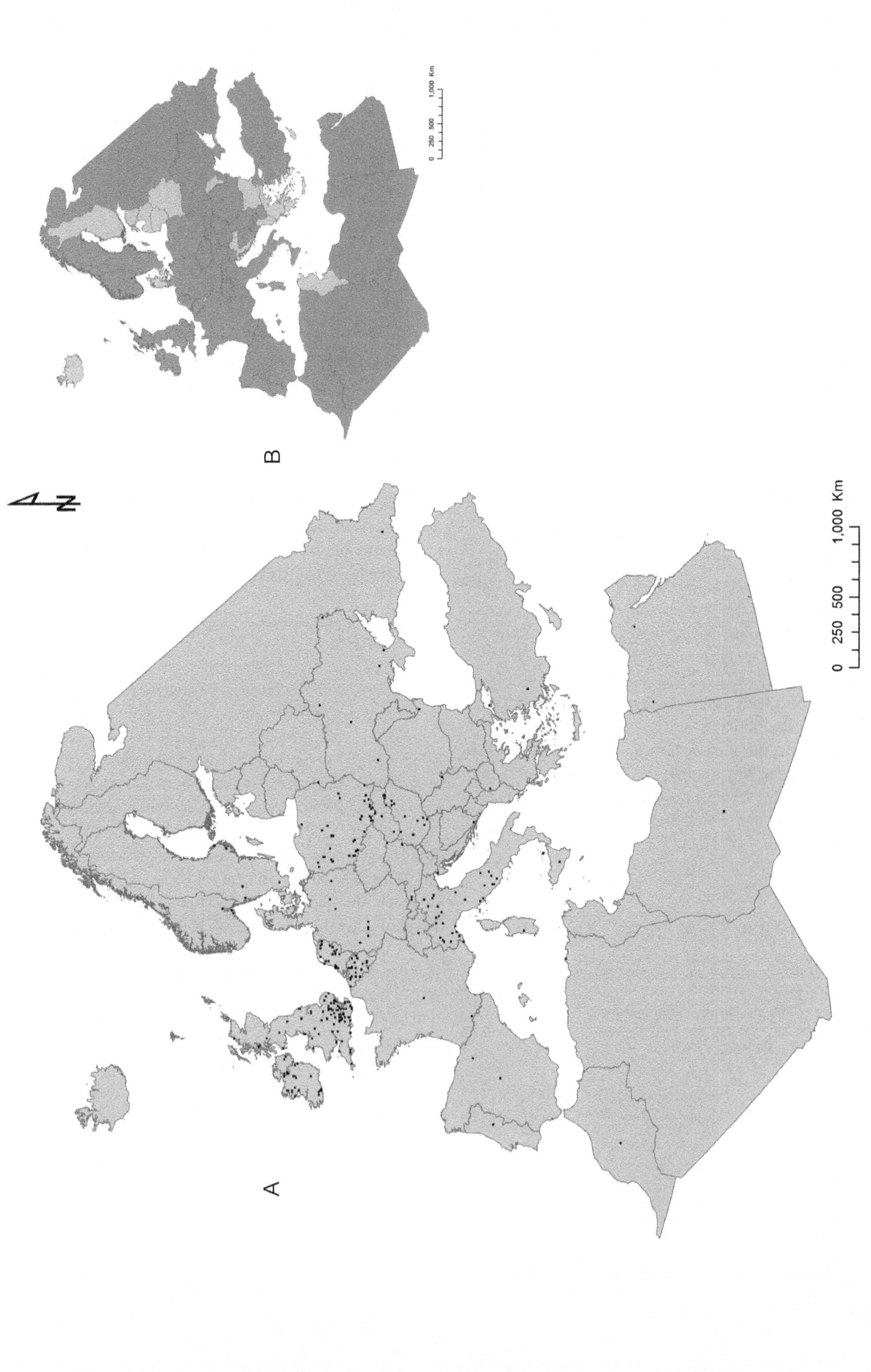

Fig. 8 **A** Distribution of *A. vespertilionis* in Europe and Northern Africa (10 × 10 km grid presence with black dots). **B** Countries where the species has been reported are marked in dark grey

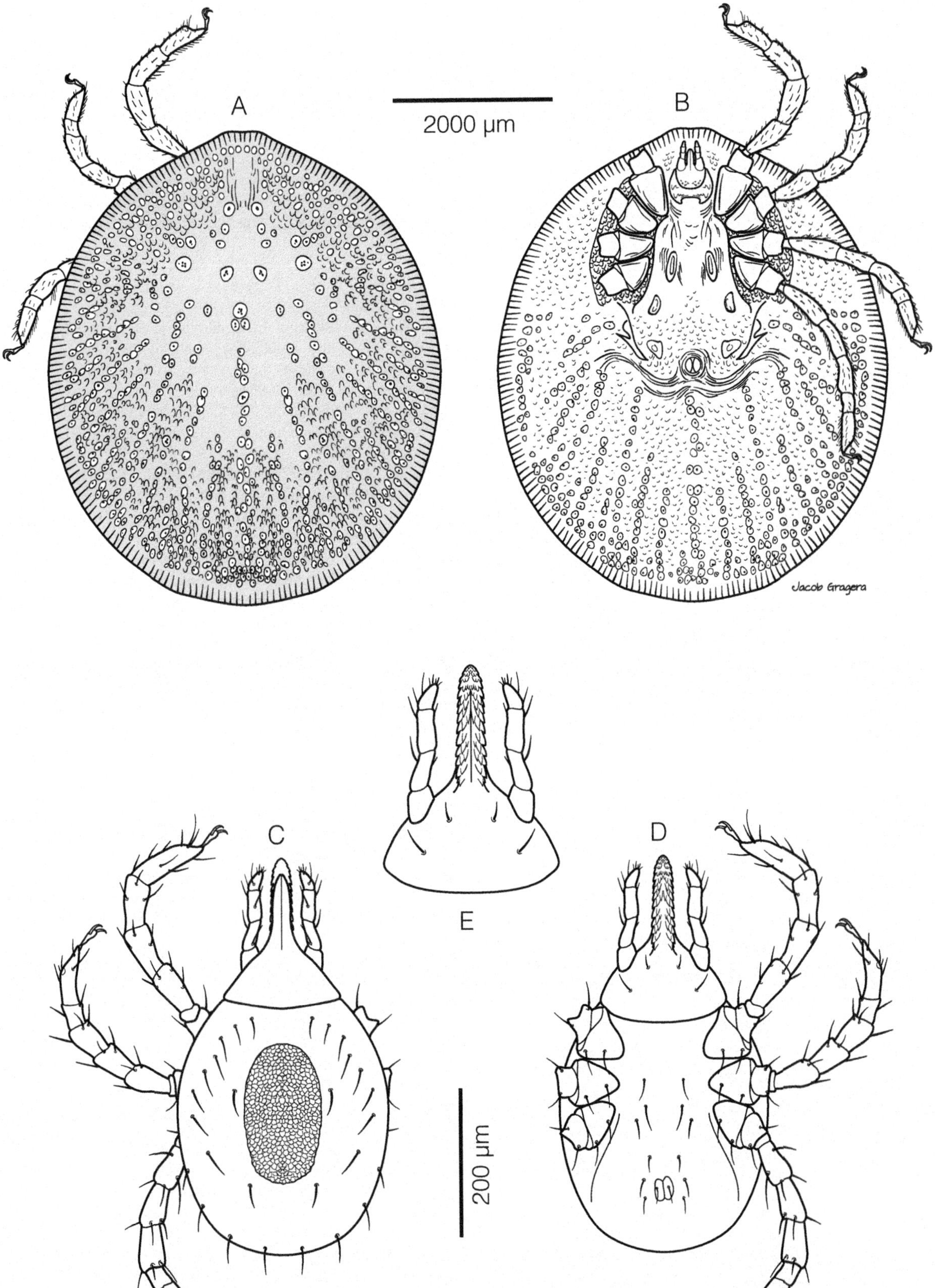

Fig. 9 **A–B** The adult of *A. vespertilionis*. **A** Dorsal, **B** ventral. **C–E** The larva of *A. vespertilionis*. **C** Dorsal, **D** ventral, **E** details of the capitulum. Illustrations from specimens collected in Spain and Italy

Switzerland, Romania, Serbia, Republic of Macedonia, Slovenia, Algeria, Morocco and Libya.

Vectorial Capacity and Pathogen Burden

When humans are bitten by *A. vespertilionis*, itching may persist for several weeks (Arthur 1963; Hosseini-Chegeni and Tavakoli 2013). Ulceration, fever and lymph node enlargement occurred in two persons bitten in their bedroom by this tick species (Jaenson et al. 1994).

Issyk-Kul and Sokuluk viruses have been isolated from this species (L'vov et al. 1973; Vermeil et al. 1996; Obsomer et al. 2013). The Q fever rickettsia, *Coxiella burnetii*, has been detected in *A. vespertilionis* ticks collected from southern Kazakhstan (Zhmaeva et al. 1966). In France, *A. vespertilionis* collected from a bat-infested attic were infected with three different bacteria: a new species of the *Ehrlichia canis* group (*Ehrlichia* sp. AvBat), a new species or subspecies of the spotted fever group (*Rickettsia* sp. AvBat) and a *Borrelia* sp. CPB1 (Socolovschi et al. 2012). The latter belongs to the relapsing fever group that caused a fatal borreliosis in a bat in the United Kingdom (Evans et al. 2009). Additionally, *B. burgdorferi* s.l. DNA has been detected in museum specimens of *A. vespertilionis* in the United Kingdom (Hubbard et al. 1998). However, this association has never been confirmed. For all of these potentially human pathogenic bacteria the vector competence of *A. vespertilionis* is unknown.

References

Arthur DR (1963) British ticks. Butterworths, London

Bursali A, Keskin A, Tekin S (2012) A review of the ticks (Acari: Ixodida) of Turkey: species diversity, hosts and geographical distribution. Exp Appl Acarol 57:91–104. https://dx.doi.org/10.1007/s10493-012-9530-4

Caeiro V (1999) General review of tick species present in Portugal. Parassitologia 41:11–15

Estrada-Peña A, Jongejan F (1999) Ticks feeding on humans: a review of records on human-biting Ixodoidea with special reference to pathogen transmission. Exp Appl Acarol 23:685–715. https://dx.doi.org/10.1023/A:1006241108739

Evans NJ, Bown K, Timofte D et al (2009) Fatal borreliosis in bat caused by relapsing fever spirochete, United Kingdom. Emerg Infect Dis 15:3–1331. https://dx.doi.org/10.3201/eid1508.090475

Filippova NA (1966) Argasid ticks (Argasidae). Fauna SSSR, Paukoobraznye 4(3)

Hoogstraal H (1958) Bat ticks of the genus *Argas* (Ixodoidea, Argasidae), 3. The subgenus *Carios*, a redescription of *A.* (*C.*) *vespertilionis* (Latreille, 1802), and variation within an Egyptian population. Ann Entomol Soc Am 51:19–26. https://dx.doi.org/10.1093/aesa/51.1.19

Hoogstraal H (1985) Argasid and nuttalliellid ticks as parasites and vectors. Adv Parasitol 24:135–238

Hoogstraal H (1956) African Ixodoidea. 1. Ticks of the Sudan. Washington, DC

Hosseini-Chegeni A, Tavakoli M (2013) *Argas vespertilionis* (Ixodida: Argasidae): a parasite of pipistrel bat in Western Iran. Persian J Acarol 2:321–330

Hubbard MJ, Baker AS, Cann KJ (1998) Distribution of *Borrelia burgdorferi* s.l. spirochaete DNA in British ticks (Argasidae and Ixodidae) since the 19th Century, assessed by PCR. Med Vet Entomol 12:89–97. https://dx.doi.org/10.1046/j.1365-2915.1998.00088.x

Imaz E, Aihartza JR, Totorika MJ (1999) Ectoparasites on bats (Gamasida, Ixodida, Diptera) in Biscay (N Iberian peninsula). Misc Zool 22:21–30

Jaenson TGT, Tälleklint L, Lundqvist L et al (1994) Geographical distribution, host associations, and vector roles of ticks (Acari: Ixodidae, Argasidae) in Sweden. J Med Entomol 31:240–256. https://dx.doi.org/10.1093/jmedent/31.2.240

Jameson LJ, Medlock JM (2011) Tick surveillance in Great Britain. Vector Borne Zoonotic Dis 11:12–403. https://dx.doi.org/10.1089/vbz.2010.0079

L'vov DK, Karas FR, Timofeev EM, et al (1973) Issyk-Kul virus, a new arbovirus isolated from bats and *Argas* (*Carios*) *vespertilionis* (Latr., 1802) in the Kirghiz S.S.R. Arch Gesamte Virusforsch 42:207–209. https://dx.doi.org/10.1007/BF01270841

Lee YT, Park CH, Cho KB, Yoon JJ (1998) Study of exoparasites, *Rickettsia* and Hantaan virus in bats. J Korean Soc Virol 28:1–9

Manzano-Román R, Díaz-Martín V, Fuente JD La, Pérez-Sánchez R (2012) Soft ticks as pathogen vectors: distribution, surveillance and control. INTECH Open Access Publisher

Mehl R (1983) Distribution and host relations of Norwegian ticks (Acari, Ixodides). Fauna Nor Ser B 30:46–51

Obsomer V, Wirtgen M, Linden A et al (2013) Spatial disaggregation of tick occurrence and ecology at a local scale as a preliminary step for spatial surveillance of tick-borne diseases: general framework and health implications in Belgium. Parasit Vectors 6:190. https://dx.doi.org/10.1186/1756-3305-6-190

Petney TN, Pfäffle MP, Skuballa JD (2012) An annotated checklist of the ticks (Acari: Ixodida) of Germany. Syst Appl Acarol 17:115–170. https://dx.doi.org/10.11158/saa.17.2.2

Shah AH, Khan MN, Iqbal Z et al (2006) Some epidemiological aspects and vector role of tick infestation on layers in the Faisalabad district (Pakistan). World's Poult Sci J 62:145–157. https://dx.doi.org/10.1079/WPS200591

Siuda K, Stanko M, Piksa K, Górz A (2009) Ticks (Acari: Ixodida) parasitizing bats in Poland and Slovakia. Wiad Parazytol 55:39–45

Socolovschi C, Kernif T, Raoult D, Parola P (2012) *Borrelia*, *Rickettsia*, and *Ehrlichia* species in bat ticks, France, 2010. Emerg Infect Dis 18:75–1966. https://dx.doi.org/10.3201/eid1812.111237

Vermeil C, Marjolet M, Chastel C (1996) *Argas* et arbovirus actualités. Bull Société Pathol Exot 89:363–365

Walter G, Rackow W (2007) Außergewöhnlich hoher Befall einer Nordfledermaus, *Eptesicus nilssonii*, mit der Lederzecke, *Argas vespertilionis* (Argasidae). Nyctalus 12:372–378

Yamaguti N, Tipton VJ, Keegan HL, Toshika S (1971) Ticks of Japan, Korea and the Ryukyu Islands. Brigham Young Univ Sci Bull 15:1–226

Zhmaeva ZM, Pchelkina AA, Belashova VS (1966) Spontaneous infection of *Argas vespertilionis* with *Rickettsia burnetii* in the south of Kazakhstan. Med Parazitol (Mosk) 35:6–595

Argas transgariepinus White, 1846 (Figs. 10 and 11)

T. N. Petney, M. P. Pfäffle, and A. Estrada-Peña

Only limited information is available on this species. Other than some records used in its original description, only a few data on its distribution are available. It is an argasid which has some peculiar morphological features, as described by Hoogstraal (1957). The larva has peritremes, a kind of elongated structure on the lateral and ventral sides of the idiosome, which are related to air exchange. These are most similar to those found in predatory and free-living mesostigmatid mites. This unusual feature, plus others found in the adult stage, provided enough background to create the subgenus *Secretargas*.

Life Cycle and Host Preferences

This is a tick found on bats, often on solitary species or in bat roosting sites. Almost all of the information comes from larvae which feed on their hosts for several days. Recorded hosts include bats from the genera *Eptesicus*, *Hypsugo*, *Myotis*, *Otonycteris*, *Pipistrellus*, *Plecotus*, *Rhinopoma*, *Rhinolophus* and *Vespertilio*. Although prevalence is low, some individual hosts may have high infestation levels. Medard et al. (2001) reported 76 larvae occurring on a single *Plecotus austriacus*. Some specimens were also found to contain the blood of geckos and in the laboratory, this species feeds on humans (Hoogstraal 1957; Hoogstraal et al. 1967; Manilla 1998; Medard et al. 2001; Frank et al. 2015).

Ecology

Only limited information is available on the ecology of this species. Hoogstraal et al. (1967) indicate that it is very hard to observe, as it is found hidden in rock crevices, at the entrance to stone wall interstices of bat caves, and because its thin body blends with its surroundings (Hoogstraal et al. 1967; Medard et al. 2001). In Egypt, it is found in rock formations in arid, desert habitats and seldom, in more humid caves used by gregarious bat species (Hoogstraal 1957). It has also been found in human habitations (Medard et al. 2001).

Distribution

Due to the migratory behaviour of some of its bat hosts, *A. transgariepinus* is distributed from Africa to middle Europe (Oliver 1989). It was originally described from South Africa and has since been recorded from Algeria, Egypt, France, Greece, Israel, Italy, Morocco, Switzerland and Spain (Beaucournu and Clerc 1968; Rageau 1972; Estrada-Peña and de la Cruz, unpublished observations).

Vectorial Capacity and Pathogen Burden

No information is available.

T. N. Petney (✉)
State Museum of Natural History, Erbprinzenstr. 13, 76133 Karlsruhe, Germany
e-mail: trevor.petney@kit.edu

M. P. Pfäffle
Department of Ecology and Parasitology, Institute of Zoology, Karlsruhe Institute of Technology, Karlsruhe, Germany
e-mail: miripfaeffle@web.de

A. Estrada-Peña
Department of Animal Health, Faculty of Veterinary Medicine, Zaragoza, Spain
e-mail: aestrada@unizar.es

A. Estrada-Peña et al. (eds.), *Ticks of Europe and North Africa*,
https://doi.org/10.1007/978-3-319-63760-0_9

Fig. 10 **A** Distribution of *A. transgariepinus* in Europe and Northern Africa (10 × 10 km grid presence with black dots). **B** Countries where the species has been reported are marked in dark grey

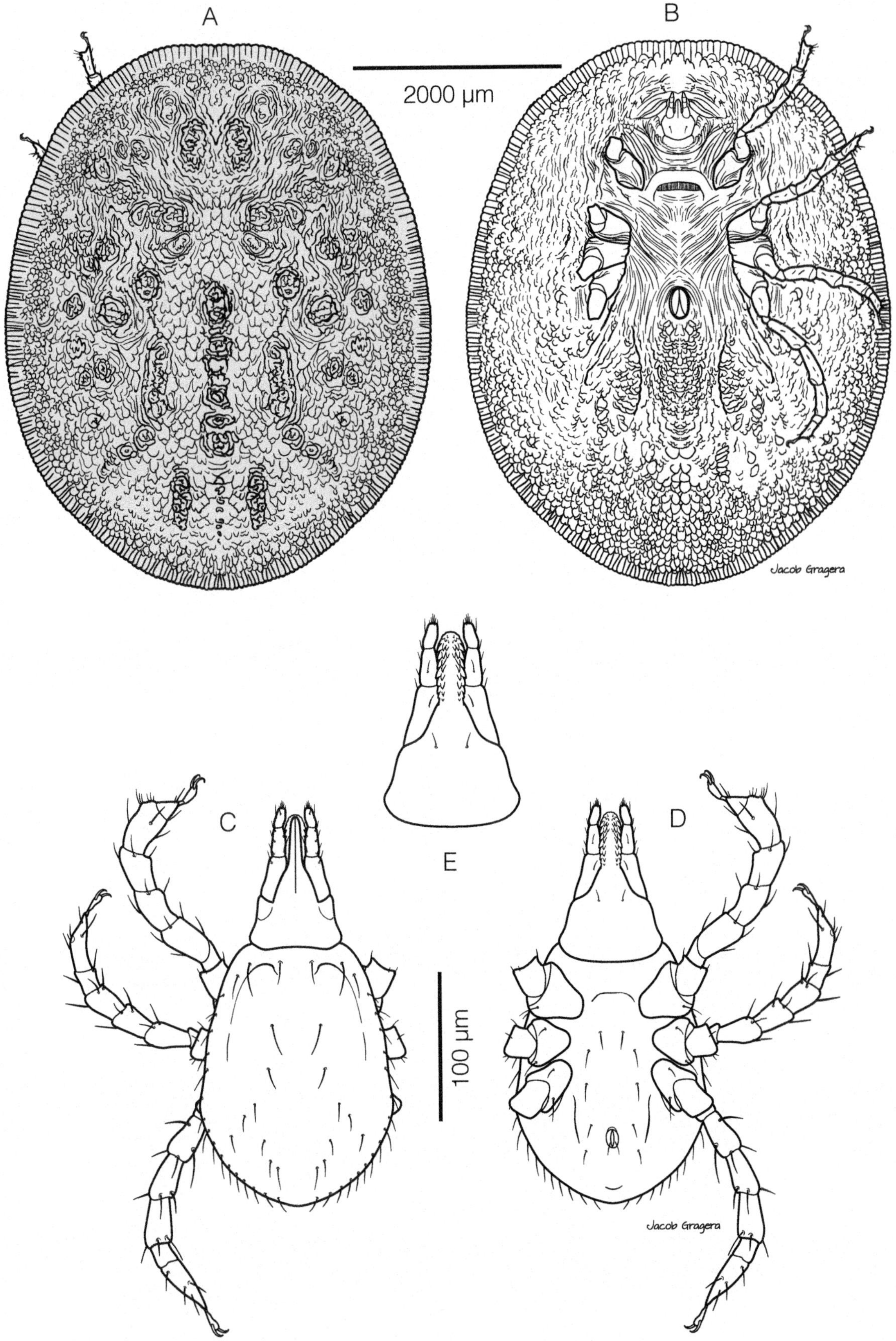

Fig. 11 **A, B** The adult of *A. transgariepinus*. **A** Dorsal, **B** ventral. **C–E** The larva of *A. transgariepinus*. **C** Dorsal, **D** ventral, **E** details of the capitulum. Illustrations from specimens collected in Spain

References

Beaucournu JC, Clerc B (1968) *Argas (Secretargas) transgariepinus* White, 1846, new tick for France and Algeria. Vie Milieu Ser C Biol Terr 19:233–236

Frank R, Kuhn T, Werblow A, Liston A, Kochmann J, Klimpel S (2015) Parasite diversity of European *Myotis* species with special emphasis on *Myotis myotis* (Microchiroptera, Vespertilionidae) from a typical nursery roost. Parasit Vectors 8:1–13. https://dx.doi.org/10.1186/s13071-015-0707-7

Hoogstraal H (1957) Bat ticks of the genus *Argas* (Ixodoidea, Argasidae) 2. *Secretargas* new subgenus and *A. transgariepinus* White, 1846, its adult and immature stages; with a definition of the subgenus *Argas*. Ann Entomol Soc Am 50:544–549. https://dx.doi.org/10.1093/aesa/50.6.544

Hoogstraal H, Uilenberg G, Blanc CP (1967) The *Argas* subgenus Secretargas (Ixodoidea, Argasidae) in Madagascar: descriptions of *A. (S.) echinops*, new species, and of the adults and nymph of *A. (S.) hoogstraali*. Ann Entomol Soc Am 60(1):62–149

Manilla G (1998) Fauna d'Italia. Acari, Ixodida. Edizioni Calderini, Bologna. 280 pp

Medard P, Guiguen C, Beaucournu JC (2001) Nouvelles récoltes d'*Argas transgariepinus* White, 1946 tique de chiroptères (Acarina - Ixodoidea - Argasidae) en France et au Maroc. Sci Rep Port-Cros Natl Park, Fr: Notes brèves 18:221–226

Oliver JH (1989) Biology and systematics of ticks (Acari: Ixodida). Annu Rev Ecol Syst 20:397–430. https://dx.doi.org/10.1146/annurev.es.20.110189.002145

Rageau J (1972) Repartation geographique et role pathogene des tiques (Acariens: Argasidae et Ixodidae) en France. Wiad Parazytol 18:707–719

Genus *Ornithodoros* Koch, 1844

A. Estrada-Peña, G. Kleinerman, and G. Baneth

The genus *Ornithodoros* was first described with the reclassification of *Ornithodoros coriaceus* and *Ornithodoros savignyi*, which had previously been attributed to the genus *Argas*. Koch translated 'Ornis', which means bird from Greek to Latin and 'doros' meaning a leather sack (Petney and Maiwald 1996). These ticks are parasites of rodents and other mammals that inhabit burrows, caves or domestic animal shelters, as well as marine birds nesting in large colonies. Several species of *Ornithodoros* are associated with bats and a few species are specific to reptiles (Hoogstraal 1985). The genus currently includes 113 species; however, there are differences of opinion between taxonomists on the systematic status of several tick species, probably because the genus is paraphyletic (Klompen and Oliver 1993; Guglielmone et al. 2010; Dantas-Torres et al. 2012). Six species have been recorded so far in the Western Palaearctic.

Morphological descriptions and illustrations of all life history stages of *O. capensis* are available in Filippova (1963, 1966). Illustrations and description of the larvae are available in Kohls et al. (1965), and adults were briefly described when compared to *O. maritimus* by Hoogstraal et al. (1976a, b).

Hoogstraal et al. (1976a) redescribed the larvae of *O. maritimus* and provided illustrations and keys for the remaining stages. They also distinguished this species from the closely related *O. coniceps* and *O. capensis* using illustrations and SEM images (Hoogstraal et al. 1976a). An adequate revision of the taxonomy of adults and immatures of *O. conciceps* was performed by Filippova (1963) and Hoogstraal et al. (1979), with comparison to other species of the *O. capensis* complex. Both references include illustrations, and Hoogstraal's report includes SEM images. Another description of the species is available in Filippova (1966).

A characterization of the larva of *O. lahorensis*, with keys and illustrations, is found in Sonenshine et al. (1966). Descriptions and illustrations of adults and larva are found in Filippova (1966). Phylogenetic analysis of nuclear rRNA sequences strongly support the placement of *O. lahorensis* within the genus *Argas* (Burger et al. 2014), in agreement with previous phylogenetic studies of the group (Black et al. 1997; Klompen et al. 2007).

Ornithodoros alactagalis Issaakjan, 1936, was described from specimens collected in burrows of rodents, badgers, foxes and hedgehogs (Filippova 1966; Sonenshine et al. 1966). Morphological characterization with illustrations and keys to larvae can be found in Sonenshine et al. (1966). Descriptions of adults and larvae, with illustrations, are available in Filippova (1966).

The larva of *O. tholozani* is described, with keys and illustrations, by Sonenshine et al. (1966). *Ornithodoros papillipes* Birula, 1895, was considered a synonym of *O. tholozani* by Hoogstraal (1985) and Sonenshine et al. (1966). It is, however, now considered a valid species (Assous and Wilamowski 2009). The adults of *O. papillipes* were adequately characterized by Pavlovsky and Skruinnik (1956). Filippova (1966) redescribed the adults and larvae of this species. Illustrations and keys are available in both references. *Ornithodoros crossi* Brumpt, 1921, was considered a synonym of *O. papillipes* by Pavlovsky (1928).

There is some confusion as to whether *O. erraticus* (Lucas 1849), *O. marocanus* Velu, 1919, *O. normandi* Larrousse, 1923 and *O. sonrai* Sautet and Witkowski, 1943 represent different species or only genetic variants of a species complex that have allopatric distributions (Hoogstraal 1985; Morel 2003; Vial et al. 2006; Trape et al. 2013). Some authors consider these species as valid and even

A. Estrada-Peña (✉)
Department of Animal Health, Faculty of Veterinary Medicine, Zaragoza, Spain
e-mail: aestrada@unizar.es

G. Kleinerman · G. Baneth
Koret School of Veterinary Medicine, Hebrew University, Rehovot, Israel
e-mail: gabriela.kleinerman@mail.huji.ac.il

G. Baneth
e-mail: gad.baneth@mail.huji.ac.il

A. Estrada-Peña et al. (eds.), *Ticks of Europe and North Africa*,
https://doi.org/10.1007/978-3-319-63760-0_10

recognize several new species based only on molecular grounds. Molecular phylogenetic analysis of 16S rRNA sequences contributed to the genetic differentiation of nine different species within the *O. erraticus* complex (Trape et al. 2013). Five new species, e.g. *O. occidentalis*, *O. costalis*, *O. rupestris*, *O. kairouanensis* and *O. merionesi*, were named (but not described), and four others identified as *O. erraticus*, *O. marocanus*, *O. sonrai* and *O. normandi* were confirmed (Trape et al. 2013) by single mutations in the 16S gene. However, these findings are not supported by a validation involving crosses and back crosses of the alleged new species, by description of immatures, nor by further molecular details. It is well known (Hoogstraal 1985) that the adults of the genus *Ornithodoros* do not have adequate characters allowing a reliable identification. The morphological analysis by Trape et al. (2013) also concluded that most of these species cannot be adequately separated by morphological features alone. However, Bouattour et al. (2010) provided reliable features to separate the adults of both *O. erraticus* and *O. normandi.* We hesitate to accept the species named by Trape et al. (2013) until further proofs accumulate, and strongly recommend the reporting of these ticks as belonging to the *O. erraticus* complex. A recent publication by Nava et al. (2016) on the South American tick *Amblyomma parvum* shows that incorrect interpretations can occur if only molecular data are used to define species. Other than the report by Bouattour et al. (2010), unpublished data on laboratory crosses strongly suggests that both *O. erraticus* and *O. sonrai* produce hybrid progeny. The best available illustrations of *O. erraticus* are those published by Hoogstraal et al. (1981) for both adults and larvae.

Key to the Species of Genus *Ornithodoros* in Europe and Northern Africa

Adults and Nymphs

Note: *Ornithodoros alactagalis* is not illustrated because of the lack of adequate material.

1a. Cheeks absent. Dorsoventral groove, pre-anal groove, and transverse postanal groove absent...***Ornithodoros lahorensis*** (Fig. 20).
1b. Cheeks present or absent. Dorsoventral groove present or absent, pre-anal groove and transverse postanal groove present...**2**.
2a. Cheeks present. Dorsoventral groove absent. Pre-anal groove and transverse postanal groove both present...**3** (**Subgenus *Alectorobius***).
2b. Cheeks present or absent. Dorsoventral groove, pre-anal groove, and transverse postanal groove present...**5** (**Subgenus *Pavlovskyella***).
3a. Mammillae of posteromedian area are evenly distributed. Discs of the posterolateral quadrant on the dorsum in several compartments...***Ornithodoros maritimus*** (Fig. 17).
3b. Mammillae of posteromedian area and discs of posterolateral quadrant not as above...**4**.
4a. Mammillae of posteromedian area arranged in chainlike rows with distinctly broad interstices...***Ornithodoros coniceps*** (Fig. 15).
4b. Discs of posterolateral quadrant in only one or two compartments...***Ornithodoros capensis*** (Fig. 13).
5a. Marginal dorsal groove well evident as a clear ridge separating the central portions of the dorsum from the lateral ones, and with very different types of mammillae on the central or the lateral portions of the dorsum...**6**.
5b. Marginal dorsal groove not apparent, with similar mammillae on the central and lateral portions of the dorsum...***Ornithodoros erraticus*** (Fig. 24).
6a. Dorsoventral groove expanded on marginal dorsal area. Small mammillae ordered in chain-like rows, often concentrated posteriorly...***Ornithodoros tholozani*** (Fig. 22).
6b. Dorsoventral groove not expanded on the dorsal part of the idiosome, rarely seen in the dorsal view. Mammillae robust, densely and more or less evenly dispersed...***Ornithodoros alactagalis***.

Larvae

Note: *Ornithodoros alactagalis* is not illustrated because of the lack of adequate material.

1a. Dorsal body surface with more than 20 pairs of dorsolateral setae. Palpal articles I through IV with setae as follows: 0, 5 5, and 10–12, respectively. Respiratory apparatus present...***Ornithodoros lahorensis*** (Fig. 20).
1b. Dorsal body surface with less than 20 pairs of dorsolateral setae. Palpal articles I through IV with a different number of setae. Respiratory apparatus absent...**2**.
2a. Palpal article I without short spines dorsally. Dorsal body surface with 10–21 pairs of dorsolateral setae and 3–5 pairs of central setae. Dorsal body setae usually pointed, barbed. Dorsal plate present. Hypostome usually pointed anteriorly, denticles throughout its length, anterior dentition: 3/3 to 5/5...3 (**Subgenus *Alectorobius***).
2b. Palpal article I with short spines dorsally. Dorsal body surface with 13 or 14 pairs of setae, usually divided into 11 or 12 dorsolateral pairs and 2 central pairs. Dorsal body setae blunt, often with a cluster of bulbous structures at their distal ends...**5** (**Subgenus *Pavlovskyella***).
3a. Hypostomal apical dentition 4/4...***Ornithodoros coniceps*** (Fig. 15).
3b. Hypostomal apical dentition 5/5 or 6/6...**4**.
4a. Dorsum of body with 15 pairs of setae. Hypostomal apical dentition 5/5...***Ornithodoros maritimus***.

4b. Dorsum of body with 22–25 pairs of setae. Hypostomal apical dentition 6/6...***Ornithodoros capensis***.

5a. Dorsal setae plumose...***Ornithodoros tholozani*** (Fig. 22).

5b. Dorsal setae of normal aspect...**6**.

6a. Post-hypostomal setae only slightly different in length... ***Ornithodoros alactagalis***.

6b. Post-hypostomal setae ph1 almost 6–8 times longer than ph2...***Ornithodoros erraticus*** (Fig. 24).

References

Assous MV, Wilamowski A (2009) Relapsing fever borreliosis in Eurasia-forgotten, but certainly not gone! Clin Microbiol Infect 15:407–414

Black WC, Klompen JSH, Keirans J (1997) Phylogenetic relationships among tick subfamilies (Ixodida: Ixodidae: Argasidae) based on the 18S nuclear rDNA gene. Mol Phylogenet Evol 7:129–144

Bouattour A, Garnier M, M'Ghirbi Y, Sarih M, Gern L, Ferquel E, Postic D, Cornet M (2010) *Borrelia crocidurae* infection of *Ornithodoros erraticus* (Lucas, 1849) in Tunisia. Vector-Borne Zoonotic Dis 10:825–830

Burger TD, Shao R, Labruna MB, Barker SC (2014) Molecular phylogeny of soft ticks (Ixodida: Argasidae) inferred from mitochondrial genome and nuclear rRNA sequences. Ticks Tick Borne Dis 5:195–207

Dantas-Torres F, Venzal JM, Bernardi LFO, Ferreira RL, Onofrio VO, Marcili A (2012) Description of a new species of bat-associated Argasid tick (Acari: Argasidae) from Brazil. J Parasitol 98:36–45

Filippova NA (1963) On ticks of the genus *Ornithodoros* Koch parasitizing birds. Trudy Zool. Inst Akad Nauk SSSR 21:16–27 (In Russian, in English NAMRU3-T124)

Filippova NA (1966) Argasid ticks (Argasidae). Fauna SSSR 4:1–255 (In Russian)

Guglielmone AA, Robbins RG, Apanaskevich DA, Petney TN, Estrada-Peña A, Horak IG, Shao R, Barker SC (2010) The Argasidae, Ixodidae and Nuttalliellidae (Acari: Ixodida) of the world: a list of valid species names. Zootaxa 2528:1–28

Hoogstraal H (1985) Argasid and nuttalliellid ticks as parasites and vectors. Adv Parasitol 24:135–238

Hoogstraal H, Kaiser MN, Easton ER (1976a) *Ornithodoros* (*Alectorobius*) *capensis* Neumann (Ixodoidea: Argasidae) parasitizing a human and birds nesting on Islands in East African Lakes. J Med Entomol 12:703–704

Hoogstraal H, Clifford CM, Keirans JE, Kaiser MN, Evanst DE (1976b) The *Ornithodoros* (*Alectorobius*) *capensis* group (Acarina: Ixodoidea: Argasidae) of the Palaearctic and oriental regions. *0*. (*A*.) *maritimus*: identity, marine bird hosts, virus infections, and distribution in Western Europe and Northwestern Africa. J Parasitol 62:799–810

Hoogstraal H, Clifford CM, Keirans JE (1979) The *Ornithodoros capensis* group (*Alectorobius*) of the Palaearctic and (Acarina: Ixodoidea: Argasidae) oriental regions. *O*. (*A*.) *coniceps* identity, bird and mammal hosts, virus infections, and distribution. J Parasitol 65:395–407

Hoogstraal H, Wassef HY, Büttiker W (1981) Ticks (Acarina) of Saudi Arabia: fam. Argasidae. Ixodidae. Fauna Saudi Arab 3:25–110

Kohls GM, Sonenshine DE, Clifford CM (1965) The systematics of the subfamily Ornithodorinae (Acarina: Argasidae). II. Identification of the larvae of the Western hemisphere and descriptions of three new species. Ann Entomol Soc Am 58:331–364

Klompen JSH, Oliver JH Jr (1993) Systematic relationships in the soft ticks (Acari: Ixodida: Argasidae). Syst Entomol 18:313–331

Klompen H, Lekvelshvili M, Black WC (2007) Phylogeny of parasitiform mites (Acari) based on rRNA. Mol Phylogenet Evol 43:936–951

Morel PC (2003) Les tiques d'Afrique et du Bassin méditerranéen (1965–1995). CIRAD-EMVT, 1342 p

Nava S, Gerardi M, Szabó MP, Mastropaolo M, Martins TF, Labruna MB, Beati L, Estrada-Peña A, Guglielmone AA (2016) Different lines of evidence used to delimit species in ticks: A study of the South American populations of *Amblyomma parvum* (Acari: Ixodidae). Ticks Tick-borne Dis 7:1168–1179

Pavlovsky EN (1928) *Ornithodorus papillipes* Birula and *O. cholodkovskyi* N.Sp. Parasitology 22:355–360

Pavlovsky EN, Skruinnik AN (1956) Contribution to the biology of the tick *Ornithodoros papillipes*. Doklady Akad Nauk S.S.S.R. 111:1403–1405 (In Russian)

Petney TP, Maiwald M (1996) Tick nomenclature. Lancet 348:1251

Sonenshine DE, Clifford CM, Kohls GM (1966) The systematics of the subfamily Ornithodorinae (Acarina: Argasidae). III. Identification of the larvae of the eastern hemisphere. Ann Entomol Soc Am 59:92–121

Trape J-F, Diatta G, Arnathau C, Bitam I, Sarih M, Belghyti D, Bouattour A, Elguero E, Vial L, Mané Y, Pugnolle F, Chauvancy G, Mahé G, Granjon L, Duplantier A-M, Durand P, Renaud F (2013) The epidemiology and geographic distribution of relapsing fever borreliosis in West and Northern Africa, with a review of the *Ornithodoros erraticus* complex (Acari: Ixodida). PLoS ONE 8:1–19

Vial L, Diatta G, Tall A, Bouganali H, Durand P, Sokhna C, Trape JF (2006) Incidence of tick-borne relapsing fever in West Africa: longitudinal study. Lancet 368:37–43

Ornithodoros (Alectorobius) capensis Neumann, 1901 (Figs. 12 and 13)

G. Kleinerman and G. Baneth

Life Cycle and Host Preferences

The life cycle of members of the *O. capensis* group may last from 100 to 195 days, and up to 1 year under laboratory conditions; adults and nymphs feed for 15–115 minutes and larvae remain attached to their host for several days. In most of the members of this group, the first instar nymphs do not feed (Hoogstraal et al. 1979; Clifford et al. 1980; Hoogstraal et al. 1985). The host range of *O. capensis* includes more than 35 seabird species in the tropics and subtropics, e.g. boobies (*Sula* spp.), terns (*Sterna* spp., *Anous* spp.), gulls (*Larus* spp.), albatrosses (*Diomedea* spp.), cormorants (*Phalacrocorax* spp.), pelicans (*Pelecanus* spp.) and penguins (*Sphenicus* spp.) (Hoogstraal et al. 1976a, b, 1985; Keirans et al. 1992; Dietrich et al. 2010; Gray et al. 2013). *Ornithodoros capensis* tends to attach to the legs of birds causing irritation, therefore high densities of ticks are known to lead to the dissertation of nests (Converse et al. 1975; Hoogstraal et al. 1976a, b; King et al. 1977; Dietrich et al. 2010). *Ornithodoros capensis* is reported to attack humans when entering into seabird nesting areas and may produce severe pruritus (Converse et al. 1975; Hoogstraal et al. 1976a, b).

Ecology

Ornithodoros capensis is one of the most widespread species of soft ticks (Dietrich et al. 2010). Its habitat is related to that of its hosts, being found in seabird colonies along shores of islands and lakes in the tropics and subtropics. The life cycle tends to couple with the breeding season of its hosts (Dietrich et al. 2010), with a single generation being completed during each breeding period, with little overlap between the different life history stages (Converse et al. 1975). As a nidiculous tick, *O. capensis* attaches to African penguins (*Sphenicus demersus*) during the night when this host is resting in its burrow; thus, it avoids the risk of being transported outside of the nesting site (Gray et al. 2013).

G. Kleinerman · G. Baneth (✉)
Koret School of Veterinary Medicine, Hebrew University, Rehovot, Israel
e-mail: gad.baneth@mail.huji.ac.il

G. Kleinerman
e-mail: gabriela.kleinerman@mail.huji.ac.il

A. Estrada-Peña et al. (eds.), *Ticks of Europe and North Africa*,
https://doi.org/10.1007/978-3-319-63760-0_11

Distribution

Ornithodoros capensis is globally distributed around the equator affecting seabird populations along the coasts and islands of the Pacific, Atlantic and Indian Oceans, the Caribbean and Coral Seas and the lakes of the eastern African Rift Valley system (Hoogstraal et al. 1976a, b; Kohls et al. 1965; Dietrich et al. 2010). A phylogenetic study that compared *O. capensis* specimens from Japan and the USA found a high degree of genetic similarity that may suggest gene flow through migratory seabirds within the Pacific or Atlantic Ocean regions (Ushijima et al. 2003). However, a high degree of genetic diversity was found in *O. capensis* ticks within the Cape Verde Archipelago, without any apparent genetic structuring, due to transoceanic colonization via host movements (Gómez-Días et al. 2012). Therefore, the effects of host dispersion on the genetic variability of *O. capensis* populations need to be further investigated.

Vectorial Capacity and Pathogen Burden

Soldado virus (*Bunyaviridae*) was the first virus found in *O. capensis*. It was collected from terns in Trinidad and later in Ethiopia, the Seychelles and the USA (Jonkers et al. 1973; Converse et al. 1975; Hoogstraal et al. 1976a, b; King et al. 1977). Naturally infected nymphs transmitted the fatal infection to immature domestic chicks (Converse et al. 1975). *Ornithodoros capensis* infected with Soldado virus is a threat to public health when seabirds nest in urban areas

Fig. 12 **A** Distribution of *O. capensis* in Europe and Northern Africa (10 × 10 km grid presence with black dots). **B** Countries where the species has been reported are marked in dark grey

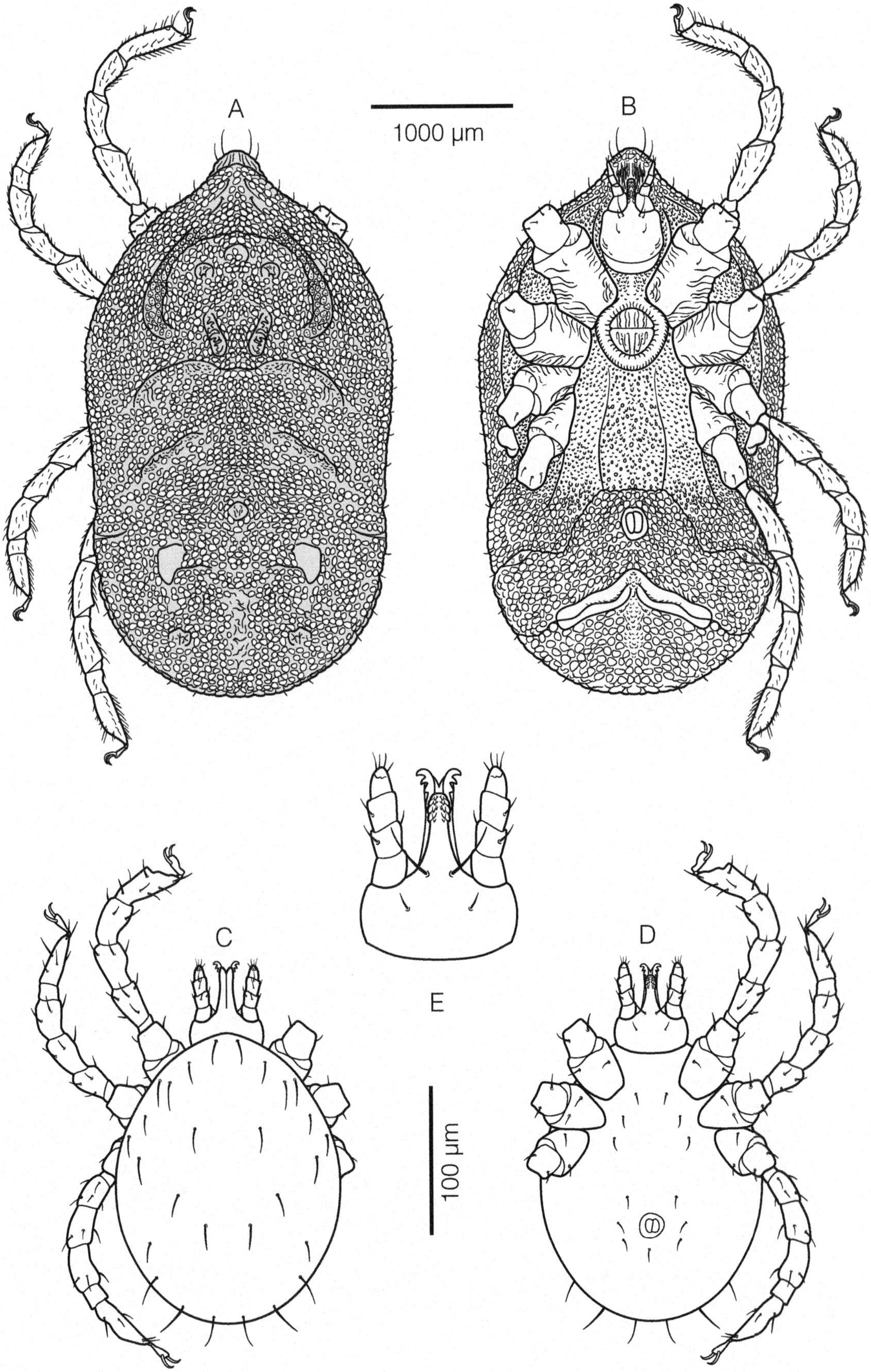

Fig. 13 **A**, **B** The adult of *O. capensis*. **A** Dorsal, **B** ventral, **C–E** the larva of *O. capensis*. **C** Dorsal, **D** ventral, **E** details of the capitulum. Illustrations from specimens collected in Azores

(Estrada-Peña and Jongejan 1999). Field-collected adult stages of *O. capensis* were able to transmit West Nile virus to naive ducklings after being experimentally infected. However, after some time, the levels of the virus decreased significantly in the positive ticks, therefore, further information is necessary in order to elucidate the actual role of this tick in the epidemiology of the disease (Hutcheson et al. 2005). Johnston Atoll virus was originally isolated from *O. capensis* from Sand Island, Johnston Atoll, Central Pacific (USA) and later from Australia, New Zealand and Hawaii (Austin 1978; Clifford 1968). The strains found in New Zealand were not pathogenic to the tick's avian hosts, but *O. capensis* was able to transmit the virus to susceptible chicks. Antibodies to Johnston Atoll virus were found in seabirds from the Coral Sea and on Australia's Great Barrier Reef (Humphery-Smith et al. 1991), but no ticks were collected in this study. There is no available information as to whether this virus is associated with human disease; however, viruses closely related to Johnston Atoll virus have been isolated from people with fever (Austin 1978). Viruses may be transported to different geographical locations through migratory seabirds harbouring infected ticks, mainly immature birds that tend to travel more. This hypothesis was proposed as ticks from seabird colonies in highly separated geographical sites have been shown to harbour related viruses (Nuttall 1984).

Ornithodoros capensis collected from nests of brown pelicans (*Pelecanus occidentalis*) in the USA were found to harbour *Rickettsia felis,* a *Borrelia* sp. similar to *Borrelia lonestari*, a *Coxiella* sp. and *Rickettsia* spp. These last two bacterial genera were found in larvae and eggs, demonstrating transovarial transmission (Reeves et al. 2006). *Rickettsia* sp. were found also in *O. capensis* ticks from Japan (Kawabata et al. 2006). *Rickettsia felis* and *Borrrelia lonestari* are known human pathogens causing murine typhus-like disease and erythema migrans, respectably (James et al. 2001; Reeves et al. 2006).

References

Austin FJ (1978) Johnston Atoll virus (quaranfil group) from *Ornithodoros capensis* (Ixodoidea: Argasidae) infesting a gannet colony in New Zealand. Am J Trop Med Hyg 27:1045–1048

Clifford CM (1968) Identification and comparison of two viruses isolated from ticks of the genus *Ornithodoros*. Am J Trop Med Hyg 17:881–885

Clifford CM, Hoogstraal H, Radovsky FR, Stiller D, Keirans E, Keirans JE (1980) *Ornithodoros (Alectorobius) amblus* (Acarina: Ixodoidea: Argasidae): identity, marine bird and human hosts, virus infections, and distribution in Peru. J Parasitol 66:312–323

Converse JD, Hoogstraal H, Moussa MI, Feare CJ, Kaiser MN (1975) Soldado virus (Hughes group) from *Ornithodoros (Alectorobius) capensis* (Ixodoidea: Argasidae) infesting sooty tern colonies in the Seychelles, Indian Ocean. Am J Trop Med Hyg 24:1010–1018

Dietrich M, Gómez-Díaz E, McCoy KD (2010) Worldwide distribution and diversity of seabird ticks: implications for the ecology and epidemiology of tick-borne pathogens. Vector-Borne Zoonotic Dis 00:1–18

Estrada-Peña A, Jongejan F (1999) Ticks feeding on humans: a review of records on human-biting Ixodoidea with special reference to pathogen transmission. Exp Appl Acarol 23:685–715

Gómez-Días E, Morris-Pocock JA, González-Solís J, McCoy KD (2012) Trans-oceanic host dispersal explains high seabird tick diversity on Cape Verde Islands. Evol Biol 11:1–4

Gray JS, Estrada-Peña AG, Vial LA (2013) Ecology of nidicolous ticks. In: Sonenshine DE, Roe RK (eds) The Biology of Ticks, vol 2. Oxford University Press, Oxford, pp 39–60

Hoogstraal H, Kaiser MN, Easton ER (1976a) *Ornithodoros (Alectorobius) capensis* Neumann (Ixodoidea: Argasidae) parasitizing a human and birds nesting on Islands in East African Lakes. J Med Entomol 12:703–704

Hoogstraal H, Clifford CM, Keirans JE (1979) The *Ornithodoros capensis* group (*Alectorobius*) of the Palaearctic and (Acarina: Ixodoidea: Argasidae) oriental regions. *O. (A) coniceps* identity, bird and mammal hosts, virus infections, and distribution. J Parasitol 65:395–407

Hoogstraal H, Clifford CM, Keirans JE, Kaiser MN, Evanst DE (1976b) The *Ornithodoros (Alectorobius) capensis* group (Acarina: Ixodoidea: Argasidae) of the Palaearctic and oriental regions. *O. (A.) maritimus*: identity, marine bird hosts, virus infections, and distribution in Western Europe and Northwestern Africa. J Parasitol 62:799–810

Hoogstraal H, Wassef HY, Hays C, Keirans JE, Keiranst JE (1985) *Ornithodoros (Alectorobius) spheniscus* n. sp. Argasidae: *Ornithodoros* [Acarina: Ixodoidea: (*Alectorobius*) *capensis* group], a tick parasite of the Humboldt penguin in Peru. J Parasitol 71:635–644

Humphery-Smith I, Cybinski DH, Byrnes KA, St George TD (1991) Seroepidemiology of arboviruses among seabirds and island residents of the Great Barrier Reef and Coral Sea. Epidemiol Infect 107:435–440

Hutcheson HJ, Gorham CH, Machain-Williams C, Loroño-Pino MA, James AM, Marlenee NL et al (2005) Experimental transmission of West Nile Virus (*Flaviviridae: Flavivirus*) by *Carios capensis* ticks from North America. Vector-Borne Zoonotic Dis 5:293–295

James MA, Liveris D, Wormser GP, Schwartz I, Montecalvo MA, Johnson BJ (2001) *Borrelia lonestari* infection after a bite by an *Amblyomma americanum* Tick. J Infect Dis 183:1810–1814

Jonkers AH, Casals J, Aitken TH, Spence L (1973) Soldado virus, a new agent from Trinidadian *Ornithodoros* ticks. J Med Entomol 10:517–519

Kawabata H, Ando S, Kishimoto T, Kurane I, Takano A, Nogami S et al (2006) First detection of Rickettsia in soft-bodied ticks associated with seabirds, Japan. Microbiol Immunol 50:403–406

Keirans JE, Hutcheson HJ, Oliver JH (1992) *Ornithodoros (Alectorobius) capensis* Neumann (Acari: Ixodoidea: Argasidae), a parasite of seabirds, established along the Southeastern Seacoast of the United States. J Med Entomol 29:371–373

King KA, Keith JO, Mitchell CA, Keirans JE (1977) Ticks as a factor in nest desertion of California brown pelicans. Condor 79:507–509

Kohls GM, Sonenshine DE, Clifford CM (1965) The systematics of the subfamily Ornithodorinae (Acarina: Argasidae). II. Identification of the larvae of the Western hemisphere and descriptions of three new species. Ann Entomol Soc Am 58:331–364

Nuttall PA (1984) Tick-borne viruses in seabird colonies. Seabird 7: 31–41

Reeves WK, Loftis A D, Sanders F, Spinks MD, Wills W, Denison A M, Gregory A, Dasch AD (2006) *Borrelia, Coxiella*, and *Rickettsia* in *Carios capensis* (Acari: Argasidae) from a Brown Pelican (*Pelecanus occidentalis*) Rookery in South Carolina, USA. Exp Appl Acarol 39:321–329

Ushijima Y, Oliver JH, Keirans JE, Tsurumit M, Kawabatat H, Watanabet H et al (2003) Mitochondrial sequence variation in *Carios capensis* (Neumann), a parasite of seabirds, collected on Torishima Island in Japan. J Parasitol 89:196–198

Ornithodoros (Alectorobius) coniceps (Canestrini, 1890) (Figs. 14 and 15)

G. Kleinerman and G. Baneth

Life Cycle and Host Preferences

The duration of the life cycle of *O. coniceps* depends on the number of its nymphal stages (3–6), which is to some extent dependent on the climatic conditions. The life cycle requires between 5 months to 2 years to be concluded (Apanaskevich and Oliver 2013); under laboratory conditions, it can be completed within 1 year (Hoogstraal et al. 1979). Each female can deposit up to 1400 eggs in several batches of around 200 eggs. Larvae hatch after 11 days at 100% relative humidity and 30 °C, attach to the host for 7 days and moult 6–8 days after to form first instar nymphs. These do not feed. Other nymphal instars and adults feed for 7–115 min and sometimes repeated feedings occur at each stage (Hoogstraal et al. 1979). *Ornithodoros coniceps* mainly infests wild and domestic pigeons (*Columba palumbus* and *C. livia*). Other hosts associated with this tick are ruddy shelducks (*Tadorna ferruginea*), swallows (*Hirundo* spp.), swifts (*Apus* spp.), sparrows (*Passer* spp.), chicken (*Gallus gallus*) and sporadically, bats (Chiroptera) (Filippova 1963; Hoogstraal et al. 1979). Attacks on humans usually occur due to entry into the pigeon or chicken nesting areas.

Ecology

Ornithodoros coniceps is the only species of its group which does not inhabit seabird nests; it is mainly associated with wild and domestic pigeons nesting in rocky areas like cliffs, wells, caves, ravines, stables, pigeon lofts or human habitations (ancient towers, bell towers, houses in ruins and attics) (Hoogstraal et al. 1979; Khoury et al. 2011). The life cycle of *O. coniceps*, as an ornithophilic species, is dependent on wild or domestic pigeon ecology (Khoury et al. 2011).

Distribution

Ornithodoros coniceps is distributed mainly in the Palaearctic region but can also be found in the Afro-Tropical and Oriental regions. It has been reported from Italy, France, Spain, United Kingdom, Israel, Turkey, Jordan, Egypt, Morocco, Tunisia, Afghanistan, Ukraine, Tajikistan, Kenya, India and Sri Lanka. In the last two locations, larvae were absent for species confirmation (Filippova 1963; Merdivenci 1968; Hoogstraal et al. 1979; Manzano-Román et al. 2012).

Vectorial Capacity and Pathogen Burden

Bites to humans have been reported in France, Spain and Israel (Estrada-Peña and Jongejan 1999; Wilamowski et al. 1999). Such bites can cause oedema, pain, chills, fever and patients sometimes require hospitalization due to toxicosis (Hoogstraal et al. 1979; Wilamowski et al. 1999).

Baku virus (*Reoviridae*) was isolated from *O. coniceps* collected in pigeon nests in the Chatkal Mountains in Uzbekistan (Labuda and Nuttall 2004). There are currently no reports of Baku virus infection in humans. Maintenance and experimental transmission of West Nile virus were proven in *O. coniceps* after feeding on infected hosts. However, no further studies were performed in order to determine the role of this tick in the epidemiology of the virus (Hoogstraal 1985).

G. Kleinerman · G. Baneth (✉)
Koret School of Veterinary Medicine, Hebrew University, Rehovot, Israel
e-mail: gad.baneth@mail.huji.ac.il

G. Kleinerman
e-mail: gabriela.kleinerman@mail.huji.ac.il

A. Estrada-Peña et al. (eds.), *Ticks of Europe and North Africa*,
https://doi.org/10.1007/978-3-319-63760-0_12

Fig. 14 **A** Distribution of *O. coniceps* in Europe and Northern Africa (10 × 10 km grid presence with black dots). **B** Countries where the species has been reported are marked in dark grey

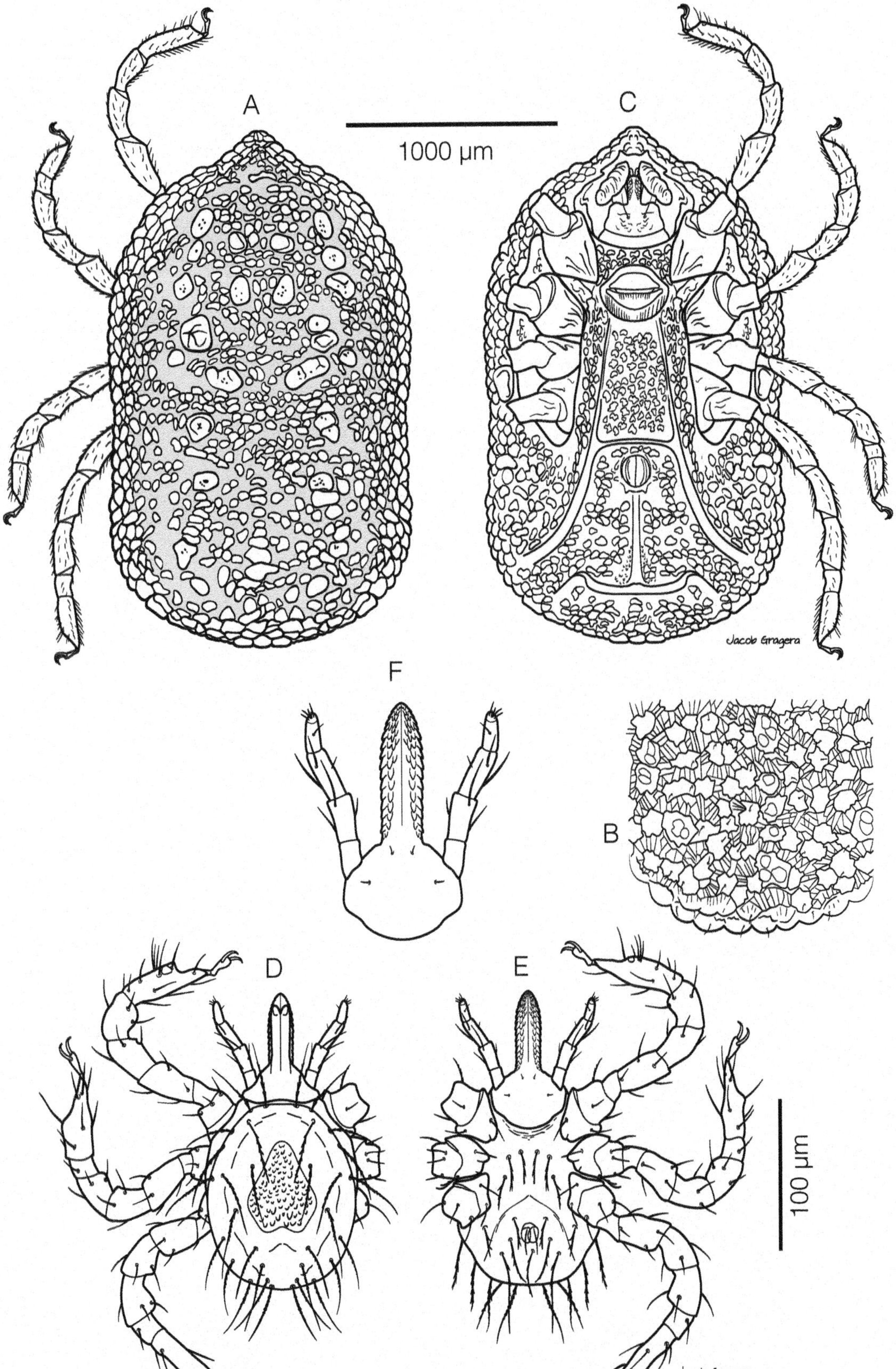

Fig. 15 **A–C** The adult of *O. coniceps*. **A** Dorsal, **B** details of dorsal cuticle, **C** ventral. **D–F** The larva of *O. coniceps*. **D** Dorsal, **E** ventral, **F** details of the capitulum. Illustrations from specimens collected in Italy

References

Apanaskevich DA, Oliver Jr JH (2013) Life cycles and natural history of ticks. In: Sonenshine DE, Roe RM (eds) Biology of ticks, vol 1, Oxford University Press, Oxford, pp 59–73

Estrada-Peña A, Jongejan F (1999) Ticks feeding on humans: a review of records on human-biting Ixodoidea with special reference to pathogen transmission. Exp Appl Acarol 23:685–715

Filippova VA (1963) On ticks of the genus *Ornithodoros* Koch parasitizing birds. Trudy Zool. Inst Akad Nauk SSSR 21:16–27 (in Russian, in English NAMRU3-T124)

Hoogstraal H (1985) Argasid and nuttalliellid ticks as parasites and vectors. Adv Parasitol 24:135–238

Hoogstraal H, Clifford CM, Keirans JE (1979) The *Ornithodoros capensis* group (*Alectorobius*) of the Palaearctic and (Acarina: Ixodoidea:Argasidae) oriental regions. *O. (A.) coniceps* identity, bird and mammal hosts, virus infections, and distribution. J Parasitol 65:395–407

Hoogstraal H, Wassef HY, Hays C, Keirans JE, Keiranst JE (1985) *Ornithodoros* (*Alectorobius*) *spheniscus* n. sp. Argasidae: *Ornithodoros* [Acarina: Ixodoidea: (*Alectorobius*) *capensis* group], a tick parasite of the Humboldt penguin in Peru. J Parasitol 71:635–644

Khoury C, Bianchi R, Andrea Massa A, Severini F, Di Luca M, Toma L (2011) A noteworthy record of *Ornithodoros* (*Alectorobius*) *coniceps* (Ixodida: Argasidae) from Central Italy. Exp Appl Acarol 54:205–209

Labuda M, Nuttall PA (2004) Tick-borne viruses. Parasitology 129: S221–S245

Manzano-Román R, Díaz-Martín V, de la Fuente J, Pérez-Sánchez R (2012) Soft ticks as pathogen vectors: distribution, surveillance and control. Parasitology 7:125–162

Merdivenci A (1968) *Ornithodoros coniceps* (Canestrini, 1890) infestation in man in Istanbul. Tip Fakültesi Mecmuasi 31:460–468

Wilamowski, A, Bromley-Schnur, HJ, Ioffe-Uspensky I et al (1999) Ticks (Ixodoidea) in Israeli towns. In: Presented at the 3rd International Conference on Urban Pests, Czech University of Agriculture, Prague

Ornithodoros (Alectorobius) maritimus Vermeil and Marguet, 1967 (Figs. 16 and 17)

G. Kleinerman and G. Baneth

Life Cycle and Host Preferences

All stages of *O. maritimus* can be found feeding on different species of seabirds, mainly on chicks and immature birds, in their nests or in nearby crevices and under stones (Hoogstraal et al. 1976; Nuttall et al. 1984; Bosch and Figuerola 1999). The seabirds associated with *O. maritimus* are terns (*Sterna hirundo*, *S. dougalli* and *S. sandvicensis*), herring gulls (*Larus argentatus*), egrets (*Egretta garzetta*), common cormorants (*Phalacrocorax carbo*), shags (*Phalacrocorax aristotelis*), razorbills (*Alca torda islandica*), common murres (*Uria aalge albionis*) and black-legged kittiwakes (*Rissa tridactyla*) (Hoogstraal et al. 1976). In the yellow-legged gull (*Larus michahellis*), infestation by *O. maritimus* can cause weight loss and shortened wing length (Bosch and Figuerola 1999). *Ornithodoros maritimus* has been reported from humans in France (Estrada-Peña and Jongejan 1999).

Ecology

Ornithodoros maritimus parasites seabirds breeding in colonies from Western Europe and northwestern Africa. Its habitat is related to the nesting preferences of the seabird species it uses as hosts, i.e. mainly sparsely vegetated, rocky coasts and cliffs (Hoogstraal et al. 1976). *Ornithodoros maritimus* is a nidicolous species that feeds for only short periods of time on its hosts (Arnal et al. 2014), therefore, it is most likely that tick infestation is restricted to the breeding season of the birds when they are on land. Dispersal may occur at local spatial scales (Dietrich et al. 2011). However, a study done on the Cape Verde Islands on *Ornithodoros* spp. from seabirds showed high local tick diversity, probably due to trans-oceanic colonization via host movements (Gómez-Días et al. 2012). Nevertheless, as most host–parasite interactions occur almost exclusively during the breeding season, the life cycle of the tick is associated with the seasonality of seabird reproduction (Dietrich et al. 2011). *O. maritimus* oviposits at the onset of spring when its hosts return to their nesting locations. This delay in oviposition may be due to the absence of available blood meals (needed to complete oogenesis) during the non-reproductive season, to a diapause mechanism based on photoperiod, or to temperature changes (Gray et al. 2013).

Distribution

Ornithodoros maritimus has been reported from France, the United Kingdom, Ireland, Tunisia, Spain, Italy, Portugal and Morocco (Hoogstraal et al. 1976; Nuttall et al. 1984; Manilla 1990; Estrada-Peña et al. 1996; Caeiro 1999; Bosch and Figuerola 1999; Dietrich et al. 2011; Arnal et al. 2014).

Vectorial Capacity and Pathogen Burden

Soldado virus (*Bunyaviridae*) has been isolated from *O. maritimus* collected in the nests of herring gulls (*Larus argentatus*) in the United Kingdom (Wales) and France (Converse et al. 1976; Chastel et al. 1984). The presence of a Soldado-like virus has also been demonstrated in *O. maritimus* from Ireland (Keirans et al. 1975). Soldado virus caused a febrile episode and persistent rhinitis in an ornithologist bitten by *O. maritimus* in a seagull colony in Morocco (Hoogstraal 1985). Soldado virus is recognized as a public health issue in urban environments where gulls feed (Estrada-Peña and Jongejan 1999). It was proven to be pathogenic to domestic chicks when transmitted by the closely related species *O. capensis* (Converse et al. 1975),

G. Kleinerman · G. Baneth (✉)
Koret School of Veterinary Medicine, Hebrew University, Rehovot, Israel
e-mail: gad.baneth@mail.huji.ac.il

G. Kleinerman
e-mail: gabriela.kleinerman@mail.huji.ac.il

A. Estrada-Peña et al. (eds.), *Ticks of Europe and North Africa*,
https://doi.org/10.1007/978-3-319-63760-0_13

Fig. 16 **A** Distribution of *O. maritimus* in Europe and Northern Africa (10 × 10 km grid presence with black dots). **B** Countries where the species has been reported are marked in dark grey

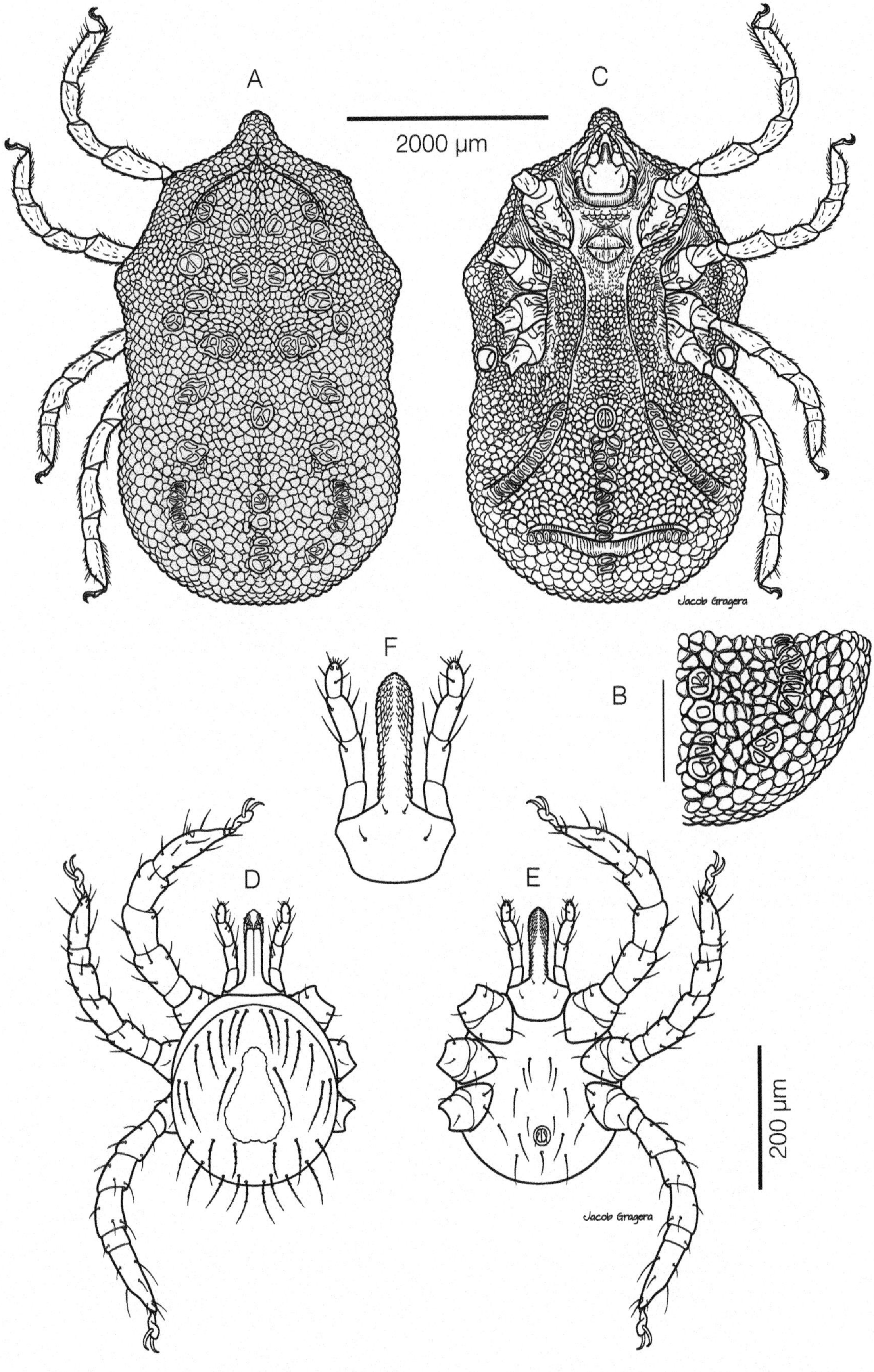

Fig. 17 A–C The adult of *O. maritimus*. **A** Dorsal, **B** details of the dorsal cuticle, **C** ventral. **D–F** The larva of *O. coniceps*. **D** Dorsal, **E** ventral, **F** details of the capitulum. Illustrations from specimens collected in Spain and Italy

however, no vectorial capacity tests have been performed on *O. maritimus*. Experimental studies show that *O. maritimus* is able to transmit West Nile virus to rodents after feeding on infected hosts (Hoogstraal et al. 1976).

A virus whose sequence was 95% similar to the Meaban virus (an RNA flavivirus) was detected in pools of *O. maritimus* sampled from yellow-legged gull chicks from the Medes Islands, Spain (Arnal et al. 2014). There is currently no information available on the zoonotic potential of Meaban or Meaban-like viruses (Arnal et al. 2014). Puffin Island virus was isolated from *O. maritimus* parasitizing a diversity of marine birds on Puffin Island, the United Kingdom (Hoogstraal 1985). Other viruses from the genus *Orbivirus* that have been isolated from *O. maritimus* ticks include Baku virus, Essaouira virus, Kala Iris virus, Great Saltee Island virus and Moellez virus. The infected ticks were found parasitizing gulls (*Larus argentatus*) in the Caspian Sea, Uzbekistan, Morocco as well as a colony of shags (*Phalacrocorax aristotelis*) in Ireland and gull and shag colonies in France (Labuda and Nuttall 2004).

References

Arnal A, Gómez-Díaz E, Cerda-Cuella M, Lecollinet S, Pearce-Duvet J, Busquets N et al (2014) Circulation of a meaban-like virus in yellow-legged gulls and seabird ticks in the western Mediterranean basin. PLoS ONE 9:1–10

Bosch M, Figuerola J (1999) Detrimental effects of ticks *Ornithodoros maritimus* on the growth of yellow-legged gull *Larus michahellis* chicks. Ardea 87:83–89

Caeiro V (1999) General review of tick species present in Portugal. Parassitologia 41:11–15

Chastel C, Guiguen C, Quillien MC, Le Lay-Roguès G, Beaucornu JC (1984) Visualization of the soldado virus by electron microscopy in the tissues of the vector tick *Ornithodoros (Alectorobius) maritimus* Vermeil and Marguet, 1967. Annales de Parasitologie Humaine et Comparee. 59:1–6

Converse JD, Hoogstraal H, Moussa MI, Evans DE (1976) Soldado virus from *Ornithodoros* (*Alectorobius*) *maritimus* (Ixodoidea: Argasidae) infesting herring gull nests on Puffin Island, Northern Wales. Acta Virol 20:243–246

Converse JD, Hoogstraal H, Moussa MI, Feare CJ, Kaiser MN (1975) Soldado virus from *Ornithodoros* (*Alectorobius*) *capensis* (Ixodoidea: Argasidae) infesting sooty tern colonies in the Seychelles, Indian Ocean. Am J Trop Med Hyg 24:1010–1018

Dietrich M, Elena Gómez-Díaz E, McCoy KD (2011) Worldwide distribution and diversity of seabird ticks: implications for the ecology and epidemiology of tick-borne pathogens. Vector-Borne Zoonotic Dis 11:453–470

Estrada-Peña A, Bosch M, Pedrocchi V (1996) First record of *Ornithodoros* (*Alectorobius*) *maritimus* Vermeil & Marguet (Acari: Argasidae) in Spain. Acarologia 37:19–21

Estrada-Peña A, Jongejan F (1999) Ticks feeding on humans: a review of records on human-biting Ixodoidea with special reference to pathogen transmission. Exp Appl Acarol 23:685–715

Gómez-Días E, Morris-Pocock JA, González-Solís J, McCoy KD (2012) Trans-oceanic host dispersal explains high seabird tick diversity on Cape Verde Islands. Evol Biol 11:1–4

Gray JS, Estrada-Peña AG, Vial LA (2013) Ecology of nidicolous ticks. In: Sonenshine DE, Roe RK (eds) The biology of ticks, vol 2, Oxford University Press, Oxford, pp 39–60

Hoogstraal H (1985) Argasid and nuttalliellid ticks as parasites and vectors. Adv Parasitol 24:135–238

Hoogstraal H, Clifford CM, Keirans JE, Makram NK, Evanst DE (1976) The *Ornithodoros* (*Alectorobius*) capensis group (Acarina: Ixodoidea: Argasidae) of the Palaearctic and Oriental regions. *O. (A.) maritimus*: identity, marine bird hosts, virus infections, and distribution in western Europe and northwestern Africa. J Parasitol 62:799–810

Keirans JE, Yunker CE, Clifford CM, Thomas LA, Walton GA, Kelly TC (1975) Isolation of a Soldado-like virus (Hughes Group) from *Ornithodorus maritimus* ticks in Ireland. Experientia 32:453–454

Labuda M, Nuttall PA (2004) Tick-borne viruses. Parasitology 129: S221–S245

Manilla G (1990) *Ornithodoros* (*Alectorobius*) *maritimus* (Acarina: Ixodoidea: Argasidae) a new species in Italy and observations on the *coniceps-capensis* group. Parassitologia 32:265–274

Nuttall PA, Kelly TC, Carey D, Moss SR, Harrap KA (1984) Mixed infections with tick-borne viruses in a seabird colony in Eire. Adv Virol 79:35–44

Ornithodoros (Pavlovskyella) alactagalis Issaakjan, 1936 (Fig. 18)

G. Kleinerman and G. Baneth

Life Cycle and Host Preferences

Under laboratory conditions (temperature of 26 °C and humidity of 90–95%), the development of *O. alactagalis* from egg to adult lasts from 5 to 10 months. All life stages feed; bloodsucking usually lasts between several minutes to 1 hour. Attachment of larvae on mice takes 1–4 h and moult to first stage nymphs occurs after 9–20 days. Female development includes 3–4 nymphal stages. For males, 2–3 nymphal stages exist. After 52–120 days of feeding, females oviposit; incubation lasts 19–32 days. Survival of wild specimens in the laboratory was reported to be longer than 6 years (Filippova 1966). *Ornithodoros alactagalis* ticks are found in burrows of animals, such as rodents (including *Psammomys obesus*, *Cricetus cricetus*, *Phodopus* spp., *Meriones* spp., *Mus musculus*, *Arvicola* spp., *Allactaga* spp.), badgers (*Meles* spp.), foxes (*Vulpes* spp.) and hedgehogs (*Hemiechinus auritus*); it was associated also with lizard and green toad burrows (*Bufo viridis*) (Filippova 1966; Soneshine et al. 1966). *Ornithodoros alactagalis* ticks may show aggressive behaviour towards humans, and researchers reported attacks when approaching animal burrows (Filippova 1966).

Ecology

The habitat of *O. alactagalis* is essentially moist burrows, which are not excessively damp. It inhabits beaches, deserts and mountainous areas covered with desert or steppe vegetation (Soneshine et al. 1966). In Azerbaijan and Armenia, larvae were observed by the end of June and beginning of July, and nymphs and adults were observed during June. Oviposition occurs from April to December (Filippova 1966).

Distribution

Ornithodoros alactagalis was reported only in the Palaearctic region, in Armenia, Azerbaijan, Georgia, Iran, Northern Caucasus, Azerbaijan and Turkey (Manzano-Román et al. 2012).

Vectorial Capacity and Pathogen Burden

A *Borrelia* sp. was reported to be transmitted by this tick to humans (Filippova 1966; Gugushvili 2013); however, no description is available on the species of *Borrelia* this tick can transmit.

G. Kleinerman · G. Baneth (✉)
Koret School of Veterinary Medicine, Hebrew University, Rehovot, Israel
e-mail: gad.baneth@mail.huji.ac.il

G. Kleinerman
e-mail: gabriela.kleinerman@mail.huji.ac.il

A. Estrada-Peña et al. (eds.), *Ticks of Europe and North Africa*,
https://doi.org/10.1007/978-3-319-63760-0_14

Fig. 18 **A** Distribution of *O. alactagalis* in Europe and Northern Africa (10 × 10 km grid presence with black dots). **B** Countries where the species has been reported are marked in dark grey

References

Gugushvili GK (2013) Duration of preservation of *Borrelia* by ticks *Ornithodoros verrucosus* Ol., Sass. et Fen. and *Ornithodoros alactagalis* Issaak. AGRIS. 185–188 (In Russian)

Filippova NA (1966) Argasid ticks (Argasidae). Fauna SSSR. 4: 1–255 (In Russian)

Manzano-Román R, Díaz-Martín V, de la Fuente J, Pérez-Sánchez R (2012) Soft ticks as pathogen vectors: distribution, surveillance and control. Parasitology 7:125–162

Sonenshine DE, Clifford CM, Kohls GM (1966) The Systematics of the Subfamily Ornithodorinae (Acarina: Argasidae). III. Identification of the larvae of the eastern hemisphere. Ann Entomol Soc Am 59:92–121

Ornithodoros (Alectorobius) lahorensis Neumann, 1908 (Figs. 19 and 20)

G. Kleinerman and G. Baneth

Life Cycle and Host Preferences

The two-host life cycle of *O. lahorensis* is unique among the Argasidae. The larvae hatch after an incubation period of 2–6 weeks, then attach to the host for 3–6 weeks and detach as third stage nymphs which moult to adults in the environment (Hoogstraal 1985). Adults attach to a new host and feed for 1–2 h. Nevertheless, if no host is available for the adult, the species becomes a one-host tick (Oliver 1989). *Ornithodoros lahorensis* is facultatively autogenous, it does not require a blood meal to oviposit and can undergo two gonotrophic cycles without taking a blood meal (Oliver 1989). Fertilized fed or unfed females deposit batches of 300–500 eggs. However, unfed females need additional blood meals to continue producing egg batches (Hoogstraal 1985). Unfed adults can live for 18 years (Hoogstraal 1985). *Ornithodoros lahorensis* infests large domestic animals; it was collected from sheep *(Ovis aries)*, camels (*Camelus dromedaries*), cattle (*Bos taurus*), goats (*Capra hircus*), horses (*Equus caballus*), donkeys (*Equus africanus asinus*) and also from dogs (*Canis familiaris*), European hares (*Lepus europaeus*) and poultry (Sonenshine et al. 1966; Telmadarraiy et al. 2004; Bursali et al. 2012). *Ornithodoros lahorensis* is reported to bite people when inhabiting human dwellings; it was collected from humans in Turkey and is a common tick parasite of man in some areas of the former Soviet Union (Parrish 1961; Sonenshine et al. 1966; Estrada-Peña and Jongejan 1999). In its animal host, it can produce anaemia, toxic reactions and paralysis (Hoogstraal et al. 1984).

G. Kleinerman · G. Baneth (✉)
Koret School of Veterinary Medicine, Hebrew University, Rehovot, Israel
e-mail: gad.baneth@mail.huji.ac.il

G. Kleinerman
e-mail: gabriela.kleinerman@mail.huji.ac.il

A. Estrada-Peña et al. (eds.), *Ticks of Europe and North Africa*,
https://doi.org/10.1007/978-3-319-63760-0_15

Ecology

The habitat of *O. lahorensis* includes stables and other animal dwellings, between bricks and stones, under plaster and in roof cracks. These ticks are found on steppe and mountain deserts from sea level to 2900 m altitude (Hoogstraal 1985). Usually, larvae attach to their hosts during the fall or winter and the final nymph instars moult to adults in the spring (Hoogstraal et al. 1984). In accordance with this, in Iran, *O. lahorensis* ticks were found to be more active during fall, late winter and spring (Telmadarraiy et al. 2004; Yakhchali and Hosseine 2006; Salari Lak et al. 2008).

Distribution

Ornithodoros lahorensis is found in the Palaearctic region, in Armenia, Kazakhstan, Russia, Kosovo, Republic of Macedonia, Syria, Turkey, Iran, Iraq, Saudi Arabia, Lebanon, Syria, Pakistan, Bulgaria, Greece, Israel, China, India and Tibet (Soneshine et al. 1966; Hoogstraal et al. 1984; Hoogstraal 1985; Manzano-Román et al. 2012).

Vectorial Capacity and Pathogen Burden

The Crimean-Congo haemorrhagic fever (CCHF) virus was isolated from mice after intraperitoneal inoculation of engorged larvae of *O. lahorensis* collected from a goat in Iran (Sureau et al. 1980). This experiment, however, does not prove transmission since the larvae did not feed on the mice. An additional study performed in Iran revealed the presence of CCHF virus in *O. lahorensis* collected from sheep and cows (Telmadarraiy et al. 2010). Again, this is not proof of a transmission of the virus, since the RNA of the virus was detected in feeding ticks. No experimental transmission was done in order to confirm the role of this tick in the epidemiology of CCHF. The presence of the virus in the

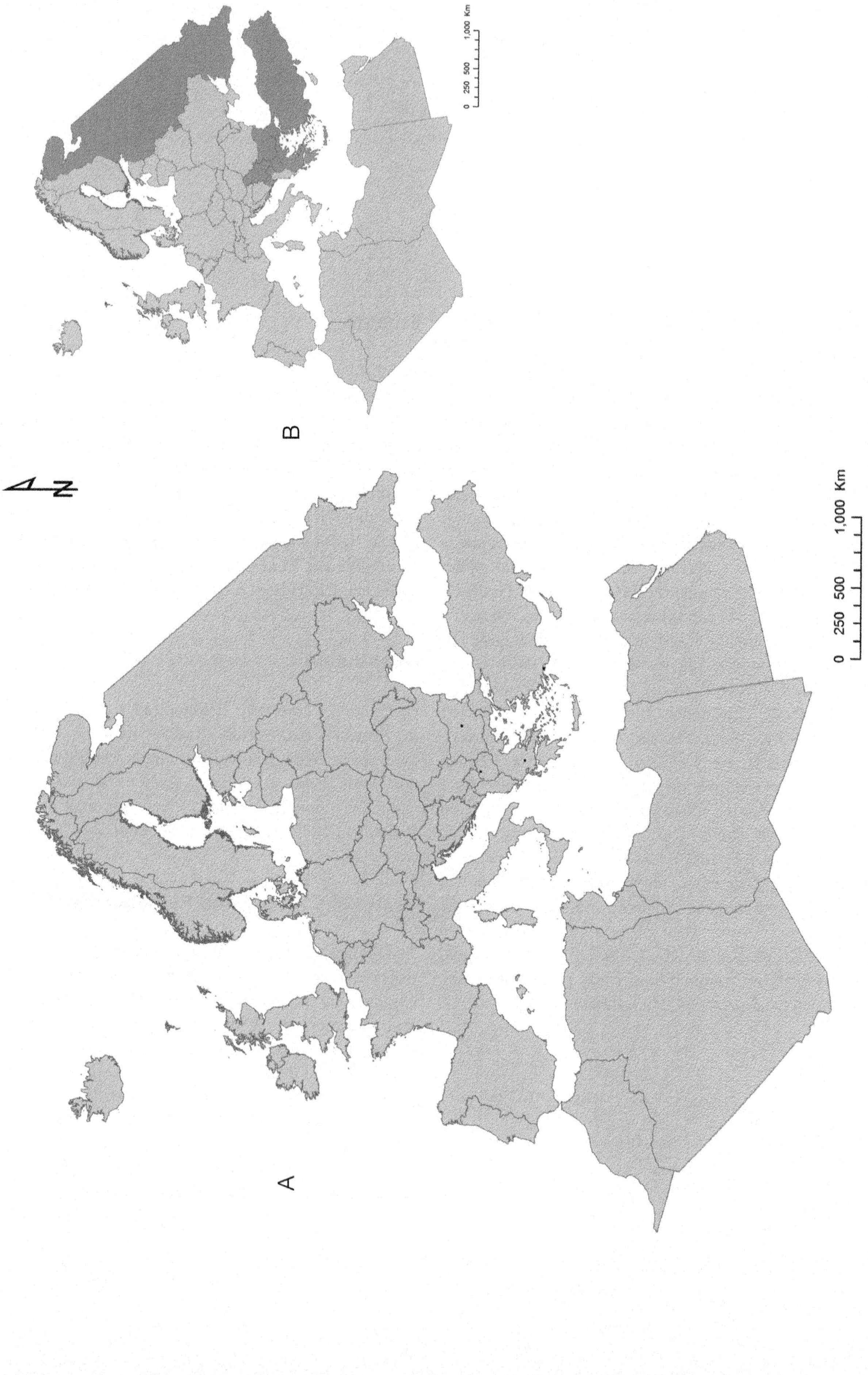

Fig. 19 A Distribution of *O. lahorensis* in Europe and Northern Africa (10 × 10 km grid presence with black dots). **B** Countries where the species has been reported are marked in dark grey

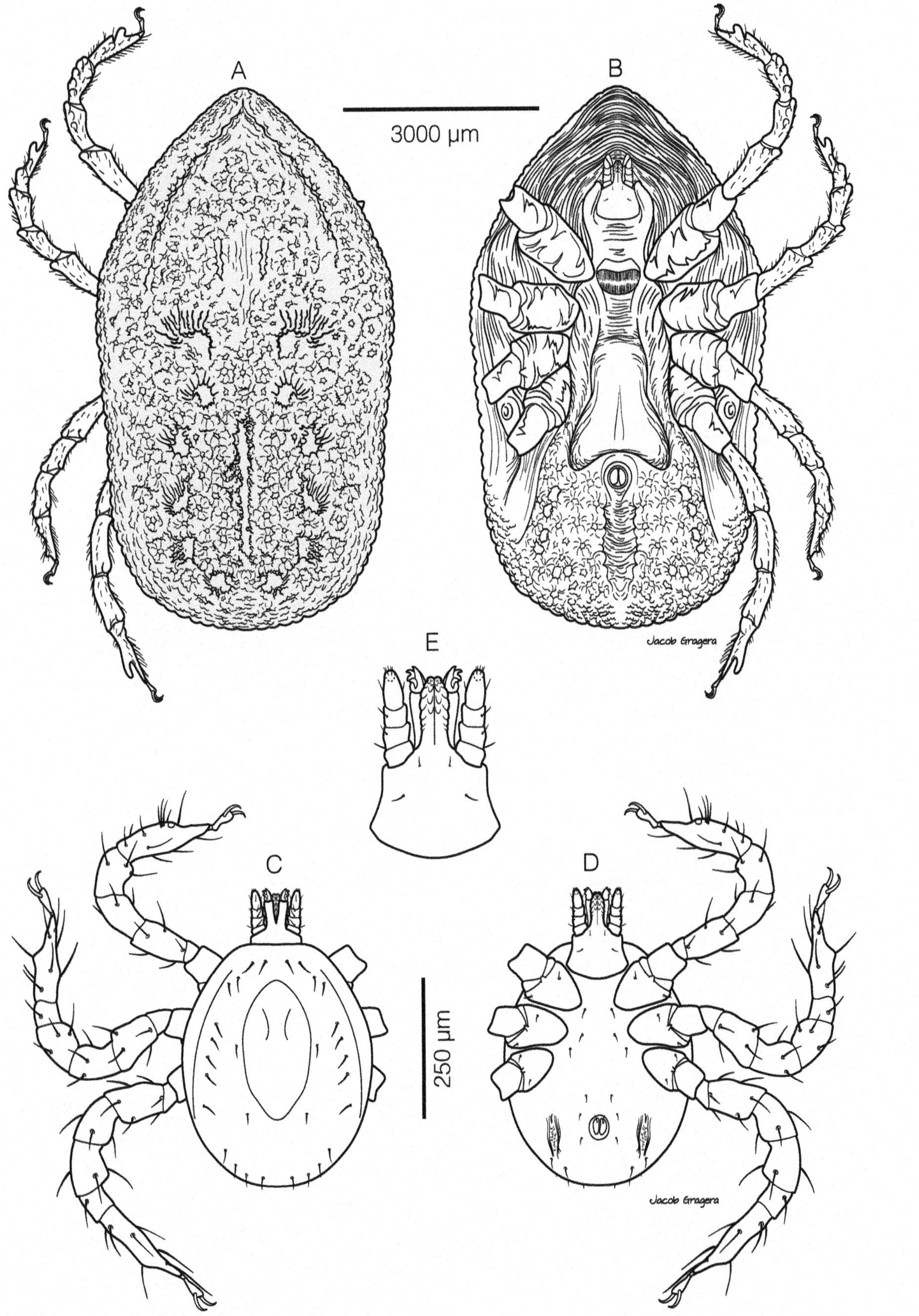

Fig. 20 **A–B** The adult of *O. lahorensis*. **A** Dorsal, **B** ventral. **C–E** The larva of *O. lahorensis*. **C** Dorsal, **D** ventral, **E** details of the capitulum. Illustrations from specimens collected in Turkey

ticks may be due to a recently ingested blood meal since CCHF virus failed to replicate in argasid ticks (Hoogstraal 1985; Labuda and Nuttall 2004) and *O. lahorensis* shares some of the same hosts with vectors of the CCHF virus (mainly *Hyalomma* spp.).

Rickettsia sibirica was shown to be experimentally transmitted by *O. lahorensis* (Estrada-Peña and Jongejan 1999) and *Rickettsia conorii* was present in tissues of *O. lahorensis,* but no transmission mechanisms were tested (Reacek 1989). Fasting *O. lahorensis* were found infected with *Brucella abortus* in areas where infected sheep had been kept one year before (Philip and Burgdorfer 1961). Hoogstraal et al. (1984) reported transmission of *Francisella tularensis* and *Coxiella burnetii* to domestic animals, and possibly to man, by *O. lahorensis* in Eurasia. However, *O. lahorensis* failed to transmit *C. burnetii* to guinea pigs after 19 months of acquiring the infection (Philip and Burgdorfer 1961).

References

Bursali A, Keskin A, Tekin S (2012) A review of the ticks (Acari: Ixodida) of Turkey: species diversity, hosts and geographical distribution. Exp Appl Acarol 57:91–104

Estrada-Peña A, Jongejan F (1999) Ticks feeding on humans: a review of records on human-biting Ixodoidea with special reference to pathogen transmission. Exp Appl Acarol 23:685–715

Hoogstraal H (1985) Argasid and nuttallielid ticks as parasites and vectors. Adv Parasitol 24:135–238

Hoogstraal H, Wassef HY, Diab FM, AI-Asgah NA, AI-Khalifa AS (1984) Acarina of Saudi Arabia *Ornithodoros (Alveonasus) lahorensis* (Fam. Argasidae) in Saudi Arabia biological, veterinary, and medical implications. Fauna Saudi Arab 6:165–169

Labuda M, Nuttall PA (2004) Tick-borne viruses. Parasitology 129: S221–S245

Manzano-Román R, Díaz-Martín V, de la Fuente J, Pérez-Sánchez R (2012) Soft ticks as pathogen vectors: distribution, surveillance and control. Parasitology 7:125–162

Oliver JH Jr (1989) Biology and systematics of ticks (Acari: Ixodida). Annu Rev Ecol Syst 20:397–430

Parrish DW (1961) The ticks (Argasidae and Ixodidae) of Turkey. J Econ Entomol 54:91–92

Philip CB, Burgdorfer W (1961) Arthropod vectors as reservoirs of microbial disease agents. Annu Rev Entomol 6:391–412

Řeáček J (1989) Ecological relationships between ticks and rickettsiae. Eur J Epidemiol 5:407–413

Salari Lak SH, Vatandoost H, Telmadarraiy Z, Entezar Mahdi R, Kia EB (2008) Seasonal activity of ticks and their importance in tick-borne infectious diseases in west Azerbaijan, Iran. Iranian J Arthropod-Borne Dis 2:28–34

Sonenshine DE, Clifford CM, Kohls GM (1966) The systematics of the subfamily Ornithodorinae (Acarina: Argasidae). III. identification of the larvae of the eastern hemisphere. Ann Entomol Soc Am 59:92–121

Sureau P, Klein JM, Casals J, Digoutte JP, Salaun JJ, Piazak N et al (1980) Isolement des virus Thogoto, Wad Medani, Wanowrie et de la Fiévre Hémorragique de Crimée-congo en Iran á partir de tiques d'animaux domestiques. Ann Virol 131:185–200

Telmadarraiy Z, Bahrami A, Vatandoost H (2004) A survey on fauna of ticks in West Azerbaijan Province, Iran. Iranian J Publ Health 33:65–69

Telmadarraiy Z, Ghiasi SM, Moradi M, Vatandoost H, Eshraghian MR, Faghihi F et al (2010) A survey of Crimean-Congo haemorrhagic fever in livestock and ticks in Ardabil Province, Iran during 2004–2005. Scandinavian J Infect Dis 41:137–141

Yakhchali M, Hosseine A (2006) Prevalence and ectoparasites fauna of sheep and goats flocks in Urmia suburb, Iran. Veterinarski Arhiv 76:431–442

Ornithodoros (Pavlovskyella) tholozani (Laboulbène and Mégnin, 1882) (Figs. 21 and 22)

G. Kleinerman and G. Baneth

Life Cycle and Host Preferences

The duration of the life cycle of *O. tholozani* is related to the presence of hosts; it can last between 7 and 12 months if hosts are abundant. However, tick stages (third stage nymphs and adults) can live for 10 years in starvation when there is irregular host availability (Avivi et al. 1973; Assous and Wilamowski 2009). Long-term survival of *O. tholozani* individuals up to 20 years was reported under laboratory conditions (Pavlovskiy and Skruinnik 1956). All stages of *O. tholozani* feed (Pavlovskiy and Skruinnik 1956) and blood sucking lasts 20–30 min (Assous and Wilamowski 2009). Normally, females feed, mate and oviposit several times in their life; each egg batch includes 100–229 eggs. However, some females may show autogenous oviposition (oviposition without a blood meal in the adult stage); this behaviour was related to a larger female size at ecdysis, but the number of eggs is smaller in autogenous batches compared to regular batches (Feldman-Muhsam and Havivi 1973). The behaviour of hyperparasitism or homovampirism where ticks feed on recently fed ticks, in order to extend their longevity was reported in *O. tholozani* (Gray et al. 2013). This conduct is probably a side effect of long starvation and not an adaptive process, where victims of homovampirism may suffer higher mortality (Gray et al. 2013). Commonly reported hosts include sheep (*Ovis aries*), goats (*Capra aegagrus hircus*), porcupines (*Hystrix* spp.), hedgehogs (*Hemiechinus* spp., *Paraechinus* spp.), foxes (*Vulpes* spp.), badgers (*Meles* spp.), jackals (*Canis aureus*), rodents (including *Rhombomys* spp., *Meriones* spp., *Pallasiomys* spp. and *Mesocricetus* spp.), camels (*Camelus* spp.) and cattle (*Bos taurus*) (Hoogstraal 1985). *Ornithodoros tholozani* can readily attack humans when the tick inhabits human habitations or when a man enters the tick environment (Sonenshine et al. 1966).

Ecology

As reported in Israel, Jordan, Syria, Lebanon, Egypt and Libya, *O. tholozani* inhabits caves, ancient ruins and archaeological sites, where it digs itself into the soil or hides in wall crevices and lives under microclimatic conditions of low temperature and high humidity (17–25 °C and 70–80%, respectively) (Avivi et al. 1973; Assous and Wilamowski 2009). It can also be found in animal shelters and burrows. At the northern limits of its distribution (southern Russia, Iran, Kashmir and Afghanistan), the tick resides in stables, barns, clay and stone fences, storerooms and human habitations (Avivi et al. 1973; Hoogstraal 1985). Its geographical habitats include semi-desert, steppe and Mediterranean climate environments (Hoogstraal 1985). *Ornithodoros tholozani* is a nidicolous tick and an indiscriminate feeder (Vial 2009). In Israel, these ticks do not migrate from their caves, even after long periods of starvation. If some ticks venture outside the refuge of the cave, it is unlikely that they will survive for more than few hours during the following day (Avivi et al. 1973). Due to the fact that this tick species is opportunistic and inhabits microhabitats where climatic conditions are buffered throughout the year, it is active during the whole year (Vial 2009; Gray et al. 2013). However, in central Asia, *O. tholozani* does not feed on cattle during winter and shows a diapause of 6 months (Vial 2009). Under laboratory conditions, *O. tholozani* is mainly a nocturnal tick; nevertheless, in caves, it is active all day (Avivi et al. 1973).

G. Kleinerman · G. Baneth (✉)
Koret School of Veterinary Medicine, Hebrew University, Rehovot, Israel
e-mail: gad.baneth@mail.huji.ac.il

G. Kleinerman
e-mail: gabriela.kleinerman@mail.huji.ac.il

A. Estrada-Peña et al. (eds.), *Ticks of Europe and North Africa*,
https://doi.org/10.1007/978-3-319-63760-0_16

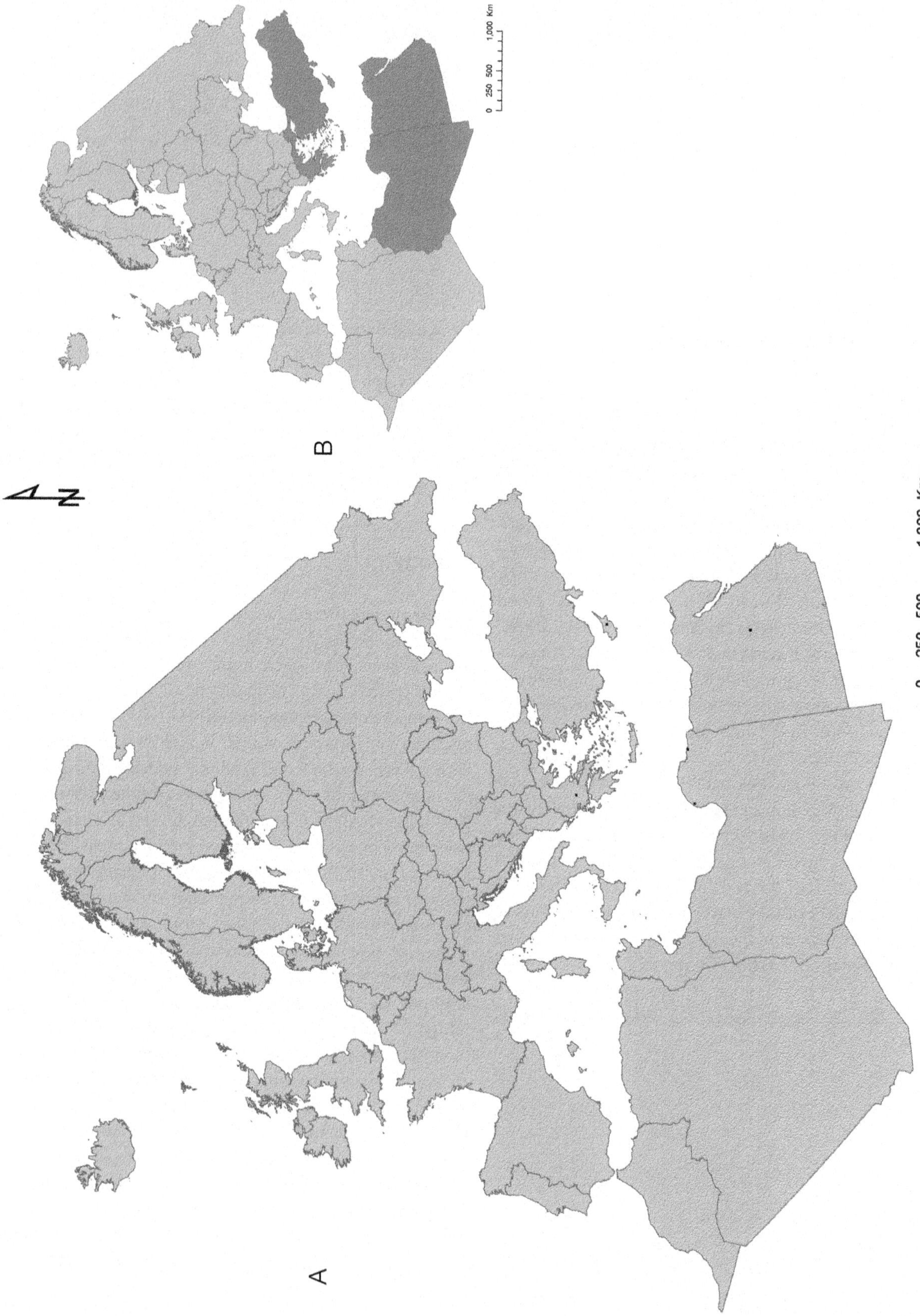

Fig. 21 **A** Distribution of *O. tholozani* in Europe and Northern Africa (10 × 10 km grid presence with black dots). **B** Countries where the species has been reported are marked in dark grey

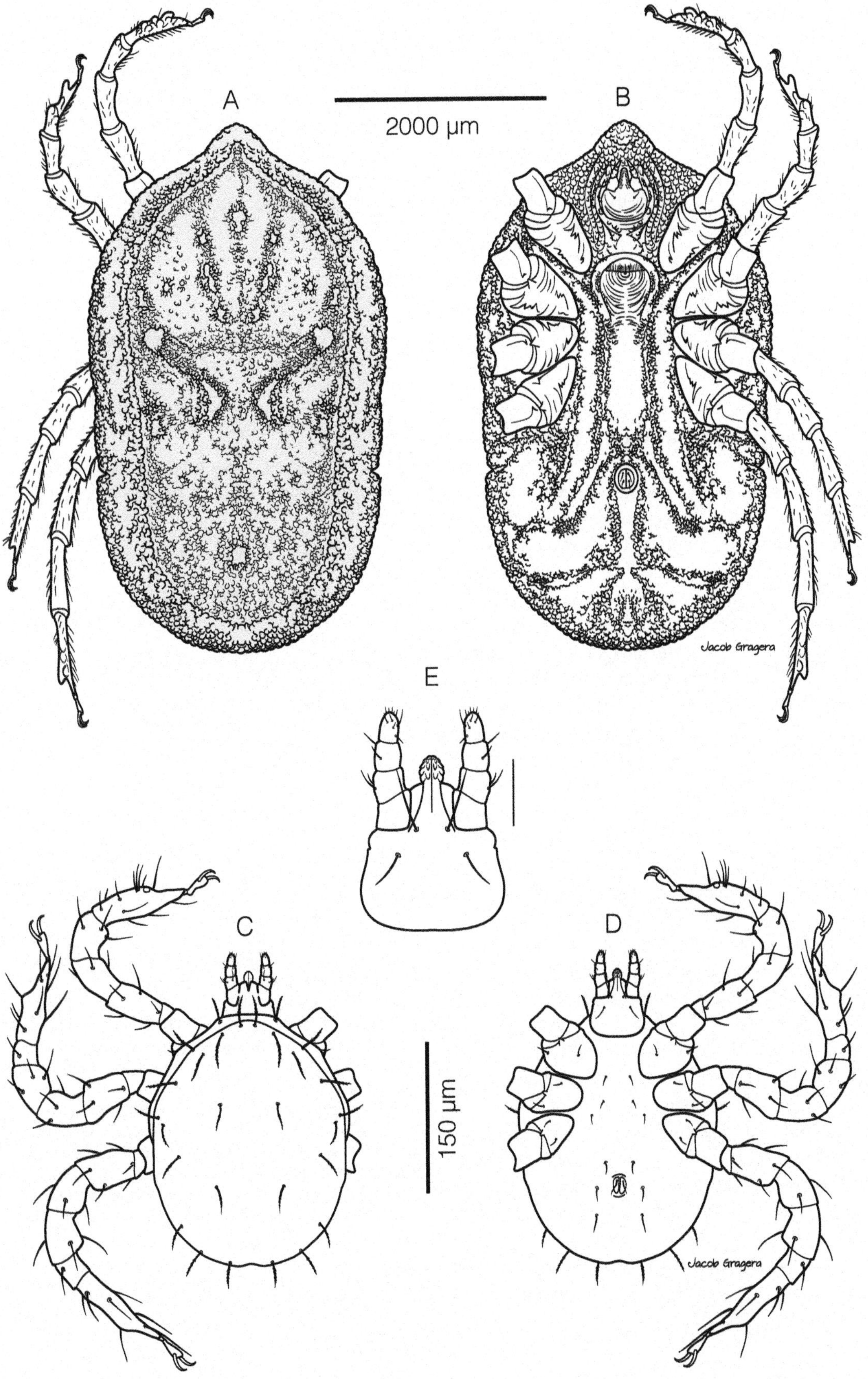

Fig. 22 **A–B** The adult of *O. tholozani*. **A** Dorsal, **B** ventral. **C–E** The larva of *O. tholozani*. **C** Dorsal, **D** ventral, **E** details of the capitulum. Illustrations from specimens collected in Turkey

Ornithodoros tholozani ticks residing in caves show a reproduction peak at the end of the summer and oviposition starts in the beginning of the fall (Avivi et al. 1973).

Distribution

Ornithodoros tholozani is widely distributed and occurs in Kashmir (India), Western China, in the central Asian republics of the former USSR (Kazakhstan, Kyrgyzstan, Tajikistan, Turkmenistan and Uzbekistan), Afghanistan, Iran, Iraq, Syria, Jordan, Turkey, Greece, Israel, Egypt, Cyprus, Libya and Lebanon (Avivi et al. 1973; Assous and Wilamowski 2009).

Vectorial Capacity and Pathogen Burden

Ornithodoros tholozani is the natural vector of *Borrelia persica,* which is the causative agent of tick-borne relapsing fever (TBRF) in humans in Eurasia. The distribution of the disease overlaps the distribution of the tick (Assous and Wilamowski 2009). The disease is characterized by recurrent periods of fever attacks accompanied by non-specific symptoms; complications are rare and include neurologic and ophthalmologic disorders (Assous and Wilamowski 2009). Molecular characterization and phylogenetic analysis of bacterial genes revealed that *B. persica* clustered separately from African and the new world TBRF, and infected humans and ticks share the same bacterial genotypes (Assous et al. 2006; Safdie et al. 2010). Transovarial transmission of *B. persica* from the females to the eggs has been reported. However, larvae do not transmit the bacterium, and only when ticks had developed to nymphs, was transmission possible to guinea pigs (Burgdorfer and Varma 1966).

Other bacteria found in *O. tholozani* include *Coxiella burnetii*, which was found to survive 5–10 years in this tick (Řeháčk 1989). DNA of *Borrelia theileri* was detected in *O. tholozani* ticks from a cave in Israel where cows were the main hosts; all positive ticks had cow blood meal, thus the presence of the bacteria may not necessarily indicate vector competence of this cattle pathogen by the tick.

West Nile virus was isolated from *O. tholozani* ticks apparently infecting birds from the southern former USSR; no transmission experiments were performed (Hoogstraal 1985). Karshi virus, a Flavivirus, was isolated from *O. tholozani* (as the synonymous *O. papillipes*) from burrows of the great gerbil, *Rhombomys opimus*, in Uzbekistan (Hoogstraal 1985). Karshi virus was proven to be pathogenic for mice and an encephalitis with fever due to this virus was reported in humans in Uzbekistan (Turell et al. 2008). Kyasanur Forest disease virus (KFVD), another Flavivirus, was proven to be transstadially and transovarially transmitted in *O. tholozani*. Furthermore, first-generation infected ticks transmitted the virus to susceptible hosts including a human (Hoogstraal 1985; Mahdi et al. 2011).

References

Assous MV, Wilamowski A (2009) Relapsing fever borreliosis in Eurasia-forgotten, but certainly not gone! Clin Microbiol Infect 15:407–414

Assous MV, Wilamowski A, Bercovier H, Marva E (2006) Molecular characterization of tickborne relapsing fever *Borrelia*. Israel Emerg Infect Dis 12:1740–1743

Avivi A, Warburg M, Galun R (1973) Ecological studies on the cave tick *Ornithodoros tholozani* and its distribution in Israel. Israel J Entomolog 3:109–1929

Burgdorfer WM, Varma GR (1966) Trans-stadial and transovarial development of disease agents in arthropods. Annu Rev Entomol 12:347–376

Feldman-Muhsam B, Havivi Y (1973) Autogeny in the tick *Ornithodoros tholozani* (Ixodoidea, Argasidae). J Med Ent 10:185–189

Gray JS, Estrada-Peña AG, Vial LA (2013) Ecology of nidicolous ticks. In: Sonenshine DE, Roe RK (eds) The biology of ticks, vol 2, Oxford University Press, Oxford, pp 39–60

Holbrook MR (2012) Kyasanur forest disease. Antiviral Res 96:353–362

Mahdi M, Erickson BR, Comer JA, Nichol ST, Rollin PE, AlMazroa MA et al. (2011) Kyasanur forest disease virus alkhurma subtype in ticks, Najran Province, Saudi Arabia. Emerg Infect Dis 17:945–947

Pavlovskiy N, Skrynnik AN (1959) Contribution to the biology of the tick *Ornithodoros papillipes* 3:1403–1405

Řeháčk J (1989) Ecological relationships between ticks and Rickettsia. Eur J Epidemiol 5:407–413

Safdie G, Farrah IY, Yahia R, Marva E, Wilamowski A, Sawalha SS et al (2010) Molecular characterization of *Borrelia persica*, the agent of tick borne relapsing fever in Israel and the Palestinian Authority. PLoS ONE 5:1–11

Sonenshine DE, Clifford CM, Kohls GM (1966) The systematics of the subfamily Ornithodorinae (Acarina: Argasidae). III. Identification of the larvae of the eastern hemisphere. Ann Entomol Soc Am 59:92–121

Turell MJ, Whitehouse CA, Butler A, Baldwin C, Hottel H, Mores CN (2008) Assay for and replication of karshi (Mammalian Tick-Borne Flavivirus Group) virus in mice. Am J Trop Med Hyg 78:344–347

Vial L (2009) Biological and ecological characteristics of soft ticks (Ixodida: Argasidae) and their impact for predicting tick and associated disease distribution. Parasite 16:191–202

Ornithodoros erraticus complex (Figs. 23 and 24)

M. M. Santos-Silva and A. Estrada-Peña

Life Cycle and Host Preferences

Ornithodoros erraticus is a complex of species morphologically, biologically and ecologically similar. As mentioned in the introductory parts of this chapter, there is confusion as for whether other populations with similar morphology and variability in small parts of a gene may represent different species or are simply variants of one species which have allopatric distributions (Morel 2003). In this way, we are reluctant to accept the species defined only on molecular phylogenetic analysis of 16S rRNA sequences (Trape et al. 2013) without further description of the larvae, and/or laboratory crosses within presumed different taxa. The extreme morphological variability of this species (or complex of species) with large variations of morphology that overlap between taxa considered as separate species is well-known. It has been also reported that the morphology of the adults is highly variable, and that only the morphological details of the larvae could help in the morphological separation. Therefore, we hesitate to accept the reports of nine different species among *O. erraticus* complex: five new species (e.g. *O. occidentalis*, *O. costalis*, *O. rupestris*, *O. kairouanensis*, and *O. merionesi*), and four other identified as *O. erraticus*, *O. marocanus*, *O. sonrai* and *O. normandi* (Trape et al. 2013). While with the available data it is impossible to ascertain the validity of the presumed new species, we think that there is not yet enough proof to promote their separation as different taxa, considering both the large morphological variability and the high degree of molecular variation found in similar studies (Vial et al. 2006). However, Bouattour et al. (2010) provided reliable features to morphologically separate the adults of both *O. erraticus* and *O. normandi*.

This group of ticks has a natural long life cycle with many nymphal stages and long periods of starvation (Gray et al. 2014). Members of the *O. erraticus* complex are haematophagous nocturnal feeders. The natural life cycle maybe as long as 2 or 3 years for ticks occurring in Spain (Oleaga-Pérez et al. 1990). Under laboratory conditions, the life cycle is completed in a maximum of 154 days (El Shoura 1987). All stages feed on various warm-blooded vertebrates, including ungulates, carnivores, insectivores, rodents among others (Encinas-Grandes et al. 1993; Dias 1994; Trape et al. 2013). The most common hosts on the Iberian Peninsula are pigs, however, this species will feed opportunistically on other vertebrate species (Encinas-Grandes et al. 1993; Wilson et al. 2013). There seem to be no differences in host preference between the adult and immature stages. This tick has been reported from humans that are considered accidental hosts (Sánchez Botija 1982).

Ecology

The *O. erraticus* complex are adapted to various ecozones. The typical habitats include mostly xerophytic ecosystems and dry woodlands (Dias 1994). Because of its status as a nidicolous tick, it can be found in diverse microhabitats, established in holes, cracks, fissures, bird nests, burrows, under stones or in the resting places of vertebrate hosts (Boinas et al. 2014; Gray et al. 2014). In the Iberian Peninsula, *O. erraticus* usually is found in walls of the buildings, particularly those used to house pigs (Boinas et al. 2014). All stages are more active during the spring and

M. M. Santos-Silva (✉)
Centro de Estudos de Vectores e Doenças Infecciosas
Dr. Francisco Cambournac (CEVDI), Instituto Nacional de Saúde
Doutor Ricardo Jorge, Águas de Moura, Portugal
e-mail: m.santos.silva@insa.min-saude.pt

A. Estrada-Peña
Department of Animal Health, Faculty of Veterinary Medicine,
Zaragoza, Spain
e-mail: aestrada@unizar.es

A. Estrada-Peña et al. (eds.), *Ticks of Europe and North Africa*,
https://doi.org/10.1007/978-3-319-63760-0_17

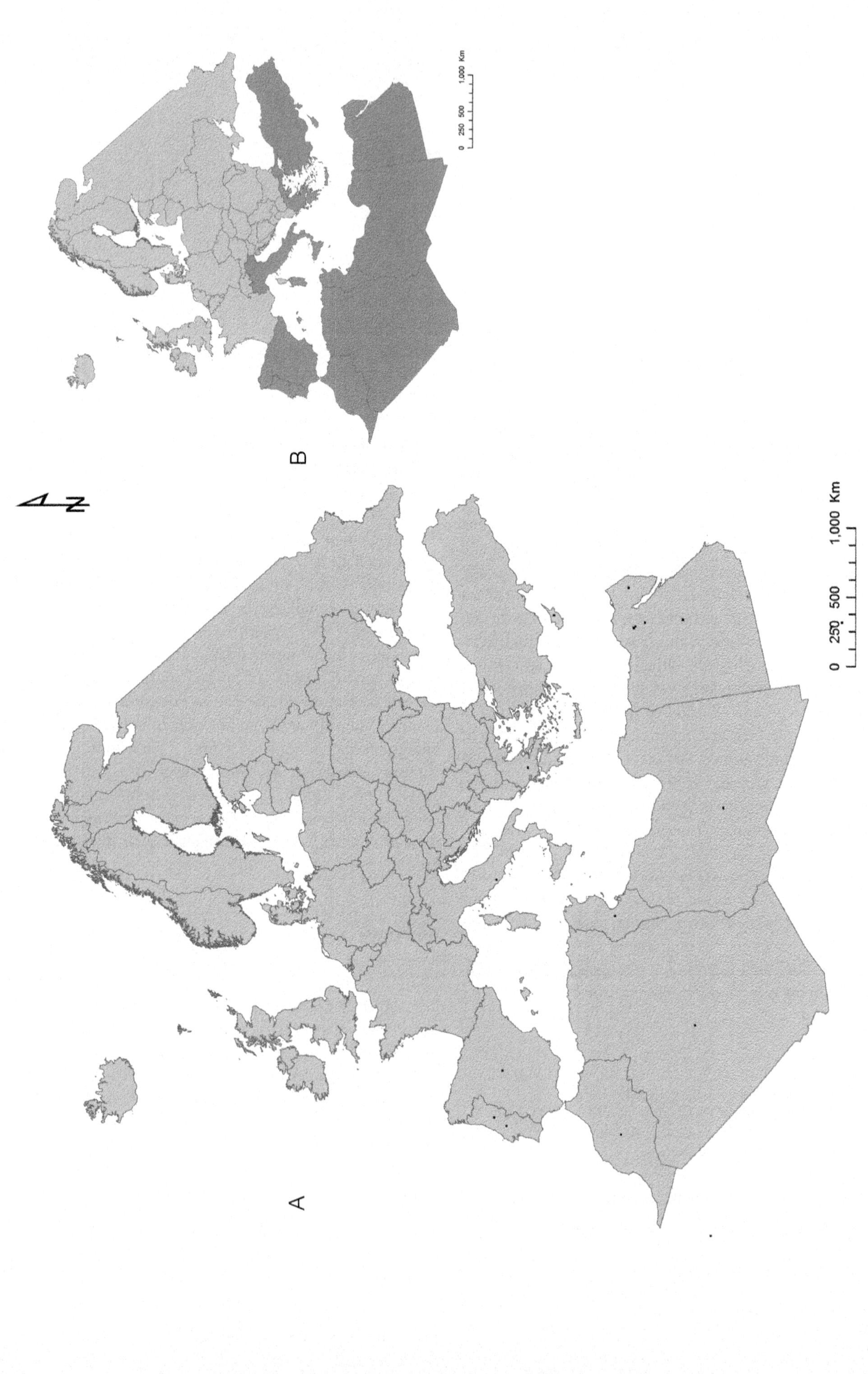

Fig. 23 **A** Distribution of *O. erraticus* in Europe and Northern Africa (10 × 10 km grid presence with black dots). **B** Countries where the species has been reported are marked in dark grey

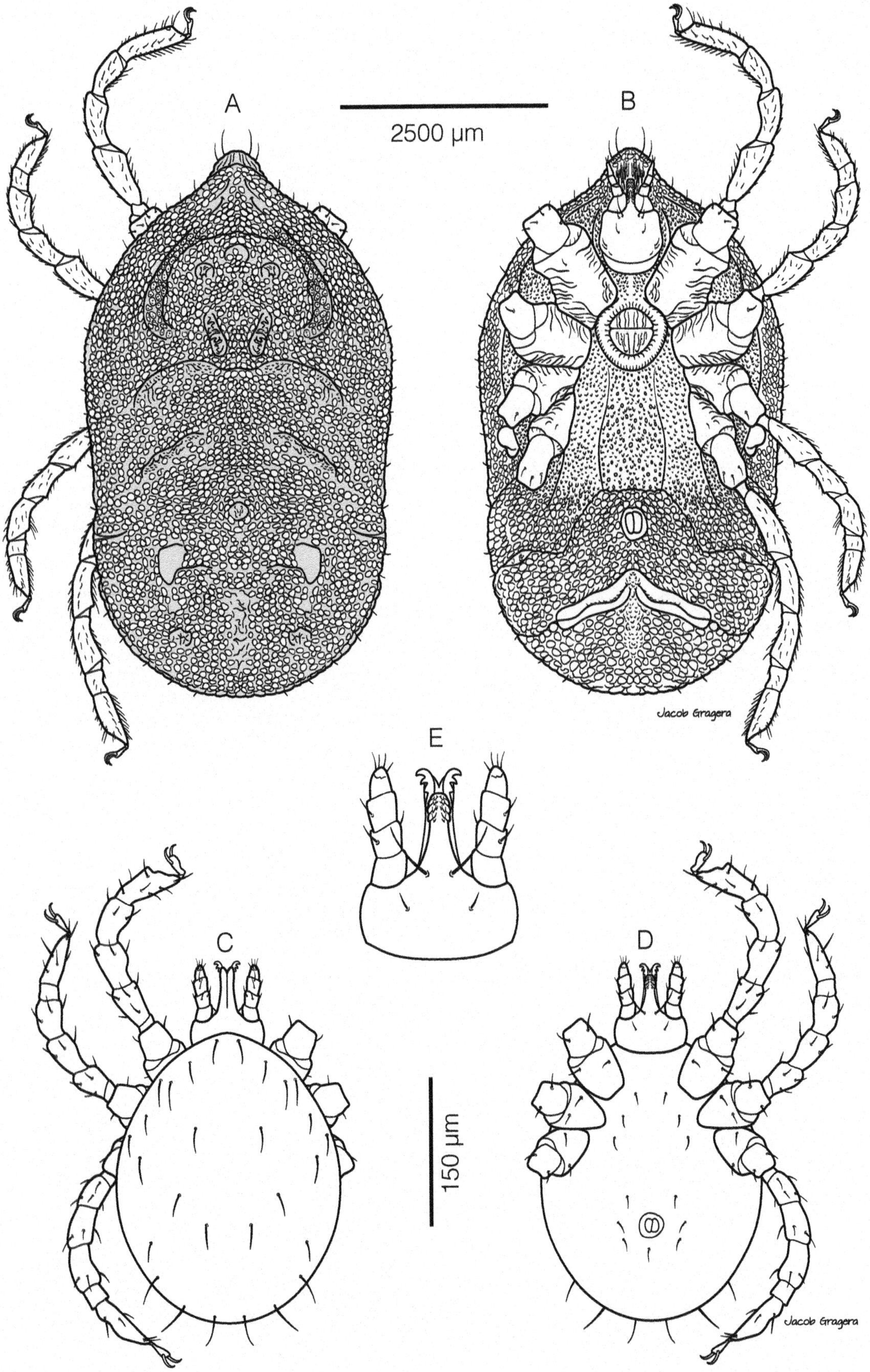

Fig. 24 A–B The adult of *O. erraticus*. **A** Dorsal, **B** ventral. **C–E** The larva of *O. erraticus*. **C** Dorsal, **D** ventral, **E** details of the capitulum. Illustrations from specimens collected in Spain and Portugal

summer months from March to September, due to their thermophilic preferences (Oleaga-Pérez et al. 1990; Gray et al. 2014). *Ornithodoros erraticus* evidences a passively and opportunistic host-seeking behaviour waiting for hosts in their environment.

Distribution

The *Ornithodoros erraticus* complex has been found in the Iberian Peninsula, north and west Africa and western Asia and includes several species most of which have relatively small geographic ranges.

Vectorial Capacity and Pathogen Burden

Experimental proof for the vectorial capacity of *O. erraticus* complex is available for African swine fever virus (Sánchez Botija 1982) and *Borrelia hispanica* (Sarih et al. 2009). Other pathogens associations are *Borrelia crocidurae* (Bouattour et al. 2010) and *Rickettsia lusitaniae* (Milhano et al. 2014). Regarding these pathogens, there is no experimental evidence for the tick's vectorial ability. The role of *O. erraticus* complex concerning other virus and bacterial pathogens remains poorly understood.

References

Boinas F, Ribeiro R, Madeira S, Palma M, Lopes de Carvalho I, Núncio S, Wilson AJ (2014) The medical and veterinary role of *Ornithodoros erraticus* complex ticks (Acari: Ixodida) on the Iberian Peninsula. J Vector Ecology 39:238–248

Botija Sánchez (1982) African swine fever. New developments. Rev Sci Tech Off Int Epiz 1:1065–1094

Bouattour A, Garnier M, M'Ghirbi Y, Sarih M, Gern L, Ferquel E, Postic D, Cornet M (2010) *Borrelia crocidurae* infection of *Ornithodoros erraticus* (Lucas, 1849) ticks in Tunisia. Vector Borne Zoonotic Dis 10:825–830

Dias JATS (1994) As carraças (Acarina-Ixodoidea) da Península Ibérica. Algumas considerações sobre a sua biogeografia e relacionamento com a ixodofauna Afropaleárctica e Afrotropical. Instituto de Investigação Cientifica Tropical, Estudos, Ensaios e Documentos 158, Lisboa

El Shoura SM (1987) The life cycle of *Ornithodoros* (*Pavlovskyella*) *erraticus* (Acari: Ixodoidea: Argasidae) in the laboratory. J Med Entomol 24:229–234

Encinas-Grandes A, Oleaga-Perez A, Pérez Sánchez R, Astigarraga A (1993) datos sobre el reservorio y vector de la peste porcina Africana, *Ornithodoros erraticus*. Anaporc 121:38–47

Gray JS, Estrada-Peña AG, Vial LA (2013) Ecology of nidicolous ticks. In: Sonenshine DE, Roe RK (eds) The biology of ticks, vol 2. Oxford University Press, Oxford, pp 39–60

Milhano N, Palma M, Marcili A, Núncio MS, de Carvalho IL, de Sousa R (2014) *Rickettsia lusitaniae* sp. nov. isolated from the soft tick *Ornithodoros erraticus* (Acarina: Argasidae). Comp Immunol Microbiol Infect Dis 37(3):189–193. https://dx.doi.org/10.1016/j.cimid.2014.01.006. Epub 30 Jan 2014

Morel PC (2003) Les tiques d'Afrique et du Bassin méditerranéen (1965–1995). CIRAD-EMVT, p 1342

Oleaga-Pérez A, Pérez-Sanchez R, Encinas-Grandes A (1990) Distribution and biology of *Ornithodoros erraticus* in parts of Spain affected by African swine fever. Vet Record 136:32–37

Sarih M, Garnier M, Boudebouch N, Bouattour A, Rihani A, Hassar M, Gern L, Postic D, Cornet M (2009) *Borrelia hispanica* relapsing fever. Morocco. Emerg Infect Dis 15:1626–1629

Trape JF, Diatta G, Arnathau C, Bitam I, Sarih M, Belghyti D, Bouattour A, Elguero E, Vial L, Mané Y, Baldé C, Pugnolle F, Chauvancy G, Mahé G, Granjon L, Duplantier J-M, Durand P, Renaud F (2013) The epidemiology and geographic distribution of relapsing fever borreliosis in West and North Africa, with a review of the *ornithodoros erraticus* complex (Acari: Ixodida). PLOS ONE 8(11):e78473, 1–19

Vial L, Durand P, Arnathau C, Halos L, Diatta G, Trape JF, Renaud F (2006) Molecular divergences of the *Ornithodoros sonrai* soft tick species, a vector of human relapsing fever in West Africa. Microbes Infect 8:2605–2611

Wilson AJ, Ribeiro R, Boinas F (2013) Use of a Bayesian network model to identify factors associated with the presence of the tick *Ornithodoros erraticus* on pig farms in southern Portugal. Prev Vet Med 110:45–53

Part II

Family Ixodidae Koch, 1844

The family Ixodidae is a large assemblage of ticks, divided in several genera and subgenera, including about 660 species. The genera can be easily separated for both adults or immatures following simple morphological details. Of importance are the presence or absence of eyes, presence or absence and position of an anal groove, the number of festoons, and the shape and relative proportions of the capitulum. Figure 25 includes a basic illustration of the morphology and dimensions of the different body portions of an ixodid tick. The illustration includes only the adults since immatures (larva and nymph) are morphologically similar to this life history stage, although they lack a genital opening. The keys below are intended to separate the genera of the family Ixodidae based on basic aspects of the morphology, and are provided separately for males, females, nymphs, and larvae.

Key to Genera of Family Ixodidae in Europe and Northern Africa

Adults

1a. Without eyes and with an anal groove surrounding the anus anteriorly...Genus ***Ixodes***
1b. With or without eyes; always without an anal groove surrounding the anus anteriorly...**2**
2a. Dorsum ornamented, with dark spots over a whitish (or pale) background...Genus ***Dermacentor***
2b. Not ornamented...**3**
3a. Eyes absent. Festoons present. Males always without adanal plates...Genus ***Haemaphysalis***
3b. Eyes and festoons present. Males with adanal plates (except in one species of *Rhipicephalus*), variously shaped...**4**
4a. Palpal segments very short, commonly shorter or as wide as the width of the basis capituli. Capitulum with an hexagonal contour...Genus ***Rhipicephalus***
4b. Palpal segments clearly longer than the width of the basis capituli, which has a rectangular or quadrangular contour...Genus ***Hyalomma***

Nymphs

1a. Anal groove surrounding the anus anteriorly in the shape of an inverted "U". Without eyes or festoons...Genus ***Ixodes***
1b. With or without eyes; always without an anal groove surrounding the anus anteriorly. Festoons present or absent...**2**
2a. With 9 festoons...Genus ***Rhipicephalus***
2b. With 11 festoons...**3**
3a. Eyes absent...Genus ***Haemaphysalis***
3b. Eyes present...**4**
4a. Hypostome with dental formula 3/3...Genus ***Dermacentor***
4b. Hypostome with dental formula 2/2...Genus ***Hyalomma***

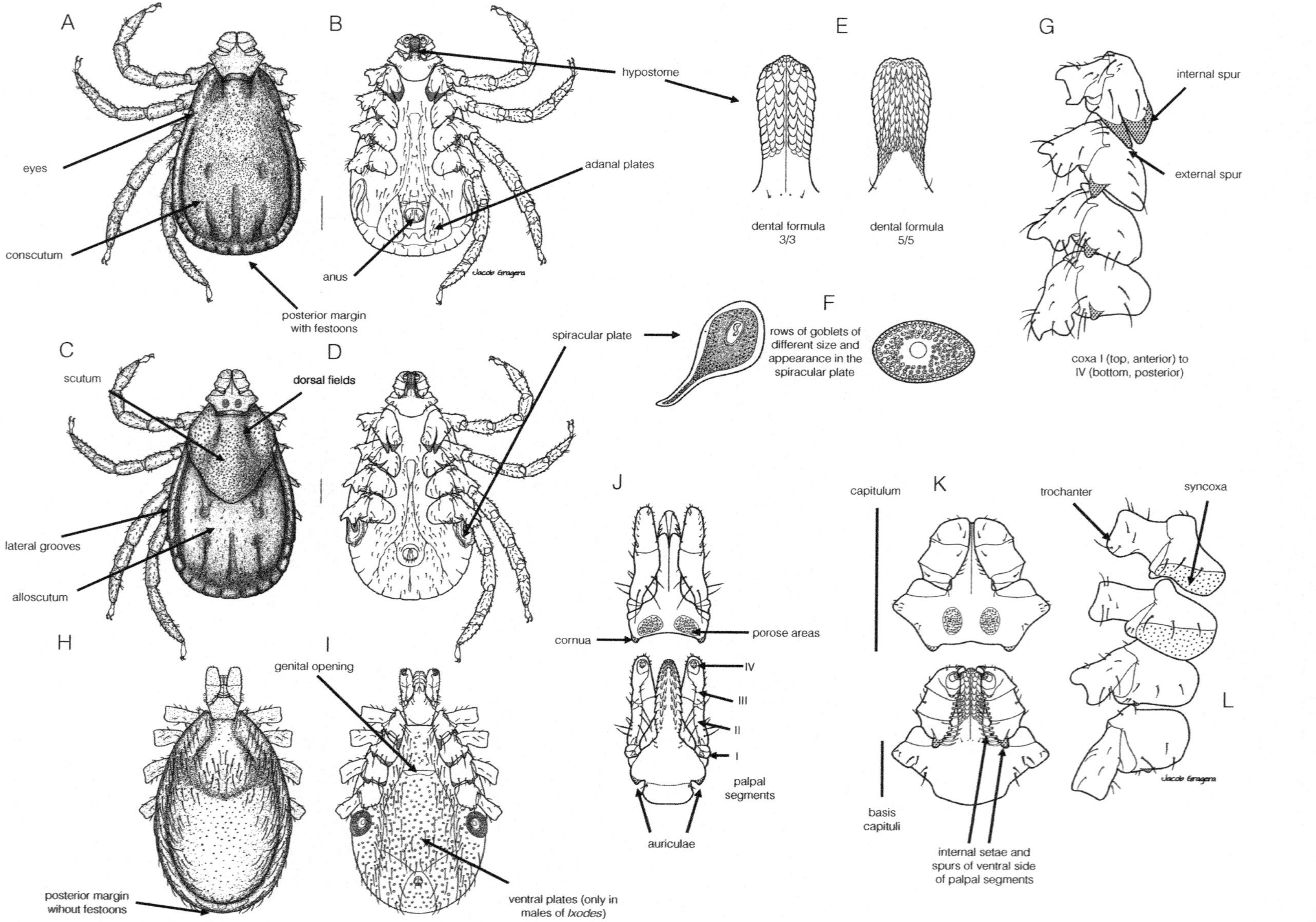

Fig. 25 The main morphological features of ticks belonging to the family Ixodidae. The illustrations show the typical features of a male (**A**: dorsal, **B**: ventral) and a female (**C**: dorsal, **D**: ventral) displaying the main morphological characteristics of the Metastriata. Further illustrations show the hypostome (**E**), the spiracular plate (**F**) and the coxae (**G**). Illustrations **H–L** show some features of prostriate ticks (genus *Ixodes*), displaying a female (**H**: dorsal, **I**: ventral), details of the capitulum and the basis capituli in Prostriata (**J**) and Metastriata (**K**), as well as details of the coxae of some *Ixodes* ticks that have syncoxae (**L**)

Larvae

1a. Anal groove surrounding the anus anteriorly in the shape of an inverted "U". With 2 pairs if hypostomal setae...Genus ***Ixodes***

1b. Anal groove absent. With 1 pair of posthypostomal setae...**2**

2a. Palpi long, with 4 articles (article 1 present)...**3**

2b. Palpi short, with 3 articles (article 1 absent)...**4**

3a. Body with 7 festoons. With 4–5 marginal dorsal setae anterior to the sensillum sagittiforme on each side...Genus ***Hyalomma***

3b. Body with 9 festoons. With 3 marginal dorsal setae anterior to the sensillum sagittiforme on each side...Genus ***Dermacentor***

4a. With eyes. With 4–5 marginal dorsal setae anterior to the sensillum sagittiforme on each side. Body without (subgenus *Boophilus*) or with 9 festoons...Genus ***Rhipicephalus***

4b. Without eyes. With 2 marginal dorsal setae anterior to the sensillum sagittiforme on each side. With 11 festoons... Genus ***Haemaphysalis***

Genus *Ixodes* Latreille, 1795

A. Estrada-Peña, M. P. Pfäffle, and T. N. Petney

There are a total of 59 species of ticks in this genus, grouped into about nine subgenera that have been reported from the Palaearctic region (Guglielmone et al. 2014). A list of synonyms was provided by Camicas et al. (1998) and further complemented by Guglielmone et al. (2009) and Guglielmone and Nava (2014). The French school of acarologists elevated most of the subgenera to the category of genera (i.e., Camicas et al. 1998). We prefer to follow the point of view of Horak et al. (2002), Guglielmone et al. (2009, 2010) and Guglielmone and Nava (2014), and consider all of the species to belong to the genus *Ixodes*, which is then divided into subgenera. A coherent view of the systematics of the subfamily Ixodinae was provided by Clifford et al. (1965). A general overview of the genus, including some species that are not recorded in the Western Palaearctic, is available in Filippova (1977). Adequate keys for all of the stages of *Ixodes*, without some species known to occur in the target area, are available in Manilla (1998). The reader should be aware that the last reference contains some inaccuracies, such as the mention to "spurs in coxae IV of larvae". In addition, most parts are in Italian, although there are keys in both Italian and English. A reliable key for the larvae of species of *Ixodes* in Europe (not the complete Western Palaearctic) was published by Márquez et al. (1992); the same key was repeated in the book by Manilla (1998).

Subgenus *Trichotoixodes* Reznik, 1961. Two species of *Ixodes* belonging to the subgenus *Trichotoixodes* are found in the Palaearctic region, namely *I. frontalis* (Panzer, 1798) and *I. turdus* Nakatsuji, 1942. Only *I. frontalis* is found in the Western Palaearctic. This species is a ubiquitous parasite on birds in the Palaearctic (Norte et al. 2012). Keys, redescriptions and illustrations for every stage of *I. frontalis* were provided by Filippova (1977) and Pérez-Eid (2007). The immatures were described by Arthur (1953) and adequate keys for immatures are available in Manilla (1998), Manilla and Iori (1992, 1993), Cordas et al. (1993), Sénevet and Rodhain (1968) as *I. pari*, which is a synonym of *I. frontalis* according to Reznik (1961). It is important to note the long history of confusion regarding the systematic status of this species. Doby (1998) still named it *I. pari*, a name that is no longer considered valid (Guglielmone et al. 2010). Authors such as Schembri (2003) and Visser et al. (2011) continued to use the name *I. pari* as valid.

Subgenus *Eschatocephalus* Frauenfeld, 1853. *Ixodes vespertilionis* Koch, 1844 and *I. simplex* Neumann, 1906 are species commonly found on bats and in the caves where bats breed in the Palaearctic and other regions. *Ixodes simplex* is a monoxenous tick of the cave bat, *Miniopterus schreibersii*, while *I. vespertilionis* has a wide host range among the Chiroptera. A new species, *I. ariadnae*, was recently described by Hornok et al. (2014). Details of its morphology were published in the original description, with additional morphological and molecular comparisons with close species in a separate paper (Hornok et al. 2015). These are three of the four species of the subgenus *Eschatocephalus*, which seems to be widely distributed, including in tropical regions and probably in Australia (Guglielmone et al. 2014). However, adequate comparisons of the populations outside the main and continuous range of these species are unavailable. Published illustrations of both *I. vespertilionis* and *I. simplex* from Japan are compatible with conspecificity with their Western Palaearctic representatives, although some molecular divergence has been found among these populations (Hornok et al. 2015). Illustrations of specimens from southern Africa lack the necessary degree of detail to determine their conspecificity. Both *I. vespertilionis* and

A. Estrada-Peña (✉)
Department of Animal Health, Faculty of Veterinary Medicine, Zaragoza, Spain
e-mail: aestrada@unizar.es

M. P. Pfäffle
Department of Ecology and Parasitology, Institute of Zoology, Karlsruhe Institute of Technology, Karlsruhe, Germany
e-mail: miripfaeffle@web.de

T. N. Petney
State Museum of Natural History, Erbprinzenstr. 13, 76133 Karlsruhe, Germany
e-mail: trevor.petney@kit.edu

A. Estrada-Peña et al. (eds.), *Ticks of Europe and North Africa*,
https://doi.org/10.1007/978-3-319-63760-0_18

I. simplex have been adequately illustrated by Yamaguti et al. (1971) and Filippova (1977). A thorough description of two of these species is available in Filippova (1972). A key for the larvae is available in Sénevet and Rodhain (1968) and for immatures in Manilla (1998), and Manilla and Iori (1992, 1993). Scanning electron microscopy (SEM) illustrations and keys can be found in Cordas et al. (1993), with further details on descriptions in Siuda (1993) and Pérez-Eid (2007).

Subgenus *Ceratixodes* Neumann, 1902. *Ixodes uriae* White, 1852 is a species widely found on sea birds. Details on its morphology, description and illustrations can be found in Filippova (1977), Yamaguti et al. (1971) and Pérez-Eid (2007). Molecular differences have been found among populations using different hosts in sea bird colonies (McCoy et al. 2003).

Subgenus *Multidentatus* Clifford, Sonenshine, Keirans and Kohls, 1973. *Ixodes rothschildi* Nuttall and Warburton 1911 is the only species in the subgenus found in the Palaearctic region. Keys for the larvae have been provided by Sénevet and Rodhain (1968). The adults of *I. rothschildi* were redescribed by Filippova (1977). Adequate morphological information and keys for closely related species are available in Pérez-Eid (2007). Arthur (1953) erroneously considered *I. rothschildi* to be a subspecies of *I. percavatus* (which is today regarded as an invalid species according to Camicas et al. 1998).

Subgenus *Scaphixodes* Schulze, 1941. Two species in the subgenus *Scaphixodes* are known to occur in the Western Palaearctic: *I. caledonicus* Nuttall, 1910 and *I. unicavatus* Neumann, 1908. Morel (2003) mentioned that *I. caledonicus* could be a synonym of *I. berlesei*, but a comparison of the types has not yet been made. We concur with the opinion by Guglielmone et al. (2014) that the two are separate species. There is adequate information for *I. berlesei* and *I. caledonicus* in Filippova (1977). Information, illustrations and keys for *I. unicavatus* and *I. caledonicus* are available in Pérez-Eid (2007). The female of *I. unicavatus* was redescribed by Filippova (1977), the male was described by Guiguen and Beaucournu (1984). The larvae of *I. unicavatus* can be separated from morphologically similar species by the keys in Sénevet and Rodhain (1968). No additional information seems to be available for these species, which are poorly known.

Subgenus *Pholeoixodes* Schulze, 1942. The species of the subgenus *Pholeoixodes* recorded from the Western Palaearctic are *I. arboricola* Schulze and Schlottke, 1929, *I. canisuga* Johnston, 1849, *I. crenulatus* Koch, 1849, *I. hexagonus* Leach, 1815, *I. kaiseri* Arthur 1957, *I. lividus* Koch, 1844, and *I. rugicollis* Schulze and Schlottke, 1930. A general review of the subgenus was provided by Emelyanova (1979).

Descriptions, keys and illustrations of *I. arboricola* are available in Filippova (1977) and Cordas et al. (1993). *Ixodes arboricola* is a species found on birds (Passeriformes and Strigiformes) that build nests in trees or walls. Haarløv (1962) and Guegan and Vermeil (1971) published accounts of the morphological variability of the species. *Ixodes arboricola* has been recorded on migratory birds and was regularly found on such species in Egypt (Clifford et al. 1965; Hoogstraal et al. 1967; Nosek and Balat 1982). There is a detailed account of its importance, distribution and ecology in Petney et al. (2012). A detailed comparison of three species of *Ixodes* commonly found on birds (*I. arboricola*, *I. frontalis*, and *I. ricinus*) was published by Heylen et al. (2014).

Ixodes rugicollis was redescribed and illustrated by Morel and Aubert (1975) and Aubert (1977). Pérez-Eid (2007) and Cordas et al. (1993) provided comparisons with morphologically similar species and keys for other species of the subgenus *Pholeoixodes*.

Ixodes hexagonus and *I. canisuga* are common species on Carnivora and, in the case of *I. hexagonus*, Western Palearctic members of the Family Erinaceidae (Order Eulipotyphla). Keys, illustrations and adequate descriptions are available in Filippova (1977), together with criteria for discriminating it from *I. lividus*, *I. arboricola* and *I. crenulatus*. Keys for immature stages can be found in Manilla (1998), Manilla and Iori (1992, 1993), and Cordas et al. (1993). Keys distinguishing between *I. hexagonus*, *I. canisuga*, *I. lividus*, *I. arboricola* and *I. rugicollis* are available (Pérez-Eid 2007). Details of the systematic history of *I. canisuga* are available in Teng (1973).

The original description of *I. kaiseri* by Arthur (1957) contains information and illustrations of the adult stages. All stages were redescribed by Filippova and Uspenskaya (1973). These authors also provided information on the taxonomic status of this species in relation to *I. crenulatus*, which was erroneously synonymized under *I. kaiseri* by Sonenshine et al. (1969). Other papers record the presence of *I. crenulatus* (compiled by Petney et al. 2012), but, as far as we know, no details are known of its ecology. Descriptions and details of the intraspecific variation in *I. crenulatus* can be found in Filippova and Panova (2000). It is possible that this species has been confused and reported as *I. canisuga* (Gothe et al. 1977).

There are adequate illustrations of *I. lividus* in Yamaguti et al. (1971). Keys, drawings and descriptions are available in Filippova (1977) and Siuda (1993) for all of the active stages. There are descriptions of the larvae of *I. lividus* in Morel and Pérez-Eid (1978a, b). SEM pictures are available in Cordas et al. (1993) and keys for the larvae in Manilla and Iori (1992) as well as Sénévet and Rodhain (1968). Keys for the nymphs are given in Manilla and Iori (1993). This species lives in the nests of and on the sand martin, *Riparia riparia*, and its distribution, although patchy, follows that of this host. *Riparia riparia* occurs throughout Europe and Asia, as well as parts of North America (Walter et al. 1979). However, surveys of the tick in colonies of *R. riparia* carried out in Portugal (Norte, personal communication) and Spain (Estrada-Peña, unpublished data) did not find this species. Pomerantzev (1950) referred to *I. plumbeus* as a valid species for what is now considered to be *I. lividus*.

Subgenus *Exopalpiger* Schulze, 1935. There are two species in this subgenus that have been recorded in the Palaearctic region: *I. trianguliceps* Birula, 1895 and *I. ghilarovi* Filippova and Panova, 1988. The former is the only one found in the Western Palaearctic where it is restricted to rodents and their nests. There are adequate illustrations and details of the morphology of *I. trianguliceps* in Filippova (1977), Morel and Pérez (1972), Cordas et al. (1993), Manilla (1998) Manilla and Iori (1992, 1993), Sénévet and Rodhain (1968), Siuda (1993) and Pérez-Eid (2007). A detailed account of its biology and ecology in England was published by Randolph (1975a, b).

Subgenus *Ixodes* Latreille, 1859. The status of *Ixodes acuminatus* Neumann, 1901 and *Ixodes redikorzevi* Olenev, 1927, is a matter of taxonomic discussion. Pomerantzev (1950) stated that both species are valid, a point of view shared by Filippova (1977). However, these authors did not describe any way to separate them. Sénévet and Rodhain (1968) provided a key for the larvae of the *I. ricinus* group, but again *I. acuminatus* is not separated from *I. redikorzevi*. It is interesting to note that Pérez-Eid (2007), when referring to *I. acuminatus*, included the drawings by Filippova (1977) for *I. redikorzevi*, with the exception of the nymph, a stage that seems to be clearly different in both taxa. The "Western" literature from Europe has numerous references, including an explicit comparison of *I. acuminatus* with other species (i.e. Manilla 1998; Manilla and Iori 1992, 1993; Morel and Pérez-Eid 1978a, b), but not with *I. redikorzevi*. Kolonin (2009) treats *I. redikorzevi* as a synonym of *I. acuminatus* based on a comparison of female specimens of *I. acuminatus* from France and Italy and *I. redikorzevi* from the former USSR, but the holotypes were not examined. There is thus a pressing need to compare both species. The types of both species are deposited in prominent collections and it should be relatively straightforward to compare these specimens. An adequate redescription would then be possible. A detailed account of the ecology of *I. acuminatus* can be found in Gilot et al. (1992).

Ixodes apronophorus Schulze, 1924, a poorly known species, is widely distributed in the target area. It is a parasite of Cricetidae, Castoridae, Soricidae, Muridae and Passeriformes (Guglielmone et al. 2014). The best available description and keys are in Filippova (1978), with additional keys and SEM illustrations in Cordas et al. (1993). Some additional illustrations can be found in Hillyard (1996). The immatures are described and compared to closely related species by Morel and Graf (1997).

Ixodes ventalloi Gil Collado, 1936, is a species which parasitizes the Mediterranean rabbit (*Oryctolagus cuniculus*). Most of the systematic problems with this species arose from two synonymies of *I. ventalloi*, namely *I. thompsoni* Arthur 1957, and *I. festai* sensu Arthur 1957. *Ixodes thompsoni* is recognised as a synonym of *I. ventalloi*. *Ixodes ventalloi* is probably restricted to the western Mediterranean, occurring as far east as Sicily. This species has also been found in Germany as an accidental introduction (Petney et al. 1996), and it seems that specimens have also been collected in the United Kingdom, although the determination was made as *I. festai* (Hillyard 1996). It is easy to determine *I. ventalloi* because of its small size and the ventrally curved auriculae of the nymphs and adults. It is difficult to state the actual range of this species, since specimens from Germany determined as *I. festai* by Walter et al. (1979) were determined using Arthur (1965), and both of these studies confused this species with *I. ventalloi* (Gilot and Perez 1978, Petney et al. 1996), as did Hillyard (1996) (see Gilot and Beaucournu 1998). *Ixodes bivari*, described by Santos-Dias from specimens collected in Portugal, was regarded as a different species because of the different appearance of the auriculae (Guglielmone et al. 2014). However, studies on a large series of specimens show that there is some variability in this character and therefore both species might be synonymous, in which case the name *ventalloi* has priority (Santos-Silva et al. 2011). Adequate keys for the separation of *I. ventalloi* from other morphologically similar species of *Ixodes* were provided by Gilot and Pérez (1978). Features adequate to separate the species from *I. ricinus* are available in Morel and Pérez-Eid (1978a, b). SEM images of the different stages of *I. ventalloi* are available in Cringoli et al. (2005). The molecular structure of close populations of *I. ventalloi* has been reported by Latrofa et al. (2017).

A special problem may exist with *I. festai* Tonelli-Rondelli, 1926. This is a poorly known species and its type is in very poor condition. The flask in the collection of the Natural History Museum of Turin, Italy, labelled as *I. festai*, contains one engorged female *Ixodes* without the capitulum, and therefore it is impossible to ascribe it to any species (see under *I. inopinatus* below). *Ixodes festai* has been reported to be a small tick that has been confused with both *I. ventalloi* and *I. ricinus*. It seems that the main hosts of this tick are birds and lizards. The specimens described by Arthur (1957, 1965) are in fact *I. ventalloi*. The larva of *I. festai* seems to be described by Sénévet and Rodhain (1968), but according to the measurements and the drawings in this publication, the specimens reported could in fact be any of several species of *Ixodes* from the Western Palaearctic. The nymph was compared with that of *I. ventalloi* by Gilot and Pérez-Eid (1978). The male of *I. festai* was described by Contini et al. (2011), but the illustrations provided in the publication are highly suggestive of *I. eldaricus*, and the authors did not include explicit ways for separating it from other related species. SEM images of the female of *I. festai* are available in Cringoli et al. (2005).

There are several species in the genus *Ixodes* which are part of the so-called "*ricinus/persulcatus* group". All of these species may share overlapping distributions and hosts and, therefore, there is a need for detailed and reliable information about them. While the adults can in some cases be reliably determined without mounting, the immatures need a careful examination and comparison with adequate descriptions or keys. *Ixodes ricinus* (Linnaeus, 1758), and *I. persulcatus* Schulze, 1930, are probably the two most widespread species of the genus *Ixodes* in the Palaearctic. The former is undoubtedly one of the most frequently collected ticks in Europe because of the frequency with which it is encountered and its importance in the transmission of pathogens to both humans and animals. Petney et al. (2012) compiled important references for understanding the general importance of this tick. Adequate descriptions, illustrations, keys and SEM figures are available in Cordas et al. (1993), Filippova (1977), Siuda (1993), Manilla and Iori (1992, 1993), Manilla (1998), Pérez-Eid (2007) and Cringoli et al. (2005). It is important to reliably determine this species and compare it with other species that might be confused with it such as *I. acuminatus, I. eldaricus, I. frontalis, I. gibbosus, I. inopinatus, I. laguri* and *I. persulcatus*. *I. persulcatus* is distributed from the eastern range of *I. ricinus* extending from approximately 30° E to Japan. However, specimens of *I. ricinus* have also been reported from southern Finland, and *I. persulcatus* has been found in both Finland, Sweden, and the Baltic countries (Jaenson et al. 2016). It is thus necessary to establish a clear separation between these two species. There is further information about the morphology of *I. persulcatus* in Yamaguti et al. (1971), Filippova (1977) and Siuda (1993). The reader should consult the study by Filippova (1985) to understand the distribution of *I. persulcatus*, which to some extent overlaps that of *I. ricinus* in Eastern Europe. Filippova (1985, 1999) states that in the sympatric zone inhabited by both species, *I. persulcatus* and *I. ricinus* inhabit different biotopes, with the latter being associated with light, broad-leaved European forests or with mostly light-leaved forests of the southern taiga. The distribution of *I. persulcatus*, however, coincides with the distribution of vegetation zones in which dark-coniferous components are predominant. Both species parasitize nearly the same range of host species. Although interspecific crosses between *I. persulcatus* and *I. ricinus* are successful, back crosses of the F1 generation with parental stock did not lead to offspring (Balashov et al. 1998). This obviously produces a bias in the reliable identification of *I. ricinus* and *I. persulcatus*, because the morphology may not totally reflect the taxonomic status of hybrid specimens (Bugmyrin et al. 2015; Kovalev et al. 2015). Additional information on this can be found in Jaenson et al. (2016).

A new species belonging to the *ricinus* group, *I. inopinatus* Estrada-Peña, Nava and Petney 2014, was recently described (Estrada-Peña et al. 2014). As far as known, it is a species that replaces *I. ricinus* in drier areas. In addition to southern Spain and a single collection from Germany (Petney et al. 2015), it has been collected in North Africa (Tunisia and Morocco), suggesting that old records of *I. ricinus* from this region might be unreliable. After its description, this tick has also been found in Algeria and some European countries (unpublished data and A. Paulauskas, personal communication). However, the presence of *I. ricinus* in northern Africa cannot be precluded because previous collections have not been examined. According to the published evidence, it seems that some reports of *I. festai* and *I. ventalloi* (Morel and Pérez-Eid 1978a, b; Cringoli et al. 2005) could in fact refer to *I. inopinatus*. However, this is based only on the examination of the illustrations in these references and not on examining the actual specimens. Estrada-Peña et al. (2014) provided illustrations and SEM images of this species, as well as comparisons with related species.

Ixodes gibbosus Nuttall, 1916 is a species with a largely eastern Mediterranean distribution, including Italy. The best reference for its determination is Saratsiotis (1970). There are adequate keys in Manilla and Iori (1992, 1993) and Manilla (1998), although the illustration of leg morphology appears to be incorrect in the latter reference. It is probable that some records of *I. ricinus* from the eastern Mediterranean are in fact *I. gibbosus* (Papadopoulos et al. 1996). This should be determined by reexamination of the collections.

Both *I. eldaricus* Dzhaparidze, 1950 and *I. laguri* Olenev, 1929 are reported from eastern parts of the Western Palaearctic. We do not currently know if they are part of the *ricinus* complex because adequate studies are missing, however, their morphology strongly suggests their inclusion in this group. Both species might have a larger distribution than currently reported. A redescription of the adults of *I. eldaricus* was carried out by Nowak-Chmura (2011). Other descriptions, illustrations and keys for every stage of *I. eldaricus* are available from Filippova (1977) and Siuda (1993). Filippova (1974) discussed the morphological variability of all stages of this species. The biology of *I. eldaricus* is poorly known and there is no information on its epidemiological importance. It is a parasite dwelling outside of the nest of the hosts, probably with a three-host developmental cycle (Siuda 1993). It has been reported as a common parasite on migratory birds from the Mediterranean Region to Central Europe (Nowak-Chmura 2011). The hitherto defined geographical distribution of this species covers Crimea, Georgia, Azerbaijan, Armenia, Kazakhstan, Turkmenistan, Kirghizia, Uzbekistan, Tajikistan, Russia (Dagestan) (Filippova 1977) and Israel (O. Erster, personal communication). On the Iraq-Turkmenistan border, the majority of records have been reported from the Kopet Dag mountain range (Filippova 1974). There have also been reports of it occurring on birds from Cyprus (Kaiser et al. 1974) and Israel, where it was described under the name *I. tatei* Arthur 1956, which is a junior synonym (Guglielmone et al. 2009). Regarding *I. laguri*, we adhere to the criteria by Guglielmone et al. (2014) who disregarded the existence of subspecies and consider the reported variations in morphology as a case of intraspecific variability. Filippova (1999), however, accepted the presence of subspecies, while Siuda (1993) did not. In any case, adequate illustrations and keys for *I. laguri* can be found in these two references.

Key to the Species of the Genus *Ixodes* in Europe and Northern Africa

Note: in these keys, we assume that *I. redikorzevi* is a synonym of *I. acuminatus*, something which has not yet been determined by examining the types of both species (Guglielmone et al. 2014).

Females

The female of *I. festai* is not included in the keys because the lack of an adequate type specimen for comparison. We did not include in these keys the females of *I. rothschildi*, *I. unicavatus* and *I. apronophorus* because the lack of reliable features to separate them from near species.

1a. Coxae I and/or II with the posterior half as syncoxae (membranous appendages)…**2**

1b. Coxae I and II without syncoxae. Coxa I with internal and external spines or spurs. Palpal segment I variably shaped but never triangular…**4**

2a. No spurs or spines on the coxae. Palpal segment I with a prominent triangular outgrowth…***Ixodes trianguliceps*** (Fig. 58).

2b. Spurs or spines on coxae. Palpal segment I without a prominent triangular outgrowth…**3**

3a. Syncoxae very small and apparent on both coxae I and II. Scutal setae not clearly longer than alloscutal setae. Alloscutal setae very numerous, covering almost completely the alloscutal surface…***Ixodes laguri*** (Fig. 83).

3b. The syncoxa is very small and apparent only on coxa I. Scutal setae clearly longer than alloscutal setae. Alloscutal setae not as above…**22**

4a. Presence of spurs directed backward on the ventral side of trochanters I–III, or at least on one of the trochanters…**5**

4b. Trochanters I–III without ventral backward-directed spurs…**7**

5a. Palpal segments II and III long, usually more than 3.5 times their width. The pre-anal groove does not reach the posterior edge of the idiosome behind the anus. Internal and external spines of coxa I of similar length. Lateral sides of the hypostome slightly convex. Setae on the middle of the alloscutum about 1.5 times longer than anterior setae on the lateral portions of the alloscutum…***Ixodes frontalis*** (Fig. 27).

5b. Palpal segments II and III relatively short, about 3 times their width. Lateral branches of the pre-anal groove reaching the posterior margin of the idiosome behind the anus. These lateral branches may be parallel, diverging or converging…**6**

6a. Second and third palpal segments widely rounded. The second segment is very narrow at its basis and then expands both medially and laterally. Auriculae as long as wide, as pointed blades. Slit of the genital opening rounded…***Ixodes caledonicus***

6b. Second and third palpal segments dorsally pear-shaped. The second palpal segment is narrow at its base and then expands gradually, especially noticeable medially. Auriculae prominent. Slit of the genital opening essentially straight… ***Ixodes berlesei***

7a. Palpi relatively long. Length of palpal segments II and III about 3.5 times their width…**8**

7b. Palpi relatively short. Length of palpal segments II and III about or less than 3 times their width…**13**

8a. Legs of normal length for the genus. Basis capituli without lateral projections…**9**

8b. Legs very long, about two or three times the total length of the idiosome in unengorged specimens. Basis capituli with side projections…**21**

9a. Tarsi of legs II to IV with a hump, very prominent mainly on tarsi I, giving the aspect of an abrupt termination near the claws, not tapering progressively…***Ixodes gibbosus*** (Fig. 77).

9b. All the tarsi tapering progressively to the claws…**10**

10a. Scutum elongate, oval, slightly longer than wide. Scutal and alloscutal setae short, those on the central scutum particularly short. Auriculae clearly visible, pointed, but not extending beyond the lateral margins of the ventral side of the basis capituli. Internal spur on coxa I essentially straight, covering the first third of coxa II…***Ixodes persulcatus*** (Fig. 71).

10b. Scutal and alloscutal setae relatively long. Setae on the central scutum not particularly shorter than the setae on the alloscutum. Scutum variously shaped. Auriculae clearly visible, rounded, and extending beyond the lateral margins of the ventral side of the basis capituli. Internal spur in coxa I variously curved…**11**

11a. Scutum slightly longer than wide. Internal spur on coxa I long, rounded, covering about a half or two thirds of coxa II in unengorged specimens. Punctations on the scutum small and homogeneously distributed. Auriculae developed as curved and distinctive horns…***Ixodes ventalloi*** (Fig. 65).

11b. Scutum slightly longer than wide. Internal spur on coxa I long, rounded, covering about half of coxa II in unengorged specimens. Punctations on the scutum small and homogeneously distributed. Auriculae variously developed….**12**

12a. Scutum slightly longer than wide. Internal spur on coxa I long, rounded, covering about half of coxa II in unengorged specimens. Punctations on the scutum small and homogeneously distributed. Auriculae variously developed but always straight, never curved as distinctive horns… ***Ixodes ricinus*** (Fig. 68).

12b. Scutum slightly wider than long. Internal spur on coxa I long, rounded, covering about the first third of coxa II in unengorged specimens. Punctations on the scutum both small and large, homogeneously distributed…***Ixodes inopinatus*** (Fig. 74).

13a. Basis capituli triangular, with posterolateral angles rounded, without cornua. All the coxae with spurs. Coxa IV with a tuft of 8–10 long setae in one row…***Ixodes simplex*** (Fig. 33).

13b. Basis capituli rectangular, with or without cornua. Coxae may have small spurs. Coxa IV with the normal complement of setae, and without the tuft of long setae in one row…**14**

14a. Scutum about twice as long as wide, piriform, with the cervical grooves narrow, well marked and reaching the posterior margin of the scutum. Alloscutal setae arranged in four bands, two medial and two lateral. Palpi broadly rounded, club-like, widest at the level of palpal segment III…***Ixodes uriae*** (Fig. 39).

14b. Without the above features…**15**

15a. Internal spur on coxa I relatively long and sharp. Coxae I–IV with short external spurs…***Ixodes hexagonus*** (Fig. 50).

15b. Internal spur on coxa I short and blunt, commonly difficult to see. Coxae I–IV without external spurs…**16**

16a. Anterior margin of the basis capituli with two rugose areas between the hypostome and the palpi. Hypostome with a similar length to that of the basis capituli. Porose areas circular and very small, separated by more than twice their diameter. Scutum as long as wide…***Ixodes rugicollis***

16b. Anterior margin of the basis capituli without rugose areas between the hypostome and the palpi. Scutum longer than wide…**17**

17a. Basis capituli, dorsally, with a clear rugose area producing two lateral fields, clearly separated in the median line of the dorsal basis capituli that encloses the porose areas. The relief reaches the anteromedial sides of the basis capituli. Porose areas with very small punctations, not confluent. Lateral grooves on the scutum distinct, deep and smooth. Lateral fields of the scutum entirely occupied by one or several crests…***Ixodes crenulatus*** (Fig. 46).

17b. Basis capituli, dorsally with or without a median rugose area separating the porose areas. If present, it is restricted only to the medial side of the basis capituli dorsally, never reaching the anterolateral margins of the basis capituli. Porose areas and pores variously shaped…**18**

18a. Porose areas with some pores confluent, giving the aspect of a fusion of some of the pores. Lateral portions of the scutum covered by longitudinal and sinuous striations… ***Ixodes arboricola***

18b. Porose areas well defined and with a clear outline, pores never confluent. Scutum may or may not be covered by striations in some parts…**19**

19a. Hypostome of a similar length to that of the basis capituli. Porose areas ovoid, placed in a depression in the middle of the basis capituli. Lateral fields of the scutum with slight striations…***Ixodes canisuga*** (Fig. 46).

19b. Hypostome clearly shorter than the length of the basis capituli…**20**

20a. Porose areas very wide, almost completely covering the dorsal portion of the basis capituli. Pores large but not fused. Lateral fields of the scutum and posterior margin with fine and very smooth striations…***Ixodes lividus*** (Fig. 54).

20b. Porose areas subcircular, restricted to the lateral two-thirds of the basis capituli with the outline clearly elevated…***Ixodes kaiseri***

21a. Palpi and legs very long. Scutum clearly longer than wide, piriform, with a row of relatively long setae arranged following the anterolateral contours of the scutum…***Ixodes vespertilionis*** (Fig. 30).

21b. Legs long but palpi relatively short. Scutum slightly longer than wide, very broad (not piriform). Only a few setae, not in a row, on the anterolateral margins of the scutum…***Ixodes ariadnae*** (Fig. 36).

22a. Internal spur on coxa I relatively long, reaching coxa II. Internal spur clearly visible on coxa II. Hypostome spear-shaped, widest at its base. Dorsal cornua long, backward-directed, very acute. Ratio of length/width of palpal segments II–III not less than 4.5. Scutal setae, especially those on the middle field of the scutum, about 1.1–1.3 times longer than alloscutal setae…***Ixodes acuminatus*** (Fig. 61).

22b. Internal spur on coxa I relatively short, never reaching coxa II. Internal spur on coxa II very short or absent, although the internal angle of the coxa II may look like a chitinous ridge. Hypostome not acutely spear-shaped, with more or less convex sides. Dorsal cornua of moderate size or slightly developed. Ratio of the length/width of palpal segments II–III never higher than 4. Scutal setae, especially those on the middle field of the scutum, about 1.5-2 times longer than alloscutal setae…***Ixodes eldaricus*** (Fig. 80).

Males

Note: the male of *I. berlesei* is unknown. *Ixodes unicavatus*, *I. rothschildi* and *I. apronophorus* have not been included because of the lack of adequate features for separation from near species.

1a. Palpal segment III conical with a sharp end pointing forward. Several tufts of flattened setae in the form of a brush on the posterior and posterolateral edges of the idiosome...***Ixodes uriae*** (Fig. 39).
1b. Palpal segment III rounded or blunt at its tip. Setae absent on the posterior and posterolateral edges of the idiosome...**2**
2a. Coxae I–III have syncoxae and lack spurs...***Ixodes trianguliceps*** (Fig. 58).
2b. Coxa III always without syncoxae...**3**
3a. Absence of cornua, auriculae and coxal spurs. Basis capituli dorsally clearly triangular. Adanal shields rectangular and not fused anteriorly to the anus...**4**
3b. Cornua, auriculae and/or coxal spurs present and variously shaped. Adanal shields fused anteriorly to the anus...**5**
4a. Legs of normal length. Palps stout with sparse, short setae. Hypostome with the tip depressed medially. Scutum without lateral grooves. Coxa IV with many long setae... ***Ixodes simplex*** (Fig. 33).
4b. Legs very long. Coxa IV with sparse and short setae...**21**
5a. Presence of strong apical spines on the tarsi of legs I–III. Hypostome with vestigial denticles. Trochanters without ventral spurs. Anal groove reaching the posterior margin of the body...***Ixodes caledonicus***
5b. Apical spines absent from the tarsi of legs I–III...**6**
6a. Trochanters with obvious ventral spurs, directed backwards. The sides of the pre-anal groove do not reach the posterior margin of the idiosome. Internal spur is present only on coxa I, with external spurs on coxae I to IV. Trochanters with small ventral spurs...***Ixodes frontalis*** (Fig. 27).
6b. Trochanters without ventral denticles. The sides of the pre-anal groove are parallel or diverging towards the rear but reach the posterior margin of the idiosome...**7**
7a. Coxa I with distinct internal and external spurs on the posterior margin. Hypostome with the tip rounded or blunt, with a more or less visible apical notch. Hypostomal teeth covering considerably more than half of the length of the hypostome, which has a short neck. Teeth usually well differentiated by size: the external are larger than the internal. Cornua present...**8**
7b. Coxa I commonly without internal or external spurs on its posterior margin. If spurs are present, then the hypostome is pointed. Hypostomal teeth variously shaped but in most cases almost unapparent...**15**
8a. Coxae I and II with distinct syncoxae. Punctations on the conscutum very large, deep, with setae on the marginal ridge of the idiosome thick and overlapping the margin of the conscutum...***Ixodes laguri*** (Fig. 83).
8b. Coxae I and II without syncoxae or only on coxa I...**9**
9a. Dorsal setae on the idiosome long, overlapping the lateral portions of the conscutum and not less than the width of the marginal groove. Dorsal cornua as distinct triangular teeth with the apex lying clearly behind the medial point of the posterior edge of the basis capituli...**10**
9b. Dorsal setae of the idiosome slightly overlapping the lateral portions of the conscutum and never as long as the width of the marginal groove. The rear edge of the basis capituli almost straight, and the cornua absent or only slightly evident. Basis capituli with auriculae variously shaped...**11**
10a. All coxae with one internal spur. Those on coxa I are short. Colour of the idiosome, basis capituli and legs light brown...***Ixodes acuminatus*** (Fig. 61).
10b. Coxae II–IV without an internal spur, although the posteromedian corner of coxa II may be slightly curved and pulled back. Internal spur of coxa I long...***Ixodes eldaricus*** (Fig. 80).
11a. Coxa I with a long and very curved internal spur, largely overlapping coxa II, coxae II–IV with internal spur. Auriculae extended into curved horns...***Ixodes ventalloi*** (Fig. 65).
11b. Coxa I with a long but variously produced internal spur that may or may not reach coxa II. Auriculae small, visible, and straight, never extend into curved horns...**12**
12a. Tarsi with a clearly visible hump. Internal spur of coxa I long but not reaching coxa II. Coxae II–IV without internal spurs...***Ixodes gibbosus*** (Fig. 77).
12b. Tarsi II–IV without a clearly visible hump, tapering gradually to the claws. Internal spur of coxa I either long or relatively short...**13**
13a. With only one row of setae between the marginal groove and the lateral margins of the idiosome...***Ixodes inopinatus*** (Fig. 74).
13b. With three or more rows of setae between the marginal groove and the lateral margin of the idiosome...**14**
14a. Internal spur of coxa I very long. Longitudinal axis of peritremal plate very much longer than the diameter of the anal valve. Area of the peritremal plate larger than the size of coxa IV. Hypostome relatively stout. Posterior margin of basis capituli essentially straight...***Ixodes ricinus*** (Fig. 68).
14b. Internal spur of coxa I relatively short. Longitudinal axis of the peritremal plate only 2.5 times longer than the longitudinal diameter of the anal valve. Hypostome very sharply narrowed at the apex...***Ixodes persulcatus*** (Fig. 71).
15a. Internal spur on coxa I sharp and clearly differentiated, commonly slightly reaching the perimeter of coxa II. Anal groove pointed anteriorly. Hypostome pointed at the apex... ***Ixodes hexagonus*** (Fig. 50).
15b. Internal spur of coxa I absent or very short, never reaching coxa II. Anterior portion of the anal groove basically straight or slightly curved, not pointed. Hypostome with a rounded tip and an apical notch...**16**
16a. Anterior margin of the basis capituli with two rugose areas between the hypostome and the palpi. Basis capituli with a polyhedral pattern engraved on its surface. Spiracular

plates small, even smaller than coxa IV, with no more than 40 goblets that are sometimes coalescent...***Ixodes rugicollis***

16b. Anterior margin of the basis capituli without any rugose areas between the hypostome and the palpi. Basis capituli without an engraved polyhedral pattern. Spiracular plates large, with 60–90 goblets...**17**

17a. Anterior portion of the anal groove curved. Conscutal surface shiny and wrinkled in some parts...***Ixodes crenulatus***

17b. The anterior portion of the anal groove is essentially straight...**18**

18a. Coxa I with chitinized internal margins. Coxae I–IV with small crests. Scutum with deep lateral grooves. Adanal plates twice as long as wide...***Ixodes canisuga*** (Fig. 46).

18b. Coxae small and without chitinized crests. Scutum with superficial lateral grooves. Adanal plates at least 3 times longer than wide...**19**

19a. The conscutal surface is smooth and shiny. Hypostome tapered, elongated. Coxa I with a small medial denticle... ***Ixodes kaiseri***

19b. The conscutal surface is wrinkled and matt. Lateral grooves with large punctations. Hypostome trapezoidal, short. Coxa I without a medial small denticle...**20**

20a. Coxa I with the posterointernal angle rounded, without any elevation of its surface (carena). Basis capituli with lateral margins parallel or slightly divergent anteriorly. Spiracular plates small, with some 45 goblets...***Ixodes arboricola***

20b. Coxa I with the posterointenal margin provided with a clear carena. Basis capituli with lateral margins clearly convergent anteriorly. Spiracular plates of medium size, with about 60 goblets...***Ixodes lividus*** (Fig. 54).

21a. Palpi ovoid, laterally curved. Palpal setae long (most with length equal to or exceeding the breadth of palpi)... ***Ixodes vespertilionis*** (Fig. 30).

21b. Lateral edge of palpi straight. Palpal setae shorter than palpal breadth (especially apically)...***Ixodes ariadnae*** (Fig. 36).

Nymphs

Note: the nymphs of *I. kaiseri, I. rugicollis, I. apronophorus, I. crenulatus, I. laguri,* and *I. rothschildi* have been omitted because the lack of suitable features for adequate separation of the species.

1a. Palpal article I very broad and fused with the basis capituli. Hypostome with dentition 2/2. Coxae I and II transformed into syncoxae. Post-hypostomal setae in the vertices of a trapezium...***Ixodes trianguliceps*** (Fig. 59).

1b. Palpal article I not broad and not fused with the basis capituli. Hypostome with basal dentition 2/2, and sub-apical 3/3. In some species this may be 4/4 in the sub-apical portion. Post-hypostomal setae variously shaped. Coxae not transformed into syncoxae...**2**

2a. Basis capituli with tectum essentially flat or only slightly oblique. Palpi commonly short...**3**

2b. Basis capituli with tectum clearly oblique. Palpi short or long...**4**

3a. Palpi with internal margins straight. Post-hypostomal setae placed in the vertices of a square. Cornua and auriculae always present. Coxae I–IV with the external spur short and robust, slightly curved. Trochanters I–III with ventral spur. Tarsi I–IV tapering gradually towards the apical end...**5**

3b. Palpi with internal margins clearly convex. Post-hypostomal setae in the corners of a trapezium. Cornua and auriculae commonly absent. Coxae I–IV with the external spurs short or absent. Trochanters without ventral spurs. Tarsi I–IV tapering abruptly towards the apical end... **6**

4a. Coxae without spines. Cornua and auriculae absent...**9**

4b. Coxae with spines. Cornua and auriculae present...**11**

5a. Idiosomal scutum wider than long. Hypostome with rounded tip. Idiosomal setae short...***Ixodes caledonicus***

5b. Idiosomal scutum longer than wide. Hypostome with pointed tip. Idiosomal setae very long...***Ixodes frontalis*** (Fig. 28).

6a. Coxa I with internal spur short and pointed. Coxae I–IV each with one very small and blunt external spur...***Ixodes hexagonus*** (Fig. 51).

6b. Coxae without spurs...**7**

7a. Cornua absent. Hypostome without a medial indentation or hollow at the tip...***Ixodes arboricola***

7b. Cornua present. The tip of the hypostome has a clear medial indentation...**8**

8a. Cornua small and pointed, posterolaterally directed. Scutum with scapular fields wrinkled...***Ixodes canisuga*** (Fig. 47).

8b. Cornua large, curved posteriorly. Scutum smooth... ***Ixodes lividus*** (Fig. 55).

9a. Legs not exceptionally long. Dorsal portion of basis capituli sub-triangular, with the posterior margins laterally rounded. Hypostome with 2/2 dentition (3/3 in the sub-apical portion). Palpi rounded apically. Palpal segment II as long as segment III. Haller's organ shorter than the maximal diameters of tarsus I...***Ixodes simplex*** (Fig. 34).

9b. Legs exceptionally long. Haller's organ longer than the maximal diameter of tarsus I...**10**

10a. Scutum rhomboidal, posteriorly narrowed. Dorsal portion of the basis capituli sub-triangular, with the posterior margins prolonged laterally as dorsoventral ridges. Hypostome with a 3/3 dentition (4/4 in the sub-apical portion). Palpi narrow, palpal article II twice the length of palpal article III...***Ixodes vespertilionis*** (Fig. 31).

10b. Scutum posteriorly broad and rounded, reverse bell-shaped, with an irregular surface, especially along the margin. Dorsal portion of basis capituli sub-triangular,

posterolaterally pointed. Hypostome with a 2/2 dentition (3/3 in the sub-apical section). Palpi broad, palpal segment II longer than segment III, but never twice its length...***Ixodes ariadnae*** (Fig. 37).
11a. Hypostome with a pointed tip, with a uniform dentition as 11 rows of 2/2. Auriculae acute and prominent...***Ixodes acuminatus*** (Fig. 62).
11b. Hypostome with tip rounded or flat. Hypostomal dentition 2/2 with 3/3 in the sub-apical portion. Auricle variously shaped or absent...**12**
12a. Auriculae rudimentary, as non-protruding ridges. Alloscutal setae thick, 4–5 times as longer as the scutal setae. Hypostome with 4–5 rows of sub-apical 3/3 dentition. Tarsi with dorsal pre-terminal hump...***Ixodes gibbosus*** (Fig. 78).
12b. Auriculae well developed and protruding. Alloscutal setae numerous, not thick. Hypostome with 7–8 rows of sub-apical 3/3 dentition. Tarsi without dorsal pre-terminal hump...**13**
13a. Auriculae produced in an acute angle, curved, medially concave. Coxa I with internal spur very long and curved. Scutum with posterolateral margins concave and posterior margin essentially straight or slightly rounded...***Ixodes ventalloi*** (Fig. 66).
13b. Auriculae produced in a straight angle. Coxa I with internal spur short or long, but always straight...**14**
14a. Scutal setae in the cervical fields very long. Internal spur of coxa I does not reach the margin of the coxa II... ***Ixodes eldaricus*** (Fig. 81).
14b. Scutal setae short. Alloscutal setae clearly longer than scutal setae. Internal spur on coxa I clearly reaching coxa II...**15**
15a. Scutal and alloscutal setae very short. Alloscutal setae as long as scutal setae in the cervical fields...***Ixodes persulcatus*** (Fig. 72).
15b. Scutal setae short. Alloscutal setae longer than scutal setae at variable ratios...**16**
16a. Alloscutal setae only 2–3 times as long as the scutal setae...***Ixodes ricinus*** (Fig. 69).
16b. Alloscutal setae at least 7–9 times as long as the scutal setae...***Ixodes inopinatus*** (Fig. 75).

Larvae

Note: the larvae of *I. uriae, I. rothschildi, I. unicavatus, I. crenulatus, I. kaiseri, I. rugicollis, I. apronophorus* and *I. laguri* have been omitted because the lack of suitable features for adequate separation of the species.

1a. With only one pair of post-hypostomal setae. Haller's organ open (but this feature appears also in other species). Palpal articles II and III fused...***Ixodes uriae*** (Fig. 40).
1b. With two pairs of post-hypostomal setae...**2**
2a. Palpal article I clearly wider than palpal article II and bearing lateral expansions that fuse with the tectum...***Ixodes trianguliceps***
2b. Palpal article I of the same length as the others and without lateral expansions...**3**
3a. Trochanter of leg I with a medial spur. Palpal articles II and III fused...**4**
3b. Trochanter of leg I without a medial spur...**6**
4a. Hypostome with two rows with a dentition of 4/4, with many small denticles 3/3. Haller's organ open. Scutum with five pairs of setae...***Ixodes rothschildi***
4b. Hypostome with only 2–3 rows of dentition 3/3, never 4/4...**5**
5a. Coxa I with a well developed medial spur of similar size to or even larger than the external spur. Haller's organ closed. Scutum with 4 pairs of setae...***Ixodes caledonicus***
5b. Coxa I without medial spur but with a chitinous protuberance near the juncture with the trochanter. Haller's organ open. Scutum with 5 pairs of setae...***Ixodes unicavatus***
6a. Without supplementary setae on the conscutum...**7**
6b. With at least one pair of supplementary setae on the conscutum. Haller's organ closed...**8**
7a. Basis capituli with a pentagonal contour. Haller's organ open. Post-hypostomal setae inserted as the four corners of a rectangle...**15**
7b. Basis capituli approximately rectangular in shape, wider than long. Haller's organ closed. Post-hypostomal setae inserted in a trapezoidal pattern, with the basis anteriorly... **16**
8a. Palpal articles I to III well separated. With 4-5 pairs of supplementary setae on the conscutum. Post-hypostomal setae inserted in a rectangular pattern. With 3 pairs of marginoventral setae...***Ixodes frontalis*** (Fig. 28).
8b. Palpal article I well separated from article II. With only one pair of supplementary setae on the conscutum. Post-hypostomal setae inserted in a trapezoidal pattern. Four pairs of marginoventral setae...**9**
9a. Hypostome with a dentition 2/2 for its complete length... **10**
9b. Hypostome with additional rows of dentition 3/3. The rest of the hypostome with a dental formula 2/2...**11**
10a. Ventral cornua with an acute angle...***Ixodes acuminatus*** (Fig. 47).
10b. Ventral cornua rounded...***Ixodes laguri*** (Fig. 84).
11a. Latero-dorsal setae measuring less than 1.5 times the length of the marginoscutal setae. Ventral cornua rounded... **12**
11b. Latero-dorsal setae measuring 2–3 times the length of the marginoscutal setae. Ventral cornua acute...**13**
12a. Basis capituli with the ventral posterior margin concave and with marginal extensions acute...***Ixodes apronophorus***

12b. Basis capituli with the ventral posterior margin straight…***Ixodes persulcatus*** (Fig. 72).

13a. Ventral cornua in a well marked, acute angle…**14**

13b. Ventral cornua rounded…***Ixodes gibbosus*** (Fig. 78).

14a. Coxa I with posterointernal spur rounded, slightly longer than the posteroexternal spur…***Ixodes ricinus*** (Fig. 69).

14b. Coxa I with posterointernal spur acute, subtriangular, at least twice as long as the posteroexternal spur…***Ixodes ventalloi*** (Fig. 66).

15a. Legs of normal length. Scutum broader than long, broadest anterior to mid-length, with a transversely straight posterior margin…***Ixodes simplex*** (Fig. 34).

15b. Legs long. Scutum variously shaped, posteriorly rounded…**22**

16a. Ventral cornua evident…**17**

16b. Ventral cornua not visible…**19**

17a. Seven pairs of marginodorsal setae. Three pairs of marginoventral setae…***Ixodes lividus*** (Fig. 55).

17b. Eight pairs of marginodorsal setae. Four pairs of marginoventral setae…**18**

18a. Marginodorsal setae of similar length to marginoscutal setae…***Ixodes kaiseri***

18b. Marginodorsal setae longer than marginoscutal setae… ***Ixodes hexagonus*** (Fig. 51).

19a. Hypostome with sub-apical rows of 3/3 dentition…**20**

19b. Hypostome with sub-apical rows of 2/2 dentition…**21**

20a. Seven pairs of marginodorsal setae…***Ixodes arboricola***

20b. Nine pairs of marginodorsal setae…***Ixodes rugicollis***

21a. Three pairs of marginoventral setae…***Ixodes crenulatus***

21b. Four pairs of marginoventral setae…***Ixodes canisuga*** (Fig. 47).

22a. Scutum broadest at mid-length. Cervical grooves reach the posterolateral margin of the scutum at the deepest point of concavity. Palpi relatively narrow, curved both medially and laterally…***Ixodes vespertilionis*** (Fig. 31).

22b. Scutum broadest anterior to mid-length. Cervical grooves reach the posterolateral margin of the scutum behind the deepest point of concavity. Palpi broad, laterally straight…***Ixodes ariadnae*** (Fig. 37).

References

Arthur DR (1953) The systematic status of *Ixodes percavatus* var. *rothschildi* Nuttall & Warburton. Parasitology 43:222–226

Arthur DR (1957) Two North African *Ixodes* ticks. *I. kaiseri* sp. nov. from Egyptian desert fox cub. A redescription of the female and a description of the male of *I. festai* Rondelli, 1926 (Ixodoidea, Ixodidae). J Parasitol 43:578–585

Arthur DR (1965) Ticks of the genus *Ixodes* in Africa. Athlone Press, London

Aubert MFA (1977) Description du mâle de *Pholeoixodes rugicollis* (Schulze and Schlottke, 1929) (Acariens Ixodina). Ann Parasitol Hum Comp 52:481–490

Balashov YS, Grigor'eva LA, Oliver JH Jr (1998) Reproductive isolation and interspecific hybridization in ixodid ticks of the *Ixodes ricinus-I. persulcatus* group (Acarina, Ixodidae). Entomol Rev 78:500–508

Bugmyrin SV, Belova OA, Ieshko EP, Bespyatova LA, Karganova GG (2015) Morphological differentiation of *Ixodes persulcatus* and *I. ricinus* hybrid larvae in experiment and under natural conditions. Ticks Tick-Borne Dis 6:129–133

Camicas JL, Hervy JP, Adam F, Morel PC (1998) Les tiques du monde. Nomenclature, stades décrits, hôtes, répartition (Acarida, Ixodida). Orstom, Paris

Clifford CM, Sonenshine DE, Keirans JE, Kohls GM (1965) Systematics of the subfamily Ixodinae (Acarina: Ixodidae). 1. The subgenera of *Ixodes*. Ann Entomol Soc Am 66:489–500

Contini C, Palmas C, Seu V, Stancampiano L, Usai F (2011) Redescription of the male of *Ixodes festai* Rondelli, 1926 (Ixodida: Ixodidae) on specimens from Sardinia (Italy). Parasite 18:235–240

Cordas T, Aeschlimann A, Morel PC (1993) Étude morphologique des Ixodidae s. str. (Schulze, 1937) de Suisse au microscope électronique à balayage. Acarologia XXXIV:21–46

Cringoli G, Iori A, Rinaldi L, Veneziano V, Genchi C (2005) "Zecche". Mappe Parassitologiche, Facoltà di Medidina Veterinaria, Naples, Italy

Doby JM (1998) Contribution à la connaissance de la biologie de *Ixodes (Trichotoixodes) pari* Leach (=*I. frontalis* (Panzer)) (Acari: Ixodidae), tique spécifique des oiseaux. Acarologia XXXIX:315–324

Emelyanova ND (1979) Taxonomic status of ixodid ticks of the genus *Pholeoixodes* within the subfamily Ixodinae and its division into subgenera. In: Efimov MV, Pronin NM (eds) Zooparasitology of Lake Baikal Bassin. Akad. Sauk SSSR, Sib. Otd. Buret. Filial, Otd. Biol., pp 5–27 (in Russian, English Translation NAMRU3 T1739)

Estrada-Peña A, Nava S, Petney T (2014) Description of all the stages of *Ixodes inopinatus* n. sp. (Acari: Ixodidae). Ticks Tick-Borne Dis 5:734–743

Filippova NA (1972) New data on ticks of the genus *Ixodes* Latreille (Ixodidea, Ixodidae) which are specific bat parasites. Zoological Institute, USSR Academy of Sciences, Leningrad, pp 279–285

Filippova NA (1974) *Ixodes eldaricus* and its distribution in southern USSR. Parazitologiya 8:504–514 (in Russian, English translation NAMRU-3, T1554)

Filippova NA (1977) Ixodid ticks (Ixodinae). Fauna USSR New Series, vol 4, pp 1–316 (in Russian)

Filippova NA (1985) The taiga tick *Ixodes persulcatus* (Acarina, Ixodidae). Morphology, systematics, ecology, medical importance. Soviet Committee for the UNESCO Programme Man and Biosphere, Leningrad, Russia, 416 pp (in Russian)

Filippova NA (1997) Fauna of Russia and neighbouring countries. Arachnoidea, Ixodid ticks of subfamily Amblyomminae, Vol IV (5), Russian Academy of Sciences, Zoological Institute, New Series 145, Nauka Publishing House, St Petersburg

Filippova NA (1999) The intraspecific taxonomic structure and identification of closely related species of ixodid ticks (Ixodidae). Parazitologiya 34:81–91 (in Russian)

Filippova NA, Panova IV (2000) Intraspecific variation of nest ambushing tick *Ixodes crenulatus* (Ixodidae). Parazitologiya 34:265–279 (in Russian)

Filippova NA, Uspenskaya IG (1973) *Ixodes kaiseri* Arthur 1957 (Ixodidae) species taxonomic status. Parazitologiya 7:297–306 (in Russian, NAMRU3 translation T636)

Gilot B, Beaucournu JC (1998) Book review: Ticks of North-West Europe by Paul D. Hillyard 1996. Parasite 5:206–208

Gilot B, Perez C (1978) Individualisation et caracterisation de deux *Ixodes* actuellement confondus: *I. festai* Rondelli, 1926, *I. ventalloi* Gil Collado, 1936 (Acarina, Ixodoidea). Revue Suisse Zool 85:143–149

Gilot B, Couatarmanach A, Guiguen C, Beaucournu J (1992) Biology and ecology of *Ixodes acuminatus* Neumann, 1901, its hosts, seasonal activity and distribution in France. Ann Parasitol Humaine Comparée 67:19–25

Gothe R, Stendel W, Holm R (1977) Zum Vorkommen von *Ixodes canisuga* Johnston, 1848, in Deutschland, ein Beitrag zur *Ixodes*-Fauna. Zeitsch. Parasitenkunde 53:123–128

Guegan J, Vermeil C (1971) *Ixodes arboricola* Schulze & Schlottke 1929. Compléments à l'étude du stade larvaire. Acarologia 13:36–46

Guglielmone AA, Robbins RG, Apanaskevich DA, Petney TN, Estrada-Peña A, Horak IG (2009) Comments on controversial tick (Acari: Ixodida) species names and species described or resurrected from 2003 to 2008. Exp Appl Acarol 48:311–327

Guglielmone AA, Robbins RG, Apanaskevich DA, Petney TN, Estrada-Peña A, Horak IG, Shao R, Barker SC (2010) The Argasidae, Ixodidae and Nuttalliellidae (Acari: Ixodida) of the world: a list of valid species names. Zootaxa 2528:1–28

Guglielmone AA, Robbins RG, Apanaskevich DA, Petney TN, Estrada-Peña A, Horak IG (2014) The hard ticks of the world (Acari: Ixodida: Ixodidae). Springer, Dordrecht, p 738

Guiguen C, Beaucournu JC (1984) Description of the male *Ixodes (Scaphixodes) unicavatus* Neumann, 1908 (Acarina, Ixodidae). Ann Parasitol Hum Comp 59:271–275

Haarløv N (1962) Variation in the ixodid tick, *Ixodes arboricola* Schulze & Schlottke 1929. Parasitology 52:425–439

Heylen D, De Coninck E, Jansen F, Madder M (2014) Differential diagnosis of three common *Ixodes* spp. ticks infesting songbirds of Western Europe: *Ixodes arboricola*, *I. frontalis* and *I. ricinus*. Ticks Tick-Borne Dis 5:693–700

Hillyard PD (1996) Ticks of north-west Europe. Field Studies Council, Shrewsbury

Hoogstraal H, Kaiser MN, Seymour C, Gaber S (1967) Noteworthy recent tick records from Egypt. 1. *Ixodes arboricola* Schulze and Schlottke infesting resident, migrant, and wintering birds in the western coastal desert. J Egyp Public Health Assoc 42:223–229

Horak IG, Camicas JL, Keirans JE (2002) The Argasidae, Ixodidae and Nuttalliellidae (Acari: Ixodida): a world list of valid tick names. Exp Appl Acarol 28:27–54

Hornok S, Kontschán J, Kováts D, Kovács R, Angyal D, Görföl T, Mihalca AD (2014) Bat ticks revisited: *Ixodes ariadnae* sp. nov. and allopatric genotypes of *I. vespertilionis* in caves of Hungary. Parasit Vectors 7:202

Hornok S, Kontschán J, Estrada-Peña A, Fernández de Mera IG, Tomanovic S, de la Fuente J (2015) Contributions to the morphology and phylogeny of the newly discovered bat tick species, *Ixodes ariadnae* in comparison with *I. vespertilionis* and *I. simplex*. Parasit Vectors 8:47

Jaenson TG, Värv K, Fröjdman I, Jääskeläinen A, Rundgren K, Versteirt V, Golovljova I (2016) First evidence of established populations of the taiga tick *Ixodes persulcatus* (Acari: Ixodidae) in Sweden. Parasit Vectors 9:1

Kaiser MN, Hoogstraal H, Watson G (1974) Ticks (Ixodoidea) on migrating birds in Cyprus, fall 1967 and spring 1968 and epidemiological considerations. Bull Entomol Res 64:97–110

Kolonin GV (2009) Fauna of ixodid ticks of the world. http://www.kolonin.org/. Accessed 15 May 2012 (Please note this page is no longer working)

Kovalev SY, Golovljova TA, Mukhacheva TA (2015) Natural hybridization between *Ixodes ricinus* and *Ixodes persulcatus* ticks evidenced by molecular genetics methods. Ticks Tick-Borne Dis 7:113–118

Latrofa MS, Giannelli A, Persichetti MF, Pennisi MG, Solona-Gallego L, Brianti E, Parisi A, Wall R, Dantas-Torres F, Otranto D (2017) *Ixodes ventalloi*: morphological and molecular support for species integrity. Parasitol Res 116:251–258. https://dx.doi.org/10.1007/s00436-016-5286-5289

Manilla G (1998) Fauna d'Italia. Acari, Ixodida. Edizioni Calderini, Bologna, 280 pp

Manilla G, Iori A (1992) Chiave illustrata delle zecche d'Italia. I. Stadi larvali delle specie della sottofamiglia Ixodinae (Acari, Ixodoidea, Ixodidae). Parassitologia 34:83–95

Manilla G, Iori A (1993) Chiave illustrata delle zecche d'Italia. II. Stadi ninfali delle specie della sottofamiglia Ixodidae (Acari, Ixodoidea, Ixoddae). Parassitologia 35:37–45

Márquez FJ, Morel P-C, Guiguen C, Beaucournu JC (1992) Clé dichotomique des Ixodidae d'Europe. I. Les larves du genre *Ixodes*. Acarologia 33:325–330

McCoy KD, Boulinier T, Tirard C, Michalakis Y (2003) Host-dependent genetic structure of parasite populations: differential dispersal of seabird tick host races. Evolution 57:288–296

Morel PC (2003) Les tiques d'Afrique et du Bassin méditerranéen (1965-1995). CIRAD-EMVT, 1342 pp

Morel PC, Aubert MFA (1975) Contribution à la connaisance de *Pholeoixodes rugicollis* (Schulze and Schlottke, 1929) (Acariens Ixodina). Cahiers ORSTOM. Ser Entomol Med Parasitol 13:99–109

Morel PC, Graf JF (1997) Morpohlogie des stases préimaginales des Ixodidae s. st. d'Europe Occidentale. VII. Description de la larve te de la nymphe *d'Ixodes apronophorus* Schulze, 1924. Acarologia, XIX:587–592

Morel P, Pérez-Eid C (1978a) Morphology of immature stages of Ixodidae s. str. from occidental Europe. 5. Larvae of *Ixodes* s. str. Acarologia 19:395–405

Morel P, Pérez-Eid C (1978b) Morphology of immature stages of Ixodidae s. str. from occidental Europe. 6. Nymphs of *Ixodes* s. str. Acarologia 19:579–586

Norte AC, Lopes de Carvalho I, Ramos JA, Gonçalves M, Gern L, Núncio S (2012) Diversity and seasonal patterns of ticks parasitizing wild birds in western Portugal. Exp Appl Acarol 58:327–339

Nosek J, Balat F (1982) Infestation of migratory birds with *Ixodes ricinus* and *I. arboricola* ticks. Biologia 37:215–219

Nowak-Chmura M (2011) *Ixodes eldaricus* Djaparidze, 1950 (Ixodidae) on migrating birds, reported first time in Poland. Vet Parasitol 186:399–402

Papadopoulos B, Morel PC, Aeschlimann A (1996) Ticks of domestic animals in the Macedonia region of Greece. Vet Parasitol 63:25–40

Pérez-Eid C (2007) Les tiques. Identification, biologie, importance médicale et vétérinaire. Editions Tec & Doc, Paris, 314 pp

Petney TN, Beichel E, Maiwald M, Hassler D (1996) *Ixodes ventalloi*: a new tick record for Germany. Appl Parasitol 37:96–98

Petney TN, Pfäffle MP, Skuballa JD (2012) An annotated checklist of the ticks (Acari: Ixodida) of Germany. Syst Appl Acarol 17:115–170

Petney TN, Moser E, Littwin N, Pfäffle M, Muders SV, Taraschewski H (2015) Additions to the "Annotated Checklist of the Ticks of Germany": *Ixodes acuminatus* and *Ixodes inopinatus*. Syst Appl Acarol 20:221–224

Pomerantzev BI (1950) Ixodid ticks (Ixodidae). Fauna SSSR, Paukoobraznye, vol 4(2), Izd. Akad. Nauk SSSR, Moscow, Leningrad, 224 pp (in Russian)

Randolph SE (1975a) Seasonal dynamics of a host-parasite system: *Ixodes trianguliceps* (Acarina: Ixodidae) and its small mammal hosts. J Animal Ecol 44:425–449

Randolph SE (1975b) Patterns of distribution of the tick *Ixodes trianguliceps* Birula on its hosts. J Animal Ecol 44:451–474

Reznik PA (1961) Contribution to study of immature stages of the tick family Ixodidae. Communication six. Description of *Ixodes frontalis* Panzer larvae and nymphs and *Ixodes redikorzev* Olenev larvae. Trudy Nauch. Islled. Protiv. Kavk. Zakavk., vol 5, pp 276–286 (in Russian, English Translation T382, NAMRU-3, Cairo, Egypt)

Santos-Silva MM, Beati L, Santos AS, De Sousa R, Núncio MS, Melo P, Bacellar F (2011) The hard-tick fauna of mainland Portugal (Acari: Ixodidae): an update on geographical distribution and known associations with hosts and pathogens. Exp Appl Acarol 55:85–121

Saratsiotis A (1970) Etude morphologique et observations biologiques sur *Ixodes gibbosus* Nuttall, 1916. Ann Parasitol Hum Comp 45:661–675

Schembri PJ (2003) Current state of knowledge of the Maltese non-marine fauna. In: Malta environment and planning authority annual report and accounts 2003. Malta Environment and Planning Authority, Floriana, pp 33–65

Sénevet G, Rodhain F (1968) Larves des principales espèces du genre *Ixodes* d'Europe Occidentale et Centrale. Description de la larva d'*Ixodes festai* Rondelli 1926. Ann Parasitol 43:513–523

Siuda K (1993) Kleszcze Polski. Monografie Parazytologiczne. Polsk. Towarzystwo Parazytol. II, p 375

Sonenshine DE, Kohls GM, Clifford CM (1969) *Ixodes crenulatus* Koch, 1844 synonomy with *Ixodes kaiseri* Arthur, 1957 and redescriptions of the male, female, nymph, and larva (Acarina: Ixodidae). Acarologia 11:193–206

Teng KF (1973) Some chinese ticks (genus *Ixodes*). Two new species (Acarina: Ixodidae). Acta Entomol Sinica 16:73–81 (In Chinese)

Visser M, Massner C, Rehbein S (2011) Massive infestation with fur mites *Lynxacarus mustelae* of a stone marten (*Martes foina*) from Tyrol. Wiener Klin Wochenschr (Supplement 1), pp 36–42

Walter G, Liebisch A, Streichert J (1979) Untersuchungen zur Biologie und Verbreitung von Zecken (Ixodoidea, Ixodidae) in Norddeutschland. I. Die Vogelcecken *Ixodes lividus* (C.L. Koch 1844) und *Ixodes arboricola* (Schulze & Schlottke 1929). Z Angew Zool 66:65–73

Yamaguti N, Tipton VJ, Keegan HL, Toshioka S (1971) Ticks of Japan, Korea, and the Ryukyu Islands. Brigham Young University, Science Bulletin, Biological Series, vol 15, 226 pp

Ixodes frontalis (Panzer, 1798) (Figs. 26–28)

M. P. Pfäffle, M. Madder, M. M. Santos-Silva, and T. N. Petney

Life-Cycle and Host Preferences

Ixodes frontalis is an exophilic, three-host tick with a life cycle of 1 year in its natural environment (Doby and Bigaignon 1997; Monks et al. 2006). All life stages infest various bird species, both cavity-nesting and open-nesting species (Heylen et al. 2014). These are mainly passerine birds, but also anseriforms, falconiforms, galliforms, gruiforms and strigiforms (Arthur 1963; Hoogstraal et al. 1963; Millán et al. 2004; Santos-Silva et al. 2011; Bona and Stanko 2013). It has been reported attaching to humans but this appears to be very rare (Gilot et al. 1997). Additionally, it can occasionally be found on other mammals, such as badgers or martens (Hillyard 1996).

Ecology

Ixodes frontalis is an ornithophilic tick that can be found in bird nests and understory vegetation (Obsomer et al. 2013). On rare occasions it can also be trapped by flagging (Schorn et al. 2011; Heylen et al. 2014; Santos-Silva et al. unpublished data). It has a preference for a variety of different habitats, such as European broadleaf forests, Atlantic mixed forests, Alpine conifer and mixed forests and English lowland beech forests (Guglielmone et al. 2014). It can be introduced into different climate zones via bird migration; however, it prefers milder climatic regions (Poupon et al. 2006). The peak activity of adult ticks is during the winter months from October to February, and for immature forms from March to October (Hillyard 1996; Santos-Silva et al. 2011; Santos-Silva et al. unpublished data).

Distribution

This tick is widely distributed in Europe, western Asia and northern Africa (Hoogstraal et al. 1963). However, in the Palaearctic it is most common in Southern and Central Europe (Arthur 1963). Records, including those from migratory birds, come from Belgium, Croatia, Cyprus, the Czech Republic, Denmark, Egypt, France, Georgia, Germany, Hungary, Italy, Moldova, the Netherlands, Norway, Poland, Portugal, Russia, Serbia, Slovakia, Spain, Sweden, Switzerland, Turkey, the United Kingdom and the Ukraine (Hoogstraal et al. 1963; Kaiser et al. 1974; Jaenson et al. 1994; Osacar-Jimenez et al. 1998; Lundqvist et al. 1998; Chastel et al. 1999; Siuda et al. 2006; Poupon et al. 2006; Monks et al. 2006; Movila et al. 2008; Jameson and Medlock 2011; Santos-Silva et al. 2011; Schorn et al. 2011; Krčmar 2012; Bursali et al. 2012; Petney et al. 2012; Obsomer et al. 2013; Bona and Stanko 2013; Jahfari et al. 2014; Rubel et al. 2014). Specimens from Finland from migratory birds (Laakkonen et al. 2009) have since been determined to be *I. ricinus* (Heylen et al. 2012).

Vectorial Capacity and Pathogen Burden

Ixodes frontalis is associated with avian tick-related syndrome (Monks et al. 2006). Since all life stages feed on birds and rarely on other vertebrates, *I. frontalis* is probably not

M. P. Pfäffle (✉)
Department of Ecology and Parasitology, Institute of Zoology, Karlsruhe Institute of Technology, Karlsruhe, Germany
e-mail: miripfaeffle@web.de

M. Madder
DG06/07 Ruisseau Creole, La Mivoie, Black River, 90625, Mauritius
e-mail: Maxime.Madder@clinglobal.com

M. M. Santos-Silva
Centro de Estudos de Vectores e Doenças Infecciosas Dr. Francisco Cambournac (CEVDI), Instituto Nacional de Saúde Doutor Ricardo Jorge, Águas de Moura, Portugal
e-mail: m.santos.silva@insa.min-saude.pt

T. N. Petney
State Museum of Natural History, Erbprinzenstr. 13, 76133 Karlsruhe, Germany
e-mail: trevor.petney@kit.edu

A. Estrada-Peña et al. (eds.), *Ticks of Europe and North Africa*,
https://doi.org/10.1007/978-3-319-63760-0_19

Fig. 26 **A** Distribution of *I. frontalis* in Europe and Northern Africa (10 × 10 km grid presence with black dots). **B** Countries where the species has been reported are marked in dark grey

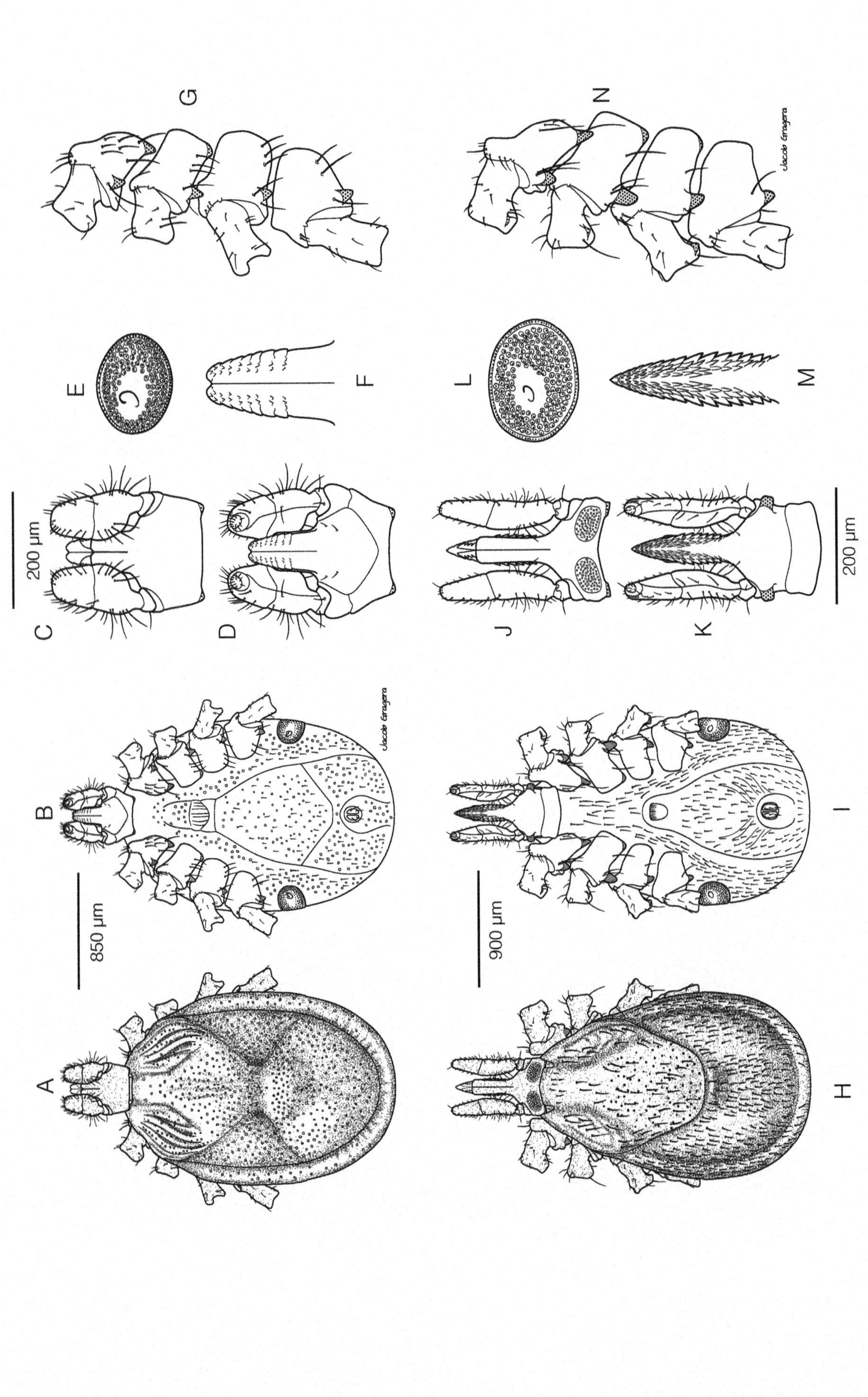

Fig. 27 **A–G** The male of *I. frontalis.* **A** Dorsal, **B** ventral, **C** capitulum, dorsal, **D** capitulum, ventral, **E** spiracular plate, **F** hypostome, **G** coxae and trochanters I–IV. **H–N** The female of *I. frontalis.* **H** Dorsal, **I** ventral, **J** capitulum, dorsal, **K** capitulum, ventral, **L** spiracular plate, **M** hypostome, **N** coxae and trochanters I–IV. Illustrations from specimens collected in Belgium, Italy, Portugal and Spain

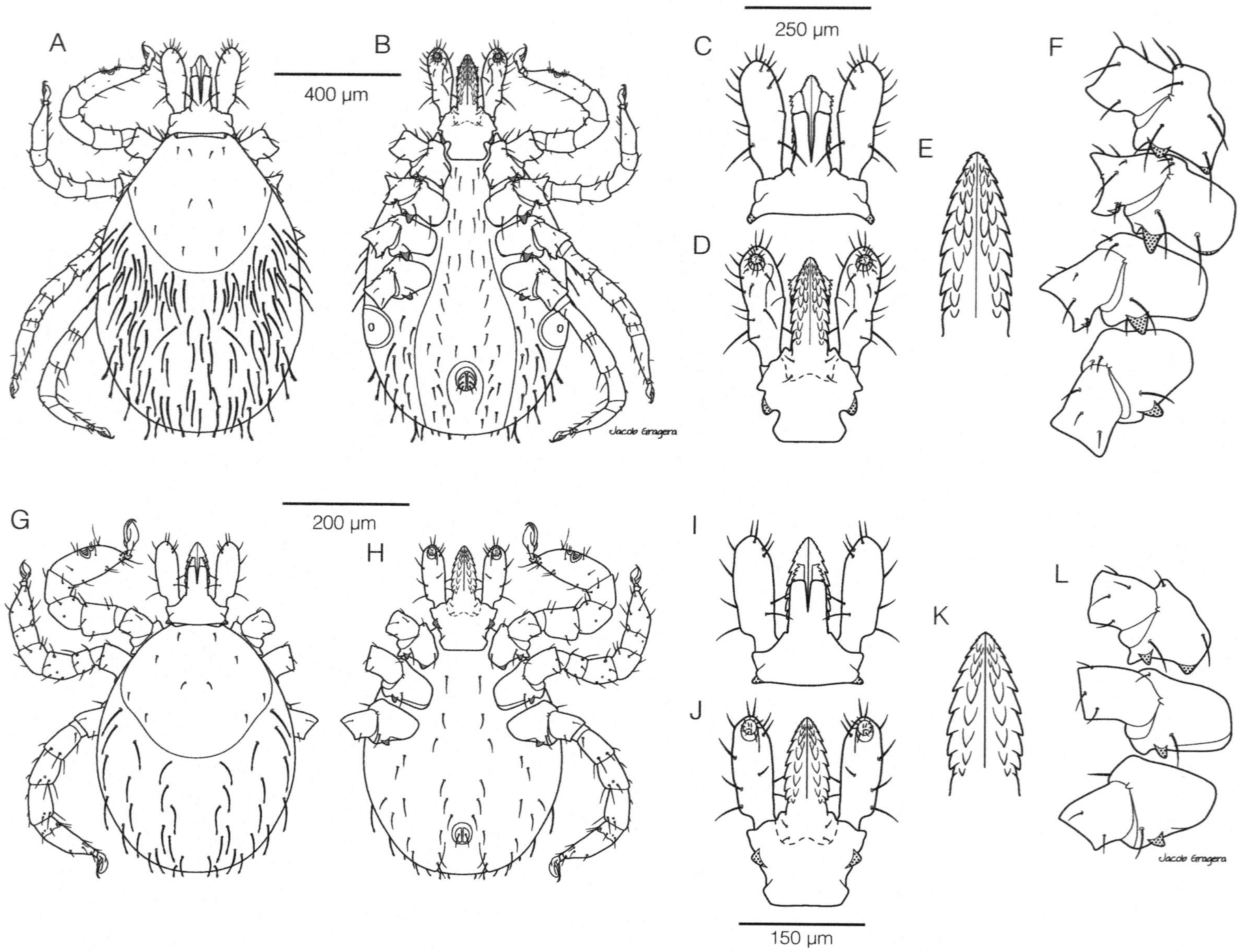

Fig. 28 **A–F** The nymph of *I. frontalis*. **A** Dorsal, **B** ventral, **C** capitulum, dorsal, **D** capitulum, ventral, **E** hypostome, **F** coxae and trochanters I–IV. **G–L** The larva of *I. frontalis*. **G** Dorsal, **H** ventral, **I** capitulum, dorsal, **J** capitulum, ventral, **K** hypostome, **L** coxae and trochanters I–III. Illustrations from larvae obtained from one engorged female collected in Spain and complemented from illustrations by Filippova (1977)

important for the direct transmission of pathogens to humans; however, it might be important for the maintenance of pathogens in a bird-tick-bird cycle. This might also be important for the further distribution of pathogens worldwide along avian migratory routes (Hoogstraal et al. 1963). It is associated with different arboviruses, such as Bahig, Chizé, Kemorov, Matruh and tick-borne encephalitis virus (Hillyard 1996; Labuda and Nuttall 2004; Obsomer et al. 2013). Although different *Borrelia* species have been found in this tick, including *B. afzelii*, *B. garinii* and *B. turdi*, its vector competence for these pathogens needs to be clarified (Estrada-Peña et al. 1995; Palomar et al. 2012; Obsomer et al. 2013). In Spain, *I. frontalis* were infected with *B. burgdorferi* s.l. only in those areas where they co-existed with *I. ricinus* (Estrada-Peña et al. 1995). *Anaplasma phagocytophilum*, *Candidatus* "Neoehrlichia mikurensis" and *Coxiella burnetii* have been isolated from this species (Hoogstraal et al. 1963; Hillyard 1996; Obsomer et al. 2013; Movila et al. 2013; Jahfari et al. 2014).

References

Arthur DR (1963) British ticks. Butterworths, London

Bona M, Stanko M (2013) First records of the tick *Ixodes frontalis* (Panzer, 1795) (Acari, Ixodidae) in Slovakia. Ticks Tick Borne Dis 4:478–481. https://dx.doi.org/10.1016/j.ttbdis.2013.06.002

Bursali A, Keskin A, Tekin S (2012) A review of the ticks (Acari: Ixodida) of Turkey: species diversity, hosts and geographical distribution. Exp Appl Acarol 57:91–104. https://dx.doi.org/10.1007/s10493-012-9530-4

Chastel C, Chandler L, Le Goff F, Chastel O, Tesh R, Shope R (1999) Chizé virus, a new phlebovirus isolated in France from *Ixodes (Trichotoixodes) frontalis*. Acta Virol 43:279–283

Doby JM, Bigaignon G (1997) Ixodid ticks parasiting birds and their pathogenic role. Rev Med Vet 148:853–860

Estrada-Peña A, Oteo JA, Estrada-Peña R, Gortazar C, Osácar JJ, Moreno JA, Castellá J (1995) *Borrelia burgdorferi* sensu lato in ticks (Acari: Ixodidae) from two different foci in Spain. Exp Appl Acarol 19:173–180. https://dx.doi.org/10.1007/BF00046289

Filippova NA (1977) Fauna of the USSR. 4. Chelicerata. Ixodid ticks of the Subfamily Ixodinae. Nauka, Leningrad (in Russian)

Filipova NA (1997) Fauna of Russia and neighbouring countries. Arachnoidea, Ixodid ticks of subfamily Amblyomminae, vol 4(5), Russian Academy of Sciences, Zoological Institute, New Series 145, Nauka Publishing House, St Petersburg

Gilot B, Beaucournu JC, Chastel C (1997) Collecte «au drapeau» et fixation sur l'homme d'*Ixodes (Trichotoixodes) frontalis* (Panzer, 1795). Parasite 4:197–199. https://dx.doi.org/10.1051/parasite/1997042197

Guglielmone AA, Robbins RG, Apanaskevich DA, Petney TN, Estrada-Peña A, Horak IG (2014) The hard ticks of the world. Springer, Heidelberg

Heylen DJ, Jacobs I, Matthysen E, Madder M (2012) Comments on article "First report of *Ixodes frontalis* (Acari: Ixodidae) in Finland, an example of foreign tick species transported by a migratory bird" (Memoranda Soc. Fauna Flora Fennica 85:16–19. 2009). Memo Soc pro Fauna Flora Fenn 88:19–20

Heylen D, De Coninck E, Jansen F, Madder M (2014) Differential diagnosis of three common *Ixodes* spp. ticks infesting songbirds of Western Europe: *Ixodes arboricola*, *I. frontalis* and *I. ricinus*. Ticks Tick Borne Dis 5:693–700. https://dx.doi.org/10.1016/j.ttbdis.2014.05.006

Hillyard PD (1996) Ticks of North-West Europe. Field Studies Council, Shrewsbury

Hoogstraal H, Kaiser MN, Traylor MA, Guindy E, Gaber S (1963) Ticks (Ixodidae) on birds migrating from Europe and Asia to Africa 1959-61. Bull World Health Organ 28:235–262

Jaenson TGT, Tälleklint L, Lundqvist L, Olsen B, Chirico J, Mejlon H (1994) Geographical distribution, host associations, and vector roles of ticks (Acari: Ixodidae, Argasidae) in Sweden. J Med Entomol 31:240–256. https://dx.doi.org/10.1093/jmedent/31.2.240

Jahfari S, Coipan EC, Fonville M, Maassen CB, van der Giessen J, Takken W, Takumi K, Sprong H (2014) Circulation of four *Anaplasma phagocytophilum* ecotypes in Europe. Parasit Vectors 7:365. https://dx.doi.org/10.1186/1756-3305-7-365

Jameson LJ, Medlock JM (2011) Tick surveillance in Great Britain. Vector Borne Zoonotic Dis 11:403–412. https://dx.doi.org/10.1089/vbz.2010.0079

Kaiser MN, Hoogstraal H, Watson GE (1974) Ticks (Ixodoidea) on migrating birds in Cyprus, fall 1967 and spring 1968, and epidemiological considerations. Bull Entomol Res 64:97–110. https://dx.doi.org/10.1017/S0007485300027024

Krčmar S (2012) Hard ticks (Acari, Ixodidae) of Croatia. Zookeys 234:19–57. https://dx.doi.org/10.3897/zookeys.234.3658

Laakkonen J, Terhivuo J, Huhtamo E, Vapalahti O, Uzcátegui NY (2009) First report of *Ixodes frontalis* (Acari: Ixodidae) in Finland, an example of foreign tick species transported by a migratory bird. Memo Soc pro Fauna Flora Fenn 85:16–19

Labuda M, Nuttall PA (2004) Tick-borne viruses. Parasitology 129: S221–S245. https://dx.doi.org/10.1017/S0031182004005220

Lundqvist L, Gray JS, Hillyard PD (1998) *Ixodes frontalis* on the Baltic island of Gotland, Sweden. Med Vet Entomol 12:215–216. https://dx.doi.org/10.1046/j.1365-2915.1998.00101.x

Millán J, Gortazar C, Martín-Mateo MP, Villafuerte R (2004) Comparative survey of the ectoparasite fauna of wild and farm-reared red-legged partridges (*Alectoris rufa*), with an ecological study in wild populations. Parasitol Res 93:79–85. https://dx.doi.org/10.1007/s00436-004-1113-9

Monks D, Fisher M, Forbes NA (2006) *Ixodes frontalis* and avian tick-related syndrome in the United Kingdom. J Small Anim Pract 47:451–455. https://dx.doi.org/10.1111/j.1748-5827.2006.00031.x

Movila A, Gatewood A, Toderas I, Duca M, Papero M, Uspenskaia I, Conovalov J, Fish D (2008) Prevalence of *Borrelia burgdorferi* sensu lato in *Ixodes ricinus* and *I. lividus* ticks collected from wild birds in the Republic of Moldova. Int J Med Microbiol 298:149–153. https://dx.doi.org/10.1016/j.ijmm.2007.12.009

Movila A, Alekseev AN, Dubinina HV, Toderas I (2013) Detection of tick-borne pathogens in ticks from migratory birds in the Baltic region of Russia. Med Vet Entomol 27:113–117. https://dx.doi.org/10.1111/j.1365-2915.2012.01037.x

Obsomer V, Wirtgen M, Linden A, Claerebout E, Heyman P, Heylen D, Madder M, Maris J, Lebrun M, Tack W, Lempereur L (2013) Spatial disaggregation of tick occurrence and ecology at a local scale as a preliminary step for spatial surveillance of tick-borne diseases: general framework and health implications in Belgium. Parasit Vectors 6:190. https://dx.doi.org/10.1186/1756-3305-6-190

Osacar-Jimenez JJ, Estrada-Peña A, Llucientes-Curdi J (1998) Ticks (Acarina: Ixodidae) of wild birds in the Ebro Middle Basin (North-East Spain). Acarologia 39:23–31

Palomar AM, Santibáñez P, Mazuelas D, Roncero L, Santibáñez S, Portillo A, Oteo JA (2012) Role of birds in dispersal of etiologic agents of tick-borne zoonoses, Spain, 2009. Emerg Infect Dis 18:1188–1191. https://dx.doi.org/10.3201/eid1807.111777

Petney TN, Pfäffle MP, Skuballa JD (2012) An annotated checklist of the ticks (Acari: Ixodida) of Germany. Syst Appl Acarol 17:115–170. https://dx.doi.org/10.11158/saa.17.2.2

Poupon M-A, Lommano E, Humair P-F, Douet V, Rais O, Schaad M, Jenni L, Gern L (2006) Prevalence of *Borrelia burgdorferi* sensu lato in ticks collected from migratory birds in Switzerland. Appl Environ Microbiol 72:976–979. https://dx.doi.org/10.1128/AEM.72.1.976-979.2006

Rubel F, Brugger K, Monazahian M, Habedank B, Dautel H, Leverenz S, Kahl O (2014) The first German map of georeferenced ixodid tick locations. Parasit Vectors 7:477. https://dx.doi.org/10.1186/s13071-014-0477-7

Santos-Silva MM, Beati L, Santos AS, De Sousa R, Núncio MS, Melo P, Santos-Reis M, Fonseca C, Formosinho P, Vilela C, Bacellar F (2011) The hard-tick fauna of mainland Portugal (Acari: Ixodidae): an update on geographical distribution and known associations with hosts and pathogens. Exp Appl Acarol 55:85–121. https://dx.doi.org/10.1007/s10493-011-9440-x

Schorn S, Schöl H, Pfister K, Silaghi C (2011) First record of *Ixodes frontalis* collected by flagging in Germany. Ticks Tick Borne Dis 2:228–230. https://dx.doi.org/10.1016/j.ttbdis.2011.07.003

Siuda K, Majszyk A, Nowak M (2006) Ticks (Acari: Ixodida) parasitizing birds (Aves) in Poland. Biol Lett 43:147–151

Ixodes vespertilionis Koch, 1844 (Figs. 29–31)

S. Hornok

Life-Cycle and Host Preferences

Ixodes vespertilionis is a three-host tick species parasitizing bats. There is no report on the laboratory rearing of this species. Based on observations in its natural habitats, the duration of the complete life cycle is estimated to be more than 15 months (Beaucournu 1967). This implies that if egg-laying starts at the beginning of the year (and may take about 45 days), the larvae will emerge after another 45 days at the beginning of April. After 10 days of maturation, larvae may engorge in 10 days before the end of bat hibernation. Nymphs interrupt development and wait until the bats return to their caves (from October), then feed for at least as long as the larvae did (Beaucournu 1967). Following a long diapause (5 months), the adults emerge in March, and females still have a chance to feed (usually for around 16 days) before the end of bat hibernation (Beaucournu 1967). The preferred hosts of *I. vespertilionis* are horseshoe bats (*Rhinolophus* spp.) that typically hibernate in caves, and sometimes other bat species which tend to use such winter roosts (*Myotis myotis*, *My. emarginatus*, *My. oxygnathus*, *Miniopterus schreibersii*) or may use such winter roosts (*My. nattereri*, *My. daubentonii*, *Plecotus auritus*) (Arthur 1956; Beaucournu 1967; Bursali et al. 2012). Rarely, *I. vespertilionis* can be found on forest bats associated with tree holes (e.g. *My. mystacinus*, several *Pipistrellus* spp.) or more urbanised bat species (e.g. *Nyctalus noctula*, *Eptesicus serotinus*) (Beaucournu 1967; Ševčík et al. 2010). *Ixodes vespertilionis* can also accidentally infest humans (Piksa et al. 2013).

Ecology

Ixodes vespertilionis is an endophilic, typically cave- dwelling tick species (Arthur 1956). Sometimes it may also occur in other cavities, such as mines, attics or cellars and tree holes (Ševčík et al. 2010; Piksa et al. 2014). In such sheltered environments, all life-cycle stages show a year-round, but changing activity. During the first half of bat hibernation (October–January), increasing numbers of questing larvae, nymphs and females can be seen on cave walls or in crevices near bats (Arthur 1956; Beaucournu 1967; Hornok et al. 2014). These will consequently attach to bats for blood-feeding. Engorged specimens may be seen between stones during this period (Arthur 1956). Males, which usually outnumber females on cave walls (Hornok et al. 2014), can also be partly or fully engorged. This may indicate blood from the nymphal stage and the simultaneous inability of males to feed on blood (Neumann 1916), or that males feed for a very short time (Nuttall et al.

S. Hornok (✉)
Department of Parasitology and Zoology, University of Veterinary Medicine, Budapest, Hungary
e-mail: hornok.sandor@univet.hu

A. Estrada-Peña et al. (eds.), *Ticks of Europe and North Africa*,
https://doi.org/10.1007/978-3-319-63760-0_20

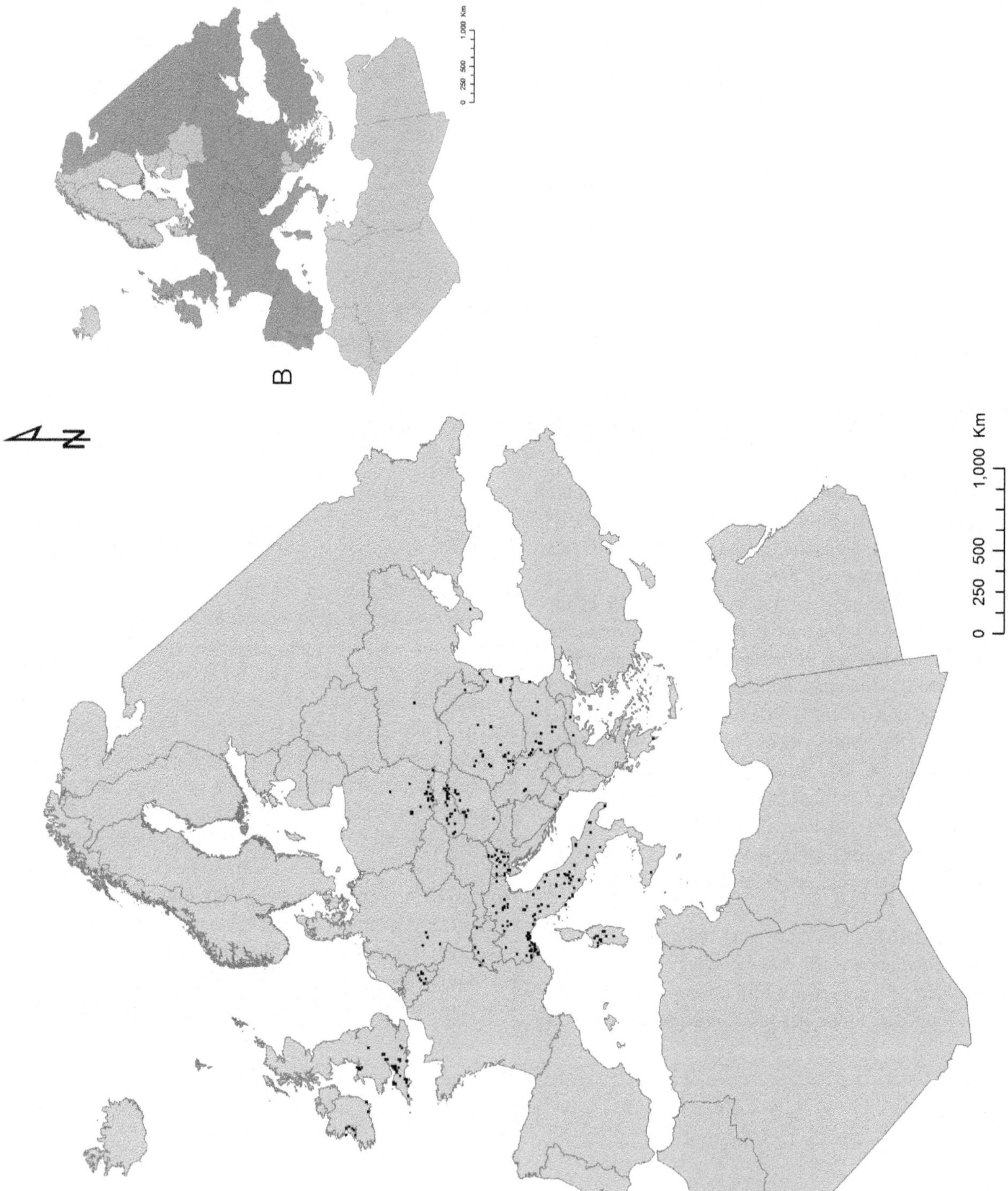

Fig. 29 **A** Distribution of *I. vespertilionis* in Europe and Northern Africa (10 × 10 km grid presence with black dots). **B** Countries where the species has been reported are marked in dark grey

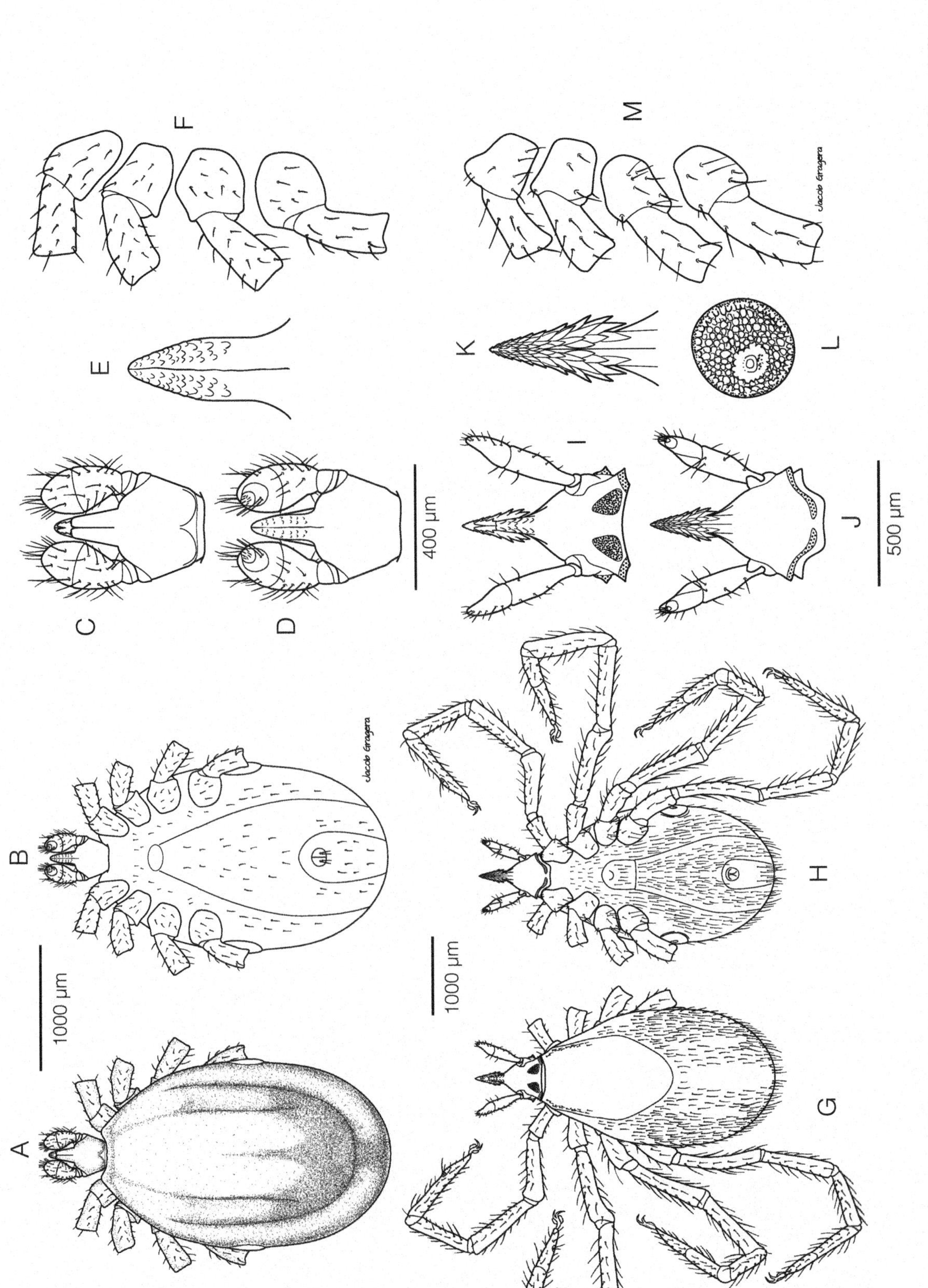

Fig. 30 **A–F** The male of *I. vespertilionis*. **A** Dorsal, **B** ventral, **C** capitulum, dorsal, **D** capitulum, ventral, **E** hypostome, **F** coxae and trochanters I–IV. **G–M** The female of *I. vespertilionis*. **G** Dorsal, **H** ventral, **I** capitulum, dorsal, **J** capitulum, ventral, **K** hypostome, **L** spiracular plate, **M** coxae and trochanters I–IV. Illustrations from specimens collected in Hungary, Morocco and Spain

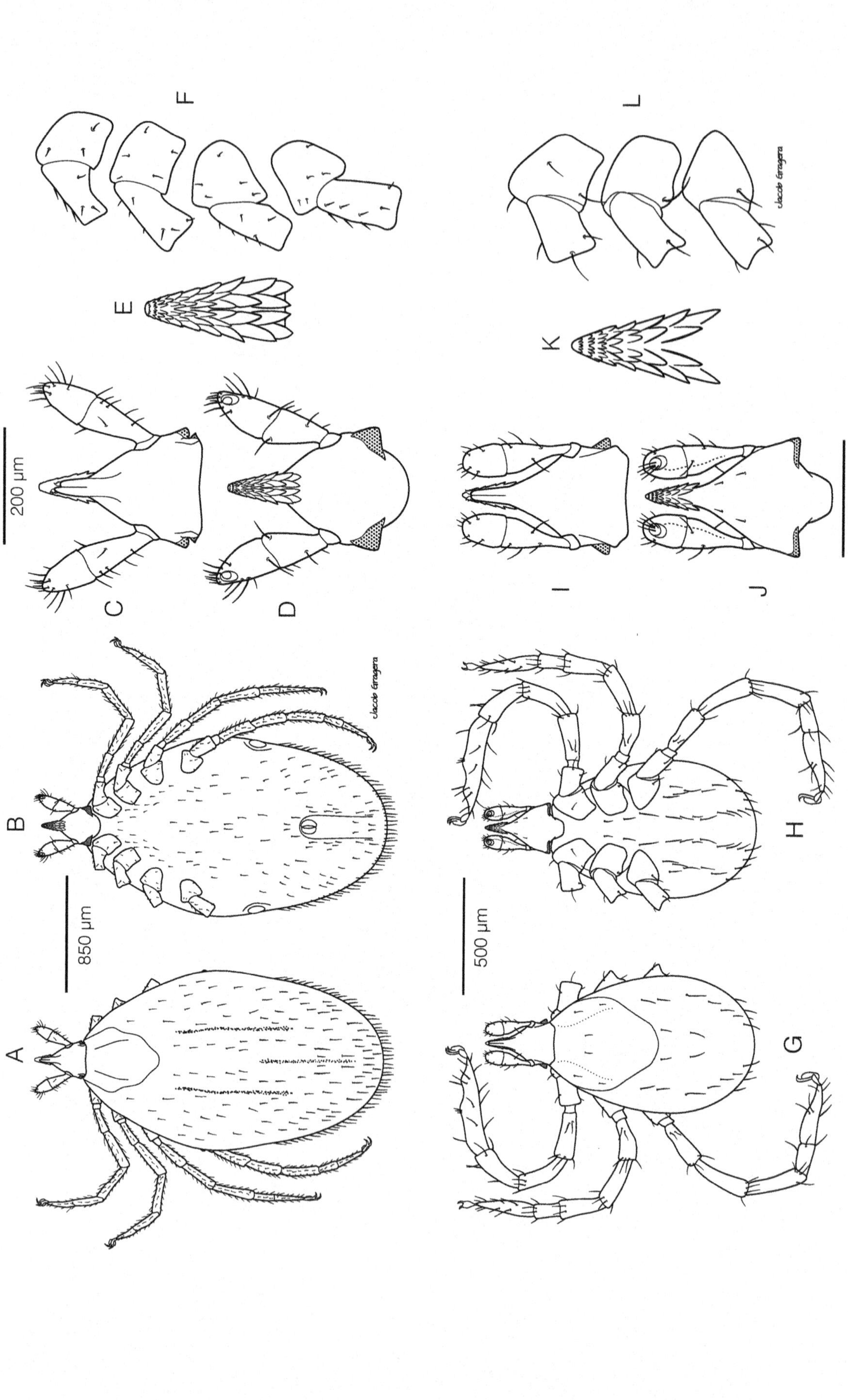

Fig. 31 **A–F** The nymph of *I. vespertilionis*. **A** Dorsal, **B** ventral, **C** capitulum, dorsal, **D** capitulum, ventral, **E** hypostome, **F** coxae and trochanters I–IV. **G–L** The larva of *I. vespertilionis*. **G** Dorsal, **H** ventral, **I** capitulum, dorsal, **J** capitulum, ventral, **K** hypostome, **L** coxae and trochanters I–III. Illustrations from specimens collected in Hungary and Morocco

1911). However, males have never been found on bats (Arthur 1956). In the spring, nymphs and females predominate on cave walls (Hornok et al. 2014), and larvae and nymphs on bats (Ševčík et al. 2010; Piksa et al. 2014). The intensity of infestation on active bats is the highest shortly after hibernation (when they form their nursery colonies), i.e. in April/May (Ševčík et al. 2010; Piksa et al. 2014), with a contemporaneously decreasing number of relevant stages in caves (Beaucournu 1967). Later (summer to early autumn) usually only males and a few unfed larvae/nymphs/females can be seen on cave walls (Arthur 1956; Beaucournu 1967; Hornok et al. 2014), with increasing numbers of larvae and nymphs on bats (Piksa et al. 2014).

Distribution

Ixodes vespertilionis has a wide distribution in Europe (except Scandinavia and the Baltic states), including Ireland, the United Kingdom, the Netherlands, Belgium, France, Spain, Portugal, Germany, Poland, Switzerland, Austria, Italy, the Czech Republic, Slovakia, Hungary, the Balkans, Greece, Turkey, Ukraine and Russia.

Vectorial Capacity and Pathogen Burden

Ixodes vespertilionis may harbour *Bartonella* species (Hornok et al. 2012), but its vectorial role is unknown. The DNA of *Babesia vesperuginis* and *B. crassa* was also shown to be present in *I. vespertilionis* (Hornok et al. 2016).

References

Arthur DR (1956) The *Ixodes* ticks of Chiroptera (Ixodoidea, Ixodidae). J Parasitol 42:180–196

Beaucournu JC (1967) Contribution a la connaissance de la biologie d'*Ixodes* (*Eschatocephalus*) *vespertilionis* Koch, 1844 et d'*Ixodes* (*Pomerantzevella*) *simplex* Neumann, 1906 (Acarina, Ixodoidea) parasites des Chiroptères. Ann Speleologie 22:543–580 (Article in French)

Bursali A, Keskin A, Tekin S (2012) A review of the ticks (Acari: Ixodida) of Turkey: species diversity, hosts and geographical distribution. Exp Appl Acarol 57:91–104

Hornok S, Kovács R, Meli ML, Kontschán J, Gönczi E, Gyuranecz M, Dán Á, Molnár V, Hofmann-Lehmann R (2012) First detection of bartonellae in a broad range of bat ectoparasites. Vet Microbiol 159:541–543

Hornok S, Kontschán J, Kováts D, Kovács R, Angyal D, Görföl T, Polacsek Z, Kalmár Z, Mihalca AD (2014) Bat ticks revisited: *Ixodes ariadnae* sp. nov. and allopatric genotypes of *I. vespertilionis* in caves of Hungary. Parasit Vectors 7:202

Hornok S, Szőke K, Kováts D, Estók P, Görföl T, Boldogh SA, Takács N, Kontschán J, Földvári G, Barti L, Corduneanu A, Sándor AD (2016) DNA of piroplasms of ruminants and dogs in ixodid bat ticks. PloS One 11:e0167735

Neumann LG (1916) Biospeologica. Ixodidei (Acariens) I. Arch Zool Exp Gen 55:515–527

Nuttall GHF, Warburton C, Cooper WF, Robinson LE (1911) Ticks: a monograph of the Ixodoidea. Part II. Ixodidae. Cambridge University Press, Cambridge, pp 1–348 (271–277)

Piksa K, Nowak-Chmura M, Siuda K (2013) First case of human infestation by the tick *Ixodes vespertilionis* (Acari: Ixodidae). Int J Acarol 38:1–2

Piksa K, Górz A, Nowak-Chmura M, Siuda K (2014) The patterns of seasonal activity of *Ixodes vespertilionis* (Acari: Ixodidae) on *Rhinolophus hipposideros* in nursery colonies. Ticks Tick-Borne Dis 5:69–74

Ševčík M, Krištofík J, Uhrin M, Benda P (2010) New records of ticks (Acari: Ixodidae) parasitising on bats in Slovakia. Vespertilio 13–14:139–147

Ixodes simplex Neumann, 1906 (Figs. 32–34)

S. Hornok

Life-Cycle and Host Preferences

Ixodes simplex is a three-host tick species parasitizing bats. There is no report on the laboratory rearing of this species, and information is also scarce on its natural life cycle. Based on some observations, the larvae and nymphs of *I. simplex* may develop over a shorter period than those of *I. vespertilionis*, but the duration of blood-sucking and diapause of these stages are most likely similar (Beaucournu 1967). The preferred host of *I. simplex* is the cave-dwelling bat *Miniopterus schreibersii* (Arthur 1956; Beaucournu 1967), but it can rarely (especially outside the range of this main host) also be found on other bat species sharing colonies with *Mi. screibersii* (Beaucournu 1967), most notably on *Myotis myotis* (Beaucournu 1966; Walter and Kock 1985), but also on *My. alcathoe* (Krištofík and Danko 2012) and *Rhinolophus* spp. (Arthur 1956; Burazerović et al. 2015).

Ecology

Ixodes simplex is an endophilic, cave-dwelling tick species (Filippova 1972). It has a year-round but changing activity pattern (Lourenço and Palmeirim 2008). The highest numbers of larvae and nymphs are seen in late spring on pregnant bats (Lourenço and Palmeirim 2008). Females can usually be collected from bats during either spring or summer when males can be seen in cave walls (Beaucournu 1966). In summer (after engorgement and copulation), females lay eggs near colonies of *Miniopterus* (Beaucournu 1967). Thus, the behaviour of *I. simplex*, based on its small size and short legs, is different from that of *I. vespertilionis* in that they wander in search of a host over shorter distances on cave walls (Beaucournu 1967). During early autumn, mating bats mostly carry immature stages of *I. simplex* (Hornok, unpublished data). Larvae and nymphs can be least frequently found on bats during their hibernation (Lourenço and Palmeirim 2008). Males probably do not feed, or they may feed very rapidly and then quickly hide themselves (Yamaguti et al. 1971).

Distribution

Ixodes simplex apparently has a decreasing south-to-north abundance in Europe, corresponding to the range of its main host. It is widespread in countries of the western and eastern Mediterranean Basin (but rare in Italy) and the Balkans (Burazerović et al. 2015). It occurs sporadically/focally in Germany, Switzerland, Poland, the Czech Republic, Slovakia and Hungary.

S. Hornok (✉)
Department of Parasitology and Zoology, University of Veterinary Medicine, Budapest, Hungary
e-mail: hornok.sandor@univet.hu

A. Estrada-Peña et al. (eds.), *Ticks of Europe and North Africa*,
https://doi.org/10.1007/978-3-319-63760-0_21

Fig. 32 **A** Distribution of *I. simplex* in Europe and Northern Africa (10 × 10 km grid presence with black dots). **B** Countries where the species has been reported are marked in dark grey

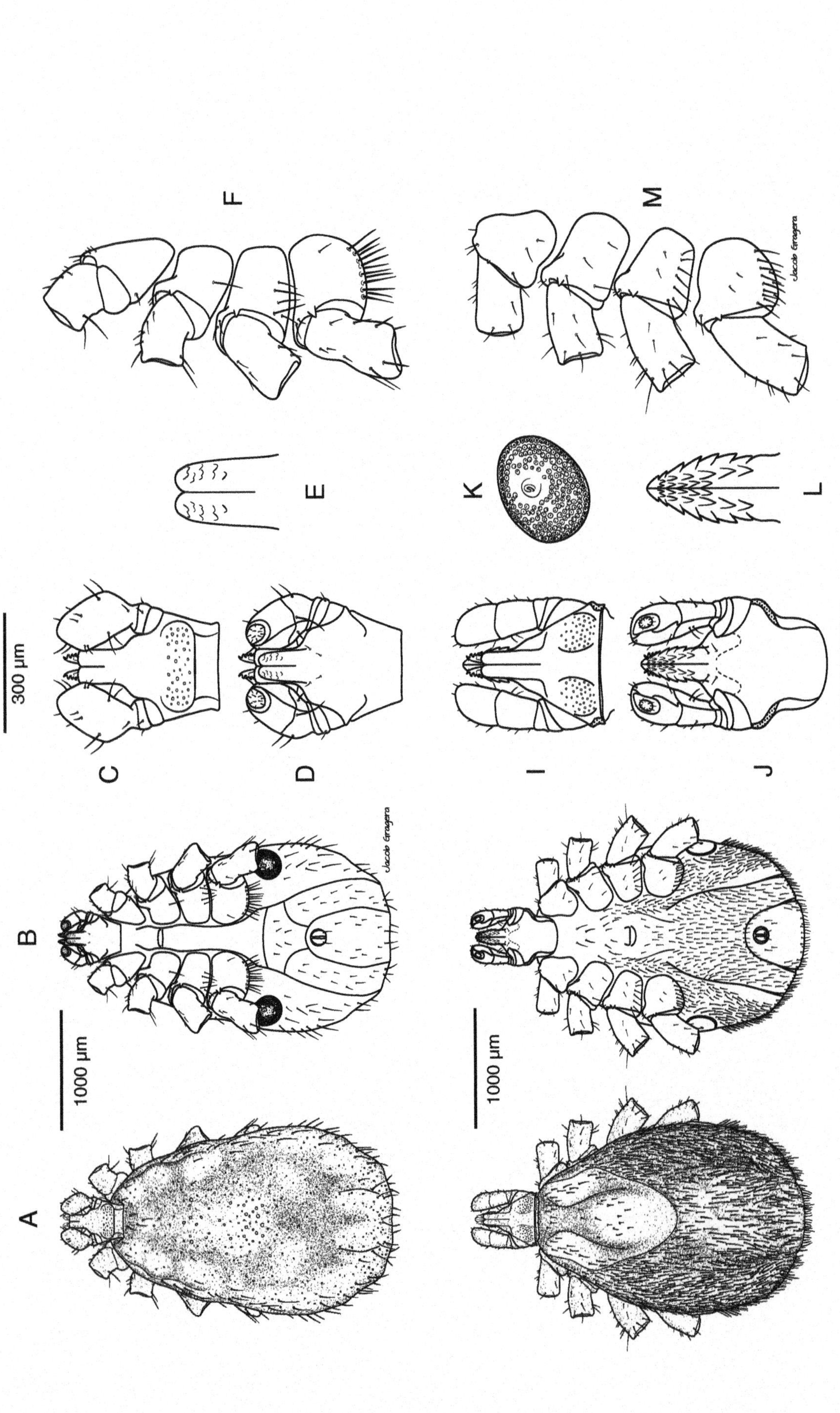

Fig. 33 **A–F** The male of *I. simplex*. **A** Dorsal, **B** ventral, **C** capitulum, dorsal, **D** capitulum, ventral, **E** hypostome, **F** coxae and trochanters I–IV. **G–M** The female of *I. simplex*. **G** Dorsal, **H** ventral, **I** capitulum, dorsal, **J** capitulum, ventral, **K** hypostome, **L** spiracular plate, **M** coxae and trochanters I–IV. Illustrations from specimens collected in Hungary

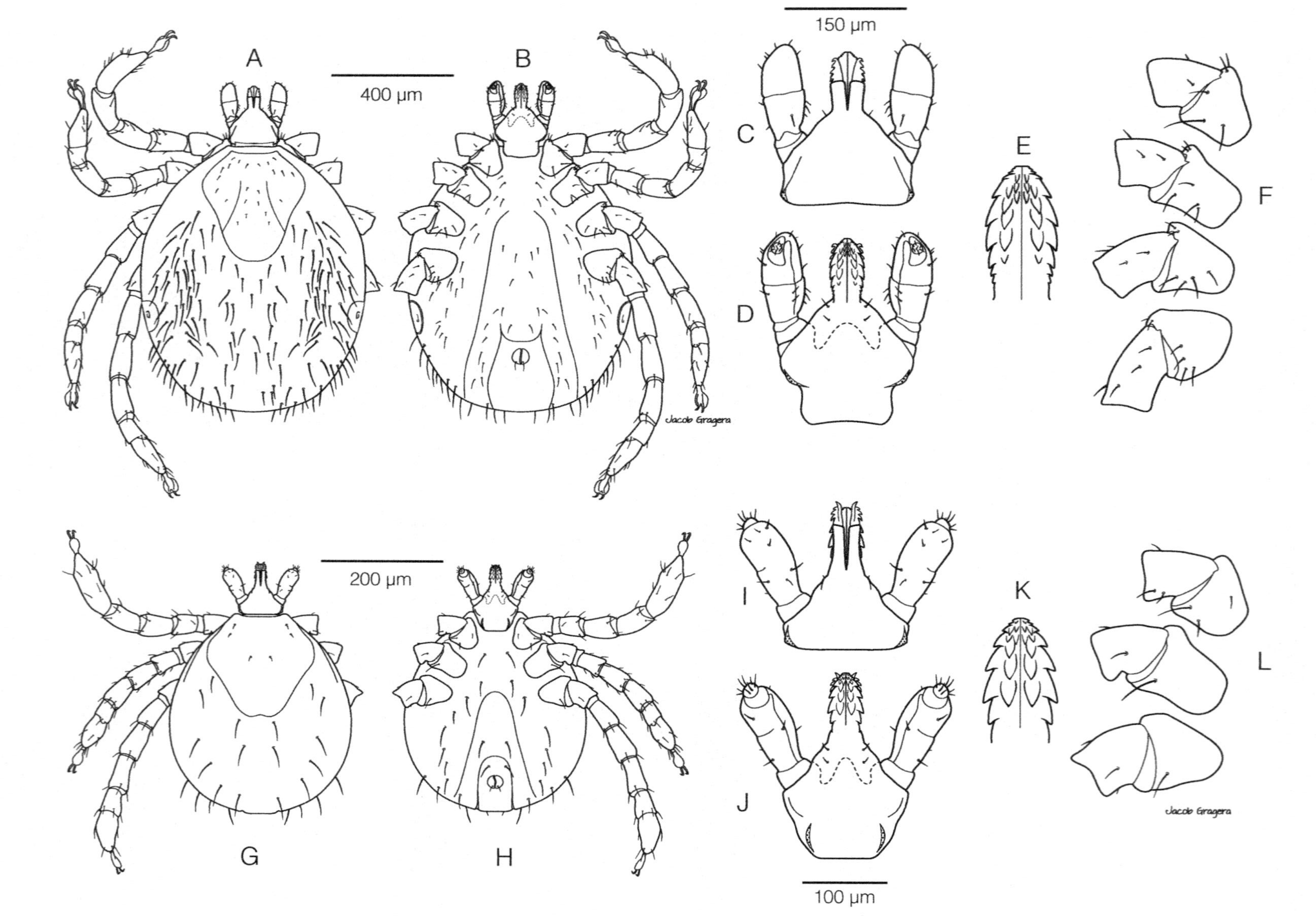

Fig. 34 **A–F** The nymph of *I. simplex*. **A** Dorsal, **B** ventral, **C** capitulum, dorsal, **D** capitulum, ventral, **E** hypostome, **F** coxae and trochanters I–IV. **G–L** The larva of *I. simplex*. **G** Dorsal, **H** ventral, **I** capitulum, dorsal, **J** capitulum, ventral, **K** hypostome, **L** coxae and trochanters I–III. Illustrations from specimens collected in Hungary

Vectorial Capacity and Pathogen Burden

There is no data on the vectorial role of *I. simplex*. However, the DNA of *Babesia crassa*, *B. venatorum*, *B. canis*, *Theileria capreoli*, *T. orientalis* and *Theileria* sp. OT3 was detected in *I. simplex* (Hornok et al. 2016).

References

Arthur DR (1956) The *Ixodes* ticks of Chiroptera (Ixodoidea, Ixodidae). J Parasitol 42:180–196

Beaucournu JC (1966) On some Palaearctic Ixodoidea (Acarina) infesting micro-Chiroptera. Ann Parasitol Hum Comp 41:495–502 (Article in French)

Beaucournu JC (1967) Contribution a la connaissance de la biologie d'*Ixodes* (*Eschatocephalus*) *vespertilionis* Koch, 1844 et d'*Ixodes* (*Pomerantzevella*) *simplex* Neumann, 1906 (Acarina, Ixodoidea) parasites des Chiroptères. Ann Speleologie 22:543–580

Burazerović J, Ćakić S, Mihaljica D, Sukara R, Ćirović D, Tomanović S (2015) Ticks (Acari: Argasidae, Ixodidae) parasitizing bats in the central Balkans. Exp Appl Acarol 66:281–291

Filippova NA (1972) New data on ticks of the genus *Ixodes* Latr. (Ixodoidea, Ixodidae) which are specific bat parasites. Entomol Rev 51:279–285

Hornok S, Szőke K, Kováts D, Estók P, Görföl T, Boldogh SA, Takács N, Kontschán J, Földvári G, Barti L, Corduneanu A, Sándor AD (2016) DNA of piroplasms of ruminants and dogs in ixodid bat ticks. PloS One 11:e0167735

Krištofík J, Danko S (2012) Arthropod ectoparasites (Acarina, Heteroptera, Diptera, Siphonaptera) of bats in Slovakia. Vespertilio 16:167–189

Lourenço S, Palmeirim JM (2008) Which factors regulate the reproduction of ectoparasites of temperate-zone cave-dwelling bats? Parasitol Res 104:127–134

Walter G, Kock D (1985) Records of *Ixodes vespertilionis*, *I. simplex* and *Argas vespertilionis* (Ixodoidea: Ixodidae, Argasidae) from German bats (Chiroptera). Z Parasitenkd 71:107–111

Yamaguti N, Tipton VJ, Keegan HL, Toshioka S (1971) Ticks of Japan, Korea and the Ryukyu Islands. Brigham Young University Science Bulletin, Biological Series, 15:155–158

Ixodes ariadnae Hornok, 2014 (Figs. 35–37)

S. Hornok

Life-Cycle and Host Preferences

Ixodes ariadnae is a three-host tick species parasitizing bats. Based on the seasonal dynamics of its developmental stages, its life cycle is most likely similar to that of *I. vespertilionis*. The preferred hosts of *I. ariadnae* appear to be *Myotis* spp. (Hornok et al. 2015a), including *My. alcathoe*, *My. blythii* (Hornok et al. 2014), *My. myotis* (Hornok et al. 2015b), *My. bechsteinii*, *My. nattereri*, *My. emarginatus*, *My. daubentonii* and *My. dasycneme* (Hornok et al. 2016), but this tick species has also been reported from *Plecotus auritus* (Hornok et al. 2014).

Ecology

All *I. ariadnae* specimens available so far were collected from the walls/floors of caves, or from bats at cave entrances. However, because the majority of known hosts of *I. ariadnae* (*P. auritus* and *Myotis* spp. excluding *My. myotis* and *My. emarginatus*) prefer tree holes as both summer and winter roosts, this species may occur in such niches. Engorged females and nymphs of *I. ariadnae* can be found in caves during spring (females in March: Hornok et al. 2014, 2015b), whereas larvae as well as nymphs are found on bats during their mating season, i.e. late August to early September, and onwards (Hornok, unpublished data).

Distribution

The known geographical range of *I. ariadnae* in Europe is limited to Hungary, Germany (Hornok et al. 2015b) and Belgium (Hornok and Krawczyk 2016).

Vectorial Capacity and Pathogen Burden

The vectorial role of *I. ariadnae* is unknown.

However, the DNA of *Babesia vesperuginis* was detected in *I. ariadnae* (Hornok et al. 2016).

S. Hornok (✉)
Department of Parasitology and Zoology, University of Veterinary Medicine, Budapest, Hungary
e-mail: hornok.sandor@univet.hu

A. Estrada-Peña et al. (eds.), *Ticks of Europe and North Africa*,
https://doi.org/10.1007/978-3-319-63760-0_22

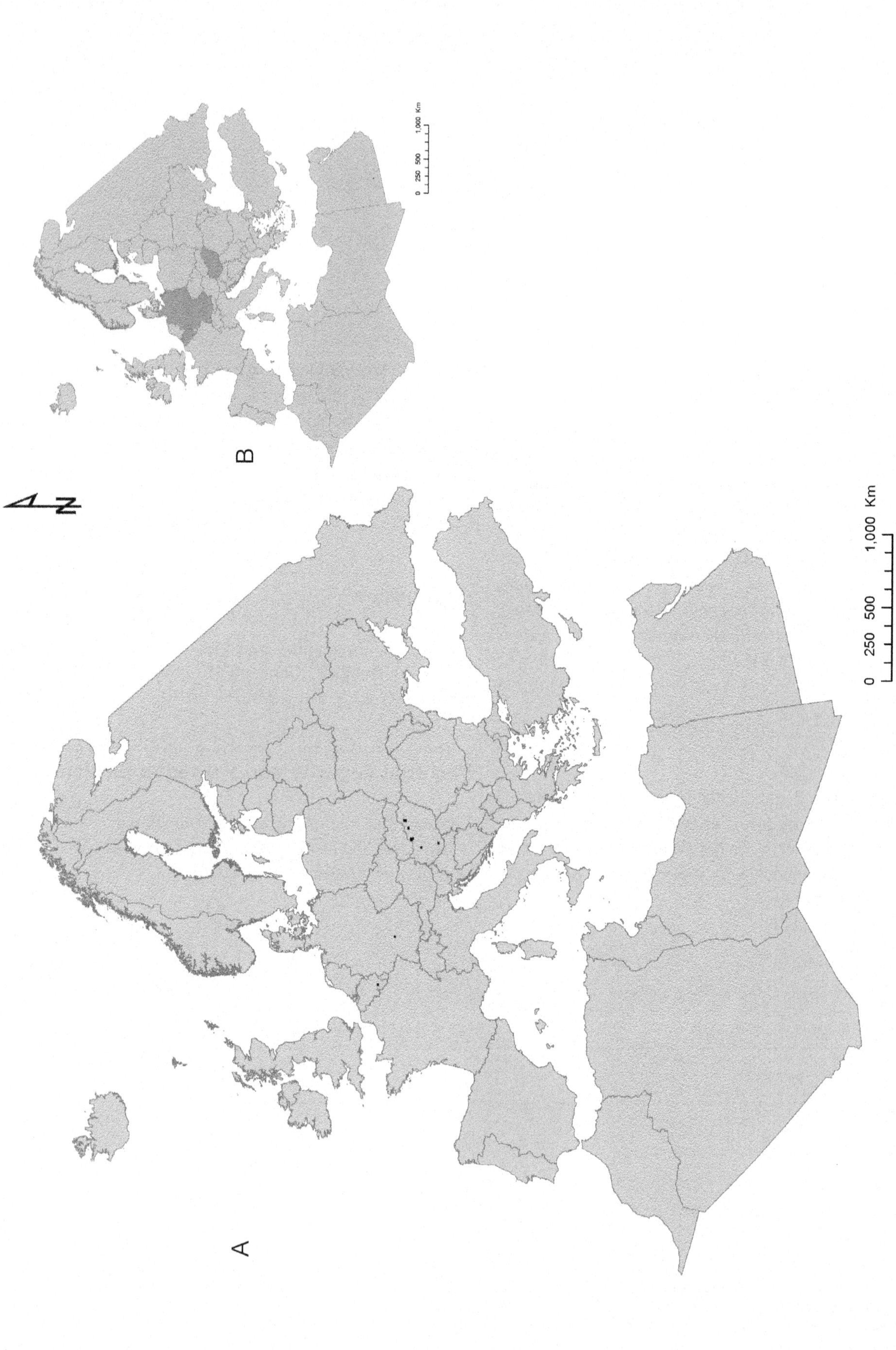

Fig. 35 **A** Distribution of *I. ariadnae* in Europe and Northern Africa (10 × 10 km grid presence with black dots). **B** Countries where the species has been reported are marked in dark grey

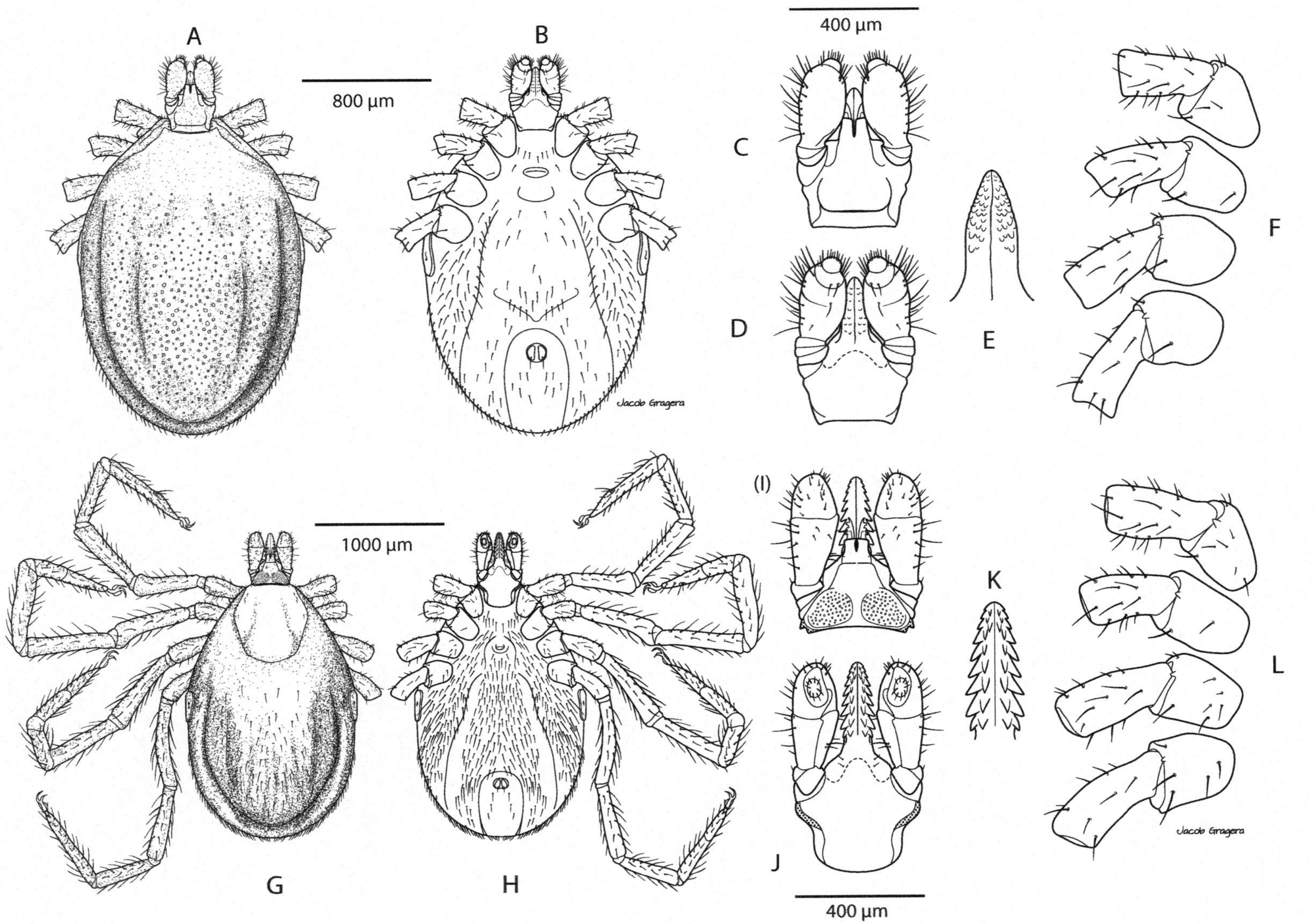

Fig. 36 **A–F** The male of *I. ariadnae*. **A** Dorsal, **B** ventral, **C** capitulum, dorsal, **D** capitulum, ventral, **E** hypostome, **F** coxae and trochanters I–IV. **G–L** The female of *I. ariadnae*. **G** Dorsal, **H** ventral, **I** capitulum, dorsal, **J** capitulum, ventral, **K** hypostome, **L** coxae and trochanters I–IV. Illustrations from specimens collected in Hungary

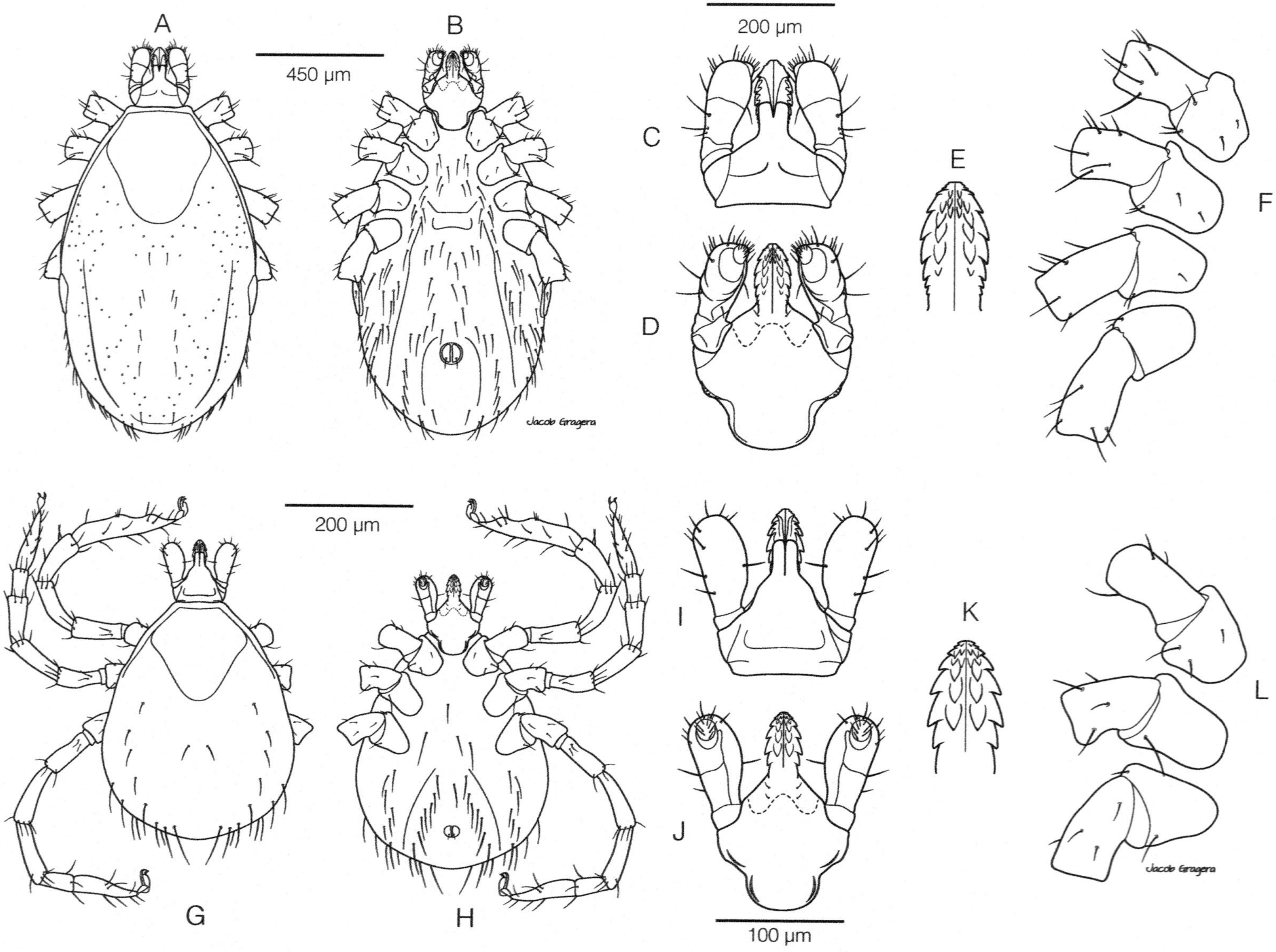

Fig. 37 **A–F** The nymph of *I. ariadnae*. **A** Dorsal, **B** ventral, **C** capitulum, dorsal, **D** capitulum, ventral, **E** hypostome, **F** coxae and trochanters I–IV. **G–L** The larva of *I. ariadnae*. **G** Dorsal, **H** ventral, **I** capitulum, dorsal, **J** capitulum, ventral, **K** hypostome, **L** coxae and trochanters I–III. Illustrations from specimens collected in Hungary

References

Hornok S, Kontschán J, Kováts D, Kovács R, Angyal D, Görföl T, Mihalca AD (2014) Bat ticks revisited: *Ixodes ariadnae* sp. nov. and allopatric genotypes of *I. vespertilionis* in caves of Hungary. Parasit Vectors 7:202

Hornok S, Estrada-Peña A, Kontschán J, Plantard O, Kunz B, Mihalca AD, Thabah A, Tomanović S, Buracerović J, Takács N, Görföl T, Estók P, Tu VT, Szőke K, Fernández de Mera IG, de la Fuente J, Takahashi M, Yamauchi T, Takano A (2015a) High degree of mitochondrial gene heterogeneity in the bat tick species *Ixodes vespertilionis, I. ariadnae* and *I. simplex* from Eurasia. Parasit Vectors 8:457

Hornok S, Takács N, Szőke K, Kunz B (2015b) First record of *Ixodes ariadnae* in Germany. Acta Vet Hung 63:347–351

Hornok S, Krawczyk A (2016) First record of *Ixodes ariadnae* in Western Europe, Belgium. Acta Vet Hung 64:467–471

Hornok S, Szőke K, Kováts D, Estók P, Görföl T, Boldogh SA, Takács N, Kontschán J, Földvári G, Barti L, Corduneanu A, Sándor AD (2016) DNA of piroplasms of ruminants and dogs in ixodid bat ticks. PLoS One 11:e0167735

Ixodes uriae White, 1852 (Figs. 38–40)

T. N. Petney and M. P. Pfäffle

It has been suggested that *Ixodes uriae* is a complex of species, although definitive taxonomic data are lacking (McCoy et al. 2005, 2012; Dietrich et al. 2013). Dietrich et al. (2014), however, suggest that evolution for host specialisation, although common and rapid, does not lead to speciation.

Life-Cycle and Host Preferences

Ixodes uriae (*Ixodes putus* was previously a commonly used junior synonym) is a three-host tick with a life cycle in the laboratory of between 9 and 12 months (Murray and Vestjens 1967). All stages feed on sea birds, including the families Spheniscidae (penguins), Diomedeidae (albatrosses), Procellariidae (petrels, prions), Stercorariidae (skuas), Phalacrocoracidae (cormorants), Laridae (gulls and kittiwakes), Alcidae (murres, razorbills, puffins), Hydrobatidae (storm petrels), Pelecanoididae (diving petrels), and Sulidae (gannets and boobies) (Dietrich et al. 2011).

Ecology

Ixodes uriae is the most common and important tick of sea birds (McCoy et al. 2001). It is found near or in seabird nesting areas, including the nests themselves (Murray and Vestjens 1967; Eveleigh and Threlfall 1975), in coastal areas and on oceanic islands, predominantly in cold temperate and polar regions, worldwide (Petney et al. 2012). This species has an extremely wide temperature tolerance ranging from −30 to +40 °C (Lee and Baust 1987), but a low dehydration tolerance combined with a high rate of water loss which confines it to hydrating microhabitats where it commonly forms aggregations (Benoit et al. 2007). When off-host, these ticks are commonly found hiding under stones (Barbosa et al. 2011). Descamps (2013) found that an increase of 1 °C in average winter temperature led to a 5% increase in the infestation of the thick-billed murre (*Uria lomvia*) by *I. uriae*, suggesting that climate change may substantially increase infestation intensities and prevalences.

Distribution

Ixodes uriae has a circumpolar distribution including cold temperate and polar areas in the Afrotropical, Australasian, Nearctic, Neotropical and Palaearctic Regions including oceanic islands.

Vectorial Capacity and Pathogen Burden

This species is either a known vector of or has been associated with a wide variety of pathogens, in particular arboviruses of which more have been isolated from *I. uriae* than any other tick species. Labuda and Nuttall (2004) list 32 species of *Reoviridae* belonging to the genus *Orbivirus*, 60 species of *Bunyaviridae* (genera *Nairovirus* and *Phlebovirus*), two *Flaviviridae* (*Flavivirus*) and three unassigned species. Recent work has added a number of new viruses to this list (Major et al. 2009). In addition to arboviruses, *I. uriae* has either been associated with or is known to transmit a variety of other pathogens including *Borrelia garinii*, which is found in this tick in both the Northern and Southern Hemispheres (Olsén et al. 1993, 1995; Gylfe et al. 1999; Smith et al. 2006).

T. N. Petney (✉)
State Museum of Natural History, Erbprinzenstr. 13, 76133 Karlsruhe, Germany
e-mail: trevor.petney@kit.edu

M. P. Pfäffle
Department of Ecology and Parasitology, Institute of Zoology, Karlsruhe Institute of Technology, Karlsruhe, Germany
e-mail: miripfaeffle@web.de

A. Estrada-Peña et al. (eds.), *Ticks of Europe and North Africa*,
https://doi.org/10.1007/978-3-319-63760-0_23

Fig. 38 **A** Distribution of *I. uriae* in Europe and Northern Africa (10 × 10 km grid presence with black dots). **B** Countries where the species has been reported are marked in dark grey

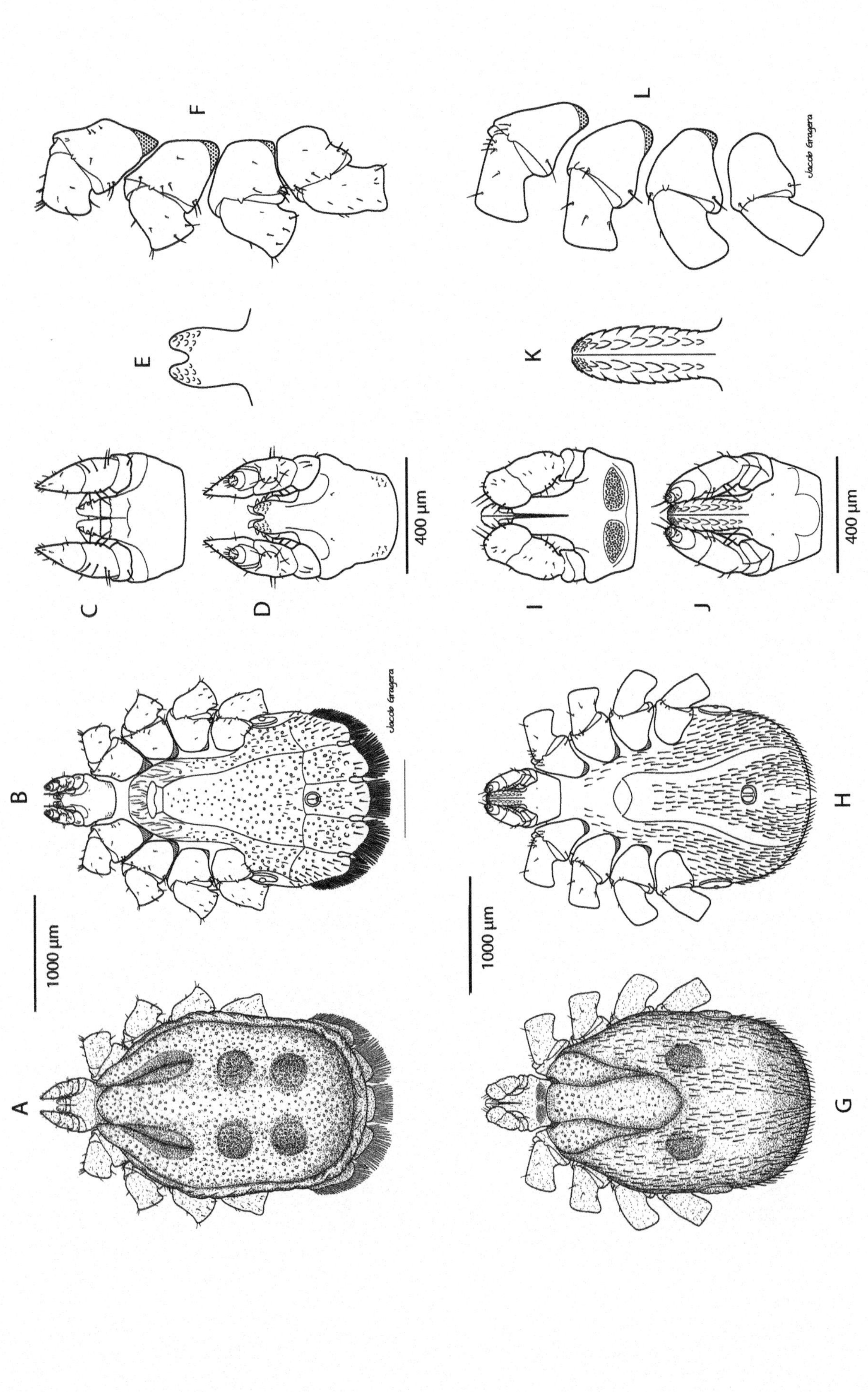

Fig. 39 **A–F** The male of *I. uriae*. **A** Dorsal, **B** ventral, **C** capitulum, dorsal, **D** capitulum, ventral, **E** hypostome, **F** coxae and trochanters I-IV. **G–L** The female of *I. uriae*. **G** Dorsal, **H** ventral, **I** capitulum, dorsal, **J** capitulum, ventral, **K** hypostome, **L** coxae and trochanters I-IV. Illustrations from specimens collected in the United Kingdom

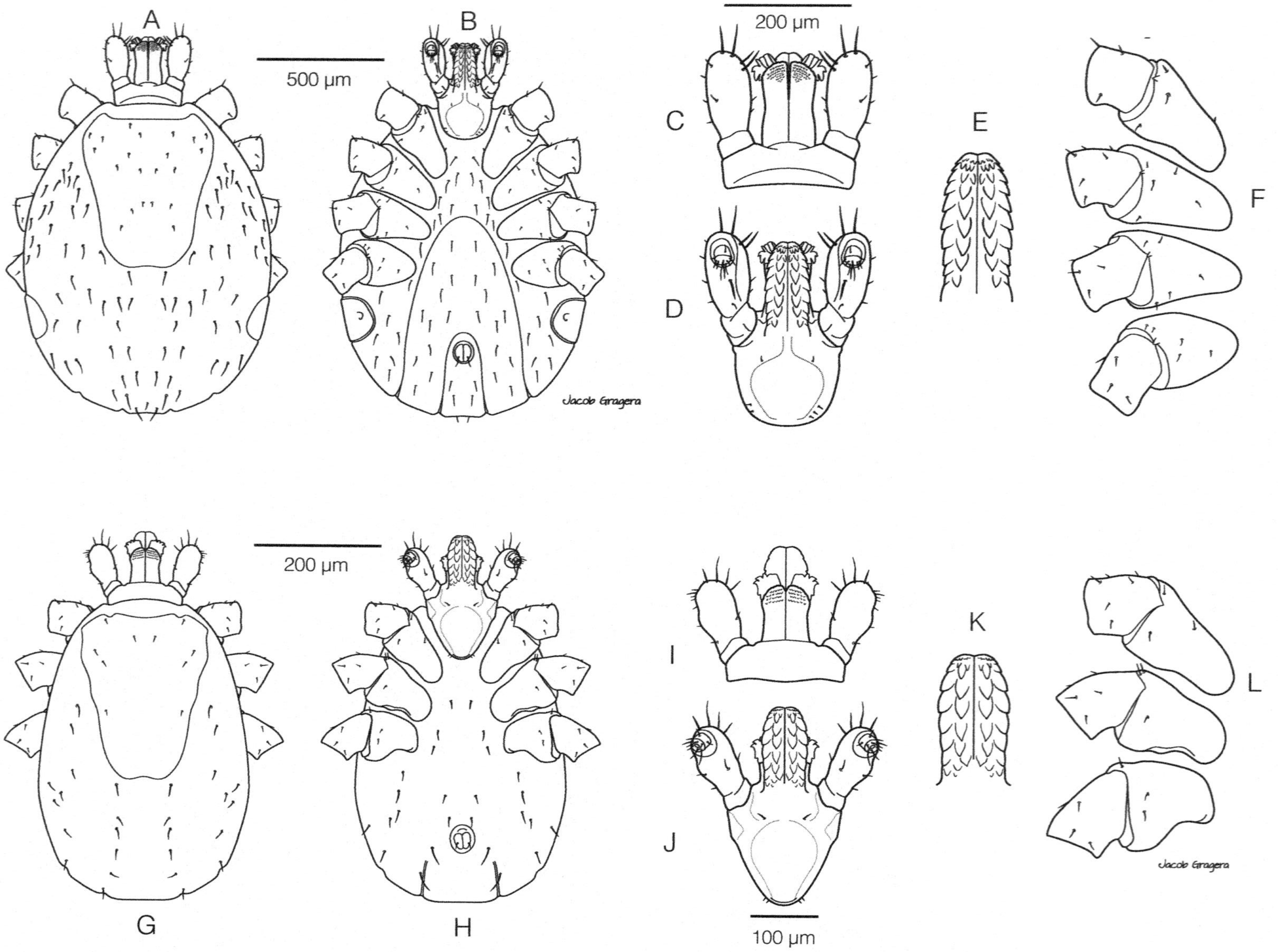

Fig. 40 **A–F** The nymph of *I. uriae*. **A** Dorsal, **B** ventral, **C** capitulum, dorsal, **D** capitulum, ventral, **E** hypostome, **F** coxae and trochanters I-IV. **G–L** The larva of *I. uriae*. **G** Dorsal, **H** ventral, **I** capitulum, dorsal, **J** capitulum, ventral, **K** hypostome, **L** coxae and trochanters I-III. Illustrations from specimens collected in the United Kingdom

References

Barbosa A, Benzal J, Vidal V, D'Amico V, Coria N, Diaz J, Motas M, Palacios MJ, Cuervo JJ, Ortiz J, Chitimia L (2011) Seabird ticks (*Ixodes uriae*) distribution along the Antarctic Peninsula. Polar Biol 34:1621–1624. https://dx.doi.org/10.1007/s00300-011-1000-7

Benoit JB, Yoder JA, Lopez-Martinez G, Elnitsky MA, Lee RE Jr, Denlinger DL (2007) Habitat requirements of the seabird tick, *Ixodes uriae* (Acari: Ixodidae), from the Antarctic Peninsula in relation to water balance characteristics of eggs, nonfed and engorged stages. J Comp Physiol B 177:205–215. https://dx.doi.org/10.1007/s00360-006-0122-7

Descamps S (2013) Winter temperature affects the prevalence of ticks in an arctic seabird. PLoS ONE 8:e65374. https://dx.doi.org/10.1371/journal.pone.0065374

Dietrich M, Gómez-Díaz E, McCoy KD (2011) Worldwide distribution and diversity of seabird ticks: implications for the ecology and epidemiology of tick-borne pathogens. Vector Borne Zoonotic Dis 11:453–470. https://dx.doi.org/10.1089/vbz.2010.0009

Dietrich M, Beati L, Elguero E, Boulinier T, McCoy KD (2013) Body size and shape evolution in host races of the tick *Ixodes uriae*. Biol J Linn Soc 108:323–334. https://dx.doi.org/10.1111/j.1095-8312.2012.02021.x

Dietrich M, Kempf F, Boulinier T, McCoy KD (2014) Tracing the colonisation and diversification of the worldwide seabird ectoparasite *Ixodes uriae*. Mol Ecol 23:3292–3305. https://dx.doi.org/10.1111/mec.12815

Eveleigh ES, Threlfall W (1975) The biology of *Ixodes (Ceratixodes) uriae* White, 1852 in Newfoundland. Acarologia 16:621–635

Gylfe A, Olsen B, Strasevicius D, Ras NM, Weihe P, Noppa L, Östberg Y, Baranton G, Bergström SI (1999) Isolation of Lyme disease *Borrelia* from puffins (*Fratercula arctica*) and seabird ticks (*Ixodes uriae*) on the Faeroe Islands. J Clin Microbiol 37:890–896

Labuda M, Nuttall PA (2004) Tick-borne viruses. Parasitology 129: S221–S245. https://dx.doi.org/10.1017/S0031182004005220

Lee REJ, Baust JG (1987) Cold-hardiness in the Antarctic tick, *Ixodes uriae*. Physiol Zool 60:499–506

Major L, La Linn M, Slade RW, Schroder WA, Hyatt AD, Gardner J, Cowley J, Suhrbier A (2009) Ticks associated with Macquarie Island penguins carry arboviruses from four genera. PLoS ONE 4: e4375. https://dx.doi.org/10.1371/journal.pone.0004375

McCoy KD, Boulinier T, Tirard C, Michalakis Y (2001) Host specificity of a generalist parasite: genetic evidence of sympatric host races in the seabird tick *Ixodes uriae*. J Evol Biol 14:395–405. https://dx.doi.org/10.1046/j.1420-9101.2001.00290.x

McCoy KD, Chapuis E, Tirard C, Boulinier T, Michalakis Y, Le Bohec C, Le Maho Y, Gauthier-Clerc M (2005) Recurrent evolution of host-specialized races in a globally distributed parasite. Proc Biol Sci 272:2389–2395. https://dx.doi.org/10.1098/rspb.2005.3230

McCoy KD, Barbosa A, Cuervo JJ (2012) Population genetic structure and colonisation of the western Antarctic Peninsula by the seabird tick *Ixodes uriae*. Mar Ecol Prog Ser 459:109–120. https://dx.doi.org/10.3354/meps09749

Murray M, Vestjens W (1967) Studies on the ectoparasites of seals and penguins. III. The distribution of the tick *Ixodes uriae* White and the flea *Parapsyllus magellanicus heardi* de Meillon on Macquarie Island. Aust. J Zool 15:715. https://dx.doi.org/10.1071/ZO9670715

Olsén B, Jaenson TG, Noppa L, Bunikis J, Bergström S (1993) A Lyme borreliosis cycle in seabirds and *Ixodes uriae* ticks. Nature 362:340–342. https://dx.doi.org/10.1038/362340a0

Olsén B, Duffy D, Jaenson T, Gylfe A, Bonnedahl J, Bergström S (1995) Transhemispheric exchange of Lyme disease spirochetes by seabirds. J Clin Microbiol 33:3270–3274

Petney TN, Pfäffle MP, Skuballa JD (2012) An annotated checklist of the ticks (Acari: Ixodida) of Germany. Syst Appl Acarol 17:115–170. https://dx.doi.org/10.11158/saa.17.2.2

Smith RP, Bin Muzaffar S, Lavers J, Lacombe EH, Cahill BK, Lubelczyk CB, Kinsler A, Mathers AJ, Rand PW (2006) *Borrelia garinii* in seabird ticks (*Ixodes uriae*), Atlantic Coast, North America. Emerg Infect Dis 12:1909–1912. https://dx.doi.org/10.3201/eid1212.060448

Ixodes rothschildi Nuttall and Warburton, 1911 (Fig. 41)

A. Estrada-Peña

Ixodes rothschildi is a poorly known species. For example, all of the published surveys have provided only specimens from the hosts. This tick has been never collected while questing for hosts.

Life-Cycle and Host Preferences

According to published data (summarized by Pérez-Eid 2007), *I. rothschildi* is a three-host species with endophilic behaviour and strong preferences for a few species of hosts, such as marine birds of the species *Puffinus puffinus*, *Larus argentatus*, *Phalacrocorax aristotelis* and *Fratercula arctica*. It seems, however, that these preferences for hosts are derived from the way in which they build their nests. Both *F. arctica* and *P. puffinus* have nests under the ground, where the ticks wait for their hosts. These nests are frequented by the hosts from approximately March to August (Pérez-Eid 2007). According to published data (Pérez-Eid and Pasquet 1983), the adults of *I. rothschildi* are first found on hosts, followed later in the year by the nymphs and the larvae. Thus, adults are likely to be the overwintering stage, and larvae the result of oviposition by the overwintering females after feeding. It is difficult to interpret the seasonal activity of the nymphs, which could be the result either of overwintering as engorged larvae or as flat nymphs or both (Pérez-Eid and Pasquet 1983; Pérez-Eid 2007).

Ecology

Few details are available on the ecology of *I. rothschildi*. Nothing can be concluded about preferences for temperature and/or relative humidity, due to its endophilic behaviour and the small distribution range.

Distribution

Ixodes rothschildi has been collected in Ireland, the United Kingdom and Brittany (Morel 1964; Pérez-Eid and Pasquet 1983; Pérez-Eid 2007).

Vectorial Capacity and Pathogen Burden

These aspects are unknown.

A. Estrada-Peña (✉)
Department of Animal Health, Faculty of Veterinary Medicine, Zaragoza, Spain
e-mail: aestrada@unizar.es

A. Estrada-Peña et al. (eds.), *Ticks of Europe and North Africa*,
https://doi.org/10.1007/978-3-319-63760-0_24

Fig. 41 **A** Distribution of *I. rothschildi* in Europe and Northern Africa (10 × 10 km grid presence with black dots). **B** Countries where the species has been reported are marked in dark grey. Specimens of this species were unavailable for illustrations

References

Morel PC (1964) Les tiques d'Afrique et du Bassin méditérranén (Ixodidea). Distribution, biologie, écologie, rôle pathogène. CD-ROM, CIRAD, 2003

Pérez-Eid C, Pasquet E (1983) Contribution à la connaissance d'une espèce de tique peu connue: *Scaphixodes rothschildi*. Prémiere decouverte des larves en France. Acarologia 24:257–260

Pérez-Eid C (2007) Les tiques. Identification, biologie, importance médicale et vétérinaire. Editions Tec & Doc, Paris, 314 pp

Ixodes caledonicus Nuttall, 1910 (Fig. 42)

A. Estrada-Peña

Ixodes caledonicus is a species with clear preferences for birds, as seems to be the rule for species of the subgenus *Scaphixodes*.

Life-Cycle and Host Preferences

Ixodes caledonicus is a three-host, endophilic tick. It is commonly found on birds that build nests in areas which are difficult to access, such as cliffs, the towers of churches, the roofs of old and abandoned buildings, or pigeon lofts. The most commonly reported hosts, summarised by Pérez-Eid (2007), are the families Procellariidae, Apodidae, and Columbidae, with a few records from Falconiformes.

Ecology

Few details can be outlined on the ecology of *I. caledonicus*. Nothing can be concluded about its preferences for temperature and/or relative humidity, due to its endophilic behaviour, but it seems that temperature, more than the relative humidity, is probably the main restricting factor for its distribution.

A. Estrada-Peña (✉)
Department of Animal Health, Faculty of Veterinary Medicine, Zaragoza, Spain
e-mail: aestrada@unizar.es

A. Estrada-Peña et al. (eds.), *Ticks of Europe and North Africa*,
https://doi.org/10.1007/978-3-319-63760-0_25

Fig. 42 **A** Distribution of *I. caledonicus* in Europe and Northern Africa (10 × 10 km grid presence with black dots). **B** Countries where the species has been reported are marked in dark grey. Specimens of this species were unavailable for illustrations

Distribution

Ixodes caledonicus has been collected in the United Kingdom, Denmark, Germany and Poland (data summarized by Pérez-Eid 2007; Siuda 1993). There is one record of the species from the island of Sardinia (Manilla 1998).

Vectorial Capacity and Pathogen Burden

These aspects are unknown.

References

Manilla G (1998) Fauna d'Italia. Acari, Ixodida. Edizioni Calderini, Bologna, 280 pp

Pérez-Eid C (2007) Les tiques. Identification, biologie, importance médicale et vétérinaire. Editions Tec & Doc, Paris, 314 pp

Siuda K (1993) Kleszcze Polski (Acari: Ixodida). II. Systematyka i rozmieszczenie. Monografie Parazyto- logiczne 12. Polskie Towarzystwo Parazytologiczne, Warszawa

Ixodes unicavatus Neumann, 1908 (Fig. 43)

T. N. Petney and M. P. Pfäffle

Life-Cycle and Host Preferences

Ixodes unicavatus is a rare species which is usually found in coastal areas where it is parasitic predominantly on colonial seabirds or birds that frequent the coast, particularly cormorants *Phalacrocorax aristotelis* and *P. carbo* (Schulze 1932; Arthur 1963; Guiguen et al. 1987; Kolonin 2008). It has also been found rarely on the Eurasian rock pipit (*Anthus petrosus*) and the gyrfalcon (*Falco rusticolus*) (Hillyard 1996).

Ecology

Ixodes unicavatus is an endophilic tick that is found in its host's nest or under stones in the vicinity of nests in rocky, marine habitats (Jaenson et al. 1994; Filippova 2007). It also occurs in Celtic broadleaf forests, English lowland beech forests, and rarely in North Atlantic moist mixed forests (Guglielmone et al. 2014). Hillyard (1996) indicates that larvae, nymphs and females have two seasonal attachment periods, from March to June and again September to November.

Distribution

Ixodes unicavatus occurs along the Atlantic coasts of Europe and the Crimea (Filippova 2007). It has been reported from France (Rageau 1972; Guiguen et al. 1987), the United Kingdom (Jameson and Medlock 2011), Ireland (Nuttall et al. 1984), the Netherlands (Obsomer et al. 2013) and Sweden (Jaenson et al. 1994).

Vectorial Capacity and Pathogen Burden

The role of *I. unicavatus* as a vector of disease agents is not known.

T. N. Petney (✉)
State Museum of Natural History, Erbprinzenstr. 13, 76133 Karlsruhe, Germany
e-mail: trevor.petney@kit.edu

M. P. Pfäffle
Department of Ecology and Parasitology, Institute of Zoology, Karlsruhe Institute of Technology, Karlsruhe, Germany
e-mail: miripfaeffle@web.de

A. Estrada-Peña et al. (eds.), *Ticks of Europe and North Africa*,
https://doi.org/10.1007/978-3-319-63760-0_26

Fig. 43 **A** Distribution of *I. unicavatus* in Europe and Northern Africa (10 × 10 km grid presence with black dots). **B** Countries where the species has been reported are marked in dark grey. Specimens of this species were unavailable for illustrations

References

Arthur DR (1963) British ticks. London, Butterworths

Filippova NA (2007) The phenomenon of morphological inversions in the ontogeny of some Palearctic species of ixodid ticks (Acari: Ixodidae). Int J Acarol 33:61–72. https://dx.doi.org/10.1080/01647950708684502

Guiguen C, Monnat JY, Launay H, Beaucournu JC (1987) Ectoparasites des oiseaux en Bretagne: 3. Ixodoidea. Cah ORSTOM Série Entomol Médicale Parasitol 73–81

Hillyard PD (1996) Ticks of north-west Europe. Field Studies Council, Shrewsbury

Guglielmone AA, Robbins RG, Apanaskevich DA, Petney TN, Estrada-Peña A, Horak, IG (2014) The hard ticks of the world. Springer, Heidelberg

Jaenson TGT, Tälleklint L, Lundqvist L, Olsen B, Chirico J, Mejlon H (1994) Geographical distribution, host associations, and vector roles of ticks (Acari: Ixodidae, Argasidae) in Sweden. J Med Entomol 31:240–256. https://dx.doi.org/10.1093/jmedent/31.2.240

Jameson LJ, Medlock JM (2011) Tick surveillance in Great Britain. Vector Borne Zoonotic Dis 11:403–12. https://dx.doi.org/10.1089/vbz.2010.0079

Kolonin G V. (2008) Birds as hosts of ixodid ticks (Acarina, Ixodidae). Entomol Rev 88:1012–1015. 10.1134/S0013873808080150

Nuttall PA, Kelly TC, Carey D, Moss SR, Harrap KA (1984) Mixed infections with tick-borne viruses in a seabird colony in Eire. Arch Virol 79:35–44. https://dx.doi.org/10.1007/BF01314301

Obsomer V, Wirtgen M, Linden A, Claerebout E, Heyman P, Heylen D, Madder M, Maris J, Lebrun M, Tack W, Lempereur L (2013) Spatial disaggregation of tick occurrence and ecology at a local scale as a preliminary step for spatial surveillance of tick-borne diseases: general framework and health implications in Belgium. Parasit Vectors 6:190. https://dx.doi.org/10.1186/1756-3305-6-190

Rageau J (1972) Repartation geographique et role pathogene des tiques (Acariens: Argasidae et Ixodidae) en France. Wiad Parazytol 18:707–719

Schulze P (1932) Die Zecken als Vogelparasiten. J Ornithol 80:318–329. https://dx.doi.org/10.1007/BF01905401

Ixodes arboricola Schulze and Schlottke, 1929 (Fig. 44)

A. D. Sándor, T. N. Petney, and M. P. Pfäffle

Life-Cycle and Host Preferences

The tree-hole tick, *Ixodes arboricola*, is a nest-dwelling species commonly hosted by hole breeding passerines. It occurs exclusively in tree holes and is thus adapted to this environment, including prolonged attachment and detachment at night to ensure that it remains only in tree-holes. It has a typical natural life cycle of 1–2 years (White et al. 2012). Under laboratory conditions, the cycle may be completed in 142–242 days (Arthur 1963), with the average attachment in one study of 8.3 (larvae), 4.4 (nymphs) or 7.5 days (females) (Heylen and Matthysen 2010). The species is a typical parasite of small woodland breeding passerines (*Parus* spp., *Sturnus vulgaris*, *Sitta europaea*, *Ficedula* spp. and hole-nesting owls); however, it has been recorded on other passerines breeding in somewhat similar circumstances, such as *Erithacus rubecula* or *Turdus merula*. Colonially nesting corvids may be an important local source of hosts (Sándor et al. 2017), while larger birds breeding in rocky niches may also be parasitized [e.g. *Falco peregrinus*, Schilling et al. (1981)]. No mammals other than tree-hole roosting bats have been found infested by this species (Arthur 1963). This tick has not been reported from humans.

A. D. Sándor (✉)
Department of Parasitology and Parasitic Diseases, University of Agricultural Sciences and Veterinary Medicine, Cluj-Napoca, Romania
e-mail: adsandor@gmail.com

T. N. Petney
State Museum of Natural History, Erbprinzenstr. 13, 76133 Karlsruhe, Germany
e-mail: trevor.petney@kit.edu

M. P. Pfäffle
Department of Ecology and Parasitology, Institute of Zoology, Karlsruhe Institute of Technology, Karlsruhe, Germany
e-mail: miripfaeffle@web.de

A. Estrada-Peña et al. (eds.), *Ticks of Europe and North Africa*,
https://doi.org/10.1007/978-3-319-63760-0_27

Ecology

This nest-dwelling tick occurs mostly in Europe, with sporadic occurrences in Central Asia, north-east China and northern Africa, where it has been found only on migratory birds. It occurs in wooded landscapes as it requires tree-holes, in a wide range of terrestrial habitats where woodlands or old solitary trees provide breeding holes for small passerines. As the species spends most of its life in these holes, it shows a reduced intrinsic mobility. *Ixodes arboricola* has a fairly long period of attachment, with larvae and nymphs able to stay on hosts for up to 20 days (Heylen and Matthysen 2010). There is a slight seasonality, with larvae being more common in winter and early spring, nymphs in spring, while adult females are observed most commonly in late winter. All stages may be observed all year-round (Heylen et al. 2014a). This species has developed specific adaptations to its nest-dwelling habit, with most individuals detaching from hosts only at night, with a prolonged attachment recorded when hosts were excluded from the typical tree-hole environment (White et al. 2012). It usually occurs in small numbers on its host and does not alter the physiological condition of its host (Heylen and Matthysen 2008); however, it was recorded to cause severe anaemia in certain cases (Schilling et al. 1981).

Distribution

Ixodes arboricola is primarily a European tick, occurring across a wide belt through the temperate region of Europe, with two small insular occurrences in Asia, in Iran and the Far East (north-east China). It has been reported in most Western Palaearctic countries (Austria, Belgium, Bulgaria, the Czech Republic, Estonia, Finland, France, Georgia, Germany, Italy, Moldova, the Netherlands, Poland, Romania, Russia, Slovakia, Slovenia, Sweden, Turkey, the Ukraine, The United Kingdom, Portugal). This species has

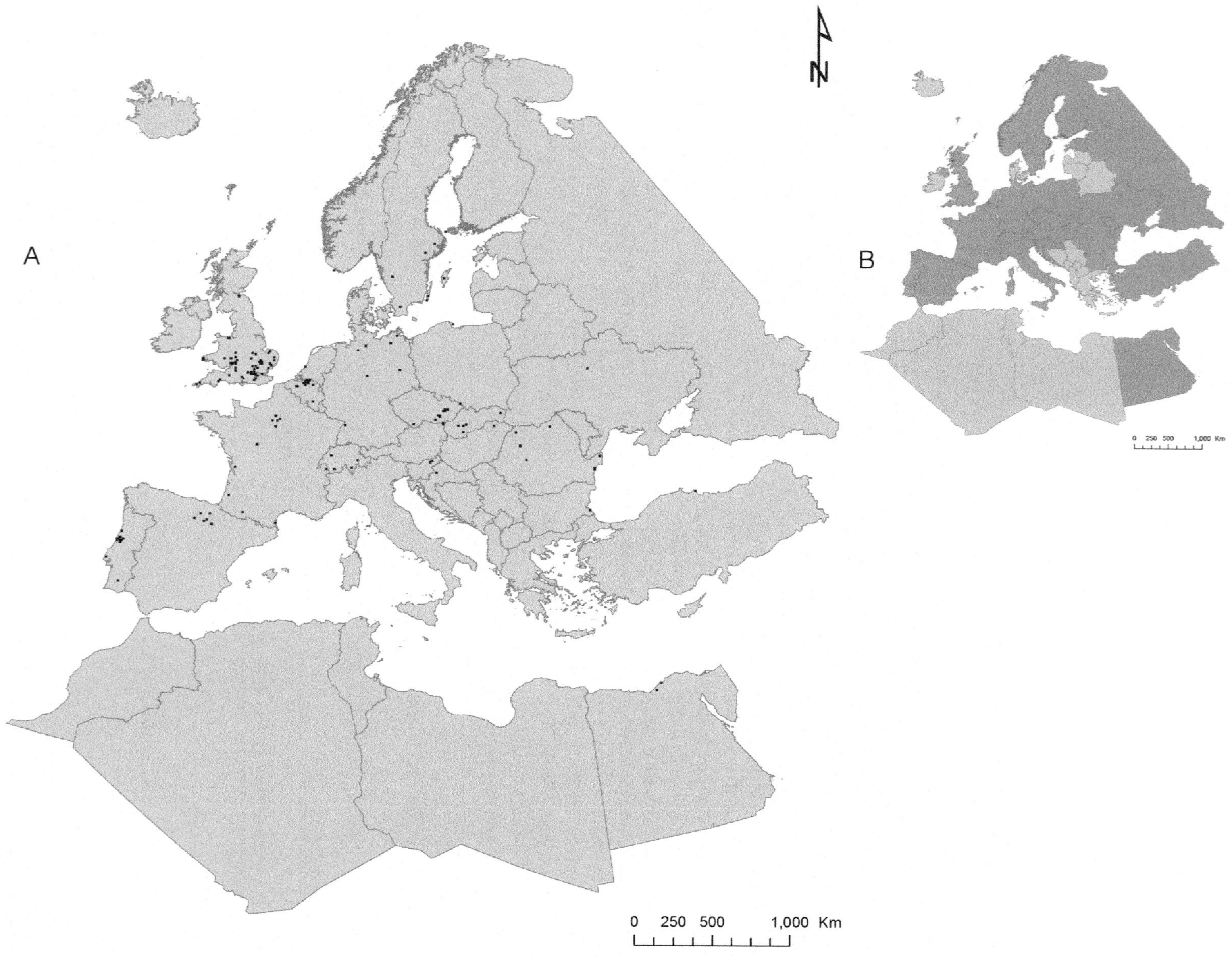

Fig. 44 **A** Distribution of *I. arboricola* in Europe and Northern Africa (10 × 10 km grid presence with black dots). **B** Countries where the species has been reported are marked in dark grey. Specimens of this species were unavialable for illustrations

been recorded on migratory birds in North Africa (Egypt) and the Middle East (Israel); however, there is no agreement as to whether it is able to maintain a full developmental cycle in these areas (Clifford and Hoogstraal 1965).

Vectorial Capacity and Pathogen Burden

The vectorial capacity of this species is poorly known. *Ixodes arboricola* was experimentally proven to play a role in the transmission of *Borrelia burgdorferi* s.l., as it was able to transmit the bacteria transstadially (Heylen et al. 2014b). This tick was also suggested to be the vector of the tick-borne encephalitis virus (Lichard and Kožuch 1967). It is suspected to have a vectorial capacity for *Rickettsia* spp. (Špitalská et al. 2011), with records of *Rickettsia helvetica*, *R. massiliae*, and *R. monacensis* (Mărcuţan et al. 2016) and *Candidatus* "Rickettsia vini" (Špitalská et al. 2011; Palomar et al. 2012; Keskin et al. 2014), although this remains to be tested.

References

Arthur DR (1963) British ticks. Butterworths, London, UK

Clifford CM, Hoogstraal H (1965) The occurrence of *Ixodes arboricola* Schulze and Schlottke (Ixodoidea: Ixodidae) in Africa on northward migrating birds. J Med Entomol 2:37–40

Heylen DJA, Matthysen E (2008) Effect of tick parasitism on the health status of a passerine bird. Funct Ecol 22:1099–1107

Heylen DJA, Matthysen E (2010) Contrasting detachment strategies in two congeneric ticks (Ixodidae) parasitizing the same songbird. Parasitology 137:661–667

Heylen DJA, Van Oosten AR, Devriendt N, Elst J, De Bruyn L, Matthysen E (2014a) Seasonal feeding activity of the tree-hole tick. *Ixodes arboricola*. Parasitology 141:1044–1051

Heylen D, Sprong H, Oers K, Fonville M, Leirs H, Matthysen E (2014b) Are the specialized bird ticks, *Ixodes arboricola* and *I. frontalis*, competent vectors for *Borrelia burgdorferi* sensu lato? Environ Microbiol 16:1081–1089

Keskin A, Koprulu TK, Bursali A, Ozsemir AC, Yavuz KE, Tekin S (2014) First record of *Ixodes arboricola* (Ixodida: Ixodidae) from Turkey with presence of *Candidatus* Rickettsia vini (Rickettsiales: Rickettsiaceae). J Med Entomol 51:864–867

Lichard M, Kožuch O (1967) Persistence of tick-borne encephalitis virus in nymphs and adults of *Ixodes arboricola* and its transmission to white mice. Acta Virol 11:480

Mărcuţan ID, Kalmár Z, Ionică AM, D'Amico G, Mihalca AD, Vasile C, Sándor AD (2016) Spotted fever group rickettsiae in ticks of migratory birds in Romania. Parasites Vectors 9:294

Palomar AM, Portillo A, Santibáñez P, Santibáñez S, García-Álvarez L, Oteo JA (2012) Genetic characterization of *Candidatus* Rickettsia vini, a new rickettsia amplified in ticks from La Rioja. Spain. Ticks Tick-Borne Dis 3:319–321

Sándor AD, Kalmár Z, Matei I, Ionică AM, Mărcuţan ID (2017) Urban breeding corvids as disseminators of ticks and emerging tick-borne pathogens. Vector-Borne Zoonotic Dis 17:152–154

Schilling F, Böttcher M, Walter G (1981) Probleme des Zeckenbefalls bei Nestlingen des Wanderfalken (*Falco peregrinus*). J Ornithologie 122(4):359–367

Špitalská E, Literák I, Kocianová E, Taragel'ová V (2011) The importance of *Ixodes arboricola* in transmission of *Rickettsia* spp., *Anaplasma phagocytophilum*, and *Borrelia burgdorferi* sensu lato in the Czech Republic, Central Europe. Vector-Borne Zoonotic Dis 11:1235–1241

White J, Heylen DJA, Matthysen E (2012) Adaptive timing of detachment in a tick parasitizing hole-nesting birds. Parasitology 139:264–270

Ixodes canisuga Johnston, 1849 (Figs. 45–47)

A. D. Sándor

Life-Cycle and Host Preferences

Ixodes canisuga is a three-host, nest-dwelling tick commonly found on domestic dogs and wild carnivores in Europe. It has a wider host range than most endophilic ticks, with mammals (primarily Carnivora, but also four other families) and birds (two families) reportedly being parasitized (Hillyard 1996). In the eastern part of its range, it is also common on most burrowing rodent and leporid species (Haixin et al. 2008). The different developmental stages do not show specific host preferences: all stages may occur at any time of the year on any individual host (Hillyard 1996). This species is not mentioned as feeding on humans (Estrada-Peña and Jongejan 1999).

Ecology

The distribution is primarily limited by the presence of its hosts. In Europe it is more common at low altitudes, and its range is linked to more humid climatic conditions (Sobrino et al. 2012). However, in its eastern range this altitudinal limitation has not been observed, and the species is common even at elevations of 4300 m a.s.l in Quinghai-Tibet Plateau (Haixin et al. 2008). It does not show a clear seasonal dynamic; however, females are more frequently found during the summer (Harris and Thompson 1978). Larvae and nymphs are active all year-round, with their development linked to temperature (Smith 1972), although no peaks are recognisable (Harris and Thompson 1978). Females lay relatively small batches of eggs (up to 400), and the life cycle in nature takes 3 years (Smith 1972).

Distribution

Ixodes canisuga has been reported in most European states, from the British Isles and Iberian Peninsula to eastern Europe, and from Finland to Italy on a north–south axis; furthermore it occurs throughout the cold steppe, boreal and alpine regions of Asia, to the Russian Far East, China and Mongolia. The limits of its distribution are between 32.5–58.6°N and 4.3°W–144.1°E (Kolonin 2009).

Vectorial Capacity and Pathogen Burden

Although it is a common tick of domestic carnivores (especially dogs), there are no studies on the vectorial capacity of this species. *Ixodes canisuga* is suspected to transmit, among other pathogens, *Theileria annae*/*Babesia microti* (Obsomer et al. 2013), *Ehrlichia canis* (Hornok et al. 2012) and *Borrelia burgdorferi* s.l. (Santos-Silva et al. 2011) without further experimental evaluation.

A. D. Sándor (✉)
Department of Parasitology and Parasitic Diseases,
University of Agricultural Sciences and
Veterinary Medicine, Cluj-Napoca, Romania
e-mail: adsandor@gmail.com

A. Estrada-Peña et al. (eds.), *Ticks of Europe and North Africa*,
https://doi.org/10.1007/978-3-319-63760-0_28

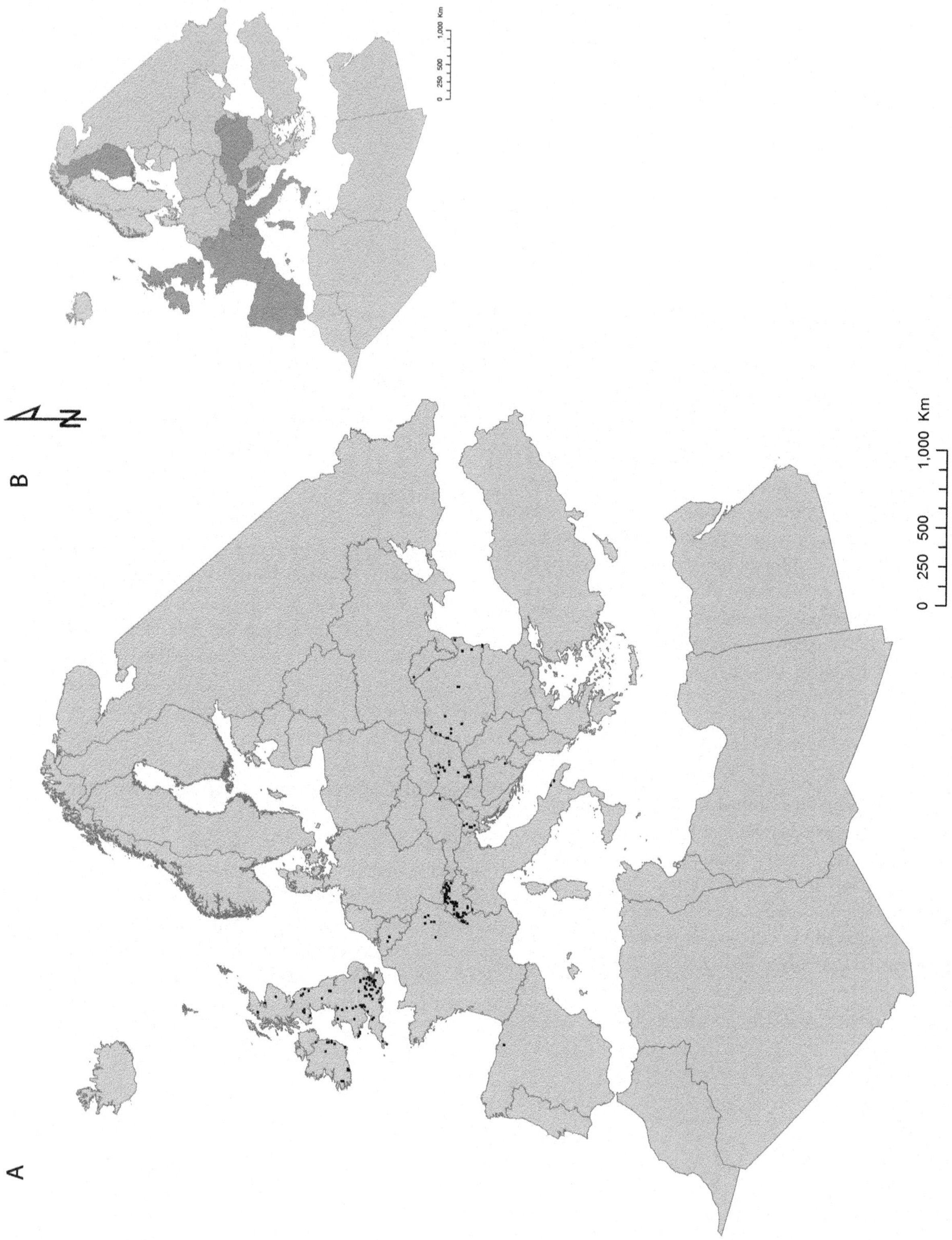

Fig. 45 **A** Distribution of *I. canisuga* in Europe and Northern Africa (10 × 10 km grid presence with black dots). **B** Countries where the species has been reported are marked in dark grey

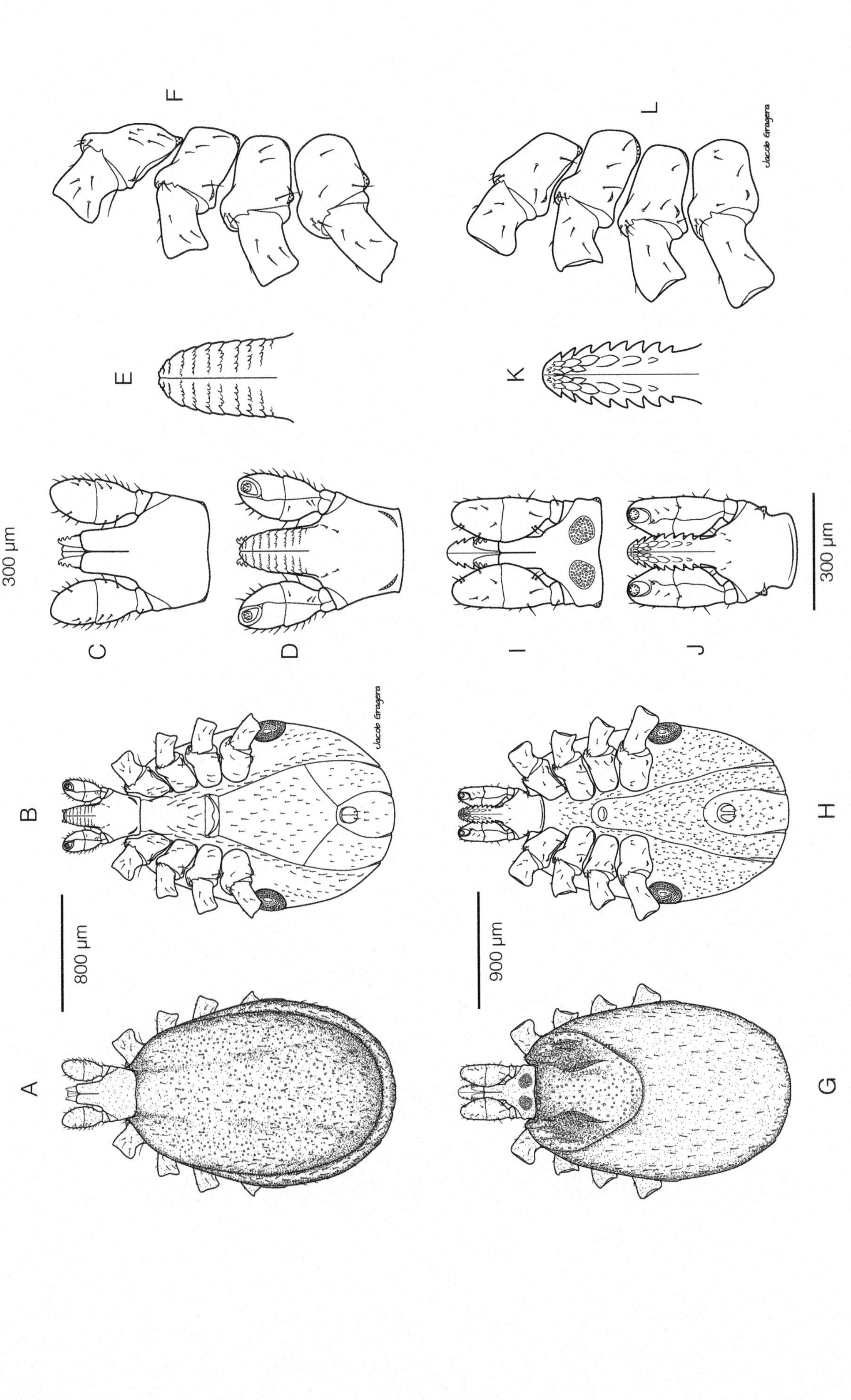

Fig. 46 **A–F** The male of *I. canisuga*. **A** Dorsal, **B** ventral, **C** capitulum, dorsal, **D** capitulum, ventral, **E** hypostome, **F** coxae and trochanters I–IV. **G–L** The female of *I. canisuga*. **G** Dorsal, **H** ventral, **I** capitulum, dorsal, **J** capitulum, ventral, **K** hypostome, **L** coxae and trochanters I–IV. Illustrations from specimens collected in France and Spain

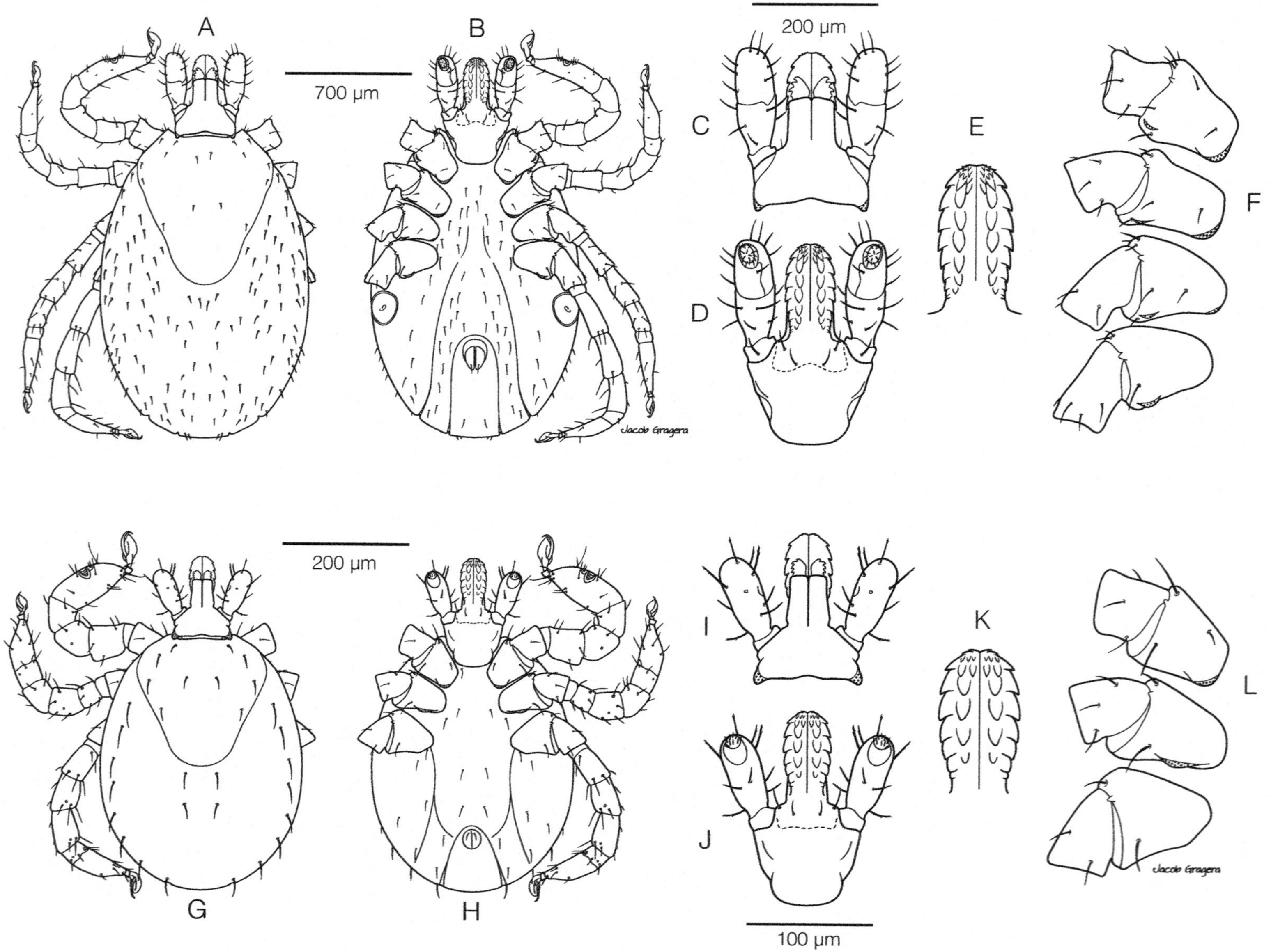

Fig. 47 **A–F** The nymph of *I. canisuga*. **A** Dorsal, **B** ventral, **C** capitulum, dorsal, **D** capitulum, ventral, **E** hypostome, **F** coxae and trochanters I–IV. **G–L** The larva of *I. canisuga*. **G** Dorsal, **H** ventral, **I** capitulum, dorsal, **J** capitulum, ventral, **K** hypostome, **L** coxae and trochanters I–III. Illustrations from specimens collected in France and Spain

References

Estrada-Peña A, Jongejan F (1999) Ticks feeding on humans: a review of records on human-biting Ixodoidea with special reference to pathogen transmission. Exp Appl Acarol 23:685–715

Haixin C, Gonghua L, Jianping S, Yifan C (2008) Host sex and ectoparasite infections of plateau pika (*Ochotona curzoniae*, Hodgson) on the Qinghai-Tibetan Plateau. Polish J Ecol 56:535–539

Harris S, Thompson GB (1978) Populations of the ticks *Ixodes* (*Pholeoixodes*) *hexagonus* and *Ixodes* (*Pholeoixodes*) *canisuga* infesting suburban foxes, *Vulpes vulpes*. J Zool 186:83–93

Hillyard, P. D. (1996). Ticks of North-West Europe. Field Studies Council, Shrewsbury

Hornok S, Fuente J, Horváth G, Fernández de Mera I, Wijnveld M, Tánczos B, Wijnveld M, Tánczos B, Farkas R, Jongejan F (2012) Molecular evidence of *Ehrlichia canis* and *Rickettsia massiliae* in ixodid ticks of carnivores from South Hungary. Acta Vet Hung 61:42–50

Kolonin GV (2009) Fauna of ixodid ticks of the world (Acari, Ixodidae). Moscow, a web-book at www.kolonin.org. Accessed on 22 Feb 2015

Obsomer V, Wirtgen M, Linden A, Claerebout E, Heyman P, Heylen D, Madder M, Maris J, Lebrun M, Tack W, Lempereur L, Hance T, Van Impe G (2013) Spatial disaggregation of tick occurrence and ecology at a local scale as a preliminary step for spatial surveillance of tick-borne diseases: general framework and health implications in Belgium. Parasites Vectors 6:190

Santos-Silva MM, Beati L, Santos AS, De Sousa R, Núncio MS, Melo P, Santos-Reis M, Fonseca C, Formosinho P, Vilela C, Bacellar F (2011) The hard-tick fauna of mainland Portugal (Acari: Ixodidae): an update on geographical distribution and known associations with hosts and pathogens. Exp Appl Acarol 55:85–121

Smith MW (1972) life history of *Ixodes canisuga* (Johnson 1849) under laboratory conditions. Ann Trop Med Parasitol 66:281–286

Sobrino R, Millán J, Oleaga Á, Gortázar C, de la Fuente J, Ruiz-Fons F (2012) Ecological preferences of exophilic and endophilic ticks (Acari: Ixodidae) parasitizing wild carnivores in the Iberian Peninsula. Vet Parasitol 184:248–257

Ixodes crenulatus Koch, 1844 (Fig. 48)

A. Estrada-Peña and T. N. Petney

Emelyanova (1979), quoting Cerný (1969), suggests that *I. crenulatus* is in fact a species group, based on morphological and ecological properties. Filippova considers the variation to involve morphotypes (Filippova and Panova 2000; Petney et al. 2012). The type specimens of this species are lost (Guglielmone et al. 2014) precluding further comparison with closely related species. It is urgent and necessary to prepare a comparative overview of the morphological details of the *Pholeoixodes* parasites of carnivores, including *I. canisuga*, *I. hexagonus*, *I. kaiseri*, *I. rugicollis* and *I. crenulatus*.

Life-Cycle and Host Preferences

Ixodes crenulatus is a typical endophilic, three-host tick species (Cerný 1969). The main hosts are foxes, badgers, hamsters, hedgehogs, mole rats and jackals (Pomerantzev 1950; Filippova and Uspenskaya 1973; Filippova 1977; Balashov 1997; Litvinov and Sapegina 2003), but dogs may also be infested (Zygner and Wedrychowicz 2006). It is occasionally found on birds, especially those contacting the burrows of the more usual hosts (Mukhammdkulov 1970). Negrobov and Borodin (1964) record small numbers of all life history stages from a badger in Germany. This record should be considered doubtful due to the poor definition of this species at the time (Sonenshine et al. 1969; Filippova and Uspenskaya 1973), and Gothe et al. (1977) consider confusion with *I. canisuga* to be likely. Gugliemone et al. (2014) cite as hosts for *I. crenulatus* species of Rodentia (several families), Carnivora (Canidae, Felidae, Mustelidae), Erinaceomorpha (Erinaceidae) and Lagomorpha (Leporidae and Ochotonidae). Other records include Bovidae for adults and nymphs, and Aves (several orders) for immatures. We must stress the difficulty in the validation of these records because of the poor morphological description of the species.

Ecology

Ixodes crenulatus is a nest-dwelling tick. No details about seasonality are available.

Distribution

Although *I. crenulatus* was supposedly described from a specimen found in Germany, it is doubtful whether this species is endemic, it is present in Poland as well as in the steppes of the Caucasus and Central Asia as far as Mongolia and China (Dash et al. 1988; Balashov 1997; Guo et al. 2002; Shi et al. 2004; Nowak-Chmura and Siuda 2012).

Vectorial Capacity and Pathogen Burden

Ixodes crenulatus appears to be involved in some areas in the sylvatic cycle of Russian tick-borne encephalitis virus (Tarasevich et al. 1971).

A. Estrada-Peña (✉)
Faculty of Veterinary Medicine, Department of Animal Health, Zaragoza, Spain
e-mail: aestrada@unizar.es

T. N. Petney
State Museum of Natural History, Erbprinzenstr. 13, 76133 Karlsruhe, Germany
e-mail: trevor.petney@kit.edu

A. Estrada-Peña et al. (eds.), *Ticks of Europe and North Africa*,
https://doi.org/10.1007/978-3-319-63760-0_29

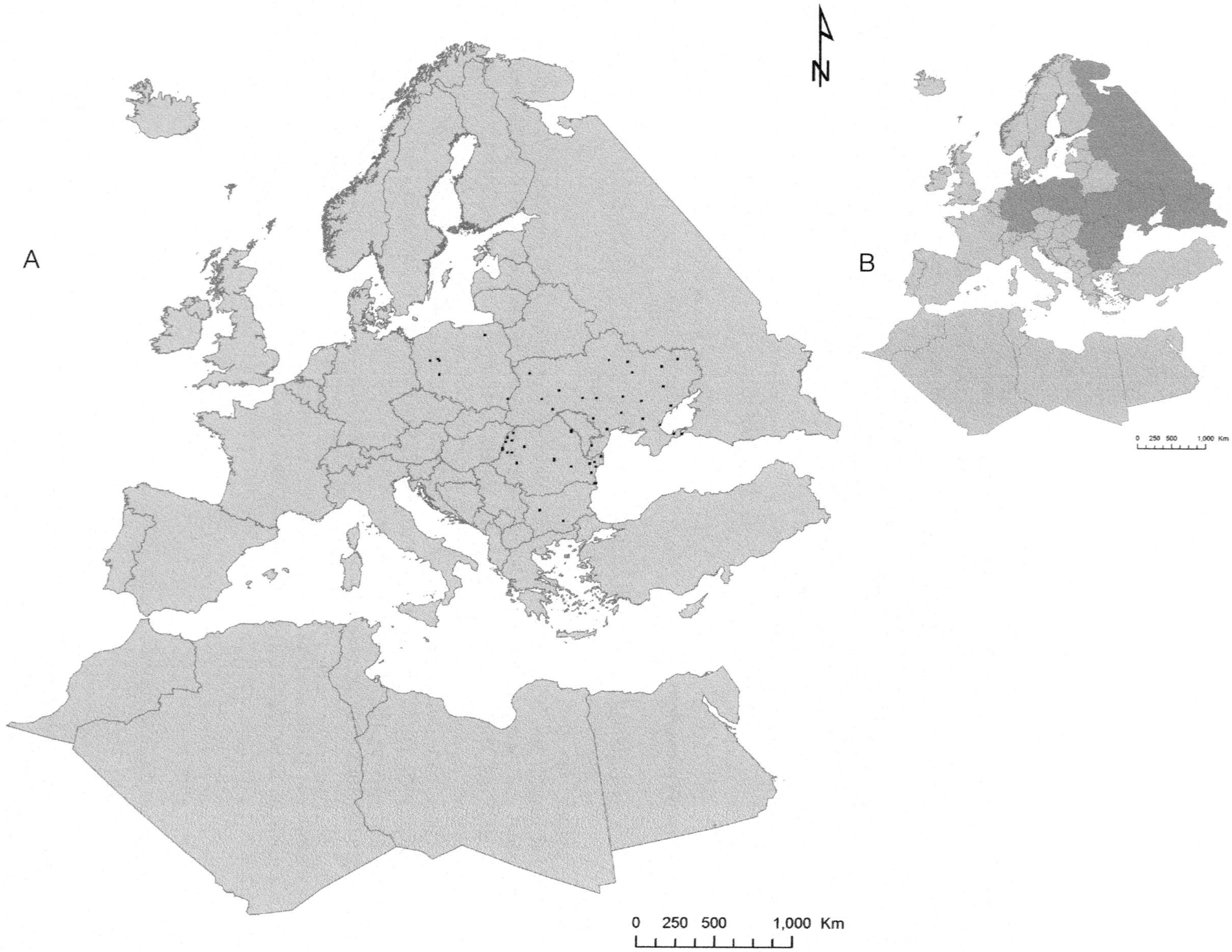

Fig. 48 **A** Distribution of *I. crenulatus* in Europe and Northern Africa (10 × 10 km grid presence with black dots). **B** Countries where the species has been reported are marked in dark grey. Specimens of this species were unavialable for illustrations

References

Balashov YS (1997) Distribution of ixodid ticks (Acarina, Ixodidae) over landscapes within their ranges. Entomol Rev 77:625–637 (in Russian)

Cerný V (1969) About some systematic and nomenclatural problems within the subgenus *Pholeoixodes* (Ixodidae). Folia Parasitologica 16:32

Dash M, Byambaa B, Neronov VM (1988) Ixodidae in Mongolia. Communication 1. Species distribution. Meditsinskaia Parazitologiia I Parazitarnye Bolezni (Moskva), vol 3, pp 37–42 (in Russian)

Emel'yanova ND (1979) Taxonomic status of ixodid ticks of the genus *Pholeoixodes* within the subfamily Ixodinae and its division into subgenera. In: Efimov MV, Pronin NM (eds) Zooparasitology of lake baikal basin. Ulan-Ude, Akad Nauk SSSR, Sib Otd Buryat Filial Otd Biol, pp 5–27

Filippova NA (1977) Ixodid ticks of the subfamily Ixodinae. Fauna SSSR, Leningrad Paukoobraznye 4. Nauka (in Russian)

Filippova NA, Panova IV (2000) Intraspecific differentiation of burrow tick *Ixodes crenulatus* (Ixodidae). Parazitologia 34:265–279

Filippova NA, Uspenskaya IG (1973) *Ixodes kaiseri* Arthur 1957 (Ixodidae) species taxonomic status. Parazitologia 7:297–306

Gothe R, Stendel W, Holm R (1977) Zum Vorkommen von *Ixodes canisuga* Johnston, 1848, in Deutschland, ein Beitrag zur *Ixodes*-Fauna. Z Parasitenkd 53:123–128

Guglielmone AA, Robbins RG, Apanaskevich DA, Petney TN, Estrada-Peña A, Horak IG (2014) The hard ticks of the world 1020 (Acari: Ixodida: Ixodidae). Springer, Dordrecht, p 738

Guo S, Yuan Z, Wu G, Wang W, Ma D, Du H (2002) Epidemiology of ovine theileriosis in Ganan region, Gansu Province, China. Parasitol Res 88 (Suppl.):36–37

Litvinov YN, Sapegina VF (2003) Ectoparasites of the Altay zokor (*Myospalax myospalax*) in northern Altay. Parazitologiya 37: 103–106

Mukhammdkulov M (1970) Wild bird hosts of argasid and ixodid ticks in northern Tadzhikistan. Izvestiya Akademii Nauk Tadzhikskoy SSR, Otdelenie Sel'skokhozyaystvennogo Biologii Nauk, vol 3, pp 87–89 (NAMRU-3 translation 591)

Negrobov VP, Borodin VS (1964) Einige seltene Zeckenfunde im mittleren Teil der DDR. Angewandte Parasitologie 5:107–111

Nowak-Chmura M, Siuda K (2012) Ticks of Poland. Review of contemporary issues and latest research. Ann Parasitol 58:125–155

Petney TN, Pfäffle MP, Skuballa JD (2012) An annotated checklist of the ticks (Acari: Ixodida) of Germany. Syst Appl Acarol 17: 115–170

Pomerantzev BI (1950) Fauna of the USSR. Arachida, vol 4, no. 2, Ixodid ticks (Ixodidae). Moscow, Academy of Sciences USSR

Shi Z, Yang Y, Li Q, Zhao H (2004) Study on species and geographic distribution of ticks in the Gansu Province. Chin J Vet Sci Technol 34:48–49

Sonenshine DE, Kohls GM, Clifford CM (1969) *Ixodes crenulatus* Koch, 1844 synonymy with *I. kaiseri* Arthur, 1957 and redescriptions of the male, female, nymph, and larva (Acarina: Ixodidae). Acarologia 11:193–206

Tarasevich LN, Tagil'tdev AA, Mal'kov GB (1971) Reisolation of tick-borne encephalitis virus from *Ixodes crenulatus* Koch ticks in southern Omsk Oblast. Tezisy Dokladov Voprosy Meditsinskoy Virusologii, Instituta Virusologii imeni Ivanovsky, D. I., Akademiya Meditsinskaya Nauk SSSR, vol 2, pp 134–135 (NAMRU-3 translation 508)

Zygner W, Wedrychowicz H (2006) Occurence of hard ticks in dogs from Warsaw area. Ann Agric Environ Med 13:355–359

Ixodes hexagonus Leach, 1815 (Figs. 49–51)

A. D. Sándor

Life-Cycle and Host Preferences

Ixodes hexagonus is a typical endophilic, three-host species with a natural life cycle of 1–2 years (Arthur 1953). Under laboratory conditions, the life cycle is completed in 142–242 days (Arthur 1951). It is a parasite of hedgehogs (*Erinaceus* spp.), with other hosts mainly being carnivorous mammals from the families Canidae, Mustelidae and Felidae. There seem to be no differences in host preferences between the adult and immature stages (Arthur 1953). This tick has been reported from humans on several occasions (Arthur 1953; Estrada-Peña and Jongejan 1999).

Ecology

Ixodes hexagonus is a nest-dwelling tick, occurring on a broad front all over Europe. It is found wherever its hosts occur (Harris and Thompson 1978). It has been found at various altitudes, ranging from 0 to 2100 m a.s.l. It does not show seasonality; however, females are more frequently found during the summer (Pfäffle et al. 2011). Larvae and nymphs are active all year-round, with their development being linked to temperature (Arthur 1951), although no peaks are recognisable (Pfäffle et al. 2011). This species has developed specific adaptations to its nest-dwelling habitat, with most individuals detaching from hosts in the evening and early morning, thus increasing the chances of successive host relocation (Matuschka et al. 1990). It may occur in high numbers on susceptible host individuals, even causing anaemia in cases of high incidence (Harris and Thompson 1978; Pfäffle et al. 2009).

Distribution

Ixodes hexagonus has been reported from most European countries, being common wherever hedgehogs and red foxes (*Vulpes vulpes*) occur. The limits of its distribution are more or less defined by the borders of the European continent, and lie between 31.3–65.8°N and 9.3°W–36.4°E (Kolonin 2009).

Vectorial Capacity and Pathogen Burden

Although it is an important tick of domestic carnivores, the vectorial capacity of this species is poorly known. *Ixodes hexagonus* was experimentally proven to successfully transmit *Borrelia burgdorferi* s.l. (Gern et al. 1991), with suspected vectorial capacity suggested for *Theileria annae* (Camacho et al. 2003), *Rickettsia helvetica* (Nijhof et al. 2007) and *Anaplasma phagocytophilum* (Skuballa et al. 2010), but without experimental proof. It is suggested to be able to transmit the tick-borne encephalitis virus (Petney et al. 2012).

A. D. Sándor (✉)
Department of Parasitology and Parasitic Diseases, University of Agricultural Sciences and Veterinary Medicine, Cluj-Napoca, Romania
e-mail: adsandor@gmail.com

A. Estrada-Peña et al. (eds.), *Ticks of Europe and North Africa*,
https://doi.org/10.1007/978-3-319-63760-0_30

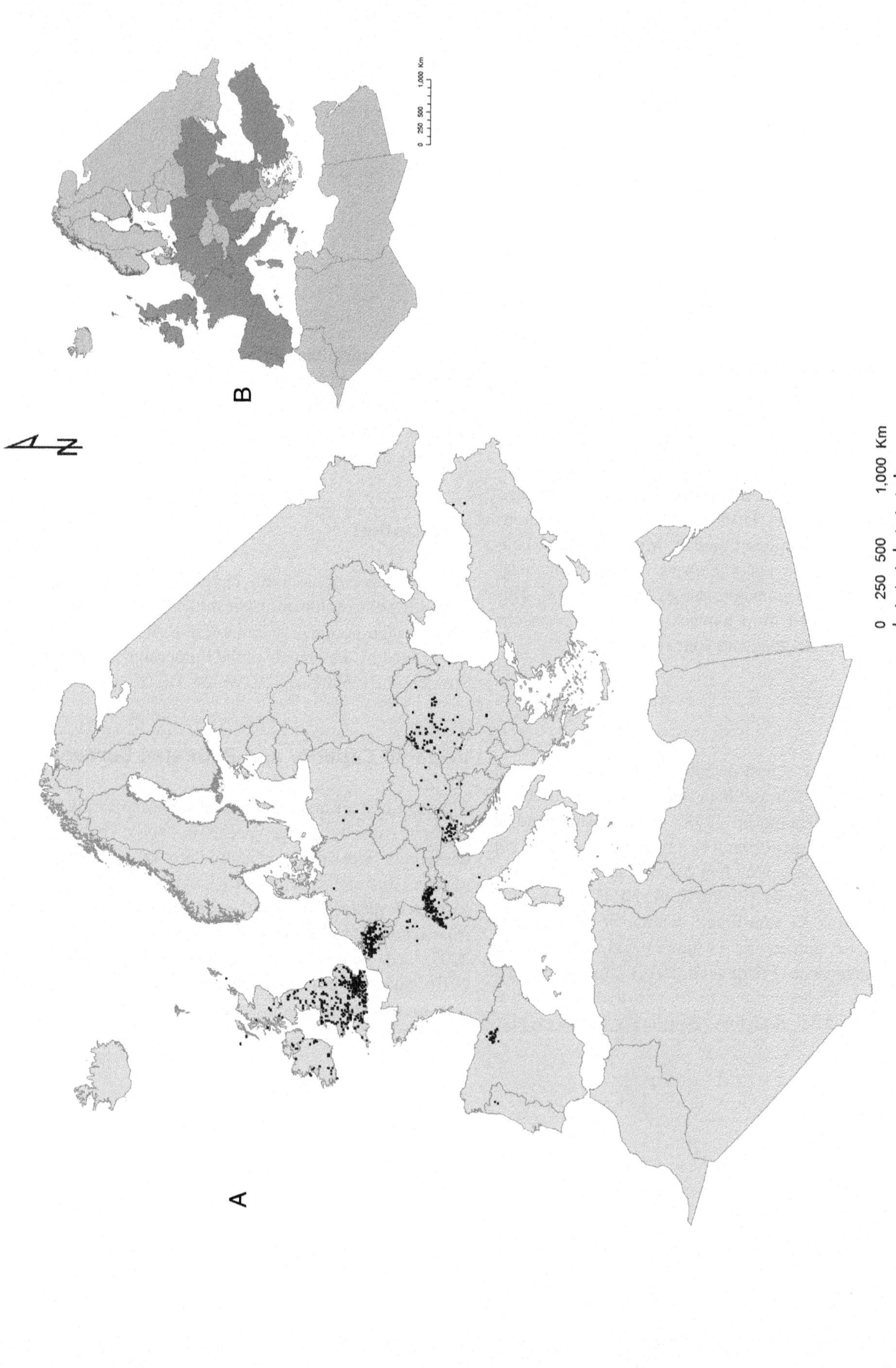

Fig. 49 **A** Distribution of *I. hexagonus* in Europe and Northern Africa (10 × 10 km grid presence with black dots). **B** Countries where the species has been reported are marked in dark grey

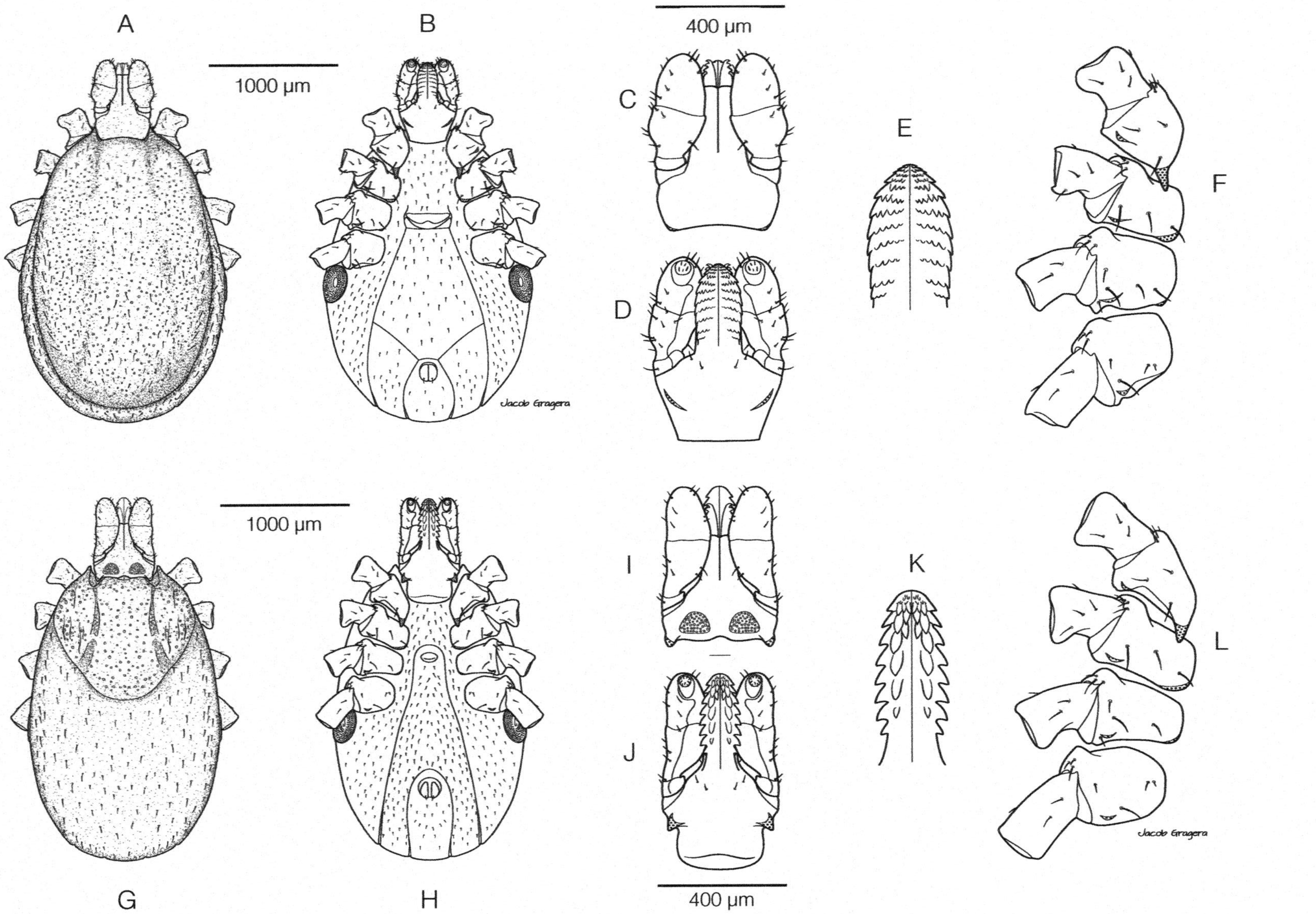

Fig. 50 **A–F** The male of *I. hexagonus*. **A** Dorsal, **B** ventral, **C** capitulum, dorsal, **D** capitulum, ventral, **E** hypostome, **F** coxae and trochanters I–IV. **G–L** The female of *I. hexagonus*. **G** Dorsal, **H** ventral, **I** capitulum, dorsal, **J** capitulum, ventral, **K** hypostome, **L** coxae and trochanters I–IV. Illustrations from specimens collected in Belgium and Spain

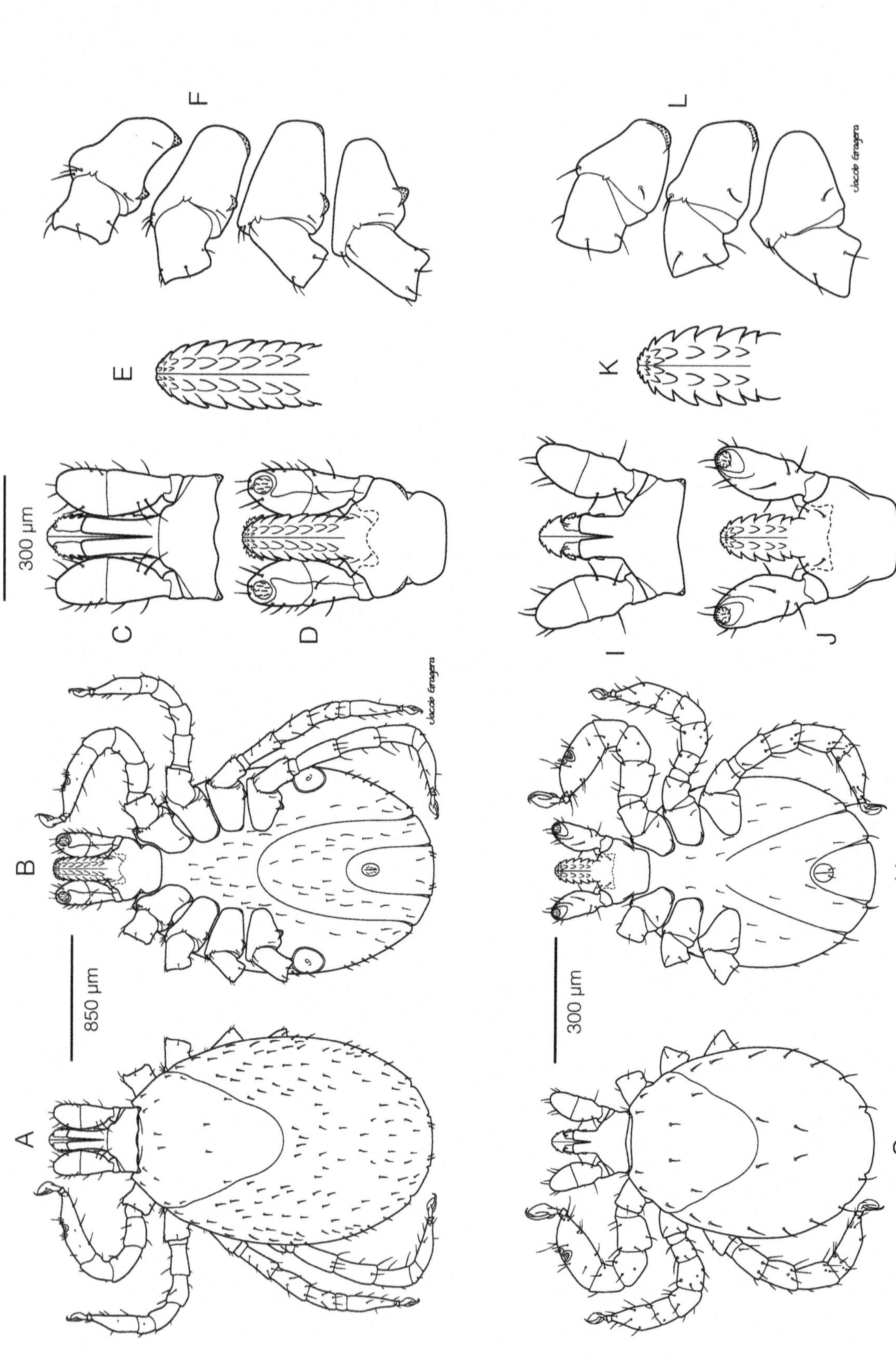

Fig. 51 **A–F** The nymph of *I. hexagonus*. **A** Dorsal, **B** ventral, **C** capitulum, dorsal, **D** capitulum, ventral, **E** hypostome, **F** coxae and trochanters I-IV. **G–L** The larva of *I. hexagonus*. **G** Dorsal, **H** ventral, **I** capitulum, dorsal, **J** capitulum, ventral, **K** hypostome, **L** coxae and trochanters I-III. Illustrations from specimens collected in Belgium and Spain

References

Arthur DR (1951) The bionomics of *Ixodes hexagonus* Leach in Britain. Parasitology 41:82–90

Arthur DR (1953) The host relationships of *Ixodes hexagonus* Leach in Britain. Parasitology 43:227–238

Camacho AT, Pallas E, Gestal JJ, Guitián FJ, Olmeda AS, Telford SR, Spielman A (2003) *Ixodes hexagonus* is the main candidate as vector of *Theileria annae* in north-west Spain. Vet Parasitol 112:157–163

Estrada-Peña A, Jongejan F (1999) Ticks feeding on humans: a review of records on human-biting Ixodoidea with special reference to pathogen transmission. Exp Appl Acarol 23:685–715

Gern L, Toutoungi LN, Hu CM, Aeschlimann A (1991) *Ixodes* (*Pholeoixodes*) *hexagonus*, an efficient vector of *Borrelia burgdorferi* in the laboratory. Med Vet Entomol 5:431–435

Harris S, Thompson GB (1978) Populations of the ticks *Ixodes* (*Pholeoixodes*) *hexagonus* and *Ixodes* (*Pholeoixodes*) *canisuga* infesting suburban foxes, *Vulpes vulpes*. J Zool 186:83–93

Kolonin GV (2009) Fauna of ixodid ticks of the world (Acari, Ixodidae). Moscow, a web-book at www.kolonin.org. Accessed on 22 Feb 2015

Matuschka FR, Richter D, Fischer P, Spielman A (1990) Nocturnal detachment of the tick *Ixodes hexagonus* from nocturnally active hosts. Med Vet Entomol 4:415–420

Nijhof AM, Bodaan C, Postigo M, Nieuwenhuijs H, Opsteegh M, Franssen L, Jebbink F, Jongejan F (2007) Ticks and associated pathogens collected from domestic animals in the Netherlands. Vector-Borne Zoonotic Dis 7:585–596

Petney TN, Pfäffle MP, Skuballa JD (2012) An annotated checklist of the ticks (Acari: Ixodida) of Germany. Syst Appl Acarol 17:115–170

Pfäffle M, Petney T, Elgas M, Skuballa J, Taraschewski H (2009) Tick-induced blood loss leads to regenerative anaemia in the European hedgehog (*Erinaceus europaeus*). Parasitology 136:443–452

Pfäffle M, Petney T, Skuballa J, Taraschewski H (2011) Comparative population dynamics of a generalist (*Ixodes ricinus*) and specialist tick (*I. hexagonus*) species from European hedgehogs. Exp Appl Acarol 54:151–164

Skuballa J, Petney T, Pfäfffie M, Taraschewski H (2010) Molecular detection of *Anaplasma phagocytophilum* in the European hedgehog (*Erinaceus europaeus*) and its ticks. Vector-Borne Zoonotic Dis 10:1055–1057

Ixodes kaiseri Arthur, 1957 (Fig. 52)

A. Estrada-Peña

The original description of *I. kaiseri* by Arthur (1957) and the redescription by Filippova and Uspenskaya (1973) provide information on the taxonomic status of the species in relation to *I. crenulatus*, which was erroneously synonymized under *I. kaiseri* by Sonenshine et al. (1969).

Life-Cycle and Host Preferences

Ixodes kaiseri is a typical endophilic, three-host tick species, with a natural life cycle of probably 1–2 years (Arthur 1965). The cycle has not yet been completed in the laboratory (Arthur 1965) and females die a few days after being collected engorged on hosts. It is a typical parasite of hedgehogs (*Erinaceus* spp.) and red foxes (*Vulpes vulpes*), with other hosts mainly being mammals from the genera *Hyaena*, *Hystrix* and *Felis*. There seem to be no differences in host preference between the adult and immature stages (Arthur 1965).

Ecology

Ixodes kaiseri is a nest-dwelling tick. No details about seasonality are known because of the scarcity of records of this species. It occurs in relatively dry and warm areas.

Distribution

Ixodes kaiseri has been reported in Egypt and Israel (Arthur 1965). The limits of its distribution are unknown and it is possible that the species has been systematically confused with *I. canisuga* (Gothe et al. 1977) or *I. hexagonus*. It seems, however, that *I. kaiseri* replaces *I. hexagonus* in warmer areas.

Vectorial Capacity and Pathogen Burden

These aspects are unknown.

A. Estrada-Peña (✉)
Faculty of Veterinary Medicine, Department of Animal Health, Zaragoza, Spain
e-mail: aestrada@unizar.es

A. Estrada-Peña et al. (eds.), *Ticks of Europe and North Africa*,
https://doi.org/10.1007/978-3-319-63760-0_31

Fig. 52 **A** Distribution of *I. kaiseri* in Europe and Northern Africa (10 × 10 km grid presence with black dots). **B** Countries where the species has been reported are marked in dark grey. Specimens of this species were unavialable for illustrations

References

Arthur DR (1965) Ticks of the genus *Ixodes* in Africa. Athlone Press, London

Gothe R, Stendel W, Holm R (1977) Zum Vorkommen von *Ixodes canisuga* Johnston, 1848, in Deutschland, ein Beitrag zur *Ixodes*-Fauna. Z Parasitenkd 53:123–128

Sonenshine DE, Kohls GM, Clifford CM (1969) *Ixodes crenulatus* Koch, 1844 synonomy with *Ixodes kaiseri* Arthur, 1957 and redescriptions of the male, female, nymph, and larva (Acarina: Ixodidae). Acarologia 11:193–206

Ixodes lividus Koch, 1844 (Figs. 53–55)

A. D. Sándor

Life-Cycle and Host Preferences

Ixodes lividus is a three-host, nest-dwelling, bird parasite typically of the sand martin (*Riparia riparia*). It is commonly found throughout this host's range. This species is typically endophilic, with a natural life cycle of 1–2 years (Ulmanen et al. 1977). Apart from its main host, in Europe it has been collected only from birds which occasionally enter sand martin nests; however, it is a fairly common parasite of house martins (*Delichon urbicum*) in Japan (Uchikawa and Sato 1969). It has not been reported from humans.

Ecology

Ixodes lividus occurs wherever its host nests (i.e. sand and loess walls). It moults and overwinters in the nest burrows of its hosts. It shows a pronounced seasonality: females are more frequently found in summer on nestling and juvenile hosts, while larvae and nymphs are found in spring on adult sand martins (Ulmanen et al. 1977). Adult males are found only in nest burrows. It may occur in high numbers on susceptible host individuals, causing anaemia and reducing breeding success and nestling condition (Szép and Moller 1999).

Distribution

The distribution of *I. lividus* overlaps with the breeding (and, to some extent, the migratory) distribution of its host species. Its range covers most of the Palaearctic, from the British Isles to Japan and the Russian Far East. The limits of its distribution are between 34.2–71.6°N and 10.4°W–145.7°E (Kolonin 2009).

Vectorial Capacity and Pathogen Burden

Although it is a tick with one dominant host species, *I. lividus* has been found to harbour several important zoonotic pathogens. While several studies have documented the occurrence of a number of viruses and bacteria, the vectorial capacity of this species is poorly known. It is suspected to transmit spotted fever *Rickettsia* sp. IXLI1 (Graham et al. 2010) and a similar *Rickettsia* sp. (Makhmetov 1968), *Borrelia burgdorferi* s.s. and *Borrelia garinii* (Movila et al 2008), together with a wide variety of viruses, such as tick-borne encephalitis, West Nile, Kama, Tyuleni-antigenic, Gadget Gully (Iakimenko et al. 1990; L'vov et al. 1998), all missing experimental transmission trials.

A. D. Sándor (✉)
Department of Parasitology and Parasitic Diseases, University of Agricultural Sciences and Veterinary Medicine, Cluj-Napoca, Romania
e-mail: adsandor@gmail.com

A. Estrada-Peña et al. (eds.), *Ticks of Europe and North Africa*,
https://doi.org/10.1007/978-3-319-63760-0_32

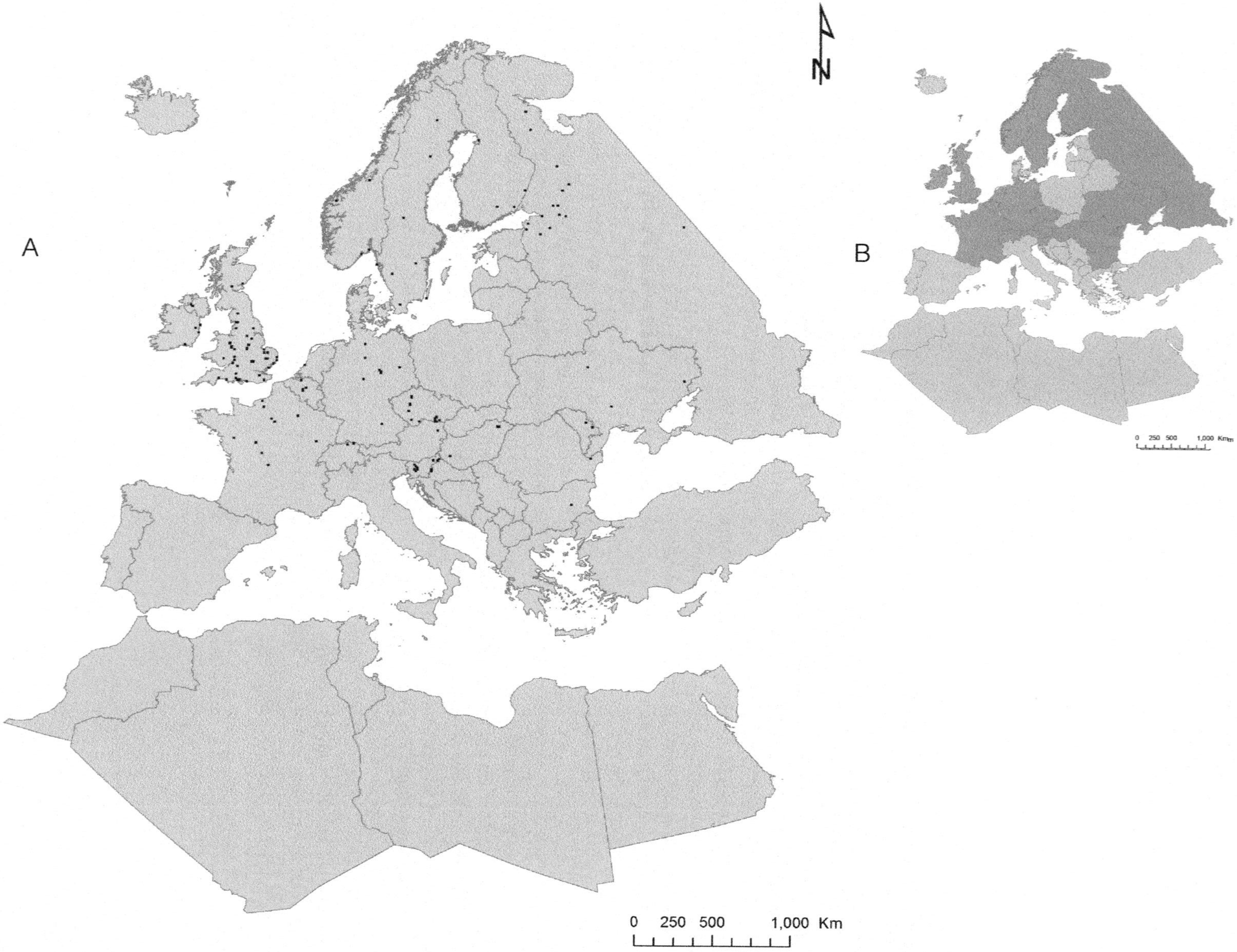

Fig. 53 **A** Distribution of *I. lividus* in Europe and Northern Africa (10 × 10 km grid presence with black dots). **B** Countries where the species has been reported are marked in dark grey

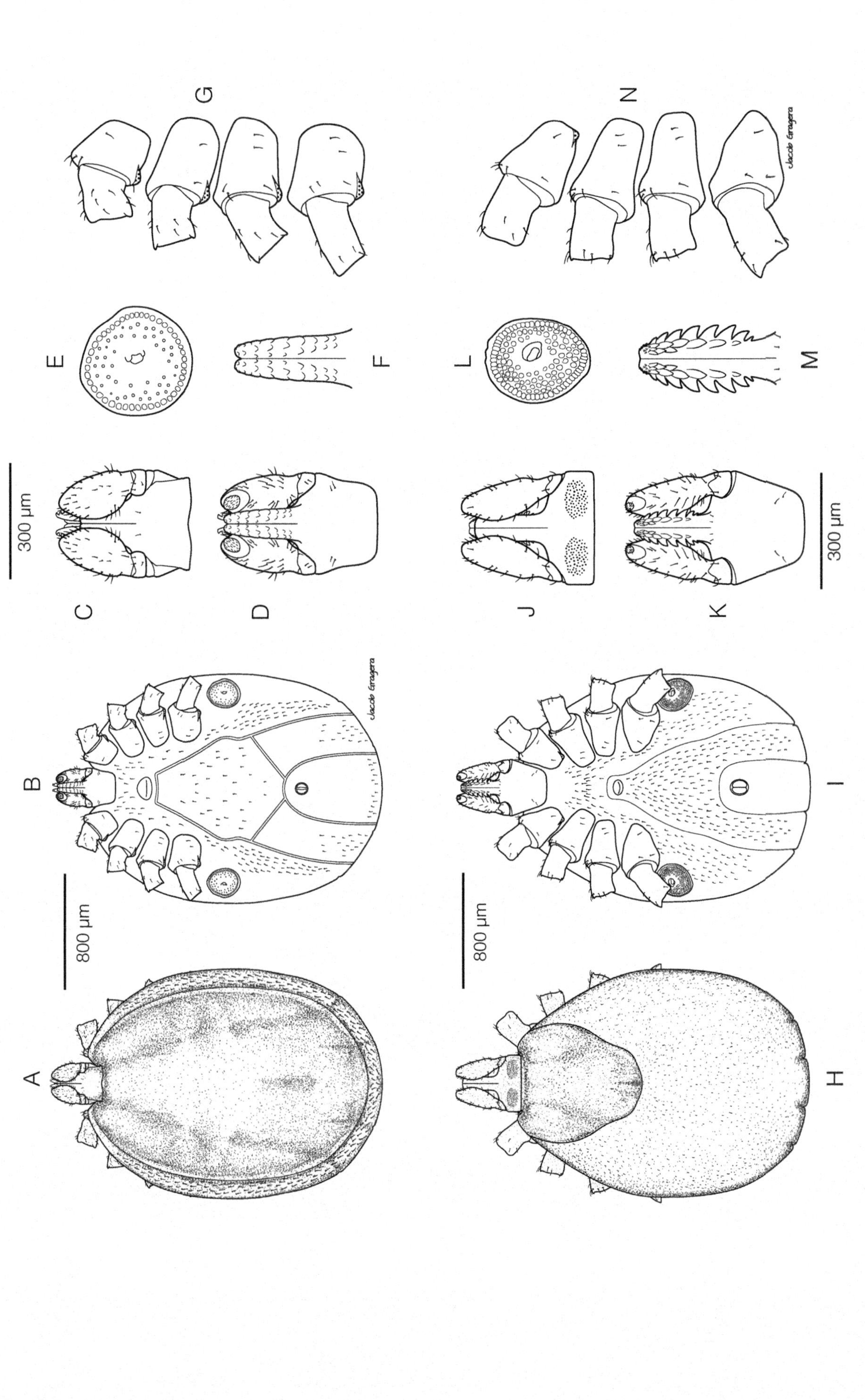

Fig. 54 **A–G** The male of *I. lividus*. **A** Dorsal, **B** ventral, **C** capitulum, dorsal, **D** capitulum, ventral, **E** spiracular plate, **F** hypostome, **G** coxae and trochanters I–IV. **H–N** The female of *I. lividus*. **H** Dorsal, **I** ventral, **J** capitulum, dorsal, **K** capitulum, ventral, **L** spiracular plate, **M** hypostome, **N** coxae and trochanters I–IV. Illustrations from specimens collected in the United Kingdom

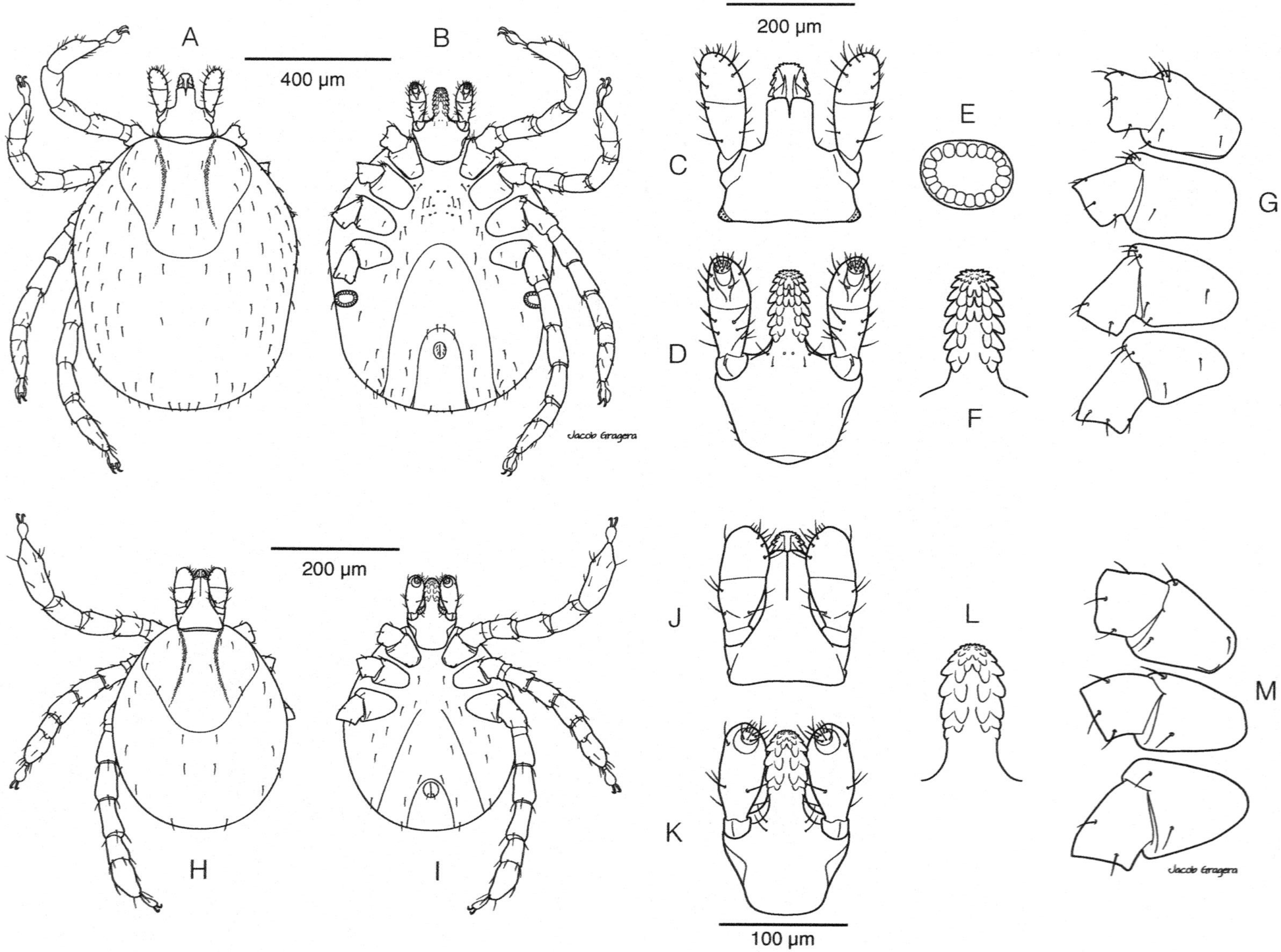

Fig. 55 **A–F** The nymph of *I. lividus*. **A** Dorsal, **B** ventral, **C** capitulum, dorsal, **D** capitulum, ventral, **E** hypostome, **F** coxae and trochanters I–IV. **G–L** The larva of *I. lividus*. **G** Dorsal, **H** ventral, **I** capitulum, dorsal, **J** capitulum, ventral, **K** hypostome, **L** coxae and trochanters I–III. Illustrations from specimens collected in the United Kingdom

References

Graham RI, Mainwaring MC, du Feu R (2010) Detection of spotted fever group *Rickettsia* spp. from bird ticks in the United Kingdom. Med Vet Entomol 24:340–343

Kolonin GV (2009) Fauna of ixodid ticks of the world (Acari, Ixodidae). Moscow, a web-book at www.kolonin.org. Accessed on 22 Feb 2015

L'vov DK, Aristova VA, Gromashevskiĭ VL, Skvortsova TM, Boĭko VA, Mel'nikova EE, Guishchina EA, Iarulova RA, Morozova TN, Petrova ES (1997) [Kama, a new virus (*Flaviviridae, Flavivirus*, Tiulenii antigenic group), isolated from *Ixodes lividus* ticks]. Voprosy virusologii 43:71–74 (in Russian)

Makhmetov MM (1968) Spontaneous infection of *Rickettsia burnetii* in ectoparasites of the sand martin. Naval Medical Research Unit No 3 Cairo (Egypt) Dept Of Medical Zoology, pp 1–7

Movila A, Gatewood A, Toderas I, Duca M, Papero M, Uspenskaia I, Conovalov J, Fish D (2008) Prevalence of *Borrelia burgdorferi* sensu lato in *Ixodes ricinus* and *I. lividus* ticks collected from wild birds in the Republic of Moldova. Int J Med Microbiol 298:149–153

Szép T, Møller AP (1999) Cost of parasitism and host immune defence in the sand martin *Riparia riparia*: a role for parent-offspring conflict? Oecologia 119:9–15

Iakimenko VV, Bogdanov II, Tagil'tsev, AA, Drokin DA, Kalmin OB (1990) [The characteristics of the relationships of arthropods of the refuge complex with the causative agents of transmissible viral infections in bird rookeries]. Parazitologiia 25:156–162 (in Russian)

Uchikawa K, Sato A (1969) The occurrence of *Argas japonicus* and *Ixodes lividus* in Nagano prefecture, Japan (Ixodoidea: Argasidae; Ixodidae). J Med Entomol 6:95–97

Ulmanen I, Saikku P, Vikberg P, Sorjonen J (1977) *Ixodes lividus* (Acari) in sand martin colonies in Fennoscandia. Oikos 1:20–26

Ixodes rugicollis Schulze and Schlottke, 1929 (Fig. 56)

M. P. Pfäffle and T. N. Petney

Life Cycle and Host Preferences

In general, very little is known about the biology of this tick. Information about the three-host life cycle comes mainly from Aubert (1981) from laboratory studies. At temperatures of 18–22 °C and a high relative humidity (vials with ticks were covered with a damp cloth) the life cycle is completed within 5.3–8.3 months, which is comparable to other related species such as *I. hexagonus* and *I. canisuga* (Aubert 1981). *I. rugicollis* uses small mustelids (e.g. *Martes foina*, *M. martes*, *Mustela nivalis*), and other carnivores, including red foxes (*Vulpes vulpes*), cats and dogs as hosts (Aubert 1976, 1981; Toutoungi et al. 1991; Kolonin 2007; Christian 2010; Siuda et al. 2010; Visser et al. 2011).

Ecology

Ixodes rugicollis is a nidicolous tick, living in the nests of its hosts (Aubert 1981). It does not seem to have altitude preferences, as it was found at various altitudes (422, 510, 1036 m) in Switzerland by Toutoungi et al. (1991). Overall information about the ecology of this tick is rare.

Distribution

This species has been reported from Austria, France, Germany, Poland, Romania and Switzerland (Aubert 1976; Toutoungi et al. 1991; Christian 2010; Siuda et al. 2010, 2006; Coipan et al. 2012; Visser et al. 2011).

Vectorial Capacity and Pathogen Burden

Almost nothing is known about the role of this species as a vector, although Aubert (1982) indicates experimental transmission of rabies virus to a ferret.

M. P. Pfäffle (✉)
Department of Ecology and Parasitology, Institute of Zoology, Karlsruhe Institute of Technology, Karlsruhe, Germany
e-mail: miripfaeffle@web.de

T. N. Petney
State Museum of Natural History, Erbprinzenstr. 13, 76133 Karlsruhe, Germany
e-mail: trevor.petney@kit.edu

A. Estrada-Peña et al. (eds.), *Ticks of Europe and North Africa*,
https://doi.org/10.1007/978-3-319-63760-0_33

Fig. 56 **A** Distribution of *I. rugicollis* in Europe and Northern Africa (10 × 10 km grid presence with black dots). **B** Countries where the species has been reported are marked in dark grey. Specimens of this species were unavialable for illustrations

References

Aubert MFA (1976) Contribution à l'étude du parasitisme du renard (*Vulpes vulpes* L.) par les Ixodidae (Acarina) dans le Nord-Est de la France. Acarologia 17:452–479

Aubert MFA (1981) Breeding of the tick *Ixodes* (*Pholeoixodes*) *rugicollis* (Acari, Ixodidae) under laboratory conditions. J Med Entomol 18:324–327

Aubert MF (1982) Experimental transmission of rabies to the ferret by an Ixodoidea: *Pholeoixodes rugicollis* (Schultze and Schlottke, 1929). Acarologia 23:125–132

Christian A (2010) Tick infestation (*Ixodes*) on feral mink (*Neovison vison*) in Central Germany. Soil Org 82:209–216

Coipan EC, Vladimirescu AF, Ciolpan O, Teodorescu I (2012) Tick species (Acari: Ixodoidea) distribution, seasonality and host associations in Romania. Trav du Muséum Natl d'Histoire Nat "Grigore Antipa" 54:301–317. https://dx.doi.org/10.2478/v10191-011-0018-y

Kolonin GV (2007) Mammals as hosts of Ixodid ticks (Acarina, Ixodidae). Entomol Rev 87:401–412. https://dx.doi.org/10.1134/S0013873807040033

Siuda K, Majszyk A, Nowak M (2006) Ticks (Acari: Ixodida) parasitizing birds (Aves) in Poland. Biol Lett 43:147–151

Siuda K, Nowak M, Gierczak M (2010) Confirmation of occurrence of *Ixodes* (*Pholeoixodes*) *rugicollis* Schulze et Schlottke, 1929 (Acari: Ixodidae) in Poland, including the morphological description and diagnostic features of this species. Wiadomości Parazytol 56:77–80

Toutoungi LN, Gern L, Aeschlimann A, Debrot S (1991) A propos du genre *Pholeoixodes*, prasite des carnivores. Acarologia 32:311–328

Visser M, Massner C, Rehbein S (2011) Massive infestation with fur mites *Lynxacarus mustelae* of a stone marten (*Martes foina*) from Tyrol. Wiener Klinische Wochenschrift (Suppl 1): 36–42. https://dx.doi.org/10.1007/s00508-011-0005-0

Ixodes trianguliceps Birula, 1895 (Figs. 57–59)

M. P. Pfäffle, T. N. Petney, and M. Madder

Life-Cycle and Host Preferences

The life cycle of *I. trianguliceps* may last from 2 to 5 years, depending on the climatic conditions (Balashov 1997). Although males are rarely found on trapped hosts, copulation seems to take place on-host, rather than off-host (Randolph 1975, 1980). *Ixodes trianguliceps* is a tick of small mammals, primarily of rodents and small insectivores (Ulmanen 1972; O'Donnell 1973; Nilsson 1974; Kolonin 2007). It is a rare, accidental parasite of humans (Hillyard 1996).

Ecology

Ixodes trianguliceps is a nidicolous species that is confined to its hosts' nests and burrows (Bown et al. 2006). Unfed nymphs and females can sometimes be found waiting for hosts on the soil surface (Graf et al. 1979; Balashov 1997). Larvae, however, are exophilic and attack their hosts in plant litter (Hillyard 1996). All life history stages of *I. trianguliceps* seem to be active throughout the year (Ulmanen 1972; Nilsson 1974). However, activity peaks seem to vary depending on the region (Graf et al. 1979). In Romania ticks were found mainly in spring and autumn (Coipan et al. 2012). Seasonal activity in north-west England seems to be triggered by rising spring temperatures, and activity peaks occur in July and October (Hussein 1980). In an earlier study from north England, O'Donnell (1973) reported the activity of larvae being the highest in mid-autumn, with the density building up in summer, while nymphs are active from late spring to autumn, with peak activities in mid-summer. Females have been reported from April to October. Other studies from Finland and the United Kingdom show that infestation of *I. trianguliceps* on rodents has two peaks, one in early summer and one in autumn, while larvae and adults show a clear bimodal activity pattern and nymphs have their highest activity in early summer (Cotton and Watts 1967; Ulmanen 1972; Nilsson 1974). Randolph (1975) reports a peak level of larval activity in autumn/winter and low larval activity through the rest of the year in the south of England. *Ixodes trianguliceps* is found in a variety of different habitats, such as moorland, meadows, peat mosses, bogs and clearings and different forest types, such as mixed forests, pine and deciduous forests, dark-coniferous forests, mountain birch forests and treeless zones at high altitude (up to 2400 m in Switzerland) (Korenberg and Lebedeva 1969; Ulmanen 1972; Nilsson 1974; Hillyard 1996). It is one of the most cold-resistent *Ixodes* species in the Palaearctic region (Nilsson 1974).

Distribution

Ixodes trianguliceps has a wide geographical distribution, occurring from Ireland to the vicinity of Lake Baikal and throughout much of Eastern Europe to the Yenisei River, which flows through Siberia from Mongolia to the Arctic Ocean (O'Donnell 1973; Balashov 1997). It has been reported from Austria, Bulgaria, Croatia, Czech Republic, Denmark, England, Finland, Germany, Hungary, Italy, Norway, Poland, Romania, Russia, Slovakia, Sweden and Switzerland (Korenberg and Lebedeva 1969; Aeschlimann et al. 1970; Nilsson 1974; Graf et al. 1979; Jaenson et al. 1994; Balashov 1997; Guryčová 1998; Nowak-Chmura and

M. P. Pfäffle (✉)
Department of Ecology and Parasitology, Institute of Zoology, Karlsruhe Institute of Technology, Karlsruhe, Germany
e-mail: miripfaeffle@web.de

T. N. Petney
State Museum of Natural History, Erbprinzenstr. 13, 76133 Karlsruhe, Germany
e-mail: trevor.petney@kit.edu

M. Madder
DG06/07 Ruisseau Creole, La Mivoie, Black River, 90625, Mauritius
e-mail: Maxime.Madder@clinglobal.com

A. Estrada-Peña et al. (eds.), *Ticks of Europe and North Africa*,
https://doi.org/10.1007/978-3-319-63760-0_34

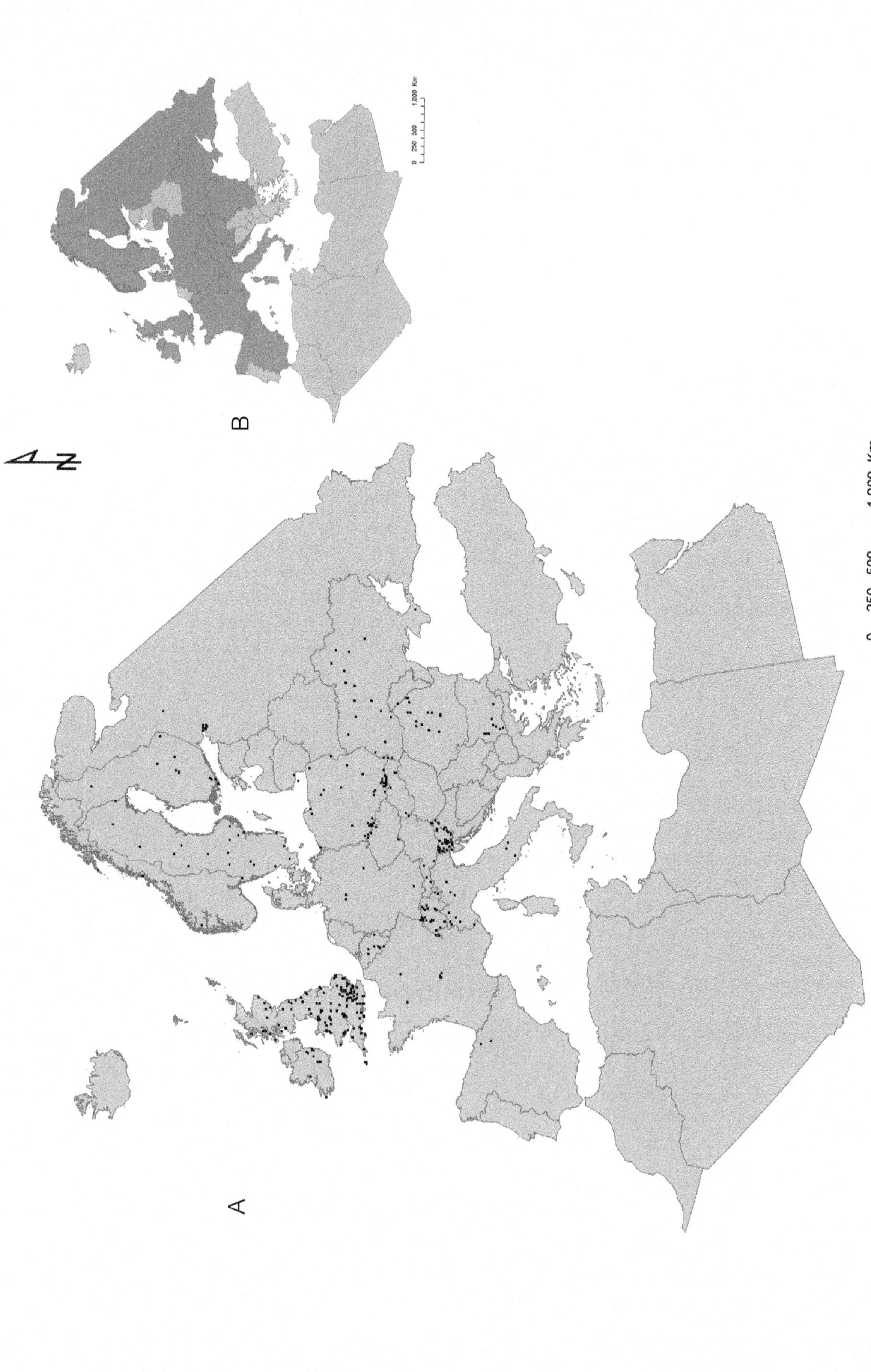

Fig. 57 **A** Distribution of *I. trianguliceps* in Europe and Northern Africa (10 × 10 km grid presence with black dots). **B** Countries where the species has been reported are marked in dark grey

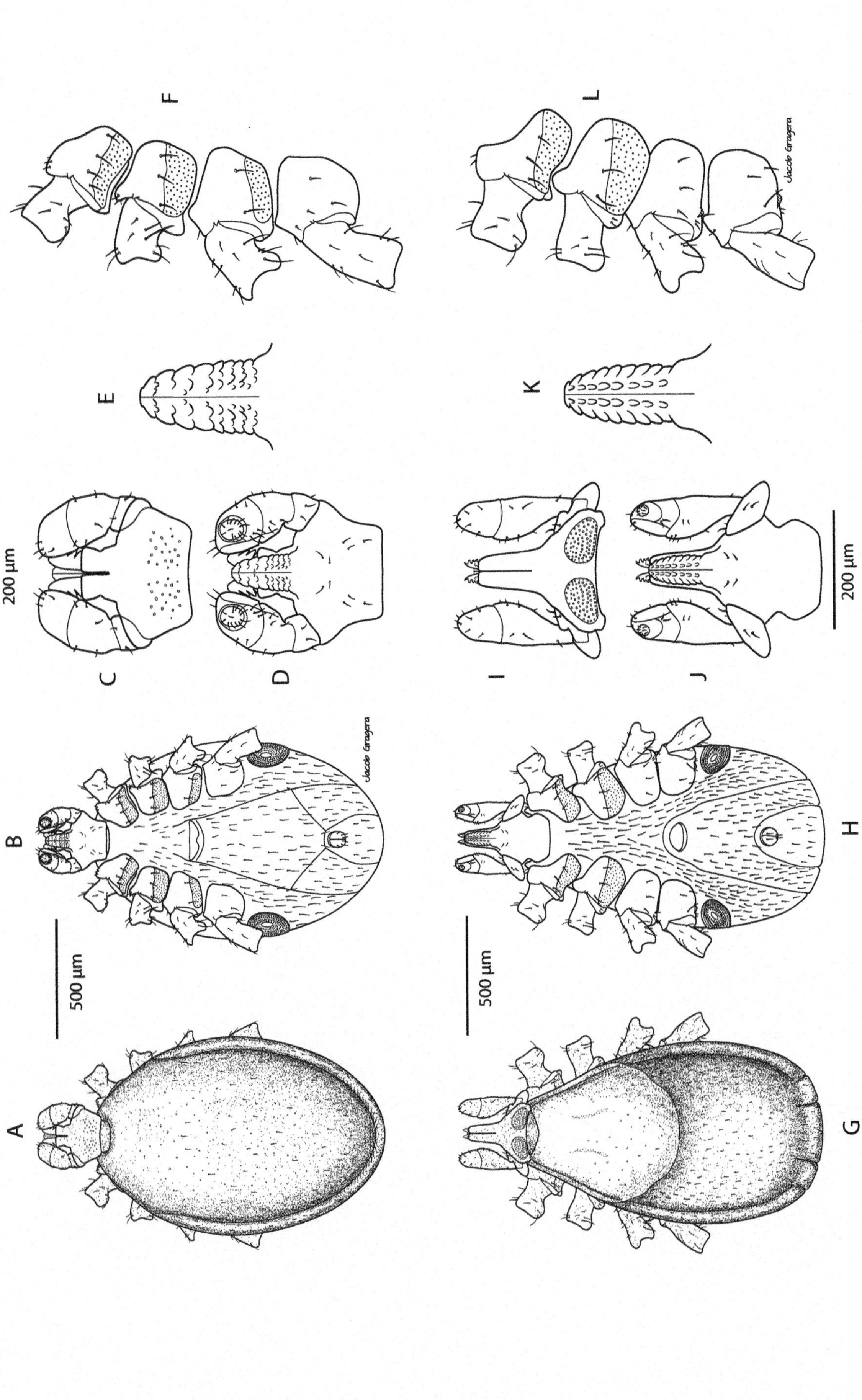

Fig. 58 **A–F** The male of *I. trianguliceps*. **A** Dorsal, **B** ventral, **C** capitulum, dorsal, **D** capitulum, ventral, **E** hypostome, **F** coxae and trochanters I–IV. **G–L** The female of *I. trianguliceps*. G Dorsal, **H** ventral, **I** capitulum, dorsal, **J** capitulum, ventral, **K** hypostome, **L** coxae and trochanters I–IV. Illustrations from specimens collected in France, Germany, and Spain

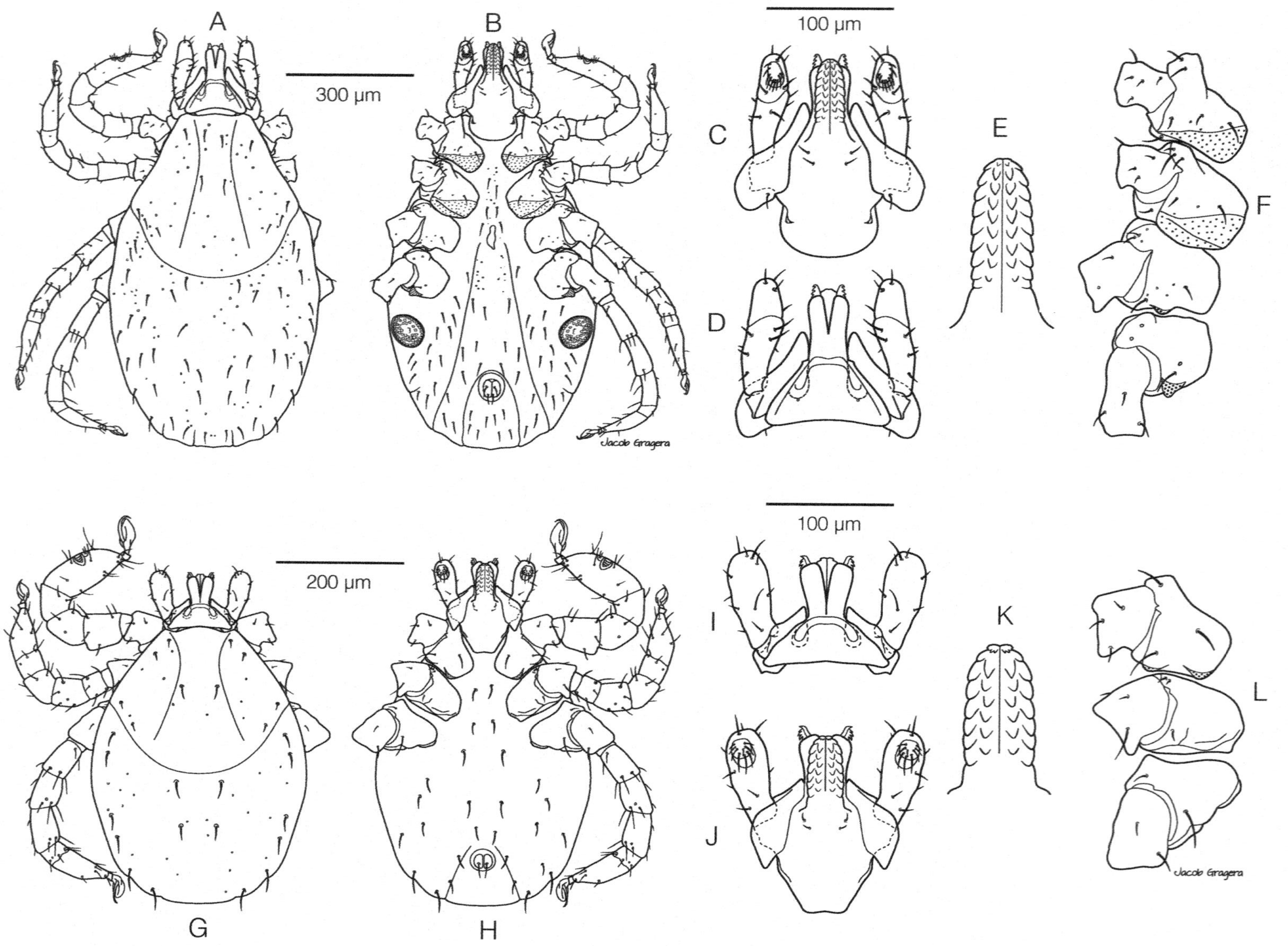

Fig. 59 **A–F** The nymph of *I. trianguliceps*. **A** Dorsal, **B** ventral, **C** capitulum, dorsal, **D** capitulum, ventral, **E** hypostome, **F** coxae and trochanters I–IV. **G–L** The larva of *I. trianguliceps*. **G** Dorsal, **H** ventral, **I** capitulum, dorsal, **J** capitulum, ventral, **K** hypostome, **L** coxae and trochanters I–III. Illustrations from specimens collected in France, Germany and Spain

Siuda 2012; Coipan et al. 2012). Its distribution is restricted to the northernmost parts of Finland, Norway and Sweden and the high mountain areas of Norway and Sweden (Nilsson 1974). With the exception of *Ixodes uriae*, it is the tick with the northernmost distribution on the Scandinavian peninsula (Jaenson et al. 1994).

Vectorial Capacity and Pathogen Burden

Ixodes trianguliceps is probably the main vector of *Babesia microti* to small rodents and is involved in the sylvatic cycle of *Anaplasma phagocytophilum* in the United Kingdom (Hussein 1980; Ogden et al. 1998; Bown et al. 2003, 2006, 2008). *Borrelia afzelii* and *B. garinii* have been isolated from this tick (Gorelova et al. 1996; Postic et al. 1997; Nefedova et al. 2005). *Francisella tularensis* has also been isolated from this species (Guryčová 1998). Aeschlimann et al. (1970) discuss its role in maintaining the enzootic cycle of tick-borne encephalitis virus in Switzerland.

References

Aeschlimann NA, Blittiker W, Diehl PA, Eichenberger G, Immler R, Weiss N (1970) Présence d'*Ixodes trianguliceps* (Birula, 1895) et d'*Ixodes apronophorus* (Schulze, 1924) en Suisse (Ixodoidea; Ixodidae). Rev Suisse Zool 77:527–536

Balashov YS (1997) Distribution of ixodid ticks (Acarina, Ixodidae) over landscapes within their ranges in Russia. Entomol Rev 76:625–637

Bown KJ, Begon M, Bennett M, Woldehiwet Z, Ogden NH (2003) Seasonal dynamics of *Anaplasma phagocytophila* in a rodent-tick (*Ixodes trianguliceps*) system, United Kingdom. Emerg Infect Dis 9:63–70

Bown KJ, Begon M, Bennett M, Birtles RJ, Burthe S, Lambin X, Telfer S, Woldehiwet Z, Ogden NH (2006) Sympatric *Ixodes trianguliceps* and *Ixodes ricinus* ticks feeding on field voles (*Microtus agrestis*): potential for increased risk of *Anaplasma phagocytophilum* in the United Kingdom? Vector Borne Zoonotic Dis 6:404–10. https://dx.doi.org/10.1089/vbz.2006.6.404

Bown KJ, Lambin X, Telford GR, et al (2008) Relative importance of *Ixodes ricinus* and *Ixodes trianguliceps* as vectors for *Anaplasma phagocytophilum* and *Babesia microti* in field vole (*Microtus agrestis*) populations. Appl Environ Microbiol 74:7118–25. https://dx.doi.org/10.1128/AEM.00625-08

Coipan EC, Vladimirescu AF, Ciolpan O, Teodorescu I (2012) Tick species (Acari: Ixodoidea) distribution, seasonality and host associations in Romania. Trav du Muséum Natl d'Histoire Nat "Grigore Antipa" 54:301–317. https://dx.doi.org/10.2478/v10191-011-0018-y

Cotton MJ, Watts CHS (1967) The ecology of the tick *Ixodes trianguliceps* Birula (Arachnida; Acarina; Ixodoidea). Parasitology 57:525–531. https://dx.doi.org/10.1017/S0031182000072401

Gorelova NB, Korenberg EI, Kovalevskiĭ IV et al (1996) Isolation of *Borrelia* from the tick *Ixodes trianguliceps* (Ixodidae) and the significance of this species in the epizootiology of ixodid tick-borne borrelioses. Parazitologiia 30:13–18

Graf J-F, Mermod C, Aeschlimann A (1979) Nouvelles données concernant la distribution, l'écologie et la biologie d'*Ixodes* (*Exopalpiger*) *trianguliceps* (Birula, 1895) en Suisse (Ixodoidea, Ixodidae). Bull Soc Neuchâtel Sci Nat 102:55–68

Guryčová D (1998) First isolation of *Francisella tularensis* subsp. *tularensis* in Europe. Eur J Epidemiol 14:797–802. https://dx.doi.org/10.1023/A:1007537405242

Hillyard PD (1996) Ticks of North-West Europe. Field Studies Council, Shrewsbury

Hussein HS (1980) *Ixodes trianguliceps*: seasonal abundance and role in the epidemiology of *Babesia microti* infection in north-western England. Ann Trop Med Parasitol 74:531–539

Jaenson TGT, Tälleklint L, Lundqvist L et al (1994) Geographical distribution, host associations, and vector roles of ticks (Acari: Ixodidae, Argasidae) in Sweden. J Med Entomol 31:240–256. https://dx.doi.org/10.1093/jmedent/31.2.240

Korenberg EI, Lebedeva NN (1969) Distribution and some general features of the ecology of *Ixodes trianguliceps* Bir. in the Soviet Union. Folia Parasitol (Praha) 16:143–152

Kolonin GV (2007) Mammals as hosts of Ixodid ticks (Acarina, Ixodidae). Entomol Rev 87:401–412. https://doi.org/10.1134/S0013873807040033

Nefedova VV, Korenberg EI, Fadeeva IA, Gorelova NB (2005) Genetic characteristics of human pathogenic *Borrelia* isolated from *Ixodes trianguliceps* Bir. and *Ixodes pavlovskyi* Pom. Med Parazitol (Mosk) 9–12

Nilsson A (1974) Distribution, host relations, and seasonal occurrence of *Ixodes trianguliceps* Birula (Acari) in Fennoscandia. Folia Parasitol (Praha) 21:233–241

Nowak-Chmura M, Siuda K (2012) Ticks of Poland. Review of contemporary issues and latest research. Ann Parasitol 58:125–155

O'Donnell TG (1973) Density and Host Relations of *Ixodes trianguliceps* Bir. in Woodland and Grassland Areas. Proc 3rd Int Congr Acarol 783–787. https://dx.doi.org/10.1007/978-94-010-2709-0

Ogden NH, Bown K, Horrocks BK, Woldehiwet Z, Bennett M (1998) Granulocytic *Ehrlichia* infection in Ixodid ticks and mammals in woodlands and uplands of the U.K. Med Vet Entomol 12:423–429

Postic D, Korenberg E, Gorelova N, et al (1997) *Borrelia burgdorferi* sensu lato in Russia and neighbouring countries: high incidence of mixed isolates. Res Microbiol 148:691–702. https://dx.doi.org/10.1016/S0923-2508(99)80068-0

Randolph SE (1975) Seasonal dynamics of a host-parasite system: *Ixodes trianguliceps* (Acarina: Ixodidae) and its small mammal hosts. J Animal Ecol 44:425–449

Randolph SE (1980) Population regulation in ticks: the effect of delayed mating on fertility in *Ixodes trianguliceps* (Acarina: Ixodidae). J Parasitol 66:287–292

Ulmanen I (1972) Distribution and ecology of *Ixodes trianguliceps* Birula (Acarina, Ixodidae) in Finland. Ann Zool Fennici 9:111–115

Ixodes acuminatus Neumann, 1901 (Figs. 60–62)

M. P. Pfäffle, T. N. Petney, and M. M. Santos-Silva

There is currently a discussion as to whether the eastern Palaearctic tick *I. redikorzevi* Olenev, 1927 is a synonym of *I. acuminatus*. According to Guglielmone et al. (2014) this can only be clarified by comparing the type specimens of both species. If the synonymy should be confirmed, the range of *I. acuminatus* would increase to include parts of eastern Europe, China, the eastern Mediterranean and the Middle East from where *I. redikorzevi* has been recorded (Petney et al. 2015).

Life-Cycle and Host Preferences

Ixodes acuminatus is a nidicolous three-host tick (Pérez-Eid 2007). No information is available about the duration of its life cycle in nature or in the laboratory. It is predominantly found on small mammals, including insectivores and rodents, but can also feed on medium-sized hosts, such as hedgehogs (Gilot et al. 1992; Kolonin 2007; Boyard et al. 2008; Gyuranecz et al. 2010; Földvári et al. 2011). Additionally, carnivores and birds are occasionally infested (Hillyard 1996; Guglielmone et al. 2014; Petney et al. 2015). Human infestation can occur (Hillyard 1996; Guglielmone et al. 2014).

M. P. Pfäffle (✉)
Department of Ecology and Parasitology, Institute of Zoology, Karlsruhe Institute of Technology, Karlsruhe, Germany
e-mail: miripfaeffle@web.de

T. N. Petney
State Museum of Natural History, Erbprinzenstr. 13, 76133 Karlsruhe, Germany
e-mail: trevor.petney@kit.edu

M. M. Santos-Silva
Centro de Estudos de Vectores e Doenças Infecciosas Dr. Francisco Cambournac (CEVDI), Instituto Nacional de Saúde Doutor Ricardo Jorge, Águas de Moura, Portugal
e-mail: m.santos.silva@insa.min-saude.pt

A. Estrada-Peña et al. (eds.), *Ticks of Europe and North Africa*,
https://doi.org/10.1007/978-3-319-63760-0_35

Ecology

Although *I. acuminatus* is a nest-dwelling species, it can occasionally be collected from the vegetation by flagging (Petney et al. 2015). A study from France showed that *I. acuminatus* is restricted to areas in which the average temperature in January does not fall below 0 °C. It is thus absent from parts of the east of France, from the northern Alps and the central Pyrenees (Gilot et al. 1992). It can be found on its hosts throughout the year, but peak activity for nymphs and adults is in spring and for larvae in summer (Gilot et al. 1992). Hillyard (1996) describes lowland habitats with an oceanic influence as preferred habitats, however, this does not coincide with the findings from Germany (Petney et al. 2015).

Distribution

Ixodes acuminatus is distributed in temperate and Mediterranean Europe including Belgium, France, Germany, the south-west corner of United Kingdom, Hungary, Italy, Portugal and Spain (Martyn 1988; Travassos Santos Dias 1995; Page and Langton 1996; Rosalino et al. 2007; Rigó et al. 2011; Kreizinger et al. 2013; Petney et al. 2015).

Vectorial Capacity and Pathogen Burden

Borrelia burgdorferi s.l. spirochaetes have been detected in *I. acuminatus* feeding on small mammals in eastern France (Doby et al. 1990). *Borrelia afzelii*, *B. spielmanii*, *B. valaisiana*, *Coxiella burnetii*, *Francisella tularensis*, *Rickettsia helvetica* and Bhanja virus have also been associated with *I. acuminatus* (Obsomer et al. 2013; Tomassone et al. 2013; Szekeres et al. 2015). No experimental studies on the vectorial ability of this species have been reported, but a recent study performed in southern Hungary suggests that *I.*

Fig. 60 **A** Distribution of *I. acuminatus* in Europe and Northern Africa (10 × 10 km grid presence with black dots). Black dots are records reported as *I. acuminatus*. Dark grey dots are records reported as *I. redikorzevi*. **B** Countries where the species has been reported are marked in dark grey

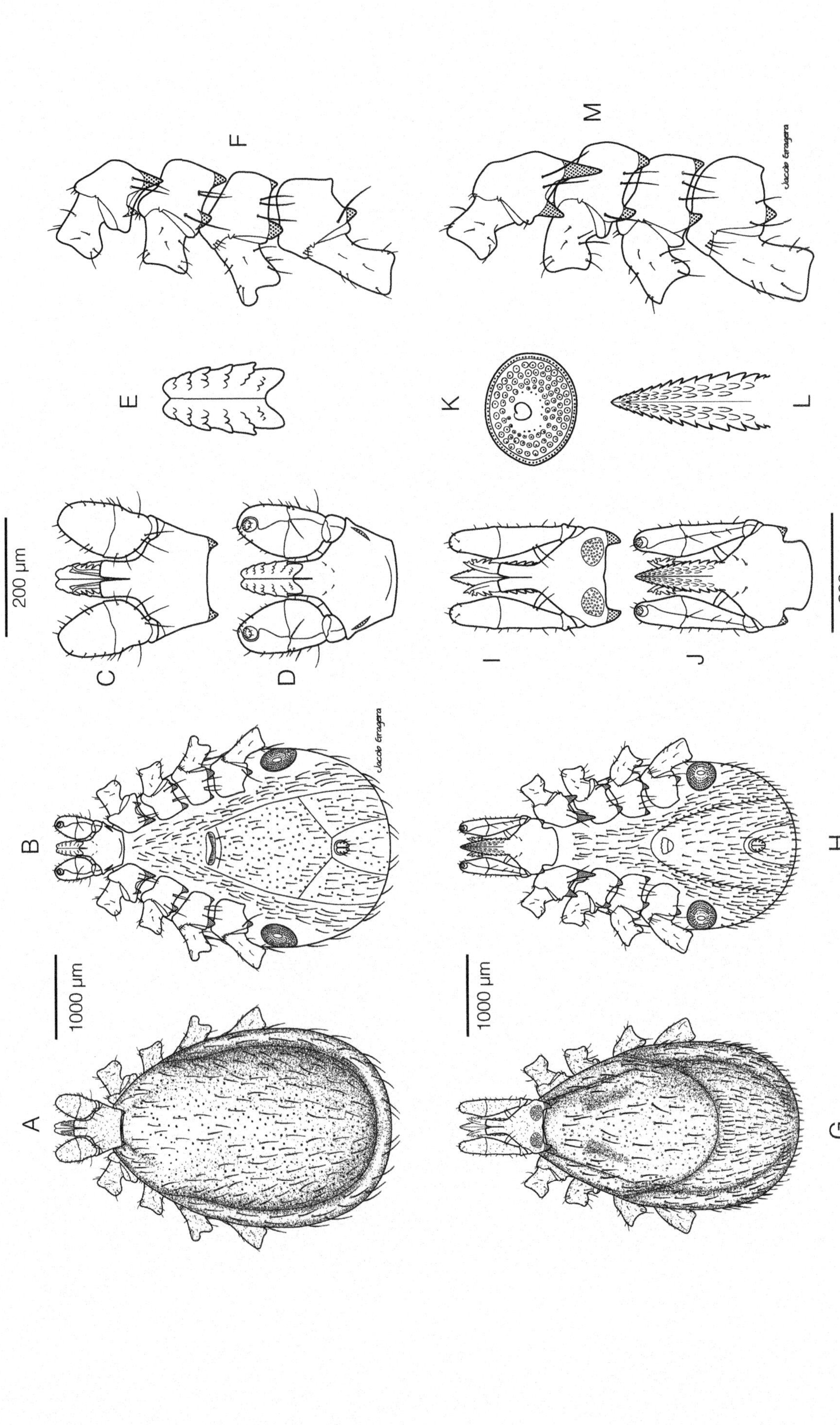

Fig. 61 **A–F** The male of *I. acuminatus*. **A** Dorsal, **B** ventral, **C** capitulum, dorsal, **D** capitulum, ventral, **E** hypostome, **F** coxae and trochanters I–IV. **G–M** The female of *I. acuminatus*. **G** Dorsal, **H** ventral, **I** capitulum, dorsal, **J** capitulum, ventral, **K** spiracular plate, **L** hypostome, **M** coxae and trochanters I–IV. Illustrations from specimens collected in France, Germany, and Spain

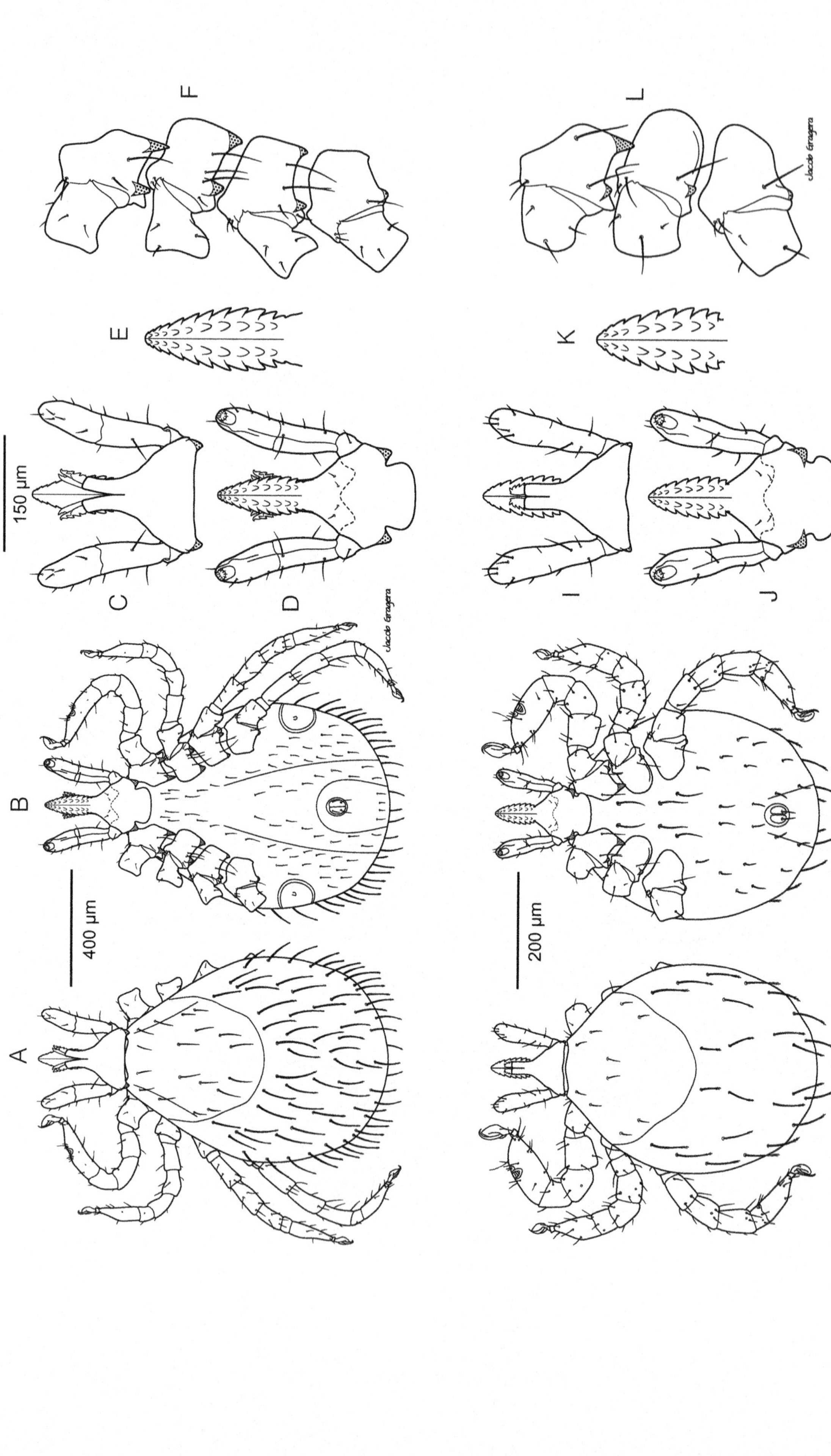

Fig. 62 **A–F** The nymph of *I. acuminatus*. **A** Dorsal, **B** ventral, **C** capitulum, dorsal, **D** capitulum, ventral, **E** hypostome, **F** coxae and trochanters I–IV. **G–L** The larva of *I. acuminatus*. **G** Dorsal, **H** ventral, **I** capitulum, dorsal, **J** capitulum, ventral, **K** hypostome, **L** coxae and trochanters I–III. Illustrations from specimens collected in France, Germany, and Spain

acuminatus might play an important role in the endophilic pathogen cycle of *B. afzelii*, similar to the role of *I. ricinus* in the exophilic pathogen cycle (Szekeres et al. 2015).

References

Boyard C, Vourc'h G, Barnouin J (2008) The relationships between *Ixodes ricinus* and small mammal species at the woodland-pasture interface. Exp Appl Acarol 44:61–76. https://dx.doi.org/10.1007/s10493-008-9132-3

Doby JM, Bigaignon G, Launay H, Costil C, Lorvellec O (1990) Presence of *Borrelia burgdorferi*, agent of tick spirochaetosis, in *Ixodes* (*Exopalpiger*) *trianguliceps* Birula, 1895 and *Ixodes* (*Ixodes*) *acuminatus* Neumann, 1901 (Acari: Ixodidae) and in *Ctenophthalmus baeticus arvernus* Jordan, 1931 and *Megabothris turbidus*. Bull Soc Française Parasitol 8:311–322

Földvári G, Rigó K, Jablonszky M et al (2011) Ticks and the city: ectoparasites of the northern white-breasted hedgehog (*Erinaceus roumanicus*) in an urban park. Ticks Tick Borne Dis 2:231–234

Gilot B, Coutarmanac'h A, Guiguen C, Beaucournu JC (1992) Bioécologie d'*Ixodes acuminatus* Neumann, 1901: hôtes et répartition spatio-temporelle de l'espèce en France. Ann Parasitol Hum comparée 67:19–25

Guglielmone AA, Robbins RG, Apanaskevich DA, Petney TN, Estrada-Peña A, Horak IG (2014) The hard ticks of the world (Acari: Ixodida: Ixodidae). Springer, Dordrecht, p 738

Gyuranecz M, Dénes B, Dán A et al (2010) Susceptibility of the common hamster (*Cricetus cricetus*) to *Francisella tularensis* and its effect on the epizootiology of tularemia in an area where both are endemic. J Wildl Dis 46:1316–20. https://dx.doi.org/10.7589/0090-3558-46.4.1316

Hillyard PD (1996) Ticks of North-West Europe. Field Studies Council, Shrewsbury

Kolonin GV (2007) Mammals as hosts of Ixodid ticks (Acarina, Ixodidae). Entomol Rev 87:401–412. https://dx.doi.org/10.1134/S0013873807040033

Kreizinger Z, Hornok S, Dán A et al (2013) Prevalence of *Francisella tularensis* and *Francisella*-like endosymbionts in the tick population of Hungary and the genetic variability of *Francisella*-like agents. Vector Borne Zoonotic Dis 13:160–163. https://dx.doi.org/10.1089/vbz.2012.1065

Martyn KP (1988) Provisional atlas of the ticks (Ixodoidea) of the British Isles. Biological Records Centre, Institute of Terrestrial Ecology

Obsomer V, Wirtgen M, Linden A et al (2013) Spatial disaggregation of tick occurrence and ecology at a local scale as a preliminary step for spatial surveillance of tick-borne diseases: general framework and health implications in Belgium. Parasit Vectors 6:190. https://dx.doi.org/10.1186/1756-3305-6-190

Page RJC, Langton SD (1996) The occurrence of ixodid ticks on wild mink *Mustela vison* in England and Wales. Med Vet Entomol 10:359–364. https://dx.doi.org/10.1111/j.1365-2915.1996.tb00757.x

Petney TN, Moser E, Littwin N et al (2015) Additions to the "Annotated Checklist of the Ticks of Germany": *Ixodes acuminatus* and *Ixodes inopinatus*. Syst Appl Acarol 20:221–224

Rigó K, Gyuranecz M, Tóth AG, Földvári G (2011) Detection of *Borrelia burgdorferi* sensu lato and *Anaplasma phagocytophilum* in small mammals and ectoparasites in Hungary. Vector Borne Zoonotic Dis 11:1499–501. https://dx.doi.org/10.1089/vbz.2011.0608

Rosalino LM, Rodrigues M, Santos-Silva M, Santos-Reis M (2007) Unusual findings on host-tick interactions through carnivore scat analysis. Exp Appl Acarol 43:293–302. https://dx.doi.org/10.1007/s10493-007-9119-5

Szekeres S, Coipan EC, Rigó K et al (2015) Eco-epidemiology of *Borrelia miyamotoi* and Lyme borreliosis spirochetes in a popular hunting and recreational forest area in Hungary. Parasit Vectors 8:309. https://dx.doi.org/10.1186/s13071-015-0922-2

Tomassone L, Grego E, Auricchio D et al (2013) Lyme borreliosis spirochetes and spotted fever group rickettsiae in ixodid ticks from Pianosa island, Tuscany Archipelago, Italy. Vector Borne Zoonotic Dis 13:84–91. https://dx.doi.org/10.1089/vbz.2012.1046

Travassos Santos Dias JA (1995) Another new tick record for the fauna of Portugal: *Ixodes acuminatus* Newman, 1901. Bol Soc Port Entomol 6:17–23

Ixodes apronophorus Schulze, 1924 (Fig. 63)

A. D. Sándor

Life-Cycle and Host Preferences

This tick is a nest-dwelling species with distinct habitat requirements. It is associated with marshes, bogs, well-vegetated wetlands or moist forests. It is a rare tick found on small mammals sharing wet habitats. Initially it was described from water voles (*Arvicola amphibius*), but it is relatively common on small mammals, including *Sorex araneus*, *S. minutus*, *Neomys anomalus*, *Microtus arvalis*, *Mi. agrestis*, *Myodes glareolus*, *My. rutilus*, *Apodemus agrarius*, *A. flavicollis*, *A. sylvaticus*, *Sciurus vulgaris*, *Tamias sibiricus*, *Cricetulus migratorius*, *Cricetus cricetus* (Schulze 1924; Serdyukova 1956; Cerný and Daniel 1960; Sharipova et al. 1960; Alifanov 1965; Aeschlimann et al. 1970; Tretyakov et al. 2012), as well as medium to large rodents such as *Ondatra zibethicus*, *Myocastor coypus* or *Castor fiber* (Negrobov and Borodin 1964; Newson and Holmes 1968; Kadulski 1998). There are occasional records on larger mammals (*Sus scrofa*, *Vulpes vulpes*) and birds (*Turdus merula*) (Turcek 1953; Grebenyuk 1966; Nosek and Sixl 1972; Gilot et al. 1976). The different developmental stages of this species do not show specific host preferences: all stages may occur at any time of the year on any individual host (Ravdonikas et al. 1971). It is not mentioned as an important species feeding on humans (Estrada-Peña and Jongejan 1999).

Ecology

The distribution is primarily limited by the presence of cold, damp and moist habitats, which are typical in the fens and bogs of boreal and tundra forests in Europe. It is common at low altitudes (in the United Kingdom, or at northern latitudes), but it also occurs in high alpine regions in Central Europe (Cerný and Daniel 1960; Mosolov 1961; Babos 1964; Mihalca et al. 2012). It does not show clear seasonal dynamics; however, larvae are more frequently found in summer (Ravdonikas et al. 1973; Sukhomlinova 1976). All forms are active all year-round (Alifanov 1965), with their development linked to temperature (Babos 1964). Population peaks vary locally (Sukhomlinova 1977; Balashov 1997; Malkova and Bogdanov 2004). Females lay relatively small batches of eggs (up to 100). The life cycle in nature is 2–3 years (Balashov 1997), while in the laboratory may take a single year (Babos 1964).

Distribution

It is found throughout a wide belt, primarily in the northern part of Eurasia. *Ixodes apronophorus* has been reported in most Western, Central and Northern European countries, from the British Isles to Eastern Europe, including the United Kingdom, France, Belgium, the Netherlands, Denmark, Switzerland, Germany, Austria, the Czech Republic, Slovakia, Hungary, Poland, Latvia, Estonia, Romania, Moldova, Russia and Ukraine (Schulze 1924; Janisch 1959; Kusilnõi 1961; Feider 1965; Aeschlimann et al. 1970; Gilot et al. 1976; Radda et al. 1986; Uspenskaia 1987; Balashov 1997; Malkova and Bogdanov 2004; Karbowiak et al. 2007; Tretyakov 2009; Nebogatkin 2012; Petney et al. 2012; Salmane 2012).

A. D. Sándor (✉)
Department of Parasitology and Parasitic Diseases, University of Agricultural Sciences and Veterinary Medicine Cluj-Napoca, Cluj-Napoca, Romania
e-mail: adsandor@gmail.com

A. Estrada-Peña et al. (eds.), *Ticks of Europe and North Africa*,
https://doi.org/10.1007/978-3-319-63760-0_36

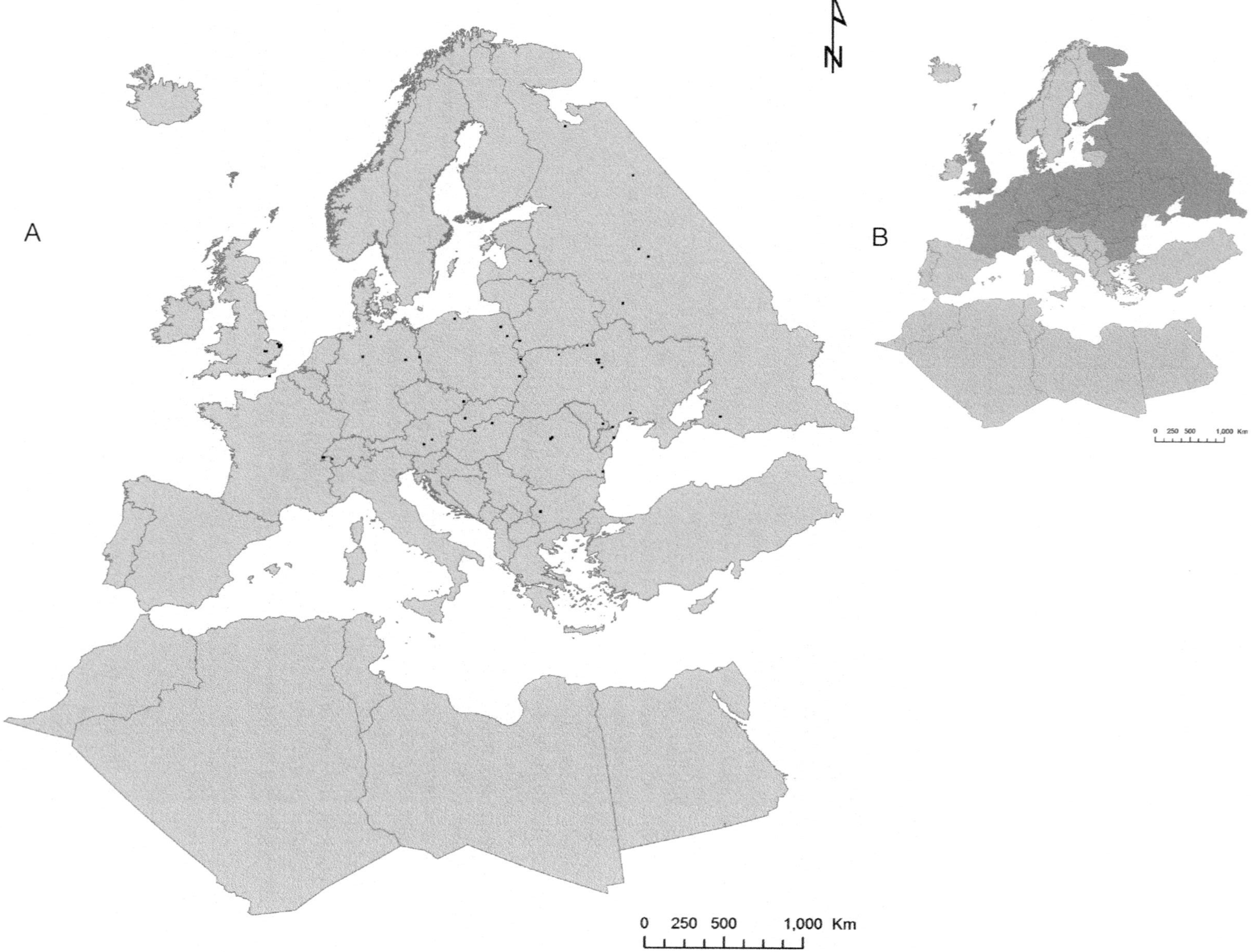

Fig. 63 **A** Distribution of *I. apronophorus* in Europe and Northern Africa (10 × 10 km grid presence with black dots). **B** Countries where the species has been reported are marked in dark grey. Specimens of this species were unavailable for illustrations

Vectorial Capacity and Pathogen Burden

Although it is a common tick of small mammals living in wetlands and along riversides in northern latitudes, there are only a limited number of studies on the vectorial capacity of this species. *Ixodes apronophorus* was proven to transmit the Omsk haemorhagic fever virus (Ravdonikas et al. 1973) and is suspected of transmitting *Francisella tularensis*, tick-borne encephalitis virus and *Babesia microti* (Mosolov 1961; Alifanov 1965; Farkas et al. 2012; Mihalca and Sándor 2013).

References

Aeschlimann A, Büttiker W, Wiehl PA, Eichenberger G, Immler R, Weiss N (1970) Présence d'*Ixodes trianguliceps* (Birula, 1895) et d'*Ixodes apronophorus* (Schulze, 1924) en Suisse (Ixodoidea, Ixodidae). Rev Suisse Zool 77:527–535

Alifanov V (1965) On the ecology and distribution of the tick *Ixodes apronophorus* P. Sch. in West Siberia in connection with its role as a vector of tularemia. Zoologicheskii Zhurnal 44:291–294

Babos S (1964) Die Zeckenfauna Mitteleuropas. Akademiai Kiado, Budapest

Balashov YS (1997) Distribution of ixodid ticks (Acarina, Ixodidae) over landscapes within their ranges. Entomol Rev 77:625–637

Cerný V, Daniel M (1960) Beitrag zur Kenntnis von Zecken und Samtmilben Rumaniens. Zoologischer Anzeiger 164:116–119

Estrada-Peña A, Jongejan F (1999) Ticks feeding on humans: a review of records on human-biting Ixodoidea with special reference to pathogen transmission. Exp Appl Acarol 23:685–715

Farkas R, Estrada-Peña A, Jaenson TG, Pascucci I, Madder M, Salman M, Tarrés-Call J (2012) Basic biology and geographical distribution of tick species involved in the transmission of animal pathogens, including zoonoses. Geographical distribution and control strategies in the Euro-Asia region. Ticks Tick-Borne Dis 6–26

Feider Z (1965) Fauna R.P. Romania. Arachnida, vol V, fasc. 2. Editura Academiei, Bucuresti

Gilot B, Moncada E, Pautou G (1976) Presence en France d'*Ixodes apronophorus* (Schulze, 1924). Ixodoidea—Ixodidae. Annales de Parasitologie Humaine Comparee 51:601–603

Grebenyuk R.V. 1966. [Ixodid ticks of Kirghizia]. Frunze Akad Nauk Kirgiz SSR Inst Biol 328 (in Russian)

Janisch M (1959) Geographical distribution of tick species in Hungary. Állattani Közlemények 47(1–2):103–110

Kadulski S (1998) Parasites of the beaver *Castor fiber* L. in Popielna. Wiadomości Parazytologiczne 44:729–736

Karbowiak G, Wieczorek M, Borowski Z, Wita I (2007) The new locality of *Ixodes apronophorus* Schulze, 1924 in Biebrza National Park, Poland. Wiad Parazytol 53:343–345

Kusilnõi AA (1961) Occurrence of *Ixodes apronophorus* on water rats in Estonia. Faunistilisi märkmeid 1(2):126

Malkova MG, Bogdanov II (2004) Parasite fauna of the water vole *Arvicola terrestris* and its nests in southern western Siberia. Parazitologiya 38:33–45

Mihalca AD, Sándor AD (2013) The role of rodents in the ecology of *Ixodes ricinus* and associated pathogens in Central and Eastern Europe. Front Cell Infect Microbiol 3:56

Mihalca AD, Dumitrache MO, Sándor AD, Magdaş C, Oltean M, Györke A, Matei IA, Ionică A, D'Amico G, Cozma V, Gherman CM (2012) Tick parasites of rodents in Romania: host preferences, community structure and geographical distribution. Parasit Vectors 5:266

Mosolov LP (1961) Records of *Ixodes apronophorus* P. Sch. in the Moscow region and observations in a natural focus of tularaemia. Med Parasitol 30:304–305

Nebogatkin I (2012). Intraurban near-water areas and their role in distribution of ticks (Acari, Ixodidae) in megapolis example of Kyiv. Vestnik zoologii 46: e-29

Negrobov VP, Borodin VS (1964) Einige seltene Zeckenfunde im mittleren Teil der DDR. Angew Parasitol 5:107–111

Newson RM, Holmes RG (1968) Some ectoparasites of the coypu (*Myocastor coypus*) in eastern England. J Anim Ecol 37:471–481

Nosek J, Sixl W (1972) Central-European ticks (Ixodoidea): key for determination. Mitteilungen Abteilung Zoologie Landesmuseum Joanneum 1:61–92

Petney TN, Pfäffle MP, Skuballa JD (2012) An annotated checklist of the ticks (Acari: Ixodida) of Germany. Syst Appl Acarol 17:115–170

Radda A, Burger I, Stanek G, Wewalka G (1986) Austrian hard ticks as vectors of *Borrelia burgdorferi*, overview. Zentralblatt Bakteriologie Mikrobiologie Hygiene. Ser A: Med Microbiol Infect Dis Virol Parasitol 263:79–82

Ravdonikas OV, Chumakov MP, Solovey EA, Ivanov DI, Korsh P (1971) Importance of the marshburrow inhabiting *Ixodes apronophorus* P. Sch. Tick in Omsk hemorrhagic fever virus circulation in a natural focus. In: Chumakov MP (ed) Viral hemorrhagic fevers. Crimean hemorrhagic fever, Omsk hemorrhagic fever with renal syndrome. Trudy Instituta Poliomielita i Virusnykh Entsefalitov. Akademii Meditsinskikh Nauk SSSR, vol 19, pp 485–491 (NAMRU-3 translation 643)

Ravdonikas OV, Chumakov MP, Ivanov DI, Korsh PV, Myasnikov YA (1973) Potentialities of accumulation of Omsk haemorrhagic fever virus in the population of Ixodid ticks of the natural focus during the epizootic season. In: Daniel M, Rosický B (eds) Proceedings of the 3rd international congress of acarology held in Prague 31 Aug–6 Sept 1971. Springer Science Business Media, Dordrecht, pp 619–626

Salmane I (2012) Ticks (Acari, Ixodida: Ixodidae, Amblyommidae) of Latvia. Latvijas Entomologs 51:153–155

Schulze P (1924) *Ixodes apronophorus* n. sp. eine neue deutsche Zecke von *Arvicola amphibius* L. Zoologischer Anzeiger 59:281–284

Serdyukova GV (1956) Ixodid Ticks of the Fauna of the USSR (Keys to the Fauna of the USSR, vol 64. Zool. Inst, Leningrad (in Russian)

Sharipova R, Lebedeva A, Grigorovich L (1960) A search for hibernation sites of forest ticks of the genus *Ixodes*. Med Parazitol 29:207–211

Sukhomlinova OI (1976) Ecology of ixodid ticks from small mammals in the Leningrad oblast. Parazitologiia 11:436–441

Tretyakov K (2009) Small mammals and their parasites (ixodid ticks) in urban forests and parks of St. Petersburg. Estonian J Ecol 58:232–239

Tretyakov KA, Medvedev SG, Apanaskevich MA (2012) Ixodid ticks in St. Petersburg: a possible threat to public health. Est J Ecol 61:215–224

Turcek F (1953) Fauno-ecological investigations on Ixodes in Slovakia in 1953. Biologia 9:464–468

Uspenskaia I (1987) [Ixodid Ticks Inhabiting the Territory Between Prut and Dnestr Rivers.] Chisinau, Republic of Moldova

Ixodes ventalloi Gil Collado, 1936 (Figs. 64–66)

T. N. Petney, D. Otranto, F. Dantas-Torres, and M. P. Pfäffle

There is some confusion concerning records of *I. ventalloi* and *I. festai* (Petney et al. 1996; Guglielmone et al. 2014). Because of this diagnostic uncertainty, no references published before Gilot and Perez (1978) will be used here.

Life-Cycle and Host Preferences

All developmental stages of *I. ventalloi* have been found on lagomorphs, carnivores and rodents. This tick is mainly a parasite of the European rabbit (*Oryctolagus cuniculus*), being popularly known as the rabbit tick (Marquez and Guiguen 1992). However, it can also be occasionally found on different carnivores such as weasel (*Mustela nivalis*), Iberian lynx (*Lynx pardinus*), red fox (*Vulpes vulpes*), cats and Egyptian mongoose (*Herpestes ichneumon*) (Dias and Santos-Reis 1989; Petney et al. 1996; Millán et al. 2007), as well as on birds such as the red-legged partridge (*Alectoris rufa*) and the chukar partridge (*Alectoris chukar*) (Estrada-Peña et al. 1984; Ioannou et al. 2009). It has been found on the crested porcupine (*Hystrix cristata*) (Mori et al. 2015), hedgehogs (*Erinaceus europaeus*) (Domínguez 2014) and humans (Gilot and Marjolet 1982).

T. N. Petney (✉)
State Museum of Natural History, Erbprinzenstr. 13, 76133 Karlsruhe, Germany
e-mail: trevor.petney@kit.edu

D. Otranto
Dipartimento di Medicina Veterinaria, Università degli Studi di Bari, Bari, Italy
e-mail: domenico.otranto@uniba.it

F. Dantas-Torres
Centro de Pesquisas Aggeu Magalhães, Fundação Oswaldo Cruz, Recife, Brazil
e-mail: filipe.vet@globo.com

M. P. Pfäffle
Department of Ecology and Parasitology, Institute of Zoology, Karlsruhe Institute of Technology, Karlsruhe, Germany
e-mail: miripfaeffle@web.de

A. Estrada-Peña et al. (eds.), *Ticks of Europe and North Africa*, https://doi.org/10.1007/978-3-319-63760-0_37

Ecology

Ixodes ventalloi is an endophilic tick, living in the nest of its main host. It is found in south-west Iberian Mediterranean sclerophyllous and mixed forests, north-west Iberian montane forest and Iberian conifer forests.

Distribution

It is a Palearctic species found in North Africa, the Iberian Peninsula and southern France. It has been recorded from Cyprus, France, the United Kingdom, Italy, Portugal and Spain (Chastel et al. 1984; Estrada-Peña et al. 1984; Dias and Santos Reis 1989; Ioannou et al. 2009; Jameson and Medlock 2011; Mori et al. 2015), and as an accidental introduction from Germany (Petney et al. 1996).

Vectorial Capacity and Pathogen Burden

Eyach and Erve viruses have been isolated from this species (Chastel et al. 1984; Faulde and Hoffmann 2001). It is known to harbour *Anaplasma phagocytophilum* (Santos et al. 2004; Márquez 2008). In addition, *Rickettsia helvetica*, *Rickettsia* sp. IRS3 (Santos-Silva et al. 2006; Márquez and Millán 2009; Tomassone et al. 2013), as well as *Coxiella burnetii* (Ioannou et al. 2009) have been detected in this tick. However, the vectorial capacity has not been proven experimentally.

Fig. 64 **A** Distribution of *I. ventalloi* in Europe and Northern Africa (10 × 10 km grid presence with black dots). **B** Countries where the species has been reported are marked in dark grey

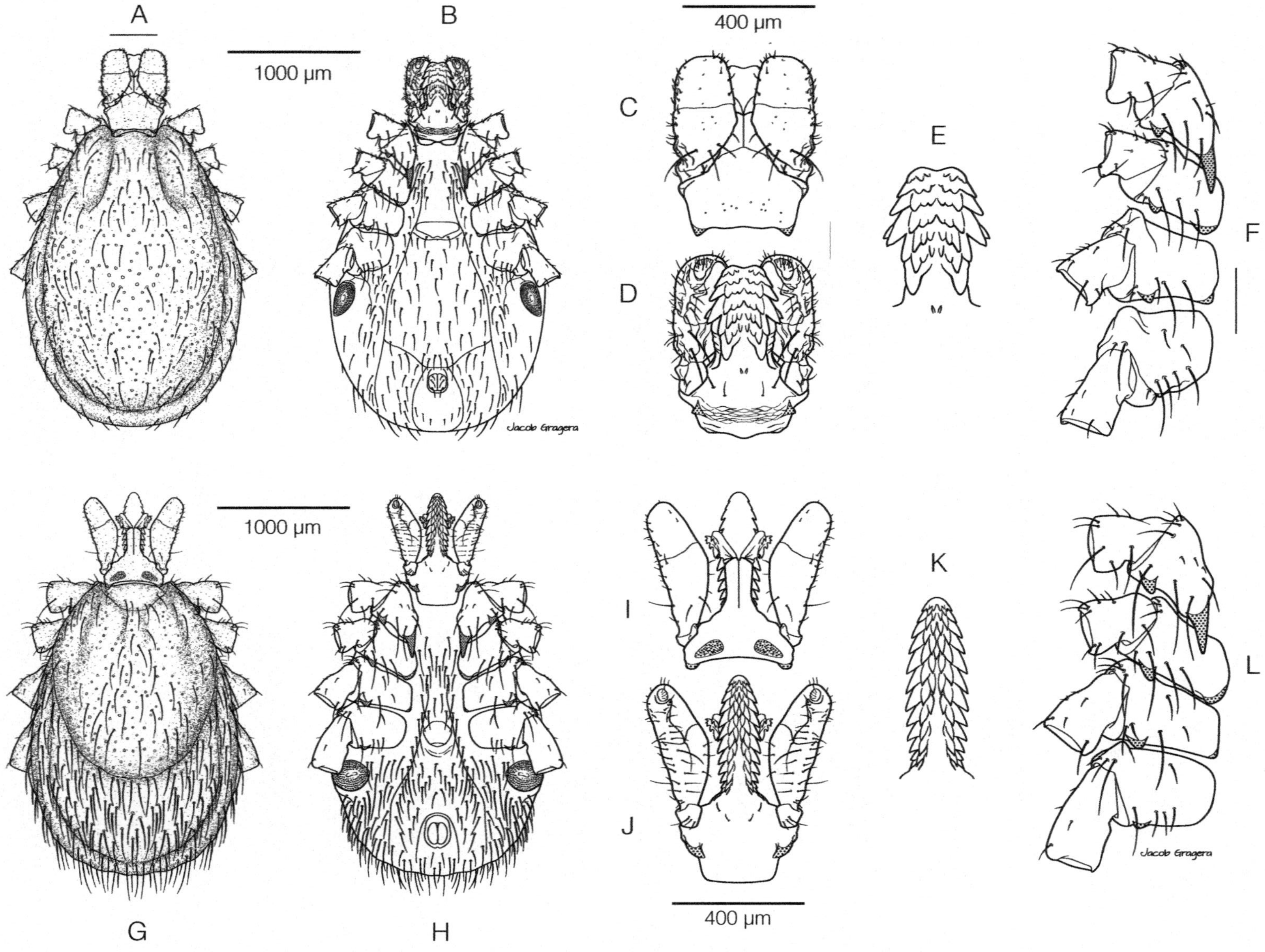

Fig. 65 **A–F** The male of *I. ventalloi*. **A** Dorsal, **B** ventral, **C** capitulum, dorsal, **D** capitulum, ventral, **E** hypostome, **F** coxae and trochanters I–IV. **G–L** The female of *I. ventalloi*. **G** Dorsal, **H** ventral, **I** capitulum, dorsal, **J** capitulum, ventral, **K** hypostome, **L** coxae and trochanters I–IV. Illustrations from specimens collected in Spain

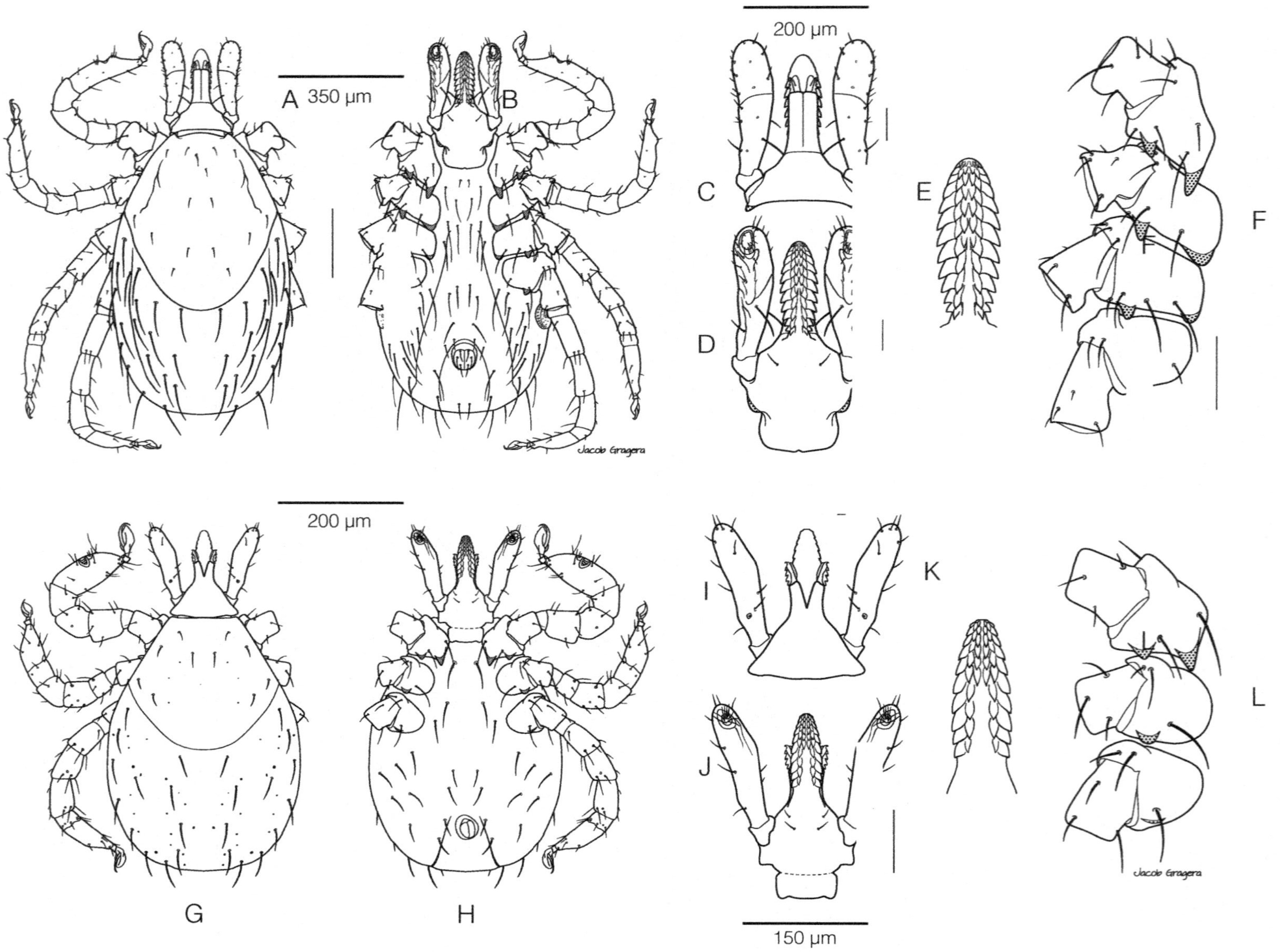

Fig. 66 **A–F** The nymph of *I. ventalloi*. **A** Dorsal, **B** ventral, **C** capitulum, dorsal, **D**: capitulum, ventral, **E** hypostome, **F** coxae and trochanters I–IV. **G–L** The larva of *I. ventalloi*. **G** Dorsal, **H** ventral, **I** capitulum, dorsal, **J** capitulum, ventral, **K** hypostome, **L** coxae and trochanters I–III. Illustrations from specimens collected in Spain

References

Chastel C, Main AJ, Couatarmanac'h A, Le Lay G, Knudson DL, Quillien MC, Beaucournu JC (1984) Isolation of Eyach virus (*Reoviridae*, Colorado tick fever group) from *Ixodes ricinus* and *I. ventalloi* ticks in France. Arch Virol 82:161–171. https://dx.doi.org/10.1007/BF01311160

Dias JATS, Santos Reis M (1989) A carraca *Ixodes ventalloi* Gil Collado, 1936 como principal ectoparasita de uma populacao de doninhas (*Mustela nivalis* Linnaeus, 1766) em Portugal. Garcia Orta 14:35–50

Domínguez G (2014) North Spain (Burgos) wild mammals ectoparasites. Parasite 11:267–272. https://dx.doi.org/10.1051/parasite/2004113267

Estrada-Peña A, Lucientes Curdi J, Castillo Hernandez JA, Sanchez Acedo C, Galmes Femenias M (1984) *Alectoris rufa*, a new host for *Ixodes ventalloi* and *Ixodes* (*Pholeoixodes*) *hexagonus* in Spain. Rev Iber Parasitol 44:215–216

Faulde M, Hoffmann G (2001) Vorkommen und Verhütung vektorassoziierter Erkrankungen des Menschen in Deutschland unter Berücksichtigung zoonotischer Aspekte. Bundesgesundheitsblatt - Gesundheitsforsch - Gesundheitsschutz 44:116–136. https://dx.doi.org/10.1007/s001030050422

Gilot B, Marjolet M (1982) Contribution à l'étude du parasitisme humain par les tiques (Ixodidae et Argasidae), plus particulièrement dans le sud-est de la France. Médecine Mal Infect 12:340–351. https://dx.doi.org/10.1016/S0399-077X(82)80014-0

Gilot B, Perez C (1978) Individualisation et caracterisation de deux *Ixodes* actuellement confondus: *I. festai* Rondelli, 1926, *I. ventalloi* Gil Collado, 1936 (Acarina, Ixodoidea)

Guglielmone AA, Robbins RG, Apanaskevich DA, Petney TN, Estrada-Peña A, Horak IG (2014) The hard ticks of the world (Acari: Ixodida: Ixodidae). Springer, Dordrecht, p 738

Ioannou I, Chochlakis D, Kasinis N, Anayiotos P, Lyssandrou A, Papadopoulos B, Tselentis Y, Psaroulaki A (2009) Carriage of *Rickettsia* spp., *Coxiella burnetii* and *Anaplasma* spp. by endemic and migratory wild birds and their ectoparasites in Cyprus. Clin Microbiol Infect 15(Suppl 2):158–160. https://dx.doi.org/10.1111/j.1469-0691.2008.02207.x

Jameson LJ, Medlock JM (2011) Tick surveillance in Great Britain. Vector Borne Zoonotic Dis 11:403–412. https://dx.doi.org/10.1089/vbz.2010.0079

Márquez FJ (2008) Spotted fever group *Rickettsia* in ticks from southeastern Spain natural parks. Exp Appl Acarol 45:185–194. https://dx.doi.org/10.1007/s10493-008-9181-7

Marquez FJ, Guiguen C (1992) Distribution sur l'hôte des ixodides parasites d'*Oryctolagus cuniculus* (L.) et facteurs qui l'affectent. Acarologia 33:141–148

Márquez FJ, Millán J (2009) rickettsiae in ticks from wild and domestic carnivores of Doñana National Park (Spain) and surrounding area. Clin Microbiol Infect 15(Suppl 2):224–226. https://dx.doi.org/10.1111/j.1469-0691.2008.02147.x

Millán J, Ruiz-Fons F, Márquez FJ, Viota M, López-Bao JV, Paz Martín-Mateo M (2007) Ectoparasites of the endangered Iberian lynx *Lynx pardinus* and sympatric wild and domestic carnivores in Spain. Med Vet Entomol 21:248–254. https://dx.doi.org/10.1111/j.1365-2915.2007.00696.x

Mori E, Sforzi A, Menchetti M, Mazza G, Lovari S, Pisanu B (2015) Ectoparasite load in the crested porcupine *Hystrix cristata* Linnaeus, 1758 in Central Italy. Parasitol Res 114:2223–2229. https://dx.doi.org/10.1007/s00436-015-4413-3

Petney TN, Beichel E, Maiwald M, Hassler D (1996) *Ixodes ventalloi*: a new tick record for Germany. Appl Parasitol 37:96–98

Santos AS, Santos-Silva MM, Almeida VC, Bacellar F, Dumler JS (2004) Detection of *Anaplasma phagocytophilum* DNA in *Ixodes* ticks (Acari: Ixodidae) from Madeira Island and Setúbal District, mainland Portugal. Emerg Infect Dis 10:1643–1648. https://dx.doi.org/10.3201/eid1009.040276

Santos-Silva MM, Sousa R, Santos AS, Melo P, Encarnação V, Bacellar F (2006) Ticks parasitizing wild birds in Portugal: detection of *Rickettsia aeschlimannii*, *R. helvetica* and *R. massiliae*. Exp Appl Acarol 39:331–338. https://dx.doi.org/10.1007/s10493-006-9008-3

Tomassone L, Grego E, Auricchio D, Iori A, Giannini F, Rambozzi L (2013) Lyme borreliosis spirochetes and spotted fever group rickettsiae in ixodid ticks from Pianosa island, Tuscany Archipelago, Italy. Vector Borne Zoonotic Dis 13:84–91. https://dx.doi.org/10.1089/vbz.2012.1046

Ixodes ricinus (Linnaeus, 1758) (Figs. 67–69)

D. Otranto, F. Dantas-Torres, and M. M. Santos-Silva

Life-Cycle and Host Preferences

Ixodes ricinus is undoubtedly the most well-known and studied European tick species. It is Europe's most common tick species and one of the most widely distributed, as well as the most important vector of many of Europe's tick-borne diseases (Petney et al. 2012). *Ixodes ricinus* is a three-host tick with catholic feeding behaviour (Nicholson et al. 2009). Its life cycle lasts from 2 to 3 years, but in some areas where the climatic conditions allow, it is active throughout the year and the life cycle is shorter (Dantas-Torres and Otranto 2013). Under laboratory conditions, the life cycle of *I. ricinus* requires 155 days to be completed (Pomerantsev 1950). Larvae and nymphs usually feed on small mammals, birds and lizards, whereas adults prefer larger mammals, such as cattle and deer (Nicholson et al. 2009; Medlock et al. 2013). It has been recorded from more than 160 species of vertebrates (Estrada-Peña and De La Fuente 2016). However, this large number of hosts may only indicate the interest of researchers working in the Western Palearctic, and therefore be biased by the large number of surveys of the tick in the target territory. *Ixodes ricinus* may show a unimodal or a bimodal temporal pattern of seasonality, but in most areas it presents a bimodal pattern with adults and nymphs peaking in spring and autumn, and adults being active most of the year. Larvae frequently peak in early summer (Petney et al. 2012). In some areas, *I. ricinus* may also present a different seasonal pattern, such as in southern Italy, were ticks can be present throughout the year, with population peaks in spring, autumn and even winter (Dantas-Torres and Otranto 2013). Details of the seasonal activity of *I. ricinus* have been adequately described and summarised by Gray (1991). The seasonality of *I. ricinus* is thought to be regulated by environmental factors involving temperature and photoperiod (Gray 1987; Randolph et al. 2002).

Ecology

There is a vast amount of ecological information available on *I. ricinus*, ranging back to the seminal studies in the 1930s and 1940s in the United Kingdom (MacLeod 1932, 1934, 1935; Milne 1947, 1949). According to Petney et al. (2012), ecological information can be found in the following publications and the references therein: Estrada-Peña (1999), (2003), Estrada-Peña et al. (2004), Perret et al. 2004, Randolph (2004), Randolph et al. (2002), Rosà et al. (2007) and Schwarz et al. (2009). *Ixodes ricinus* is an exophilic tick, commonly found in deciduous and coniferous woodland and mixed forests. It is sensitive to climatic conditions and depends on a moist microclimate, being less able to tolerate excessive heat and desiccation (Bowman and Nuttall 2008). Thus, *I. ricinus* requires a relative humidity of at least 80% to survive during its off-host period, being therefore restricted to areas of moderate to high rainfall with vegetation that retains a high humidity (Medlock et al. 2013). Most of the impact of the climate trends on ticks has been checked surveying the past and current distribution of this tick (Jaenson et al. 2011, 2012).

D. Otranto (✉)
Dipartimento di Medicina Veterinaria, Università degli Studi di Bari, Bari, Italy
e-mail: domenico.otranto@uniba.it

F. Dantas-Torres
Centro de Pesquisas Aggeu Magalhães, Fundação Oswaldo Cruz, Recife, Brazil
e-mail: filipe.vet@globo.com

M. M. Santos-Silva
Centro de Estudos de Vectores e Doenças Infecciosas
Dr. Francisco Cambournac (CEVDI), Instituto Nacional de Saúde
Doutor Ricardo Jorge, Águas de Moura, Portugal
e-mail: m.santos.silva@insa.min-saude.pt

A. Estrada-Peña et al. (eds.), *Ticks of Europe and North Africa*,
https://doi.org/10.1007/978-3-319-63760-0_38

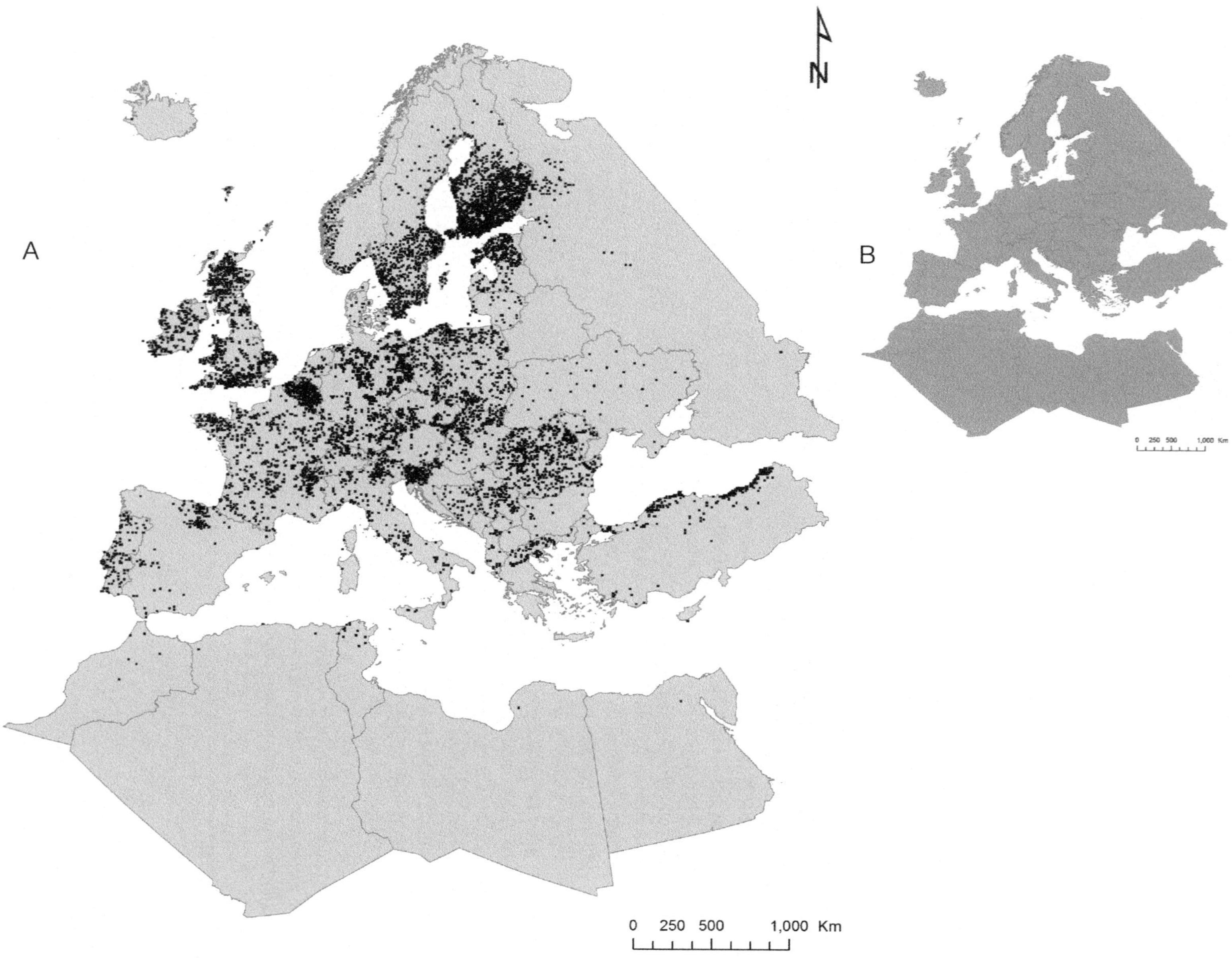

Fig. 67 **A** Distribution of *I. ricinus* in Europe and Northern Africa (10 × 10 km grid presence with black dots). **B** Countries where the species has been reported are marked in dark grey

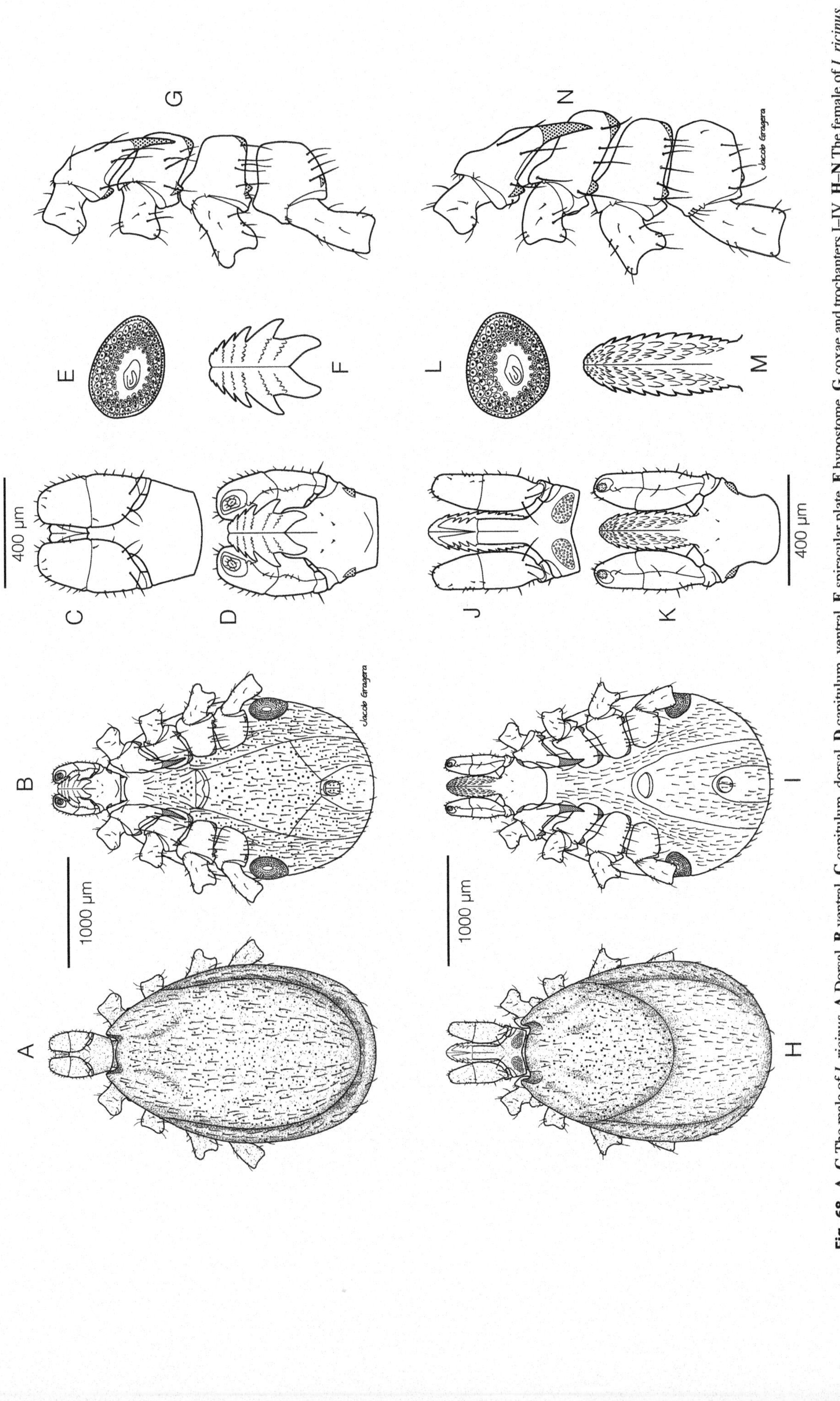

Fig. 68 **A–G** The male of *I. ricinus*. **A** Dorsal, **B** ventral, **C** capitulum, dorsal, **D** capitulum, ventral, **E** spiracular plate, **F** hypostome, **G** coxae and trochanters I–IV. **H–N** The female of *I. ricinus*. **H** Dorsal, **I** ventral, **J** capitulum, dorsal, **K** capitulum, ventral, **L** spiracular plate, **M** hypostome, **N** coxae and trochanters I–IV. Illustrations from specimens collected in Belgium, France, Germany, Spain, Sweden, and the United Kingdom

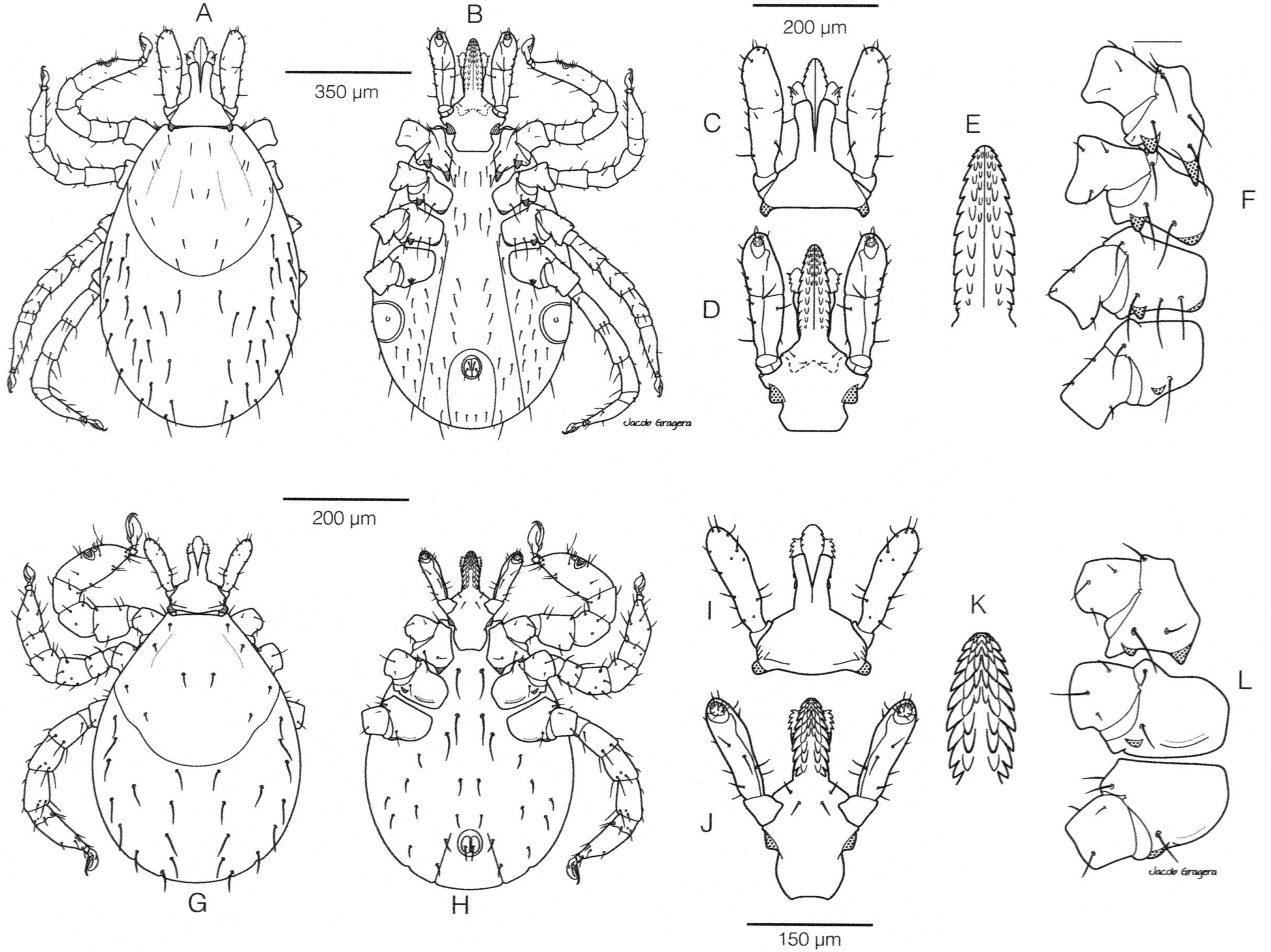

Fig. 69 **A–F** The nymph of *I. ricinus*. **A** Dorsal, **B** ventral, **C** capitulum, dorsal, **D** capitulum, ventral, **E** hypostome, **F** coxae and trochanters I–IV. **G–L** The larva of *I. ricinus*. **G** Dorsal, **H** ventral, **I** capitulum, dorsal, **J** capitulum, ventral, **K** hypostome, **L** coxae and trochanters I–III. Illustrations from specimens collected in Belgium, France, Germany, Spain, Sweden, and the United Kingdom

Distribution

Ixodes ricinus is a species widely distributed in the Western Palearctic. It occurs from the United Kingdom extending eastwards to the Volga River and Turkey and from southern Finland in the north to southern Italy and parts of North Africa in the south (Arthur 1963, 1965; Balashov 1997; Filippova 1999). Current information indicates that its range is expanding both northwards as well as in altitude in mountainous areas (Lindgren et al. 2000; Materna et al. 2008), which may be due to global climate change (but see Randolph 2004). It can be common in cities. We do not currently know whether *I. ricinus* exists in northern Africa. Some of the collections of the tick in that region have been identified as a new species (see *I. inopinatus*). It is thus necessary to clearly establish if the tick is present in the mountain ranges of Morocco, Algeria and Tunisia.

Vectorial Capacity and Pathogen Burden

The tick-borne encephalitis (TBE) complex of viruses has several tick species as vectors and predominantly mammals as natural hosts (Gaunt and Gould 2005; Gould and Solomon 2008). The status of TBE is currently controversial, with one study suggesting its division into 3 subtypes (European, Far Eastern and Siberian; Fauquet 2005), while the other suggests that TBE and louping-ill belong to the same species with four subgroups: western TBE, eastern TBE, Turkish sheep encephalitis and louping-ill virus (Grard et al. 2007). Non-viremic transmission with infection occurring between cofeeding ticks has been demonstrated (Labuda et al. 1993a, b). Louping-ill virus (LIV) occurs in the United Kingdom and Ireland, where it is a pathogen predominantly of sheep and red grouse, *Lagopus lagopus scoticus*, although it may also affect cattle, pigs, roe deer (*Capreolus capreolus*), red deer (*Cervus elaphus*), the field vole (*Microtus agrestis*) and humans (Walton and O'Donnell 1967). Sheep suffering from encephalomyelitis in both Norway and Spain have been shown to be infected with LIV and a related virus, respectively (Gao et al. 1993; Marin et al. 1995). It is potentially important that louping ill virus can be infective for birds, which may introduce the virus into new areas (Reid 1975; Hudson 1992).

Eyach virus, which is related to Colorado tick fever virus, was isolated from *I. ricinus* near the Eyach River in Baden, and was recently rediscovered in this state after more than 25 years (Hassler et al. 2003). Tribec virus was first isolated from *I. ricinus* in Slovakia in 1963 (Gresíkova et al. 1965). It is a member of the Kemerovo serogroup. Both of these viruses have been implicated as causes of human disease (Malkova et al. 1980; Charrel et al. 2004). Lipovnik, Erve and Uukuniemi viruses have also been isolated from *I. ricinus* ticks.

Ixodes ricinus is the major vector of *Borrelia burgdorferi* s.l. in its sylvatic cycles and in terms of transmission to humans in Europe (Rauter and Hartung 2005; Skuballa et al. 2007, 2012; Fingerle et al. 2008; Margos et al. 2009). It has also been suggested to be the major vector of the relapsing fever spirochete, *Borrelia miyamotoi*, in Western Europe. This species has been shown to be pathogenic to humans (Platonov et al. 2011). *Rickettsia slovaca* and *R. helvetica* have been reported from *I. ricinus* (Rehacek 1984; Nilsson et al. 1999). *Rickettsia monacensis* has been also associated with *I. ricinus* in Germany (Schorn et al. 2011). *Anaplasma phagocytophilum* is a significant human and animal pathogen (Hartelt et al. 2004; Silaghi et al. 2008). *Bartonella henselae* has been recovered from *I. ricinus* in Central Europe (Schouls et al. 1999; Sanogo et al. 2003; Halos et al. 2005), and recently Cotté et al. (2008) demonstrated the transmission of this pathogen by this tick species. *Francisella tularensis*, causing tularemia, has also been recovered from *I. ricinus*, although *Dermacentor reticulatus* (Fabricius 1794) appears to be the dominant vector (Hubalek et al. 1996; Gurycova 1998).

Babesia divergens, which is a pathogen of cattle, is transmitted by *I. ricinus* (Kjemtrup and Conrad 2000; Hartelt et al. 2004). This species can cause disease in immunocompromised humans, for example following splenectomy (Kjemtrup and Conrad 2000). *Babesia venatorum* (*Babesia* sp. EU1), a recently discovered member of this genus, is associated with *I. ricinus* and is thought to use deer as hosts (Herwaldt et al. 2003; Duh et al. 2005). Walter (1981) demonstrated that *I. ricinus* nymphs collected from a natural focus in Germany were infected with *Babesia microti*. Later, Walter and Weber (1981) demonstrated transstadial transmission from larvae to nymphs but not to adults. Transovarial transmission was not observed. A recent study in Poland has shown very low prevalences of *B. microti* in *I. ricinus* (0.6%) (Sinski et al. 2006).

References

Arthur DR (1963) British ticks. London, Butterworths
Arthur DR (1965) Ticks of the genus *Ixodes* in Africa. Athlone Press, London
Balashov YS (1997) Distribution of ixodid ticks (Acarina, Ixodidae) over landscapes within their ranges. Entomol Rev 77:625–637 (in Russian)
Bowman A, Nuttall P (2008) Ticks: biology, disease and control. Cambridge University Press, Cambridge
Charrel RN, Attoui H, Butenko AM, Clegg JC, Deubel V, Frolova TV, Gould EA, Gritsun TS, Heinz FX, Labuda M, Lashkevich VA, Loktev V, Lundkvist A, Lvov DV, Mandl CW, Niedrig M, Papa A, Petrov VS, Plyusnin A, Randolph S, Süss J, Zlobin VI, de Lamballerie X (2004) Tick-borne virus diseases of human interest in Europe. Clin Microbiol Infect 10:1040–1055

Cotté V, Bonnet S, Le Rhun D, Le Naour E, Chauvin A, Boulouis HJ, Lecuelle B, Lilin T, Vayssier-Taussat M (2008) Transmission of *Bartonella henselae* by *Ixodes ricinus*. Emerg Infect Dis 14:1074–1080

Dantas-Torres F, Otranto D (2013) Seasonal dynamics of *Ixodes ricinus* on ground level and higher vegetation in a preserved wooded area in southern Europe. Vet Parasitol 192:253–258

Duh D, Petrovec M, Avsic-Zupanc D (2005) Molecular characterization of human pathogen *Babesia* EU1 in *Ixodes ricinus* ticks in Slovenia. J Parasitol 91:463–465

Estrada-Peña A (1999) Geostatistics as predictive tools to estimate *Ixodes ricinus* (Acari: Ixodidae) habitat suitability in the western Palearctic from AVHRR satellite imagery. Exp Appl Acarol 23:337–349

Estrada-Peña A (2003) The relationships between habitat topology, critical scales of connectivity and tick abundance *Ixodes ricinus* in a heterogeneous landscape in northern Spain. Ecography 26:661–671

Estrada-Peña A, De La Fuente J (2016) Species interactions in occurrence data for a community of tick-transmitted pathogens. Sci data 3:160056

Estrada-Peña A, Quilez J, Sanchez Acedo C (2004) Species composition, distribution, and ecological preferences of the ticks of grazing sheep in north-central Spain. Med Vet Entomol 18:123–133

Fauquet CM (2005) Virus taxonomy: VIII report of the international committee on the taxonomy of viruses. Elsevier Academic Press, Amsterdam, pp 981–988

Filippova NA (1999) Systematic relationships of the *Ixodes ricinus* species complex in the Palearctic Faunal Region. In: Needham GR, Mitchell R, Horn DJ, Welbourn WC (eds) Acarology IX, vol 2. Ohio, Ohio Biological Survey, pp 355–361

Fingerle V, Schulte-Spechtel UC, Ruzic-Sabljic E, Leonhard S, Hofmann H, Weber K, Pfister K, Strle F, Wilske B (2008) Epidemiological aspects and molecular characterization of *Borrelia burgdorferi* s.l. from southern Germany with special respect to the new species *Borrelia spielmanii* sp. nov. Int J Med Microbiol 298:279–290

Gao GF, Jiang WR, Hussain MH, Venugopal K, Gritsun TS, Reid HW, Gould EA (1993) Sequencing and antigenic studies of a Norwegian virus isolated from encephalomyelitic sheep confirm the existence of louping ill virus outside Great Britain and Ireland. J Gen Virol 74:109–114

Gaunt MW, Gould EA (2005) Rapid subgroup identification of the flaviviruses using degenerate primer E-gene RT-PCR and site specific restriction enzyme analysis. J Virol Methods 128:113–127

Gould EA, Solomon T (2008) Pathogenic flaviviruses. Lancet 371:500–509

Grard G, Moureau G, Charrel RN, Lemasson JJ, Gonzales JP, Gallian P, Gritsun TS, Holmes EC, Gould EA, de Lamballerie X (2007) Genetic characterization of tick-borne flaviviruses: new insights into evolution, pathogenic determinants and taxonomy. Virology 361:80–92

Gresíkova M, Ernek E, Kozuch O, Nosek J (1965) Some ecological aspects on Tribic virus. Folia Parasitol 17:379–382

Gray JS (1987) Mating and behavioural diapause in *Ixodes ricinus* L. Exp Appl Acarol 3:61–71

Gray JS (1991) The development and seasonal activity of the tick *Ixodes ricinus*: a vector of Lyme borreliosis. Rev Med Vet Entomol 79(6):323–333

Gurycova D (1998) First isolation of *Francisella tularensis* subsp. *tularensis* in Europe. Eur J Epidemiol 14:797–802

Halos L, Jamal T, Maillard R, Beugnet F, Le Menach A, Boulouis HJ, Vayssier-Taussat M (2005) Evidence of *Bartonella* sp. in questing adult and nymphal *Ixodes ricinus* ticks from France and coinfection with *Borrelia burgdorferi* sensu lato and *Babesia* sp. Vet Res 36:79–87

Hartelt K, Oehme R, Frank H, Brockmann SO, Hassler D, Kimmig P (2004) Pathogens and symbionts in ticks: prevalence of *Anaplasma phagocytophilum* (*Ehrlichia* sp.), *Wolbachia* sp., *Rickettsia* sp., and *Babesia* sp. in southern Germany. Int J Med Microbiol 293(Suppl 37):86–92

Hassler D, Oehme R, Kimmig P, Dobler G (2003) Eyach virus: first detection from ticks after more than 25 years in south-western Germany. Deutsche Medizinische Wochenschrift 128:1874

Herwaldt BL, Caccio S, Gherlinzoni F, Aspöck H, Slemenda SB, Piccaluga PP, Martinelli G, Edelhofer R, Hollenstein U, Poletti G, Pampiglione S, Löschenberger K, Tura S, Pieniazek NJ (2003) Molecular characterization of a non-*Babesia divergens* organism causing babesiosis in Europe. Emerg Infect Dis 9:942–948

Hubalek Z, Treml F, Halouzka J, Juricova Z, Hunady M, Janik V (1996) Frequent isolation of *Francisella tularensis* from *Dermacentor reticulatus* ticks in an enzootic focus of tularemia. Med Vet Entomol 10:241–246

Hudson PJ (1992) Grouse in space and time. The Game Conservancy Trust, Fordingbridge

Jaenson TG, Lindgren E (2011) The range of *Ixodes ricinus* and the risk of contracting Lyme borreliosis will increase northwards when the vegetation period becomes longer. Ticks Tick-Borne Dis 2:44–49

Jaenson TG, Jaenson DG, Eisen L, Petersson E, Lindgren E (2012) Changes in the geographical distribution and abundance of the tick *Ixodes ricinus* during the past 30 years in Sweden. Parasit Vectors 5:8

Kjemtrup AM, Conrad PA (2000) Human babesiosis: an emerging tick-borne disease. Int J Parasitol 30:1323–1337

Labuda M, Jones LD, Williams T, Danielova V, Nuttall PA (1993a) Efficient transmission of tickborne encephalitis virus between cofeeding ticks. J Med Entomol 30:295–299

Labuda M, Nuttall PA, Kozuch O, Eleckova E, Williams T, Zuffova E, Sabo A (1993b) Non-viremic transmission of tick-borne encephalitis virus: a mechanism for arbovirus survival in nature. Experimentia 49:802–805

Lindgren E, Tälleklint L, Polfeldt T (2000) Impact of climatic change on the northern latitude limit and population density of the disease-transmitting European tick *Ixodes ricinus*. Environ Health Perspect 108:119–123

Malkova D, Holubova J, Kolman JM (1980) Antibodies against some arboviruses in persons with vari- ous neuropathies. Acta Virol 24:298

Materna J, Daniel M, Metelka L, Harcarik J (2008) The vertical distribution, density and the development of the tick *Ixodes ricinus* in mountain areas influenced by climate changes (the Krkonose Mts., Czech Republic). Int J Med Microbiol 298(Suppl 1):25–37

MacLeod J (1932) The bionomics of *Ixodes ricinus* L., the "sheep tick" of Scotland. Parasitology 24:382–400

MacLeod J (1934) *Ixodes ricinus* in relation to its physical environment: influence of climate on development. Parasitology 26:282–305

MacLeod J (1935) *Ixodes ricinus* in relation to its physical environment. II. The factors governing survival and activity. Parasitology 27:123–144

Margos G, Vollmer SA, Cornet M, Garnier M, Fingerle V, Wilske B, Bormane A, Vitorino L, Collares-Pereira M, Drancourt M, Kurtenbach K (2009) A new *Borrelia* species defined by multilocus sequence analysis of housekeeping genes. Appl Environ Microbiol 75:5410–5416

Marin MS, McKenzie J, Guo GF, Reid HW, Antoniadis A, Gould EA (1995) The virus causing encephalomyelitis in sheep in Spain: a new member of the tick-borne encephalitis group. Res Vet Sci 58:11–13

Medlock JM, Hansford KM, Bormane A, Derdakova M, Estrada-Peña A, George JC, Golovljova I, Jaenson TG, Jensen JK, Jensen PM,

Kazimirova M, Oteo JA, Papa A, Pfister K, Plantard O, Randolph SE, Rizzoli A, Santos-Silva MM, Sprong H, Vial L, Hendrickx G, Zeller H, Van Bortel W (2013) Driving forces for changes in geographical distribution of *Ixodes ricinus* ticks in Europe. Parasit Vectors 2(6):1. https://dx.doi.org/10.1186/1756-3305-6-1

Milne A (1947) The ecology of the sheep tick, *Ixodes ricinus* L.: the infestation of hill sheep. Parasitology 38:34–50

Milne A (1949) The ecology of the sheep tick, *Ixodes ricinus* L.: spatial distribution. Parasitology 40:35–45

Nicholson WL, Sonenshine DE, Lane RS, Uilenberg G (2009) Ticks (Ixodida). In: Mullen GR, Durden LA (eds) Medical and veterinary entomology, 2nd edn. Academic Press, San Diego, pp 493–541

Nilsson K, Lindquist O, Liu AJ, Jaenson TGT, Friman G, Pahlson C (1999) *Rickettsia helvetica* in *Ixodes ricinus* ticks in Sweden. J Clin Microbiol 37:400–403

Perret J-L, Rais O, Gern, L (2004) Influence of climate on the proportion of *Ixodes ricinus* nymphs and adults in a tick population. J Med Entomol 41:361–365

Petney TN, Pfäfffie MP, Skuballa JD (2012) An annotated checklist of the ticks (Acari: Ixodida) of Germany. Syst Appl Acarol 17:115–170

Platonov AE, Karan LS, Kolyasnikova NM, Makhneva NA, Toporkova MG, Maleev VV, Fish D, Krause PJ (2011) Humans infected with relapsing fever spirochete *Borrelia miyamotoi*, Russia. Emerg Infect Dis 17:1816–1823

Pomerantsev BI (1950) Ixodid ticks (Ixodidae). Fauna SSSR, Paukoobraznye, n. s., (41), 4,224 p (English translation by Elbl A., edited by Anastos G. The American Institute of Biological Sciences, Washington)

Randolph SE, Green RM, Hoodless AN, Peacey MF (2002) An empirical quantitative framework for the seasonal population dynamics of the tick *Ixodes ricinus*. Int J Parasitol 32:979–989

Randolph SE (2004) Evidence that climate change has caused 'emergence' of tick-borne diseases in Europe? Int J Med Microbiol 297(Suppl 37):5–15

Rauter C, Hartung T (2005) Prevalence of *Borrelia burgdorferi* sensu lato genospecies in *Ixodes ricinus* ticks in Europe: a metaanalysis. Appl Environ Microbiol 71:7203–7216

Reid HW (1975) Experimental infection of red grouse with louping-ill virus. I. The viraemia and antibody response. J Comp Parasitol 85:223–229

Rehacek J (1984) *Rickettsia slovaca*, the organism and its ecology. Acta Scientarum Naturalium Academiae Scientiarum Bohemicae Brno 18:1–50

Rosà R, Pugliese A, Ghosh M, Perkins SE, Rizzoli A (2007) Temporal variation of *Ixodes ricinus* intensity on the rodent host *Apodemus flavicollis* in relation to local climate and host dynamics. Vector-Borne Zoonotic Dis 7:285–295

Sanogo YO, Zeaiter Z, Caruso G, Merola F, Shpynov S, Brouqui P, Raoult D (2003) *Bartonella henselae* in *Ixodes ricinus* ticks (Acari: Ixodida) removed from humans, Belluno Province, Italy. Emerg Infect Dis 9:329–332

Schwarz A, Maier WA, Kistemann T, Kampen H (2009) Analysis of the distribution of the tick *Ixodes ricinus* L. (Acari: Ixodidae) in a nature reserve of western Germany using geographic information systems. Int J Hyg Environ Health 212:87–96

Schorn S, Pfister K, Reulen H, Mahling M, Silaghi C (2011) Occurrence of *Babesia* spp., *Rickettsia* spp. and *Bartonella* spp. in *Ixodes ricinus* in Bavarian public parks, Germany. Parasit Vectors 4(1):135

Schouls LM, Van de Pol I, Rijpkema SG, Schot CS (1999) Detection and identification of *Ehrlichia, Borrelia burgdorferi* sensu lato, and *Bartonella* species in Dutch *Ixodes ricinus* ticks. J Clin Microbiol 37:2215–2222

Silaghi C, Gilles J, Höhle M, Fingerle V, Just FT, Pfister K (2008) *Anaplasma phagocytophilum* infection in *Ixodes ricinus*, Bavaria, Germany. Emerg Infect Dis 14:972–974

Sinski E, Bajer A, Welc R, Pawelczyk A, Ogrzewalska M, Behnke JM (2006) *Babesia microti*: prevalence in wild rodents and *Ixodes ricinus* ticks from the Mazury Lake District of north-eastern Poland. Int J Med Microbiol 296:137–143

Skuballa J, Oehme R, Hartelt K, Petney TN, Bücher T, Kimmig P, Taraschewski H (2007) European hedgehogs as hosts for *Borrelia* spp. Germany. Emerg Infect Dis 13:952–953

Skuballa J, Petney T, Pfäffle M, Oehme R, Hartelt H, Fingerle V, Kimmig P, Taraschewski H (2012) Occurence of different *Borrelia burgdorferi* sensu lato genospecies including *B. afzelii, B. bavariensis* and *B. spielmanii* in hedgehogs (*Erinaceus* spp.) in Europe. Ticks Tick Borne Dis 3:8–13

Walter G (1981) Isolierung von *Babesia microti* (Franca 1912) aus freilebenden Nymphen von *Ixodes ricinus* (Linnaeus 1758). Acta Trop 38:187–188

Walter G, Weber G (1981) Untersuchungen zur Übertragung (transstadial, transovarial) von *Babesia microti*, Stamm "Hannover 1," in *Ixodes ricinus*. Trop Parasitol 32:228–230

Walton GA, O'Donnell TG (1967) Observations on *Ixodes ricinus* and louping ill in Ireland. In: Evans GO (ed) Proceedings of the 2nd international congress of acarology, Sutton Bonington (England) 19th–25th July, 1967, Akademiai Kiado, Budapest, pp 601–608

Ixodes persulcatus Schulze, 1930 (Figs. 70–72)

M. P. Pfäffle, T. N. Petney, and T. G. T. Jaenson

Life Cycle and Host Preferences

Ixodes persulcatus, the taiga tick, is a three-host tick. Under laboratory conditions (20 °C, 12 h light/dark cycle) the life cycle takes about 250 days (Konnai et al. 2008). It is a generalist tick, feeding on a variety of different animal species, ranging from reptiles to birds to mammals. The immature developmental stages prefer to feed on small and medium-sized mammals and birds, while the females parasitize large mammals, such as deer, cattle, sheep or hares (Gray 1998). In contrast to *I. ricinus,* in which the nymphal stage is the most important stage feeding on and transmitting tick-borne pathogens to humans, in *I. persulcatus* the adult females are usually found on humans (Estrada-Peña and Jongejan 1999; Korenberg et al. 2002). The nymphs of *I. persulcatus* rarely attach to humans (Korenberg et al. 2002).

Ecology

Ixodes persulcatus is an exophilic tick with all three developmental stages searching for hosts by questing on the vegetation. During this questing phase, as well as during the off-host developmental phases, the tick is vulnerable to desiccation. Due to this risk, *I. persulcatus* needs habitats with a dense vegetation cover and a mat of leaf litter so that the relative humidity at the base of the vegetation remains above 80% throughout the driest times of the year, usually the summer (Gray 1998). It occurs in several ecoregions, being associated with broad-leaved and spruce forests and mixed forests of European type, and can be found up to 3000 m a.s.l. (Shpynov et al. 2003; Guglielmone et al. 2014). The northern limit of the reproductive distribution of *I. persulcatus* is determined by the 1600° isotherm of effective accumulated temperature during the days when the mean daily temperature is ≥ 10 °C. However, specimens can occasionally be found further north in localities where the tick is unable to reproduce (Uspensky et al. 2003; Yasyukevich et al. 2009). In Russia, the steppe and forest-steppe are considered to be the southern limit to the distribution of this species (Yasyukevich et al. 2009). Korenberg (2000) describes four major activity patterns in *I. persulcatus* adults: (i) very early onset (late March) and very late cessation of activity (mid-September to early November), which occur in habitats offering optimal warmth and moisture such as mixed broad-leaved-coniferous forests; (ii) relative early onset (April) and relative late cessation of activity, occurring in secondary forests and the coniferous forest zone of eastern Europe and the pre-Ural region, as well as small-leaved and mountain light coniferous forests of west and central Siberia; (iii) relatively early onset (April) and early cessation of activity (July) occurring in European mixed and small-leaved forests of the forest-steppe type and Asian light coniferous forests undergoing transformation into steppe; (iv) late onset (May) and early cessation of activity (July) occurring in dark-coniferous forests, European oak forests, and western Siberian small-leaved forests. The activity of immatures reaches a peak in late spring to early summer, usually 10–20 days later than that of adult ticks (Korenberg 2000). The different activity patterns probably depend on the climate and the weather that occur in those regions where certain vegetation types occur, rather than the vegetation type.

M. P. Pfäffle
Department of Ecology and Parasitology, Institute of Zoology, Karlsruhe Institute of Technology, Karlsruhe, Germany
e-mail: miripfaeffle@web.de

T. N. Petney (✉)
State Museum of Natural History, Erbprinzenstr. 13, 76133 Karlsruhe, Germany
e-mail: trevor.petney@kit.edu

T. G. T. Jaenson
Medical Entomology Unit, Department of Organismal Biology, Evolutionary Biology Centre, Uppsala University, Uppsala, Sweden
e-mail: thomas.jaenson@ebc.uu.se

A. Estrada-Peña et al. (eds.), *Ticks of Europe and North Africa*,
https://doi.org/10.1007/978-3-319-63760-0_39

Fig. 70 **A** Distribution of *I. persulcatus* in Europe and Northern Africa (10 × 10 km grid presence with black dots). **B** Countries where the species has been reported are marked in dark grey

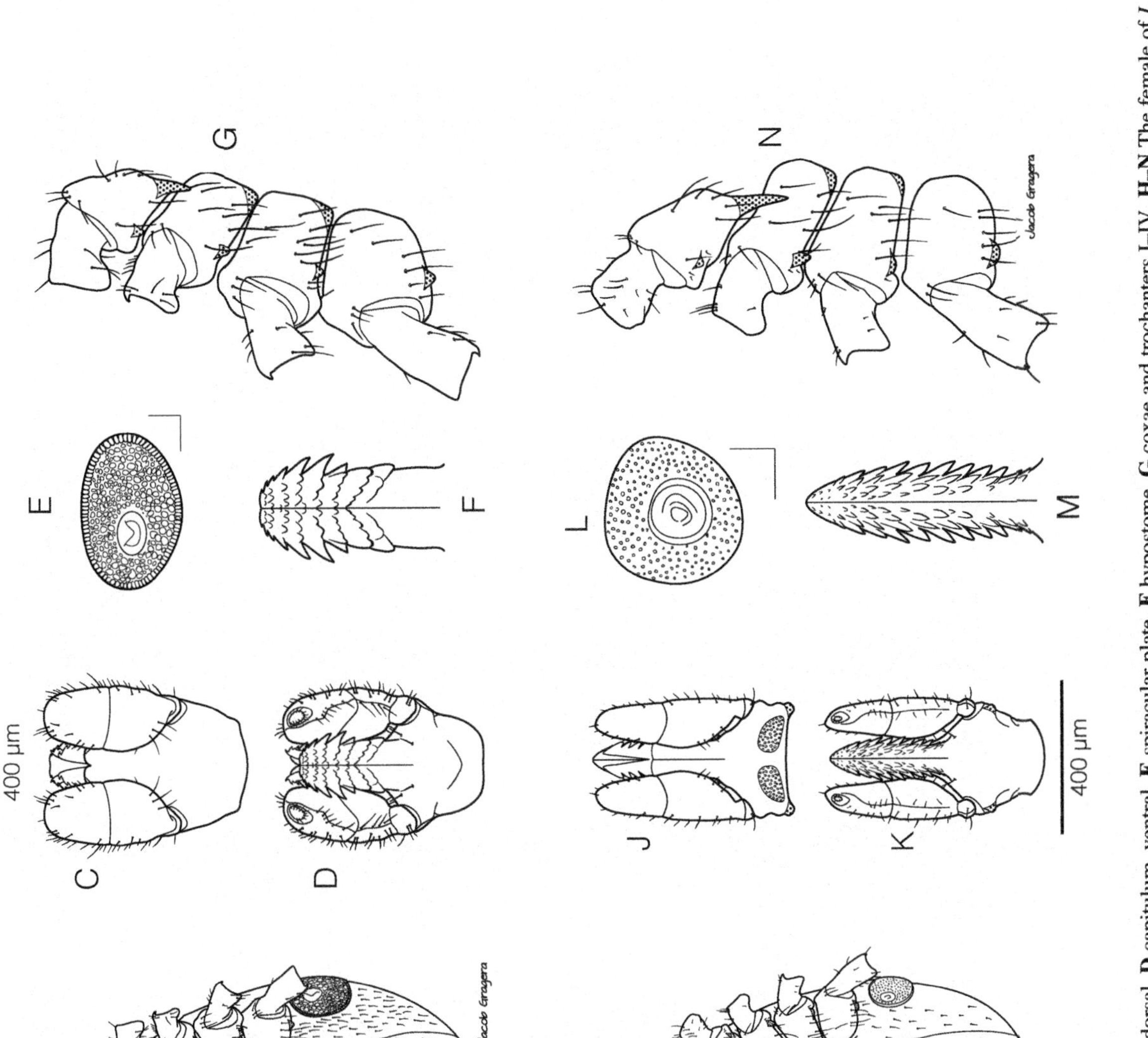

Fig. 71 **A–G** The male of *I. persulcatus*. **A** Dorsal, **B** ventral, **C** capitulum, dorsal, **D** capitulum, ventral, **E** spiracular plate, **F** hypostome, **G** coxae and trochanters I–IV. **H–N** The female of *I. persulcatus*. **H** Dorsal, **I** ventral, **J** capitulum, dorsal, **K** capitulum, ventral, **L** spiracular plate, **M** hypostome, **N** coxae and trochanters I–IV. Illustrations from specimens collected in Finland

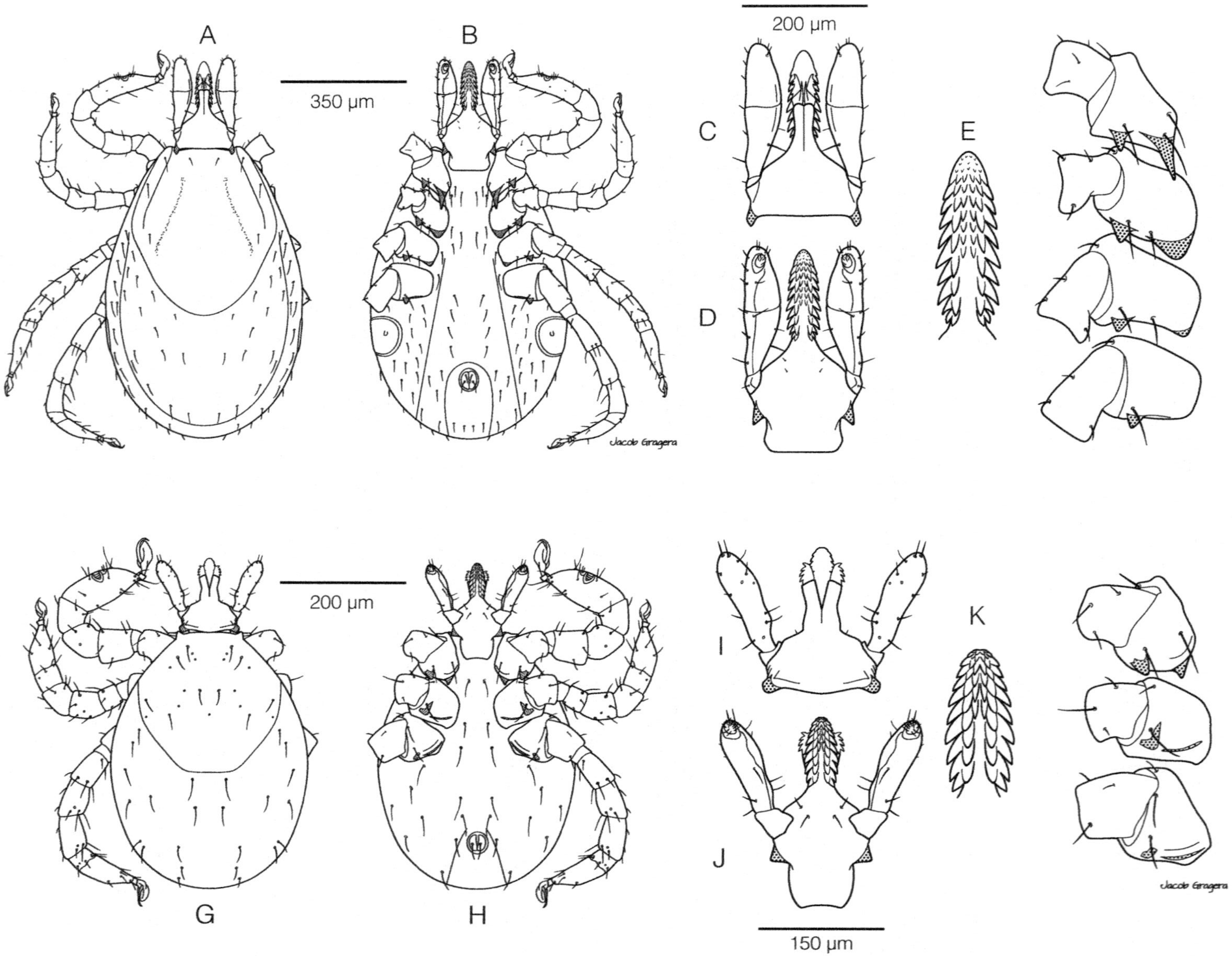

Fig. 72 **A–F** The nymph of *I. persulcatus*. **A** Dorsal, **B** ventral, **C** capitulum, dorsal, **D** capitulum, ventral, **E** hypostome, **F** coxae and trochanters I–IV. **G–L** The larva of *I. persulcatus*. **G** Dorsal, **H** ventral, **I** capitulum, dorsal, **J** capitulum, ventral, **K** hypostome, **L** coxae and trochanters I–III. Illustrations from specimens collected in Finland

Distribution

Ixodes persulcatus is the most common tick species in north-eastern Europe and northern Asia. Amongst others, it has been reported from China, Estonia, Japan, Korea, Kyrgyzstan, Latvia, Lithuania (one record), Poland, Russia, Slovakia and Turkey (Kriuchechnikov et al. 1988; Jaenson et al. 1994; Fukunaga et al. 1995; Postic et al. 1997; Sun and Xu 2003; Kim et al. 2003; Bormane et al. 2004; Labuda and Nuttall 2004; Bursali et al. 2010). It is occasionally introduced to western and northern Europe by spring-migrating birds (Jaenson et al. 1994; Petney et al. 2012).

Vectorial Capacity and Pathogen Burden

Ixodes persulcatus belongs to the group of the most important vectors of the *Borrelia* species causing Lyme disease in Eastern Europe and Asia. It is known to transmit *Borrelia afzelii* and *B. garinii* (Sun and Xu 2003; Masuzawa et al. 2008). *Borrelia burgdorferi* s.s. has been detected in this tick (Alekseev et al. 2004). Additionally, it is a vector of *Borrelia miyamotoi* (Fukunaga et al. 1995). The taiga tick also transmits various *Rickettsia* spp., including, *R. helvetica, R. japonica, R. sibirica*, and *Candidatus* "Rickettsia tarasevichiae" (Shpynov et al. 2003; Parola et al. 2005; Inokuma et al. 2007; Ando et al. 2010). *Ixodes persulcatus* acts as a vector for *Anaplasma phagocytophilum, Ehrlichia muris*, and *Candidatus* "Neoehrlichia mikurensis" (Kim et al. 2003; Cao et al. 2003; Alekseev et al. 2004; Inokuma et al. 2007; Masuzawa et al. 2008; Silaghi et al. 2012), as well as for *Babesia microti* and *Bartonella henselae* (Alekseev et al. 2004; Morozova et al. 2004). *Ehrlichia chaffeensis* has been detected in this tick (Kim et al. 2003). It is known to transmit Kemorovo virus, Omsk hemorrhagic fever virus, Powassan virus, and tick-borne encephalitis virus (Nuttall et al. 1994; Labuda and Nuttall 2004; Dantas-Torres et al. 2012).

References

Alekseev AN, Dubinina HV, Jushkova OV (2004) First report on the coexistence and compatibility of seven tickborne pathogens in unfed adult *Ixodes persulcatus* Schulze (Acarina: Ixodidae). Int J Med Microbiol Suppl 293:104–108. https://dx.doi.org/10.1016/S1433-1128(04)80015-9

Ando S, Kurosawa M, Sakata A et al (2010) Human *Rickettsia heilongjiangensis* infection, Japan. Emerg Infect Dis 16:1306–1308. https://dx.doi.org/10.3201/eid1608.100049

Bormane A, Lucenko I, Duks A et al (2004) Vectors of tick-borne diseases and epidemiological situation in Latvia in 1993–2002. Int J Med Microbiol Suppl 293:36–47. https://dx.doi.org/10.1016/S1433-1128(04)80007-X

Bursali A, Tekin S, Orhan M et al (2010) Ixodid ticks (Acari: Ixodidae) infesting humans in Tokat Province of Turkey: species diversity and seasonal activity. J Vector Ecol 35:180–186. https://dx.doi.org/10.1111/j.1948-7134.2010.00075.x

Cao W-C, Zhao Q-M, Zhang P-H et al (2003) Prevalence of *Anaplasma phagocytophila* and *Borrelia burgdorferi* in *Ixodes persulcatus* ticks from north-eastern China. Am J Trop Med Hyg 68:547–550

Dantas-Torres F, Chomel BB, Otranto D (2012) Ticks and tick-borne diseases: a One Health perspective. Trends Parasitol 28:437–446. https://dx.doi.org/10.1016/j.pt.2012.07.003

Estrada-Peña A, Jongejan F (1999) Ticks feeding on humans: a review of records on human-biting Ixodoidea with special reference to pathogen transmission. Exp Appl Acarol 23:685–715. https://dx.doi.org/10.1023/A:1006241108739

Fukunaga M, Takahashi Y, Tsuruta Y et al (1995) Genetic and phenotypic analysis of *Borrelia miyamotoi* sp. nov., isolated from the ixodid tick *Ixodes persulcatus*, the vector for Lyme disease in Japan. Int J Syst Bacteriol 45:804–810. https://dx.doi.org/10.1099/00207713-45-4-804

Gray JS (1998) The ecology of ticks transmitting Lyme borreliosis. Exp Appl Acarol 22:249–258. https://dx.doi.org/10.1023/A:1006070416135

Guglielmone AA, Robbins RG, Apanaskevich DA, Petney TN, Estrada-Peña A, Horak IG (2014) The hard ticks of the world (Acari: Ixodida: Ixodidae). Springer, Dordrecht, p 738

Inokuma H, Ohashi M, Tanabe S, Miyahara K (2007) Prevalence of tick-borne *Rickettsia* and *Ehrlichia* in *Ixodes persulcatus* and *Ixodes ovatus* in Tokachi District, Eastern Hokkaido, Japan. J Vet Med Sci 69:661–664. https://dx.doi.org/10.1292/jvms.69.661

Jaenson TGT, Tälleklint L, Lundqvist L et al (1994) Geographical distribution, host associations, and vector roles of ticks (Acari: Ixodidae, Argasidae) in Sweden. J Med Entomol 31:240–256. https://dx.doi.org/10.1093/jmedent/31.2.240

Kim C-M, Kim M-S, Park M-S et al (2003) Identification of *Ehrlichia chaffeensis, Anaplasma phagocytophilum*, and *A. bovis* in *Haemaphysalis longicornis* and *Ixodes persulcatus* ticks from Korea. Vector Borne Zoonotic Dis 3:17–26. https://dx.doi.org/10.1089/153036603765627424

Konnai S, Saito Y, Nishikado H et al (2008) Establishment of a laboratory colony of taiga tick *Ixodes persulcatus* for tick-borne pathogen transmission studies. Jpn J Vet Res 55:85–92

Korenberg EI (2000) Seasonal population dynamics of *Ixodes* ticks and tick-borne encephalitis virus. Exp Appl Acarol 24:665–681. https://dx.doi.org/10.1023/a:1010798518261

Korenberg EI, Gorelova NB, Kovalevskii YI (2002) Ecology of *Borrelia burgdorferi* sensu lato in Russia. In: Gray JS, Kahl O, Lane RS, Stanek G (eds) Lyme borreliosis biology epidemiology and control, 1st edn. CABI Publishing, Oxon, pp 175–200

Kriuchechnikov VN, Korenberg EI, Shcherbakov S V et al (1988) Identification of *Borrelia* isolated in the USSR from *Ixodes persulcatus* Schulze ticks. Zh Mikrobiol Epidemiol Immunobiol 12:41–44

Labuda M, Nuttall PA (2004) Tick-borne viruses. Parasitology 129: S221–S245. https://dx.doi.org/10.1017/S0031182004005220

Masuzawa T, Kharitonenkov IG, Okamoto Y et al (2008) Prevalence of *Anaplasma phagocytophilum* and its coinfection with *Borrelia afzelii* in *Ixodes ricinus* and *Ixodes persulcatus* ticks inhabiting Tver Province (Russia)—a sympatric region for both tick species. J Med Microbiol 57:986–991. https://dx.doi.org/10.1099/jmm.0.47721-0

Morozova DOV, Cabello FC, Dobrotvorsky AK (2004) Semi-nested PCR detection of *Bartonella henselae* in *Ixodes persulcatus* ticks from Western Siberia, Russia. Vector Borne Zoonotic Dis 4:306–310

Nuttall PA, Jones LD, Labuda M, Kaufman WR (1994) Adaptations of arboviruses to ticks. J Med Entomol 31:1–9. https://dx.doi.org/10.1093/jmedent/31.1.1

Parola P, Davoust B, Raoult D (2005) Tick- and flea-borne rickettsial emerging zoonoses. Vet Res 36:469–492. https://dx.doi.org/10.1051/vetres:2005004

Petney TN, Pfäffle MP, Skuballa JD (2012) An annotated checklist of the ticks (Acari: Ixodida) of Germany. Syst Appl Acarol 17:115–170. https://dx.doi.org/10.11158/saa.17.2.2

Postic D, Korenberg E, Gorelova N et al (1997) *Borrelia burgdorferi* sensu lato in Russia and neighbouring countries: high incidence of mixed isolates. Res Microbiol 148:691–702. https://dx.doi.org/10.1016/S0923-2508(99)80068-0

Shpynov S, Fournier PE, Rudakov N, Raoult D (2003) '*Candidatus* Rickettsia tarasevichiae' in *Ixodes persulcatus* ticks collected in Russia. Ann N Y Acad Sci 990:162–172. https://dx.doi.org/10.1111/j.1749-6632.2003.tb07358.x

Silaghi C, Woll D, Mahling M et al (2012) *Candidatus* Neoehrlichia mikurensis in rodents in an area with sympatric existence of the hard ticks *Ixodes ricinus* and *Dermacentor reticulatus*, Germany. Parasit Vectors 5:285. https://dx.doi.org/10.1186/1756-3305-5-285

Sun Y, Xu R (2003) Ability of *Ixodes persulcatus*, *Haemaphysalis concinna* and *Dermacentor silvarum* ticks to acquire and transstadially transmit *Borrelia garinii*. Exp Appl Acarol 31:151–160. https://dx.doi.org/10.1023/B:APPA.0000005119.30172.43

Uspensky I, Garruto RM, Goldfarb L (2003) The taiga tick *Ixodes persulcatus* (Acari: Ixodidae) in the Sakha Republic (Yakutia) of Russia: distributional and reproductive ranges. J Med Entomol 40:119–122. https://dx.doi.org/10.1603/0022-2585-40.1.119

Yasyukevich VV, Kazakova EV, Popov IO, Semenov SM (2009) Distribution of *Ixodes ricinus* L., 1758 and *Ixodes persulcatus* Schulze, 1930 (Parasitoformes, Ixodidae) in Russia and adjacent countries in view of observable climate changes. Dokl Earth Sci 427:1030–1034. https://dx.doi.org/10.1134/S1028334X09060312

Ixodes inopinatus Estrada-Peña, Nava and Petney, 2014 (Figs. 73–75)

A. Estrada-Peña

Ixodes inopinatus was described from specimens collected in Spain, Portugal, Germany, Morocco and Tunisia, mostly in areas where it replaces *I. ricinus*. Since its original description, this species has also been collected in Austria and Romania. In the latter country, it has been recorded in sympatry with *I. ricinus* (unpublished records).

Life-Cycle and Host Preferences

Ixodes inopinatus is an exophilous tick. The only recorded hosts for immatures are lizards of several species. The most common hosts for adults are the red fox, *Vulpes vulpes*. Several females have been found feeding on sheep in southern Germany (Estrada-Peña et al. 2014).

Ecology

Ixodes inopinatus mainly inhabits deciduous and coniferous forests in Mediterranean-type habitats. It has been collected in forests of *Pinus* spp. and *Quercus* spp., with around 400–700 mm of rainfall per year, and an altitude up to 800 m a.s.l. Since it was only recently described, no details about its ecology or seasonality are known. Immatures have been collected on hosts and by flagging in spring.

Distribution

It has been found in Spain, Portugal, Germany, Morocco, Tunisia, Austria and Romania, but the full range of its distribution remains to be determined.

Vectorial Capacity and Pathogen Burden

These aspects are unknown.

Reference

Estrada-Peña A, Nava S, Petney T (2014) Description of all the stages of *Ixodes inopinatus* n. sp. (Acari: Ixodidae). Ticks Tick-Borne Dis 5:734–743

A. Estrada-Peña (✉)
Faculty of Veterinary Medicine, Department of Animal Health, Zaragoza, Spain
e-mail: aestrada@unizar.es

A. Estrada-Peña et al. (eds.), *Ticks of Europe and North Africa*,
https://doi.org/10.1007/978-3-319-63760-0_40

Fig. 73 **A** Distribution of *I. inopinatus* in Europe and Northern Africa (10 × 10 km grid presence with black dots). **B** Countries where the species has been reported are marked in dark grey

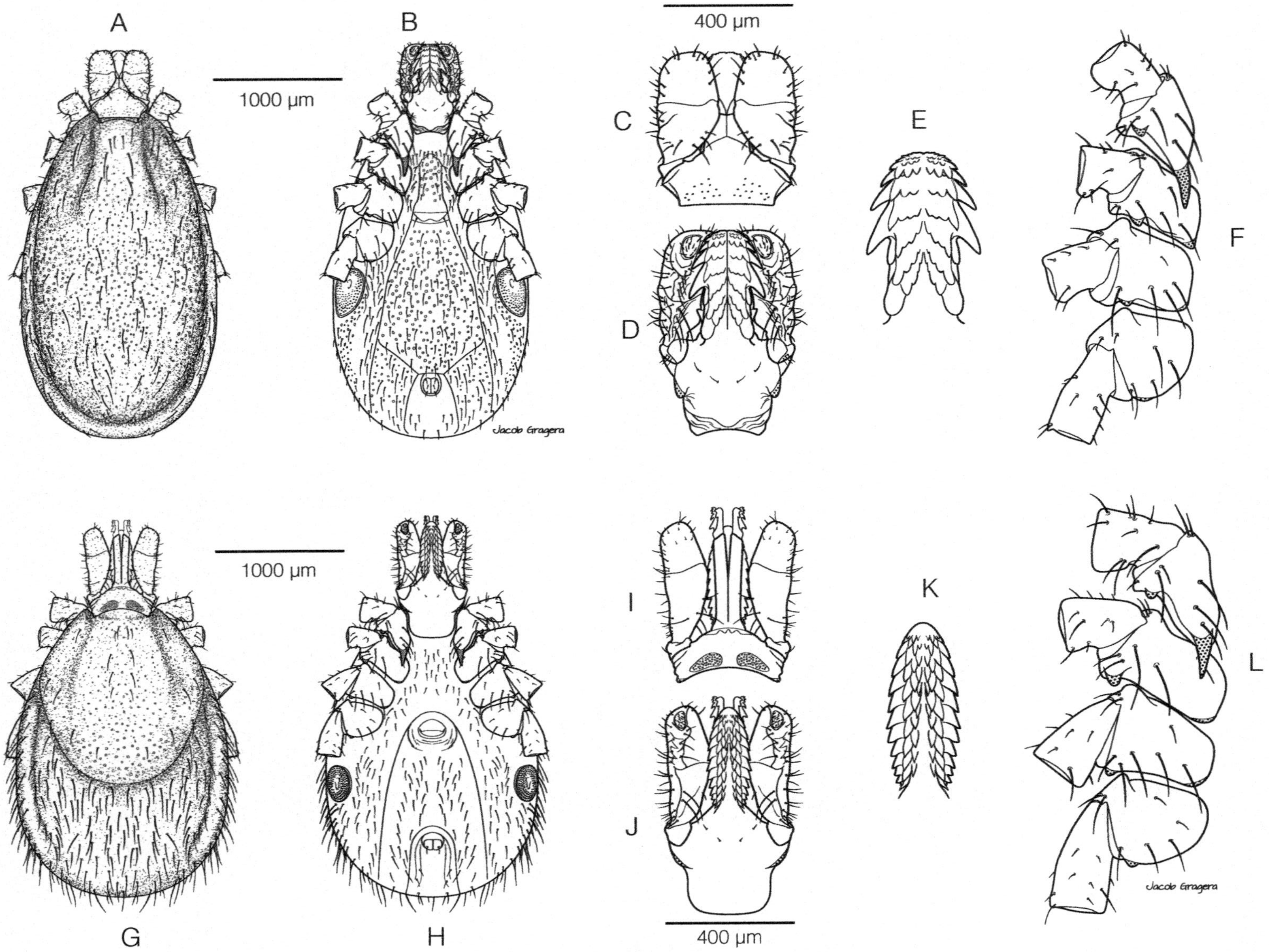

Fig. 74 **A–F** The male of *I. inopinatus*. **A** Dorsal, **B** ventral, **C** capitulum, dorsal, **D** capitulum, ventral, **E** hypostome, **F** coxae and trochanters I–IV. **G–L** The female of *I. inopinatus*. **G** Dorsal, **H** ventral, **I** capitulum, dorsal, **J** capitulum, ventral, **K** hypostome, **L** coxae and trochanters I–IV. Illustrations from specimens collected in Spain and Tunisia

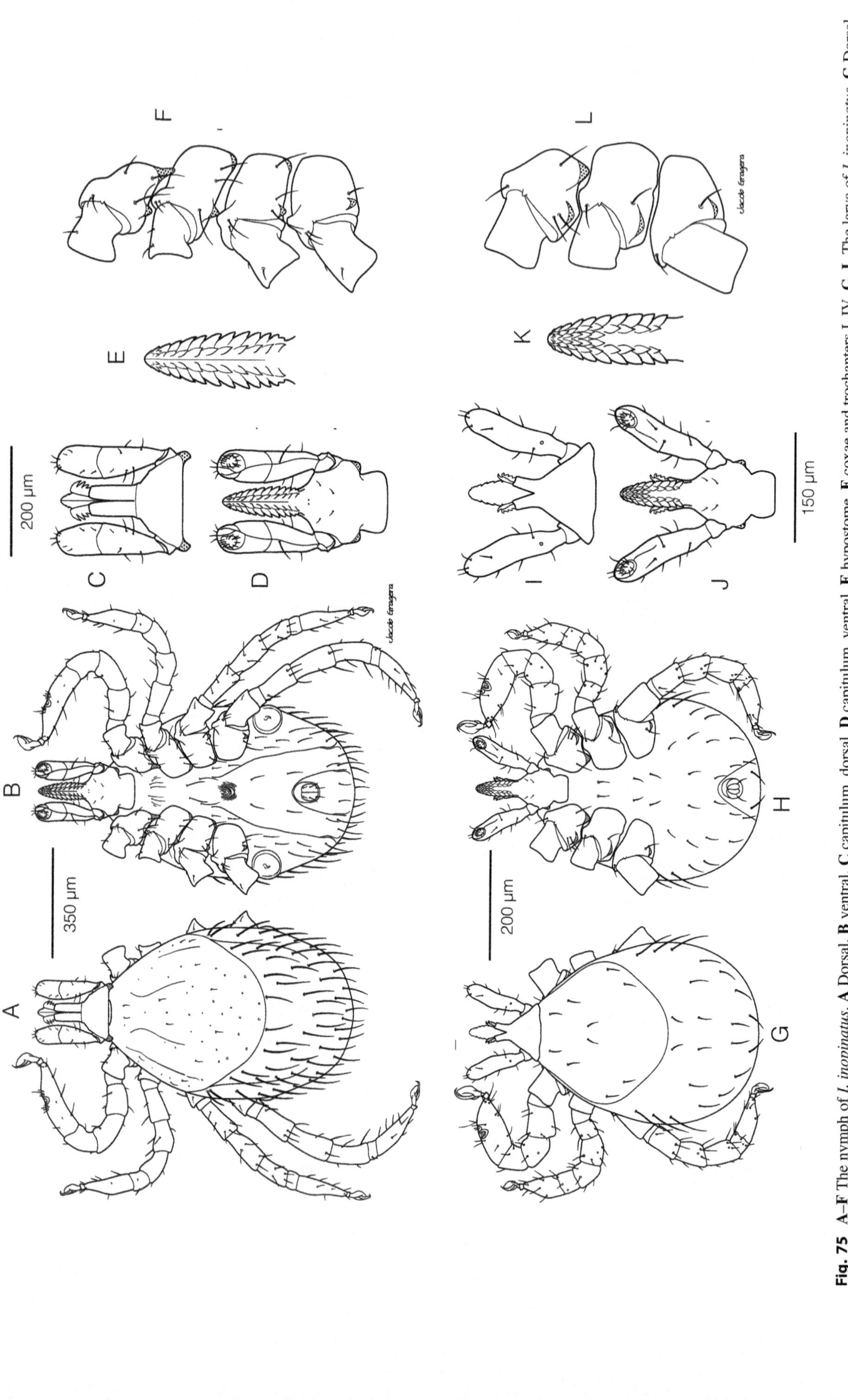

Fig. 75 **A–F** The nymph of *I. inopinatus*. **A** Dorsal, **B** ventral, **C** capitulum, dorsal, **D** capitulum, ventral, **E** hypostome, **F** coxae and trochanters I–IV. **G–L** The larva of *I. inopinatus*. **G** Dorsal, **H** ventral, **I** capitulum, dorsal, **J** capitulum, ventral, **K** hypostome, **L** coxae and trochanters I–III. Illustrations from specimens collected in Spain and Tunisia

Ixodes gibbosus Nuttall, 1916 (Figs. 76–78)

A. Estrada-Peña

Ixodes gibbosus is a relatively common but poorly known species, so far reported from parts of the Mediterranean Basin. It is a species close to *l. ricinus* and *l. persulcatus*, but is distributed in areas considered too warm and dry to support permanent populations of *I. ricinus* which it replaces in several Mediterranean countries. Nevertheless, it has been collected sympatrically with *I. ricinus* in some areas of Greece (Anna Papa, personal communication).

Life-Cycle and Host Preferences

Ixodes gibbosus is an exophilic, three-host tick. Many details of its life cycle are unknown. From records in nature, it is expected that the life cycle is similar to that of *I. ricinus*, but shorter because the higher temperature in the places where *I. gibbosus* exists. It is active during the winter and spring. Cattle, sheep, goats, dogs and wild ungulates can be infested with this tick. On both cattle and sheep, it prefers to feed on the ears, eyes and neck. Passeriform birds have been recorded as hosts for immatures. According to Guglielmone et al. (2014), *I. gibbosus* has been recorded on Equidae, Camelidae, Carnivora (unknown species), Erinaceidae, Leporidae and unknown orders of Aves (Turdidae is reported for nymphs, but the record is unverified). It has been found feeding on humans (Saratisotis 1970; Morel 2003).

Ecology

I. gibbosus mainly inhabits Mediterranean forests, woodlands and scrub.

Distribution

It occurs in the eastern part of the Mediterranean region (Greece, Turkey, Israel), and has also been recorded as far west as Italy, which appears to marks its western limit. It has been suggested that records of *I. ricinus* from southern Italy and Sicily may actually be *I. gibbosus*. However, Dantas-Torres and Otranto (2013) have reported and described the phenology of a permanent populations of *I. ricinus* in southern Italy, which probably represents the most permanent southerly population of this species. It is also present in the countries of the former Yugoslavia, where it seems to coexist with *I. ricinus*. As indicated, it also coexists with *I. ricinus* in Eastern parts of Greece.

Vectorial Capacity and Pathogen Burden

These aspects are unknown.

A. Estrada-Peña (✉)
Faculty of Veterinary Medicine, Department of Animal Health, Zaragoza, Spain
e-mail: aestrada@unizar.es

A. Estrada-Peña et al. (eds.), *Ticks of Europe and North Africa*,
https://doi.org/10.1007/978-3-319-63760-0_41

Fig. 76 **A** Distribution of *I. gibbosus* in Europe and Northern Africa (10 × 10 km grid presence with black dots). **B** Countries where the species has been reported are marked in dark grey

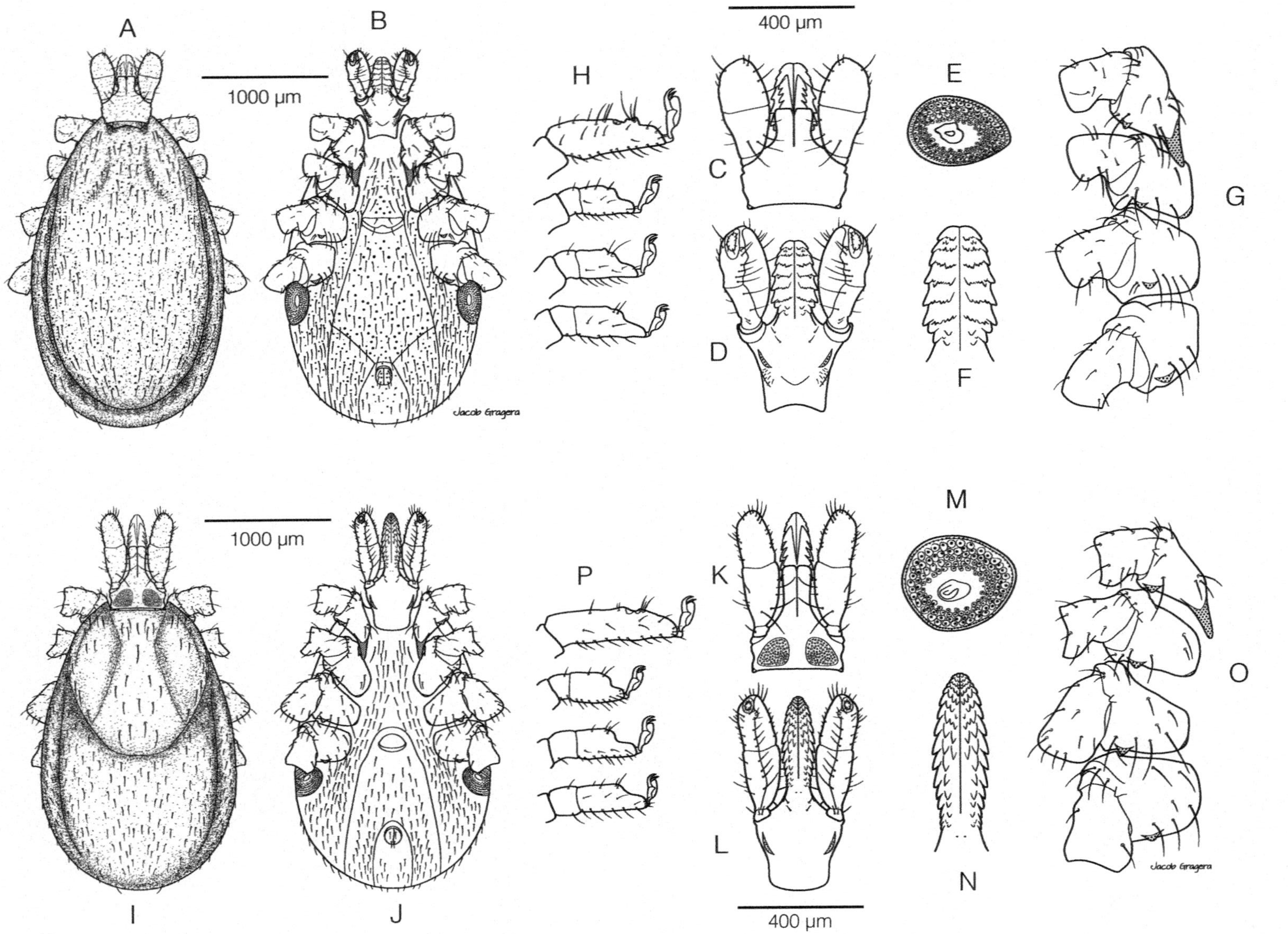

Fig. 77 **A–H** The male of *I. gibbosus*. **A** Dorsal, **B** ventral, **C** capitulum, dorsal, **D** capitulum, ventral, **E** spiracular plate, **F** hypostome, **G** coxae and trochanters I–IV, **H** tarsi I–IV. **I–P** The female of *I. gibbosus*. **I** Dorsal, **J** ventral, **K** capitulum, dorsal, **L** capitulum, ventral, **M** spiracular plate, **N** hypostome, **O** coxae and trochanters I–IV, **P** tarsi I–IV. Illustrations from specimens collected in Greece

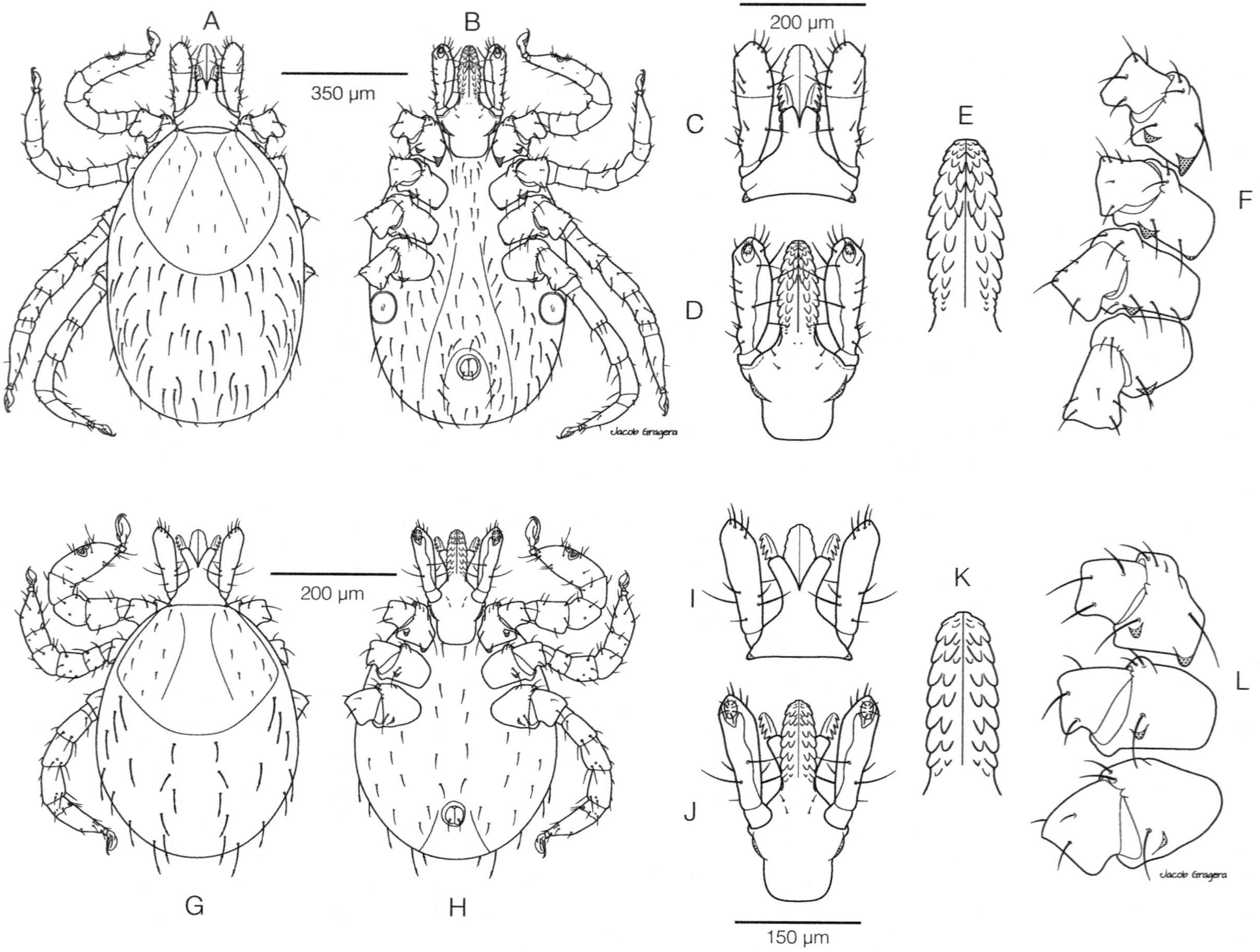

Fig. 78 **A–F** The nymph of *I. gibbosus*. **A** Dorsal, **B** ventral, **C** capitulum, dorsal, **D** capitulum, ventral, **E** hypostome, **F** coxae and trochanters I–IV. **G–L** The larva of *I. gibbosus*. **G** Dorsal, **H** ventral, **I** capitulum, dorsal, **J** capitulum, ventral, **K** hypostome, **L** coxae and trochanters I–III. Illustrations from specimens collected in Greece

References

Dantas-Torres F, Otranto D (2013) Seasonal dynamics of *Ixodes ricinus* on ground level and higher vegetation in a preserved wooded area in southern Europe. Vet Parasitol 192:253–258

Guglielmone AA, Robbins RG, Apanaskevich DA, Petney TN, Estrada-Peña A, Horak IG (2014) The hard ticks of the world (Acari: Ixodida: Ixodidae). Springer, Dordrecht, p 738

Morel PC (2003) Les tiques d'Afrique et du Bassin méditerranéen (1965-1995). CIRAD-EMVT, 1342 pp

Saratisotis A (1970) Étude morphologique et observations biologiques sûr *Ixodes gibbosus* Nuttall. Ann Parasitol Hum Comp 45:661–675

Ixodes eldaricus Dzhaparidze, 1950 (Figs. 79–81)

A. Estrada-Peña

Ixodes eldaricus is a poorly known species, so far reported from parts of Russia, the Near East and Central Asia, with populations in Eastern Europe, detected on migratory hosts. It is a species which has probably been confused with *I. frontalis*, *I. acuminatus* and *I. ricinus*, with which it has morphological similarities.

Life-Cycle and Host Preferences

Ixodes eldaricus is an exophilous tick. All stages are parasites of ground feeding birds (Nowak-Chmura 2011). Mice have been recorded as hosts for larvae and nymphs. Females have been collected in February–June and September–December, and more abundantly in April–May and October–November. Nymphs have been found in January, March–May, July, and October–December and larvae in April–June and October–December (Filippova 1974).

Ecology

Ixodes eldaricus mainly inhabits mountainous coniferous and deciduous scrub forests in mountain river valleys up to an altitude of ca. 1800 m a.s.l.

Distribution

The hitherto defined geographical distribution of *I. eldaricus* covers the southern Ukraine (Crimea), Georgia, Azerbaijan, Armenia, Kazakhstan, Turkmenistan, Kirghizia, Uzbekistan, Tajikistan, and Russia (Dagestan) (Filippova 1974). There have also been reports of this species on birds in Cyprus (Kaiser et al. 1974) and Israel, where it was described under the name *Ixodes tatei* Arthur 1956, a junior synonym of *I. eldaricus*. Permanent populations have been found in Israel (Erster unpublished data).

Vectorial Capacity and Pathogen Burden

These aspects are unknown.

A. Estrada-Peña (✉)
Faculty of Veterinary Medicine, Department of Animal Health, Zaragoza, Spain
e-mail: aestrada@unizar.es

A. Estrada-Peña et al. (eds.), *Ticks of Europe and North Africa*,
https://doi.org/10.1007/978-3-319-63760-0_42

Fig. 79 **A** Distribution of *I. eldaricus* in Europe and Northern Africa (10 × 10 km grid presence with black dots). **B** Countries where the species has been reported are marked in dark grey

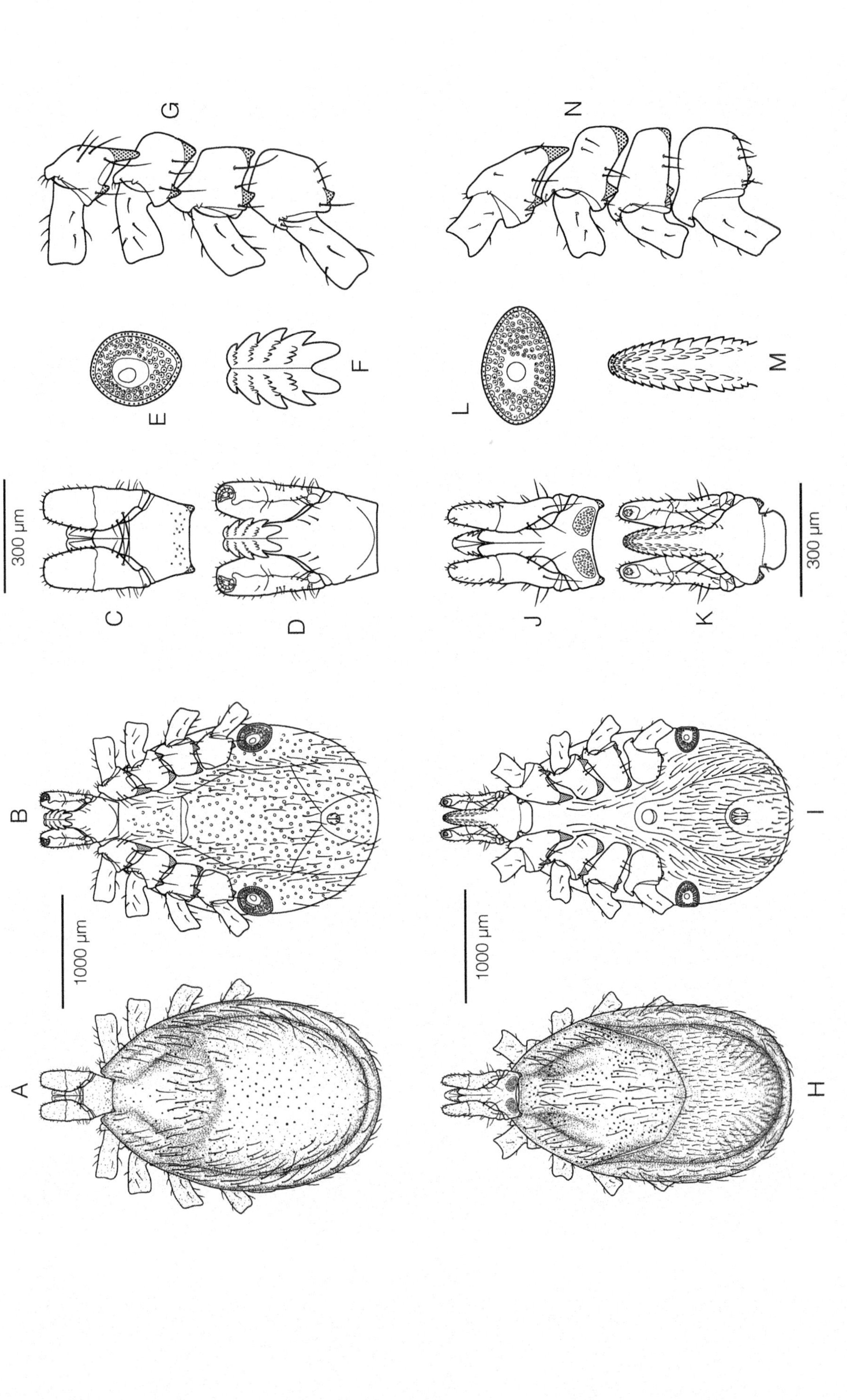

Fig. 80 **A–G** The male of *I. eldaricus*. **A** Dorsal, **B** ventral, **C** capitulum, dorsal, **D** capitulum, ventral, **E** spiracular plate, **F** hypostome, **G** coxae and trochanters I–IV. **H–N** The female of *I. eldaricus*. **H** Dorsal, **I** ventral, **J** capitulum, dorsal, **K** capitulum, ventral, **L** spiracular plate, **M** hypostome, **N** coxae and trochanters I–IV. Illustrations from specimens collected in Israel

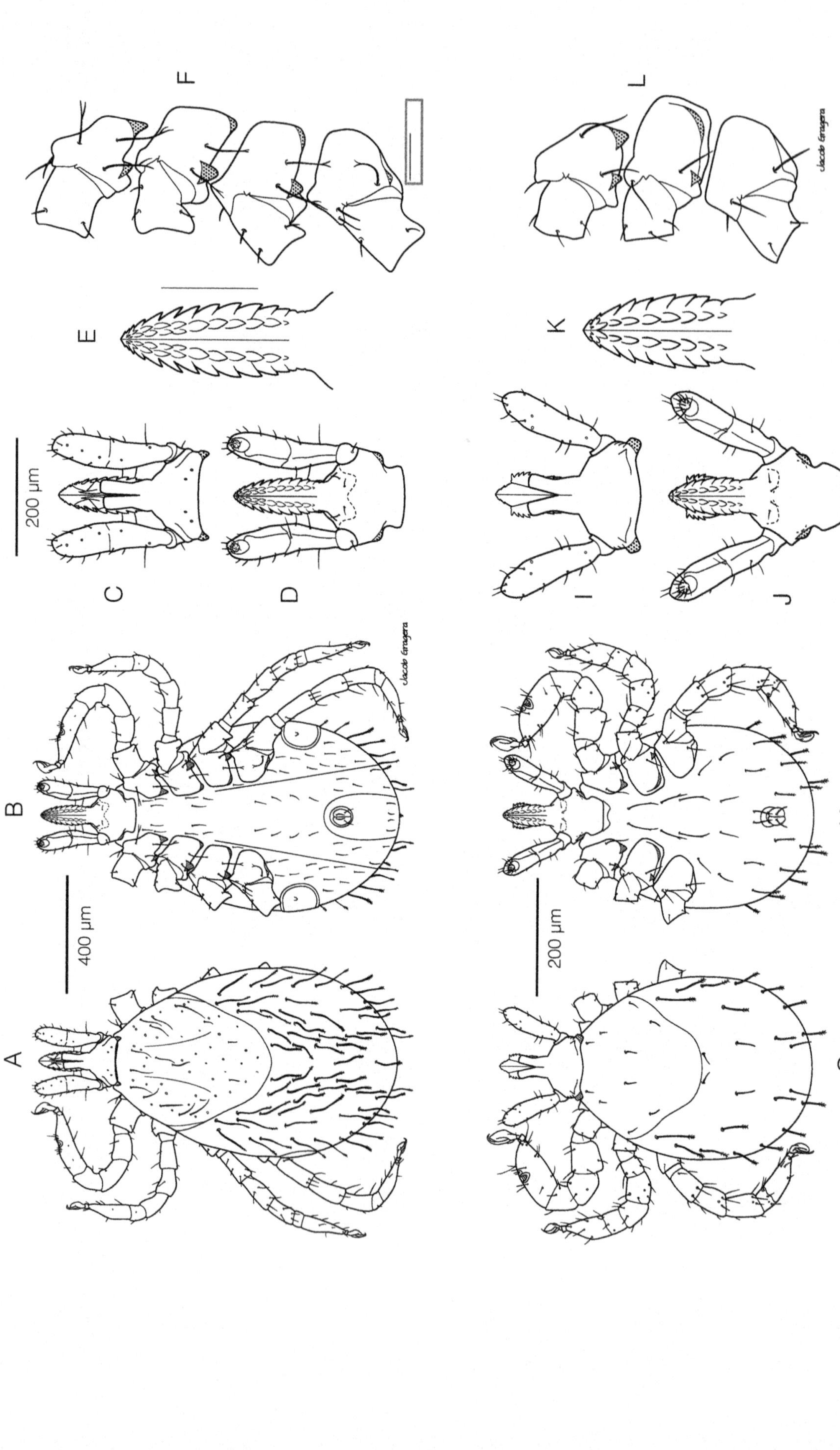

Fig. 81 **A–F** The nymph of *I. eldaricus*. **A** Dorsal, **B** ventral, **C** capitulum, dorsal, **D** capitulum, ventral, **E** hypostome, **F** coxae and trochanters I–IV. **G–L** The larva of *I. eldaricus*. **G** Dorsal, **H** ventral, **I** capitulum, dorsal, **J** capitulum, ventral, **K** hypostome, **L** coxae and trochanters I–III. Illustrations from specimens collected in Israel

References

Berdyev A (1973) O nakhozhdenii kleshcha *Ixodes eldaricus* Djab., 1950 na gryzunakh v Kopetdage (Ixodoidea, Ixodidae). Parazitologiya 7:183–184

Filippova NA (1974) *Ixodes eldaricus* i ego rasprostranenie na yuge SSSR. Parazitologiya 8:504–514

Kaiser MN, Hoogstraal H, Watson GE (1974) Ticks (Ixodoidea) on migrating birds in Cyprus, fall 1967 and spring 1968, and epidemiological considerations. Bull Entomol Res 64:97–110

Nowak-Chmura M (2011) *Ixodes eldaricus* Djaparidze, 1950 (Ixodidae) on migrating birds. Reported first time in Poland. Vet Parasitol 186(3):399–402

Ixodes laguri Olenev, 1929 (Figs. 82–84)

A. D. Mihalca and G. D'Amico

Life-Cycle and Host Preferences

This is a three-host tick species with a natural life cycle of 2–3 years (Honzáková et al. 1979). Under experimental conditions, the life cycle can be completed on laboratory mice and lasts between 96 and 192 days (Pomerantzev 1959). The typical hosts are burrowing rodents, mainly Cricetidae, Sciuridae and Gliridae, but occasional reports are also found for other rodent families, as well as for hedgehogs and small carnivores (Pomerantzev 1959). No difference in the host preferences of immatures and adults are known. However, according to Pomerantzev (1959), adult males are not parasitic. The tick has been reported on humans (Bursali et al. 2010).

Ecology

Although closely dependent on the relatively constant microclimate of the host's burrow, it seems that the seasonal activity of this tick is variable according to the geographical region. Generally, all stages are active from March to January, with two peaks: one higher in April–May and the second, less intense in July–August (Pomerantzev 1959; Feider 1965). This species overwinters mainly as females and nymphs, and rarely as larvae (Pomerantzev 1959; Honzáková et al. 1979). Males seem to be present only in late spring–early summer (Feider 1965). A high humidity is required for development, with the eggs being particularly sensitive to dessication (Feider 1965). Its distribution is confined to humid lowlands rather than arid habitats. The tick is typically nidiculous, although occasionally it has been found questing in very close proximity to the burrows of its hosts (Mihalca, personal observation).

Distribution

It has been reported in Armenia, Austria, Azerbaijan, Belarus, Bulgaria, Dagestan, Estonia, Georgia, Hungary, Kazakhstan, Latvia, Lithuania, Mongolia, Moldova, Romania, Russia, Slovakia, Turkey, Turkmenistan, Ukraine and Uzbekistan (Arthur 1957; Pomerantzev 1959; Honzáková et al. 1979; Filippova 2008; Bursali et al. 2010; Kiefer et al. 2010; Suppan 2013).

Vectorial Capacity and Pathogen Burden

The vectorial role of *I. laguri* is mostly unknown. Bozhenko and Shevchenko (1956) suggested it as a vector for *Francisella tularensis*. An uncertain species of *Rickettsia* (described as "*R. pavlovskii*") causing human "infectious nephroso-nephritis" has been isolated from ticks of this species in the former USSR (Philip and Burgdorfer 1961). No further aspects are known about its epidemiological significance.

A. D. Mihalca · G. D'Amico (✉)
Department of Parasitology and Parasitic Diseases, University of Agricultural Sciences and Veterinary Medicine, Cluj-Napoca, Romania
e-mail: gianluca.damico@usamvcluj.ro

A. D. Mihalca
e-mail: amihalca@usamvcluj.ro

A. Estrada-Peña et al. (eds.), *Ticks of Europe and North Africa*,
https://doi.org/10.1007/978-3-319-63760-0_43

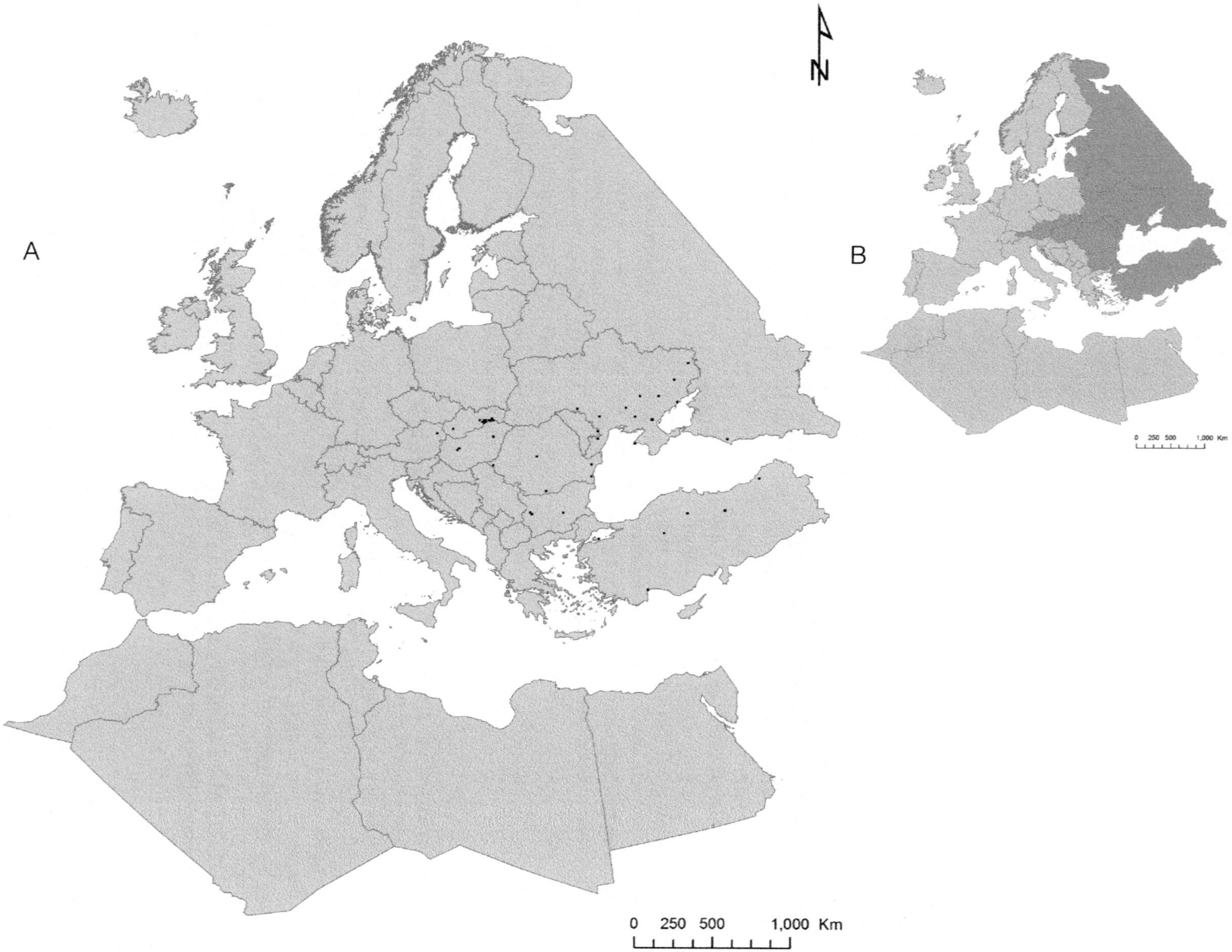

Fig. 82 **A** Distribution of *I. laguri* in Europe and Northern Africa (10 × 10 km grid presence with black dots). **B** Countries where the species has been reported are marked in dark grey

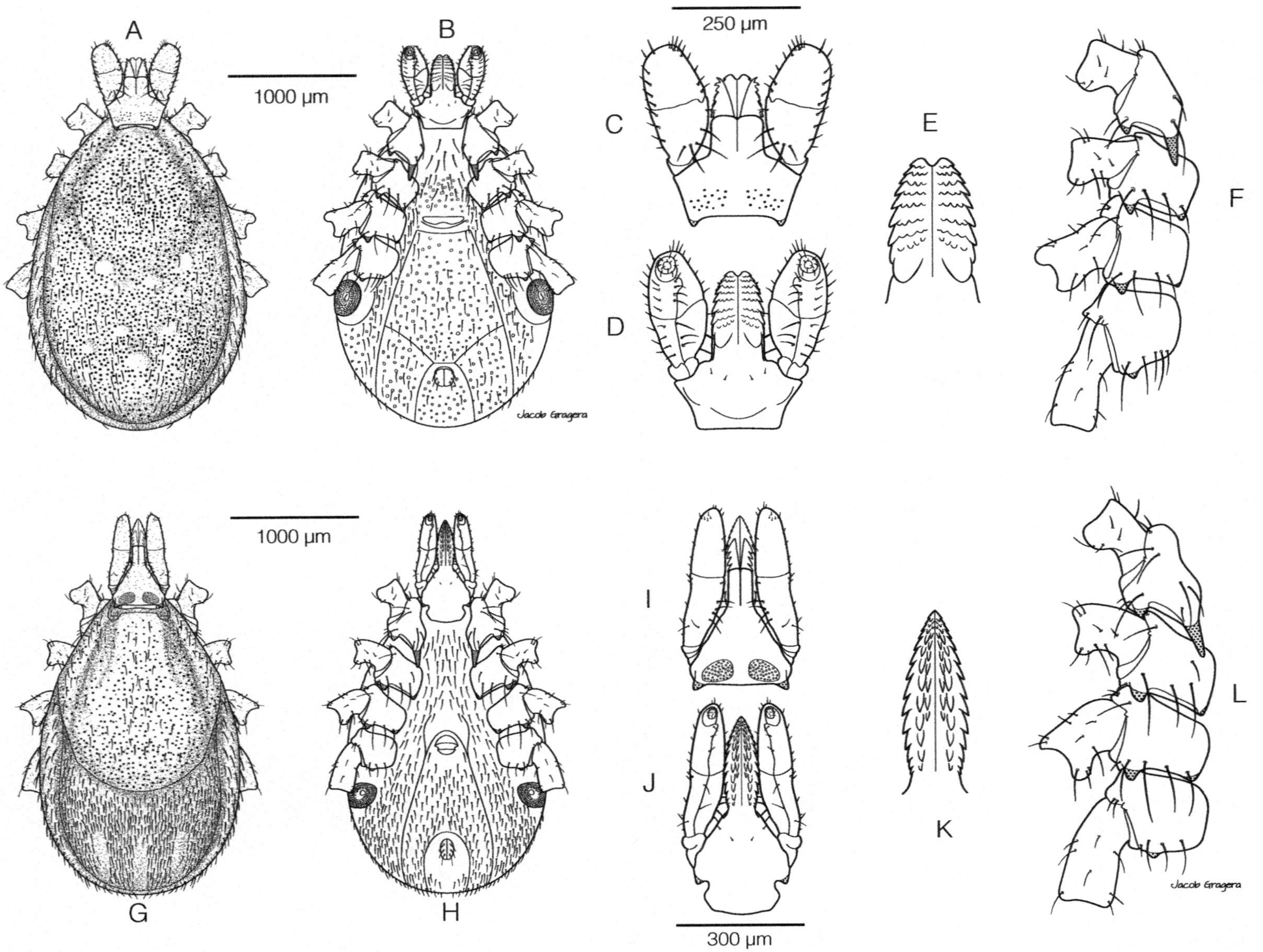

Fig. 83 **A–F** The male of *I. laguri*. **A** Dorsal, **B** ventral, **C** capitulum, dorsal, **D** capitulum, ventral, **E** hypostome, **F** coxae and trochanters I–IV. **G–L** The female of *I. laguri*. **G** Dorsal, **H** ventral, **I** capitulum, dorsal, **J** capitulum, ventral, **K** hypostome, **L** coxae and trochanters I–IV. Illustrations from specimens collected in Slovakia

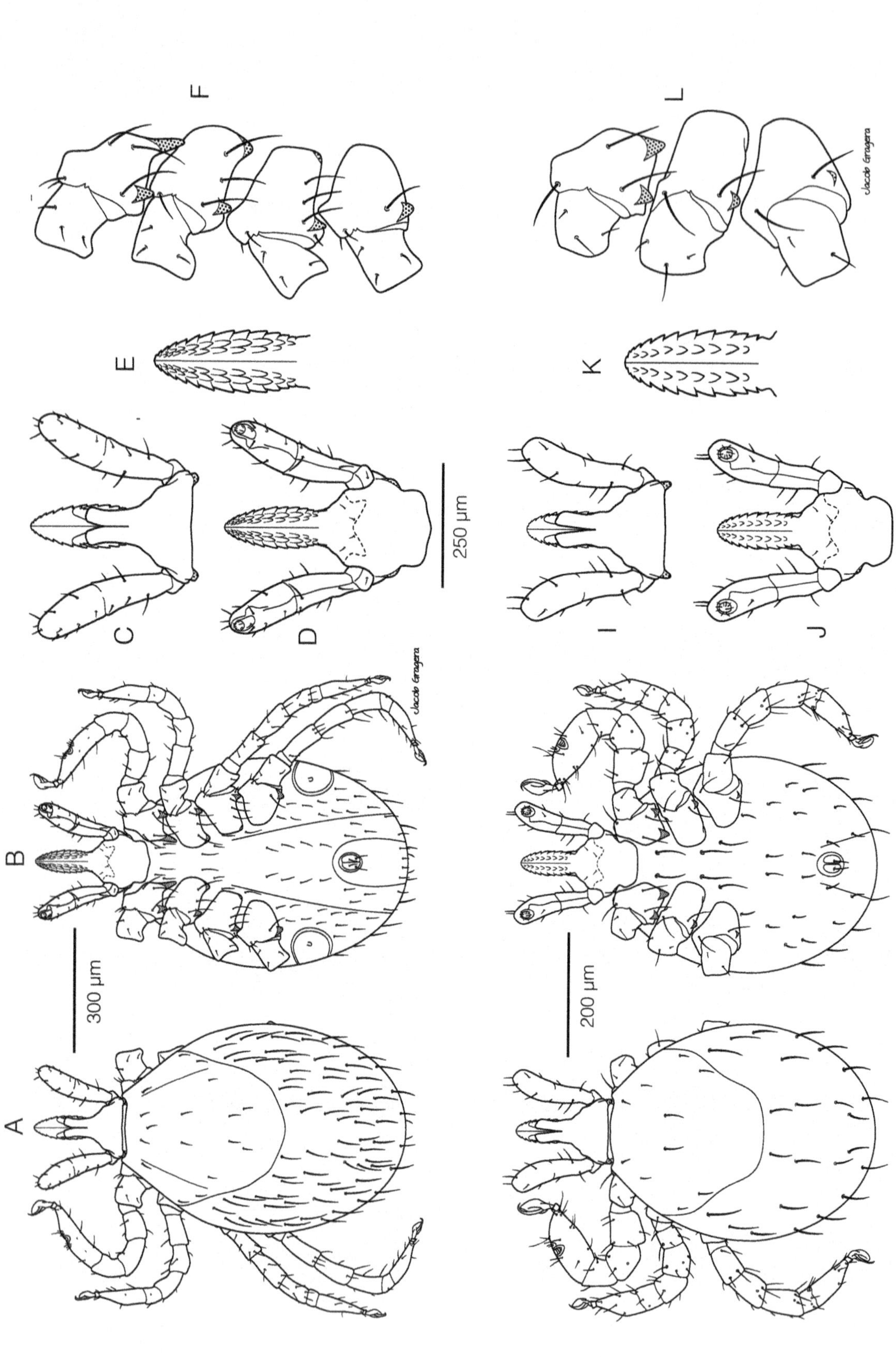

Fig. 84 **A–F** The nymph of *I. laguri*. **A** Dorsal, **B** ventral, **C** capitulum, dorsal, **D** capitulum, ventral, **E** hypostome, **F** coxae and trochanters I–IV. **G–L** The larva of *I. laguri*. **G** Dorsal, **H** ventral, **I** capitulum, dorsal, **J** capitulum, ventral, **K** hypostome, **L** coxae and trochanters I–III. Illustrations from specimens collected in Slovakia

References

Arthur DR (1957) Studies on exotic *Ixodes* ticks (Ixodoidea, Ixodidae) from United States navy and army activities. J Parasitol 43:681–694

Bozhenko VP, Shevchenko SF (1956) To the ecology of the tick *Ixodes laguri laguri* Ol. in connection with its significance in the maintenance of some natural foci of tularemia. Zool Zh 35:837–842

Bursali A, Tekin S, Orhan M, Keskin A, Ozkan M (2010) Ixodid ticks (Acari: Ixodidae) infesting humans in Tokat Province of Turkey: species diversity and seasonal activity. J Vector Ecol 35:180–186

Feider Z (1965) Arachnida. Acaromorpha, Suprafamily Ixodoidea (Ticks), Fauna of the Peoples Republic of Romania [in Romanian]. Bucharest: Editura Academiei Republicii Populare Române, 404 pp

Filippova NA (2008) Type specimens of argasid and ixodid ticks (Ixodoidea: Argasidae, Ixodidae) in the collection of the Zoological Institute, Russian Academy of Sciences (St. Petersburg). Entomol Rev 88:1002–1011

Honzáková E, Cerný V, Daniel M, Dusbabek F (1980) Development of the tick *Ixodes laguri* Ol. in the nests of the European suslik *Citellus citellus* (L.). Folia Parasitologica 27:71–75

Kiefer D, Pfister K, Tserennorov D, Bolormaa G, Otgonbaatar D, Samjaa R, Kiefer MS (2010) Current state of Ixodidae research in Mongolia. Erforsch Biol Res Mongolei 11:405-418

Philip CB, Burgdorfer W (1961) Arthropod vectors as reservoirs of microbial disease agents. Annu Rev Entomol 6:391–412

Pomerantzev BI (1959) Fauna of USSR Arachnida. Ixodid Ticks (Ixodidae) 4(2)

Suppan J (2013) Structure and possible functions of the spiracle glands in *Ixodes ricinus* L. Doctoral dissertation, Uniwien

Genus *Haemaphysalis* Koch, 1844

A. Estrada-Peña, M. P. Pfäffle, and T. N. Petney

Haemaphysalis Koch, 1844, is a widely distributed genus of ticks with 167 accepted species recognized worldwide (Guglielmone et al. 2014). The larvae, nymphs and at least one adult stage are described for only 98 (59%) of all the known species. Nevertheless, most parasitic stages of species within the Palaearctic are known (Guglielmone et al. 2014). There are at least eight species of the genus *Haemaphysalis* known that occur in the Western Palaearctic, namely *Haemaphysalis inermis* Birula, 1895, *Haemaphysalis punctata* Canestrini and Fanzago, 1878, *Haemaphysalis sulcata* Canestrini and Fanzago, 1878, *Haemaphysalis caucasica* Olenev, 1928, *Haemaphysalis concinna* Koch, 1844, *Haemaphysalis hispanica* Gil Collado, 1938, *Haemaphysalis erinacei* Pavesi, 1884 and *Haemaphysalis parva* (Neumann 1897). An almost complete list of synonymies of these species can be found in Camicas et al. (1998). All stages of the species recorded in the Western Palaearctic have been adequately described and illustrated.

Haemaphysalis inermis was redescribed with keys and illustrations by Filippova (1997) and Manilla (1998). There are adequate illustrations and a reliable redescription in Hoogstraal (1969) and Hoogstraal and Kim (1985) which were repeated to some extent by Siuda (1993) and Pérez-Eid (2007). Basic SEM pictures of both the male and female are available in Cringoli et al. (2005).

A. Estrada-Peña (✉)
Department of Animal Health, Faculty of Veterinary Medicine, Zaragoza, Spain
e-mail: aestrada@unizar.es

M. P. Pfäffle
Department of Ecology and Parasitology, Institute of Zoology, Karlsruhe Institute of Technology, Karlsruhe, Germany
e-mail: miripfaeffle@web.de

T. N. Petney
State Museum of Natural History, Erbprinzenstr. 13, 76133 Karlsruhe, Germany
e-mail: trevor.petney@kit.edu

A. Estrada-Peña et al. (eds.), *Ticks of Europe and North Africa*,
https://doi.org/10.1007/978-3-319-63760-0_44

Haemaphysalis punctata was redescribed, illustrated, compared with close species and its individual variation assessed by Pomerantzev (1950) and Filippova (1997). Additional illustrations are available in Hoogstraal and Kim (1985), which are repeated in Siuda (1993). Keys and further comments on its morphology and biology are available in Pérez-Eid (2007). SEM pictures are available in Cringoli et al. (2005). Additional keys for every stage are available in Manilla (1998).

Haemaphysalis sulcata was redescribed, illustrated, compared with closely related species and the individual variation assessed by Filippova (1997). Additional illustrations and keys are available in Pérez-Eid (2007). SEM pictures are available in Cringoli et al. (2005). Additional keys for every stage are available in Manilla (1998). *Haemaphysalis cretica* Sénevet and Caminopteros, 1936 (originally named *Hae. punctata cretica*) and used in Feldman-Muhsam (1951) and Yeruham et al. (1996), is a synonym of *Haemaphysalis sulcata* as stated in Camicas et al. (1998). There is a sequence of the 16S rDNA mitochondrial gene in GenBank (L34308) deposited by Black and Piesman (1994) under the name *Hae. cretica.* This name is erroneously repeated every time that this sequence is used for comparative molecular studies, generating additional and unnecessary noise. *Haemaphysalis cholodkovskyi* Olenev, 1928 is also a synonym of *Hae. sulcata* according to Guglielmone and Nava (2014).

Haemaphysalis concinna was adequately redescribed, illustrated and keys provided by Filippova (1997). There are also illustrations of every life history stage in Siuda (1993) and Yamaguti et al. (1971). SEM pictures of both the male and female are included in Cringoli et al. (2005). There is an extensive review of the ecology of this species in Nosek (1971) and Lebedeva and Korenberg (1981). *Haemaphysalis concinna* was incorrectly subdivided into three subspecies by Neumann (1911). Two of them (*Hae. concinna concinna* and *Hae. concinna kochi*) are synonyms of *Hae. concinna*, but the third (*Hae. concinna longicornis*) is actually *Hae. longicornis* Neumann 1901, which occurs in the Eastern Palaearctic.

Haemaphysalis caucasica is a poorly known and rare species, which has even been considered as an endangered species (Durden and Keirans 1996). It is a tick restricted to desert and xeric shrublands, usually a parasite of Loporidae. However, the species has been also found on a large variety of mammals and birds (Guglielmone et al. 2014). The best redescription and illustrations, with keys, are available from Filippova (1997).

Haemaphysalis erinacei is the valid name for what has been called *Hae. numidiana* by several authors. *Hae. erinacei* was mistakenly named *Hae. numidiana* by Neumann (1905), and *Hae. parva* was incorrectly reported as *Hae. otophila* Schulze, 1919. The subspecies reported for *Hae. erinacei*, namely *Hae. erinacei ornata* Feldman-Muhsam, 1956, *Hae. e. turanica* Pospelova-Shtrom, 1940 and *Hae. e. taurica* Pospelova-Shtrom, 1940, are considered synonyms of *Hae. erinacei*, according to Guglielmone and Nava (2014). On some occasions, these subspecies have been cited as species (i.e. Camicas et al. 1998). There are good redescriptions and illustrations in Filippova et al. (1993), still including all three subspecies. We consider that the illustrations by Morel et al. (1977) are not adequate to determine this species. SEM micrographs of the adults are available in Cringoli et al. (2005).

Haemaphysalis hispanica is a poorly known species that is restricted to the European rabbit, *Oryctolagus cuniculus*, and its burrows. Sometimes, this species can be collected on predators of rabbits, such as foxes and birds of prey. The best redescription and illustrations of the species are available in Hoogstraal and Morel (1970). A general review of its ecology is available in Márquez and Guiguen (1992). This species seems to be a relict parasite restricted to most Western populations of *O. cuniculus*, since it has not been recorded from other countries where the host was introduced or is naturally present.

Haemaphysalis parva was adequately redescribed and illustrated by Filippova (1997). SEM pictures for adults and nymphs are available in Cringoli et al. (2005).

Key to the Species of Genus *Haemaphysalis* in Europe and Northern Africa

Females

1a. Palpi long and narrow, cylindrical with rounded tips. The posterolateral margins of article II do not project laterally at the basis capituli; article III without ventral spur; suture between articles II and III not noticeable; dorsolateral margins of basis capituli connected to the posterior margin by anteriorly curved folds; dorsal side of the idiosome without marginal grooves…***Haemaphysalis inermis*** (Fig. 86).

1b. Palpi short, conical or triangular…**2**.

2a. Palpi short, large, conical, slanting medially onto the tectum and with the posterolateral margins of article I slightly projecting from the lateral margin of the basis capituli. Palpal article III with a ventral spur and fused with article II by a marked suture; posterior margin of the basis capituli with lateral spiniform or rounded extensions; cornua absent. Marginal grooves present…**3**.

2b. Palpi short, not large, from conical to triangular, with median margins not slanting towards the tectum and with the posterolateral margins of article II clearly projecting from the lateral margin of the basis capituli…**4**.

3a. Posterior margin of the basis capituli with lateral extensions that form spurs with the lateral margins; porose areas quite large, subcircular, close together. Article II of the palpi with posterolateral margins forming rounded angles that project slightly from the margin of base. Palpal article III with a small ventral spur not extending to the suture with article II. Marginal groove extending continuously from one subscapular portion to another. Coxa IV with a short internal spur…***Haemaphysalis punctata*** (Fig. 89).

3b. Posterior margin of the basis capituli extending somewhat, broadly rounded at both ends. Porose areas rather large, oval, separated. Article II of the palpi with posterolateral margins projecting from the base of the capitulum with sharp angles. Article III of the palpi with a large ventral spur that almost reaches the suture with article II. Marginal grooves extending so as to overlap, on each side, the first 2 of the 11 festoons. Coxa IV with a short inner spur with the tip strongly curved laterally…***Haemaphysalis sulcata*** (Fig. 92).

4a. Palpi conical, broad, with posterolateral margins of article II projecting for half of their length from the lateral margin of the basis capituli and forming sharp, anteriorly produced angles. Palpal article II with rounded tip. Cornua large, robust. Porose areas large. Scutum variously shaped. Posterolateral margins of idiosome with 9 or 11 festoons…**5**.

4b. Palpi clearly triangular, widely projected in the basal portion of palpal segment II, with sharp angles…**6**.

5a. Ventral margins of articles II and III of the palpi with setae with hair-like projections; article III with a rounded tip and wide ventral spur not extending to the suture with article II. Cornua wide, sub-triangular, with rounded tips. Porose areas as narrow ovals. Scutum cordiform, longer than wide. Marginal grooves short, not reaching the nine festoons on the margins of the idiosome. Coxae I–IV with small, almost equal internal spurs…***Haemaphysalis parva*** (Fig. 101).

5b. Ventral margins of articles II and III of the palpi with setae without hair-like projections. Palpal article III with a rounded tip and ventral spur not extending to the suture with article II. Cornua wide, triangular, pointed. Porose areas as broad ovals. Scutum short, wider than long; lateral grooves long, reaching the first 2 of the 11 festoons; coxa I with a long, pointed, thin inner spur; coxae II–IV with reduced inner spurs...***Haemaphysalis concinna*** (Fig. 98).

6a. Punctations on the scutal shield few, regular and homogeneously distributed over its surface. The basis capituli is rectangular in its dorsal view, with the anterior margin sub-equal to the posterior margin...***Haemaphysalis caucasica*** (Fig. 95).

6b. Punctations on the scutal shield abundant, of various sizes and not homogeneously distributed over its surface, with larger punctations over the lateral sides of the scutal shield. The basis capituli is trapezoidal dorsally, the anterior margin longer than posteriorly...**7**.

7a. Palpal shape triangular, palpi short. Posterior margin of palpal article II essentially straight. The basis capituli short, much wider than long. Posterior margin of palpal article II without spurs. Cornua short...***Haemaphysalis erinacei*** (Fig. 95).

7b. Palpi bell-shaped. Posterior margin of palpal article II rounded, with clearly round spurs. The basis capituli essentially trapezoidal, almost as wide as long. Cornua long...***Haemaphysalis hispanica*** (Fig. 104).

Males

1a. Palpi long and narrow, with convex median margins and rounded tips, with article II not extending beyond the lateral margin of the basis capituli the posterior margin of which is narrow and rounded at both ends. Article III of the palpi without a ventral spur. The suture between articles II and III not noticeable. Cornua absent. Conscutum without lateral grooves. Tarsi without apical, ventral, small spurs. Hypostome with 2/2 dentition...***Haemaphysalis inermis*** (Fig. 86).

1b. Palpi short, with article II extending beyond the lateral margin of the basis capituli, which is always rectangular, without extensions on its posterior margin and with marked cornua. Palpal article III with a ventral spur and connected to article II by a noticeable suture. Conscutum with lateral grooves. Tarsi with ventral, apical, small spurs. Hypostome with 4/4 or 5/5 dentition. Palpi variously shaped...**2**.

2a. Palpi robust, conical, with median margin of article II inclined on tectum...**3**.

2b. Palpi not robust, variously shaped and with the median article of article II not inclined on tectum...**4**.

3a. Posterolateral margins of article II of the palpi rounded and projecting slightly beyond the lateral margin of the basis capituli. Palpal article II with a wide ventral spur extending beyond the suture with article III. Basis capituli clearly wider than long, with short, wide cornua with broadly rounded tips. Lateral grooves wide and overlapping the first three festoons, which total 11. Coxa IV with a very long, thin, medially concave, internal spur...***Haemaphysalis punctata*** (Fig. 89).

3b. Posterolateral margins of article II of the palpi with a rounded tip and projecting slightly beyond the lateral margin of the basis capituli. Palpal article II with a wide and pointed ventral spur extending to the suture with article II. Basis capituli clearly longer than wide, with long, curved cornua with rounded tips. Lateral grooves overlapping only the first of the 11 festoons. Coxa IV with a robust, rather long internal, laterally concave, spur...***Haemaphysalis sulcata*** (Fig. 92).

4a. Palpi pincer- or acorn-shaped, projecting variously from the basis capituli. Cornua robust. Idiosome with 9 or 11 festoons...**5**.

4b. Palpi triangular or bell-shaped, with the posterolateral margins of article II extending for over three-fourth of their length from the lateral margin of the basis capituli, forming an anteriorly produced acute angle. Palpal article II with a punctiform tip and a long, narrow, pointed ventral spur that extends beyond the suture with article II...**6**.

5a. Palpi acorn-shaped, with the posterolateral margins of article II extending for less than their length beyond the lateral margin of the basis capituli, forming an angle with a rounded tip; article III conical with a rounded apex and overlapped by article IV, which is exceptionally and typically protuberant, with a short ventral spur not extending beyond the suture with article II. Hypostome with 10 rows of 4/4 dentition. Basis capituli longer that wide, with robust, short, wide cornua with rounded tips. Lateral grooves short and not overlapping any of the nine festoons. Coxae I–IV with short, almost equal internal spurs...***Haemaphysalis parva*** (Fig. 101).

5b. Palpi pincer-shaped, with the posterolateral margins of article II extending for half their length beyond the lateral margin of the basis capituli, converging in a short extension that is rounded dorsally and punctiform ventrally. Article II with a strongly concave inner margin and wide ventral spur with a rounded tip extending beyond the suture with article II. Hypostome with 10–11 rows of 6/6 dentition. Basis capituli wider than long, with robust, triangular, pointed cornua. Lateral grooves overlapping only the first of the 11 festoons. Coxa I with a long internal spur. Coxae II–IV with inner spurs gradually reducing in size...***Haemaphysalis concinna*** (Fig. 98).

6a. Dorsal punctations few, very large and heterogeneously distributed over the surface of the conscutum. Palpal segment II very wide and laterally produced into wide and round flaps. Cornua long. The ventral spur on palpal segment III reaches the anterior margin of palpal segment II...***Haemaphysalis hispanica*** (Fig. 107).

6b. Dorsal punctations abundant, relatively small and homogeneously distributed over the scutum. Ventral spur on palpal segment III barely reaching the anterior margin of palpal segment II...**7**.

7a. Cornua long. Palpal segment II laterally produced into large and pointed flaps with posterior projections... ***Haemaphysalis caucasica*** (Fig. 95).

7b. Cornua short. Palpal segment II laterally produced into large and pointed flaps without posterior projections... ***Haemaphysalis erinacei*** (Fig. 104).

Nymphs

1a. Palpi cylindrical, with rounded tips and posterior margins of article II not projecting from the lateral margins of the capitulum base; ventral internal margins of articles II and III without fimbriate setae; basis capituli with rounded lateral extensions; idiosoma without marginal grooves...***Haemaphysalis inermis*** (Fig. 87).

1b. Palpi variously shaped, with rounded or pointed spurs. Ventral internal margins of articles II and III with fimbriate setae. Idiosoma with marginal grooves, either complete or incomplete...**2**.

2a. Coxae without spurs or with very small ones. Article II of palpi having posterolateral margins with the pointed end projecting somewhat from the margin of the base; article III with a rounded tip, with a small ventral spur; posterior margin of the base with rounded lateral extensions; each side of the marginal grooves reaching only the first 2 of the 11 festoons...***Haemaphysalis sulcata*** (Fig. 93).

2b. Coxae with spurs clearly visible...**3**.

3a. Article II of the palpi with rounded posterolateral margins slightly projecting from the base; article III with a rounded tip and small ventral spur not extending to the suture with article II; posterior margin of the base of the capitulum with spiniform lateral extensions; no break in the marginal grooves from one subscapular portion to another ...***Haemaphysalis punctata*** (Fig. 90).

3b. Article II of the palpi with rounded or pointed posterolateral margins clearly projecting from the base...**4**.

4a. Palpi with rounded tips. Base of the capitulum with internally curved, pointed cornua; marginal grooves short, reaching the first festoon...***Haemaphysalis parva*** (Fig. 102).

4b. Palpi with narrow tips. Base of capitulum with triangular, pointed cornua; marginal grooves short or long...**5**.

5a. Palpal segment II not extended posteriorly...**6**.

5b. Palpal segment II extended posteriorly...**7**.

6a. Dorsal scutum rounded, wider than long, with short cervical grooves; marginal grooves short and without any of the 11 festoons; stigmata large. Only one fimbriate seta on the internal margin of palpal segment II on its ventral side ... ***Haemaphysalis concinna*** (Fig. 99).

6b. Dorsal scutum narrow, longer than wide, with long cervical grooves; marginal grooves long, reaching the first pair of 11 festoons. Three fimbriate setae on the internal margin of the palpal segment II on its ventral side... ***Haemaphysalis erinacei*** (Fig. 105).

7a. Internal margin of dorsal palpal segment II truncated into an angle of 90°. Coxal spurs long...***Haemaphysalis caucasica*** (Fig. 96).

7b. Internal margin of dorsal palpal segment II essentially straight. Coxal spurs relatively short...***Haemaphysalis hispanica*** (Fig. 108).

Larvae

1a. Palpi elongate, with lateral margins not projecting from the sides of capitulum base; articles II and III fused... ***Haemaphysalis inermis*** (Fig. 87).

1b. Palpi short, with posterolateral margins of article II more or less projecting from the base of the capitulum...**2**.

2a. Cornua absent; idiosoma with nine festoons...***Haemaphysalis punctata*** (Fig. 90).

2b. Cornua absent or present; idiosoma with 11 festoons...**3**.

3a. Cornua absent. Article II of palpi with posterolateral margins converging to a sharp tip; article III with a pointed ventral spur extending to the suture with article II. All of the coxae with spurs...***Haemaphysalis sulcata*** (Fig. 93).

3b. Cornua present or absent. Palpi triangular or bell-shaped, with or without posterior projections...**4**.

4a. Cornua present. Posterior margin of palpal article II with a ventral spur projecting posteriorly...**5**.

4b. Cornua present or absent. Posterior margin of palpal article II without a ventral spur projecting posteriorly...**6**.

5a. Coxal spurs only on coxa I, with some chitinous ridges on coxae II and III. Cornua present and long...***Haemaphysalis parva*** (Fig. 102).

5b. Coxal spurs on every coxa. Cornua present and rounded...***Haemaphysalis caucasica*** (Fig. 96).

6a. Cornua absent. Coxal spurs only on coxae I, with small chitinous ridges on coxae II and III. Article I of palpi with a backwards directed spur on its ventral surface...***Haemaphysalis erinacei*** (Fig. 105).

6b. Cornua present or absent. Article III of palpi with a spur on its ventral surface, slightly reaching the suture with palpal article II. Article I without a spur in its ventral surface...**7**.

7a. External contour of the palpi concave, with a clear lateral prominence on the basis of second palpal segment, then narrowing to palpal segment III. Ventral spur on palpal segment III short. Cornua present, small and widely rounded. Coxal spur evident on coxa I, very small ridges on coxae II and III...***Haemaphysalis concinna*** (Fig. 99).

7b. External contour of the palpi convex, campanulate, with a clear lateral prominence on the basis of second palpal segment, then straight or slightly convex to palpi III. Cornua present, relatively long and wide. Coxal spurs evident on the three coxae...***Haemaphysalis hispanica*** (Fig. 108).

References

Black WC, Piesman J (1994) Phylogeny of hard- and soft-tick taxa (Acari: Ixodidae) based on mitochondrial 16S rDNA sequences. Proc Nat Acad Sci 91:10034–10038

Camicas JL, Hervy JP, Adam F, Morel PC (1998) Les tiques du monde. Nomenclature, stades décrits, hôtes, répartition (Acarida, Ixodida). Orstom, Paris

Cringoli G, Iori A, Rinaldi L, Veneziano V, Genchi C (2005) Mappe parassitologiche - Zecche. Series Edit, Giuseppe Cringoli, Napoli

Durden LA, Keirans JE (1996) Host-parasite coextinction and the plight of tick conservation. Am Entomol 42:87–91

Feldman-Muhsam B (1951) A note on east Mediterranean species of the genus *Haemaphysalis*. Bull Res Council Israel 1:96–107

Filippova NA (1997) Ixodid ticks of the subfamily Amblyomminae. Fauna Russia Neighbouring Countries 4:1–436 (in Russian)

Filippova NA, Panova IV, Musatov SA (1993) The taxonomic pattern of the polytipic species *Haemaphysalis erinacei*. Parazitologiya 27:193–215 (in Russian)

Guglielmone AA, Nava S (2014) Names for Ixodidae (Acari: Ixodoidea): valid, synonyms, *incertae sedis, nomina dubia, nomina nuda, lapsus*, incorrect and suppressed names, with notes on confusions and misidentifications. Zootaxa 3767:1–256

Guglielmone AA, Robbins RG, Apanaskevich DA, Petney TN, Estrada-Peña A, Horak IG (2014) The hard ticks of the world (Acari: Ixodida: Ixodidae). Springer, Dordrecht, p 738

Hoogstraal H (1969) *Haemaphysalis (Alloceraea) kitaokai* sp. n. of Japan, and keys to species in the structurally primitive subgenus *Alloceraea* Schulze of Eurasia (Ixodoidea, Ixodidae). J Parasitol 55:211–221

Hoogstraal H, Kim K-C (1985) Tick and mammal coevolution, with emphasis on *Haemaphysalis*. In: Kim K-C (ed) Coevolution of parasitic arthropods and mammals. Wiley, New York and other cities, pp 505–568

Hoogstraal H, Morel PC (1970) *Haemaphysalis (Rhipistoma) hispanica* Gil Collado, a parasite of the european rabbit, re-description of adults and description of immature stages (Ixodoidea: Ixodidae). J Parasitol 56:813–822

Lebedeva NN, Korenberg EI (1981) Distribution of *Haemaphysalis concinna* in the Soviet Union and some general features of its ecology. Folia Parasitol 28:249–261

Manilla G (1998) Fauna d'Italia. Acari, Ixodida. Edizioni Calderini, Bologna, p 280

Márquez FJ, Guiguen C (1992) Distribution sur l'hôte des ixodides parasites d'*Oryctolagus cuniculus* (L.) et facteurs qui l'affectent. Acarologia 33:141–148

Morel PC, Sobrero L, Totaro S (1977) Ospiti naturali d'*Haemaphysalis erinacei* Pavesi, 1884. Riv Parassitol 38:93–95

Neumann LG (1905) Notes sur les ixodidés. III. Arch Parasitol 9:225–241

Neumann LG (1911) Ixodidae. Das Tierreich (26):169

Nosek J (1971) The ecology, bionomic and behaviour of *Haemaphysalis concinna* tick. Z Parasitenk 36:233–241

Pérez-Eid C (2007) Les tiques. Identification, biologie, importance médicale et vétérinaire. Editions Tec & Doc, Paris, p 314

Pomerantzev BI (1950) Ixodid ticks (Ixodidae). Fauna SSSR, Paukoobraznye, vol 4, no 2, Izd. Akad. Nauk SSSR, Moscow, Leningrad, p 224 (in Russian)

Siuda K (1993) Kleszcze Polski (Acari: Ixodida). II. Systematyka irozmieszczenie. Monografie Parazyto- logiczne 12. Polskie Towar-zystwo Parazytologiczne, Warszawa

Yamaguti N, Tipton VJ, Keegan HL, Toshioka S (1971) Ticks of Japan, Korea, and the Ryukyu Islands. Brigham Young Univ Sci Bull Biol Ser 15:1–226

Yeruham I, Hadani A, Galker F, Rosen S (1996) The seasonal occurrence of ticks (Acari: Ixodidae) on sheep and in the field in the Judean area of Israel. Exp Appl Acarol 20:47–56

Haemaphysalis inermis Birula, 1895 (Figs. 85–87)

D. Otranto, F. Dantas-Torres, M. M. Santos-Silva, and Z. Vatansever

Life Cycle and Host Preferences

The winter tick, *Haemaphysalis inermis*, is an exophilous three-host tick with a peculiar life cycle in that immature forms feed on a host and fall off within a few hours (Pomerantzev 1950). The life cycle of this species may be completed in 1 year, while in non-optimal conditions it can take 2–3 years. Under laboratory conditions, the life cycle can be completed in 165 days (Nosek et al. 1967). The adults of this species feed on domestic and wild ungulates, while immature stages prefer small mammals rather than lizards and birds. These ticks can parasitize a range of mammals including cattle, horses, sheep, deer, dogs, foxes and hedgehogs. Human infestation has also been reported (Pomerantzev 1950; Bursali et al. 2012).

Ecology

Haemaphysalis inermis is generally found in temperate broadleaf and mixed forests (Guglielmone et al. 2014). Immature forms are active during the dry season from June to September (Nosek et al. 1981). Larval stages are active during spring and summer on small mammals, infrequently on lizards and birds, whereas nymphs are present from late spring to early autumn (Manilla 1998). Adults are active during the humid season from October until May (Nosek et al. 1981), and they peak in November and December in the Mediterranean regions (Hornok 2009; Dantas-Torres and Otranto 2013).

Distribution

Haemaphysalis inermis is sparsely distributed, frequently well adapted to different biotopes and usually accompanied by *I. ricinus*. This tick species has been reported in Eurasia including Albania, Austria, Bulgaria, Czech Republic, France, Greece, Hungary, Iran, Italy, Poland, Portugal, Romania, Slovakia, Spain and Turkey (Manilla 1998; Hornok and Farkas 2009). During the last decades, in some European countries such as Hungary (Hornok and Farkas 2009) and Spain (García-Sanmartín et al. 2008; Portillo et al. 2008), *Hae. inermis* has recently shown a rather high abundance and occurrence which suggests that it might be expanding its range slightly (Santos-Silva et al. 2011).

Vectorial Capacity and Pathogen Burden

Experimental proof for the vectorial capacity of *Hae. inermis* is available for tick-borne encephalitis virus (Nosek et al. 1981). *Haemaphysalis inermis* has been implicated as a putative vector of pathogens to humans (e.g. *Rickettsia aeschlimannii, Rickettsia helvetica*) in field-collected ticks in Spain and Hungary (Portillo et al. 2008; Hornok et al. 2010). Other reports involve *Babesia bigemina* (García-Sanmartín et al. 2008). The actual vector role of *Hae. inermis* for the transmission of these pathogens remains poorly known.

D. Otranto (✉)
Dipartimento di Medicina Veterinaria, Università degli Studi di Bari, Bari, Italy
e-mail: domenico.otranto@uniba.it

F. Dantas-Torres
Centro de Pesquisas Aggeu Magalhães, Fundação Oswaldo Cruz, Recife, Brazil
e-mail: filipe.dantas@cpqam.fiocruz.br

M. M. Santos-Silva
Centro de Estudos de Vectores e Doenças Infecciosas Dr. Francisco Cambournac (CEVDI), Instituto Nacional de Saúde Doutor Ricardo Jorge, Águas de Moura, Lisboa, Portugal
e-mail: m.santos.silva@insa.min-saude.pt

Z. Vatansever
Department of Parasitology, Faculty of Veterinary Medicine, Kafkas University, Kars, Turkey
e-mail: zativet@gmail.com

A. Estrada-Peña et al. (eds.), *Ticks of Europe and North Africa*,
https://doi.org/10.1007/978-3-319-63760-0_45

Fig. 85 **A** Distribution of *Hae. inermis* in Europe and Northern Africa (10 × 10 km grid presence with black dots). **B** Countries where the species has been reported are marked in dark grey

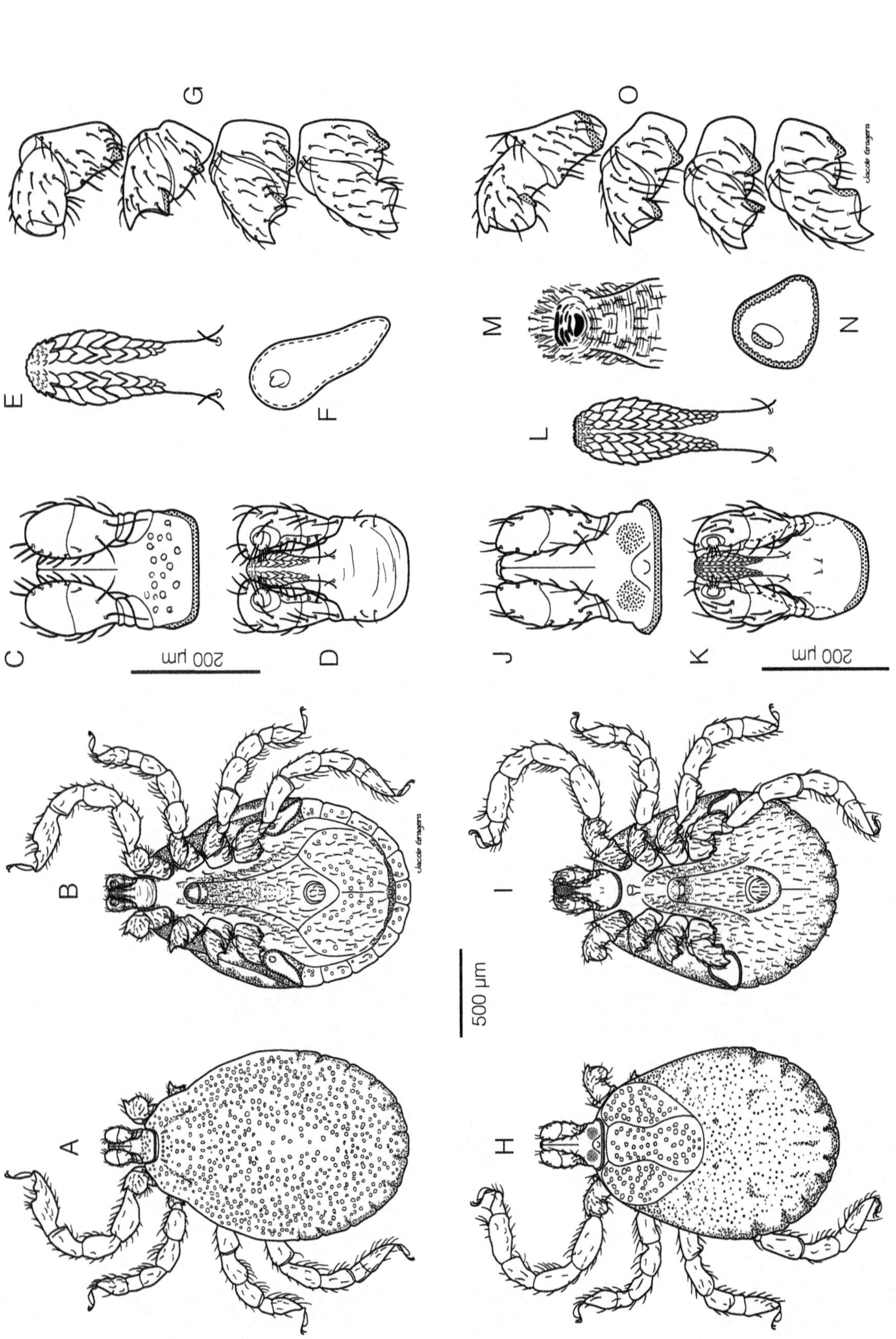

Fig. 86 **A–G** The male of *Hae. inermis*. **A** Dorsal, **B** ventral, **C** capitulum, dorsal, **D** capitulum, ventral, **E** hypostome, **F** spiracular plate, **G** coxae and trochanters I–IV. **H–O** The female of *Hae. inermis*. **H** Dorsal, **I** ventral, **J** capitulum, ventral, **K** capitulum, dorsal, **L** hypostome, **M** genital opening, **N** spiracular plate, **O** coxae and trochanters I–IV. Illustrations from specimens collected in northern Spain and northern France, complemented with illustrations by Filippova (1997) and Manilla (1998) and scanning electron micrographs by Cringoli et al. (2005)

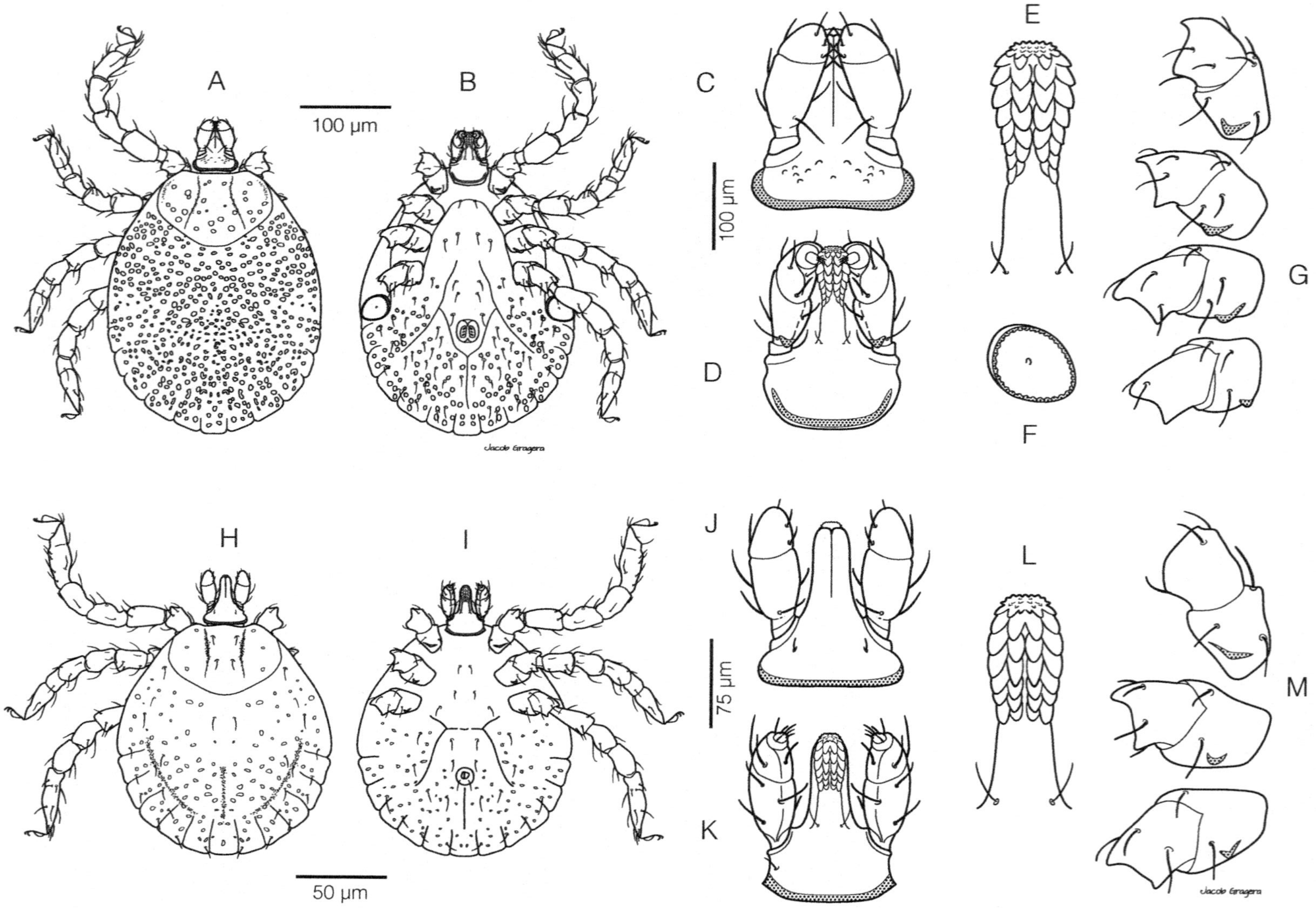

Fig. 87 **A–G** The nymph of *Hae. inermis*. **A** Dorsal, **B** ventral, **C** capitulum, dorsal, **D** capitulum, ventral, **E** hypostome, **F** spiracular plate, **G** coxae and trochanters I–IV. **H–M** The larva of *Hae. inermis*. **H** Dorsal, **I** ventral, **J** capitulum, dorsal, **K** capitulum, ventral, **L** hypostome, **M** coxae and trochanters I–III. Illustrations from larvae obtained from one engorged female collected in northern Spain and complemented from illustrations by Filippova (1997) and Manilla (1998) and scanning electron micrographs by Cringoli et al. (2005)

References

Bursali A, Keskin A, Tekin S (2012) A review of the ticks (Acari: Ixodida) of Turkey: species diversity, hosts and geographical distribution. Exp Appl Acarol 57:91–104

Cringoli G, Iori A, Rinaldi L, Veneziano V, Genchi C (2005) Mappe parassitologiche—Zecche. Series Edit, Giuseppe Cringoli, Napoli

Dantas-Torres F, Otranto D (2013) Species diversity and abundance of ticks in three habitats in southern Italy. Ticks Tick Borne Dis 4:251–255

Filippova NA (1997) Ixodid ticks of subfamily Amblyomminae. In the fauna of Russia and neighboring countries. New Series no. 145: Arachnoidea, vol 4(5), St. Petersburg, Nauka (in Russian)

García-Sanmartín J, Barandika JF, Juste RA, García-Pérez AL, Hurtado A (2008) Distribution and molecular detection of *Theileria* and *Babesia* in questing ticks from northern Spain. Med Vet Entomol 22:318–325

Guglielmone AA, Robbins RG, Apanaskevich DA, Petney TN, Estrada-Peña A, Horak IG (2014) The hard ticks of the world (Acari: Ixodida: Ixodidae). Springer, Dordrecht, p 738

Hornok S (2009) Allochronic seasonal peak activities of *Dermacentor* and *Haemaphysalis* spp. under continental climate in Hungary. Vet Parasitol 163:366–369

Hornok S, Farkas R (2009) Influence of biotope on the distribution and peak activity of questing ixodid ticks in Hungary. Med Vet Entomol 23:41–46

Hornok S, Meli ML, Perreten A, Farkas R, Willi B, Beugnet F, Lutz H, Hofmann-Lehmann R (2010) Molecular investigation of hard ticks (Acari: Ixodidae) and fleas (Siphonaptera: Pulicidae) as potential vectors of rickettsial and mycoplasmal agents. Vet Microbiol 140:98–104

Manilla G (1998) Fauna d'Italia. Acari: Ixodida. Ed Calderini, Bologna (in Italian)

Nosek J, Lichard M, Sztankay M (1967) The ecology of ticks in the Tribec and Hronský Inovec Mountains. Bull World Health Organ 36:49–59

Nosek J, Kozuch O, Lysy J (1981) The survival of the tick-borne encephalitis (TBE) virus in nymphs of *Haemaphysalis inermis* tick and its transmission to pygmy mouse (*Micromys minutus*). Cah. O.R.S.T.O.M., sér. Ent Méd. Et Parasitol XIX:67–69

Pomerantzev BI (1950) Ixodid ticks (Ixodidae). Fauna SSSR, Paukoobraznye, n. s., (41), 4, p 224 (English translation by Elbl A., edited by Anastos G. The American Institute of Biological Sciences, Washington)

Portillo A, Santibáñez P, Santibáñez S, Pérez-Martínez L, Oteo JA (2008) Detection of *Rickettsia* spp. in *Haemaphysalis* ticks collected in La Rioja, Spain. Vector Borne Zoono Dis 8:653–658

Santos-Silva MM, Beati L, Santos AS, De Sousa R, Núncio MS, Melo P, Santos-Reis M, Fonseca C, Formosinho P, Vilela C, Bacellar F (2011) The hard-tick fauna of mainland Portugal (Acari: Ixodidae): an update on geographical distribution and known associations with hosts and pathogens. Exp Appl Acarol 55:85–121. https://dx.doi.org/10.1007/s10493-011-9440-x

Haemaphysalis punctata Canestrini and Fanzago, 1878 (Figs. 88–90)

M. P. Pfäffle, M. M. Santos-Silva, T. G. T. Jaenson, Z. Vatansever, and T. N. Petney

Life Cycle and Host Preferences

Haemaphysalis punctata is a three-host tick with a natural life cycle of 1–3 years (Nosek et al. 1967; Nosek 1971; Liebisch et al. 1976; Farkas et al. 2012). Under laboratory conditions, the life cycle can be completed in 165–229 days (Nosek 1971; Liebisch et al. 1976). Small mammals, hares, hedgehogs, birds and even lizards are hosts of the immature stages, while adults mainly feed on wild and domestic ungulates, particularly cattle, sheep and goats (Arthur 1963; Nosek 1971; Liebisch et al. 1976; Gilot 1985; Guglielmone et al. 2014). Other hosts for the adult ticks include domestic dogs, red foxes, mustelids, lagomorphs and hedgehogs (Arthur 1963; Nosek 1971; Hillyard 1996). Humans are bitten infrequently (Arthur 1963; Nosek 1971).

M. P. Pfäffle (✉)
Department of Ecology and Parasitology, Institute of Zoology, Karlsruhe Institute of Technology, Karlsruhe, Germany
e-mail: miripfaeffle@web.de

M. M. Santos-Silva
Centro de Estudos de Vectores e Doenças Infecciosas
Dr. Francisco Cambournac (CEVDI), Instituto Nacional de Saúde Doutor Ricardo Jorge, Águas de Moura, Portugal
e-mail: m.santos.silva@insa.min-saude.pt

T. G. T. Jaenson
Medical Entomology Unit, Department of Organismal Biology, Evolutionary Biology Centre, Uppsala University, Uppsala, Sweden
e-mail: thomas.jaenson@ebc.uu.se

Z. Vatansever
Department of Parasitology, Faculty of Veterinary Medicine, Kafkas University, Kars, Turkey
e-mail: zativet@gmail.com

T. N. Petney
State Museum of Natural History, Erbprinzenstr. 13, 76133 Karlsruhe, Germany
e-mail: trevor.petney@kit.edu

A. Estrada-Peña et al. (eds.), *Ticks of Europe and North Africa*,
https://doi.org/10.1007/978-3-319-63760-0_46

Ecology

In general, *Hae. punctata* is ecologically very adaptable and tolerates different climatic environments. It can be found in a very wide variety of habitats from cold to mild and humid climates to drier biotopes (Pomerantzev 1950; Nosek 1971; Estrada-Peña et al. 2004). It inhabits pastures, forest margins, forest steppes, brush areas, limestone pastures, artificial conifer forests, oak forests with scarce undercover and, rarely, even evergreen oak forests (Nosek 1971; Tälleklint 1996; Curioni et al. 2004). In Italy, this species has been found up to 1300 m a.s.l. (Curioni et al. 2004). In Slovakia, it is restricted to areas with annual isotherms of 7–9 °C and isohyets of 650–1000 mm (Nosek 1971). Depending on the geographic region, the seasonal activity of *Hae. punctata* is variable. Larvae are active in the summer months from June to September (Nosek 1971; Liebisch et al. 1976; Gilot 1985; Estrada-Peña et al. 1990; Curioni et al. 2004). Nymphs show a bimodal activity pattern with peaks from April to October and from December to February (Nosek et al. 1967; Nosek 1971; Liebisch et al. 1976; Gilot 1985; Estrada-Peña et al. 1990). The adults, like the nymphs have a bimodal activity pattern with peaks from March to June and from September to November (Pomerantzev 1950; Nosek et al. 1967; Nosek 1971; Liebisch et al. 1976; Gilot 1985; Estrada-Peña et al. 1990; Curioni et al. 2004). In Turkey, Hoogstraal (1959) found adult *Hae. punctata* occasionally feeding in winter.

Distribution

Haemaphysalis punctata is a widely distributed species in the Palaearctic region. It has been recorded from Albania, Belgium, Bulgaria, Croatia, Cyprus, the Czech Republic, Denmark, France, Germany, United Kingdom, Greece, Hungary, Italy, Iran, Moldova, the Netherlands, Poland, Portugal, Romania, Russia, Serbia, Slovakia, Spain, south-eastern Sweden, Switzerland, Turkey, the Ukraine,

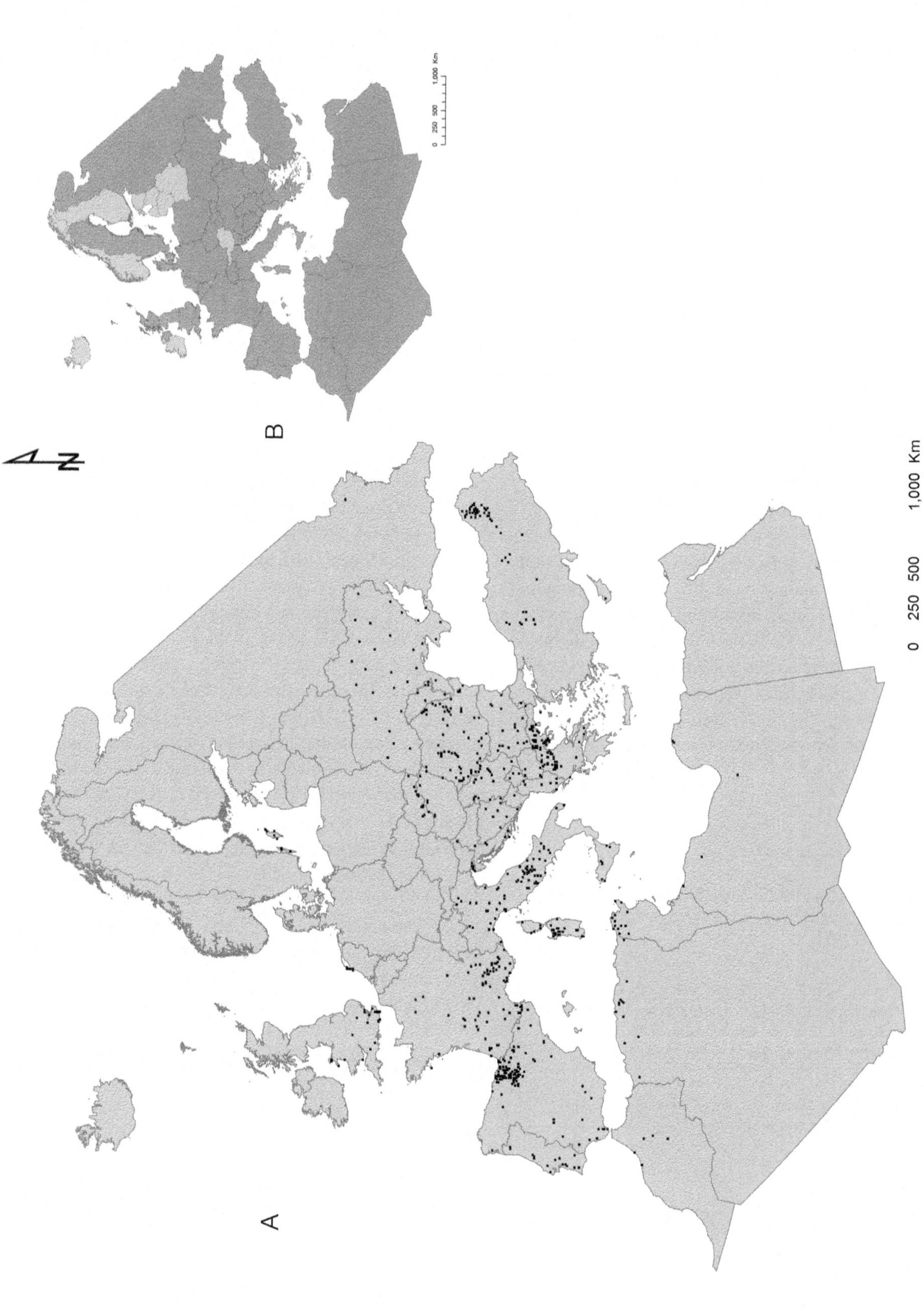

Fig. 88 **A** Distribution of *Hae. punctata* in Europe and Northern Africa (10 × 10 km grid presence with black dots), **B** countries where the species has been reported are marked in dark grey

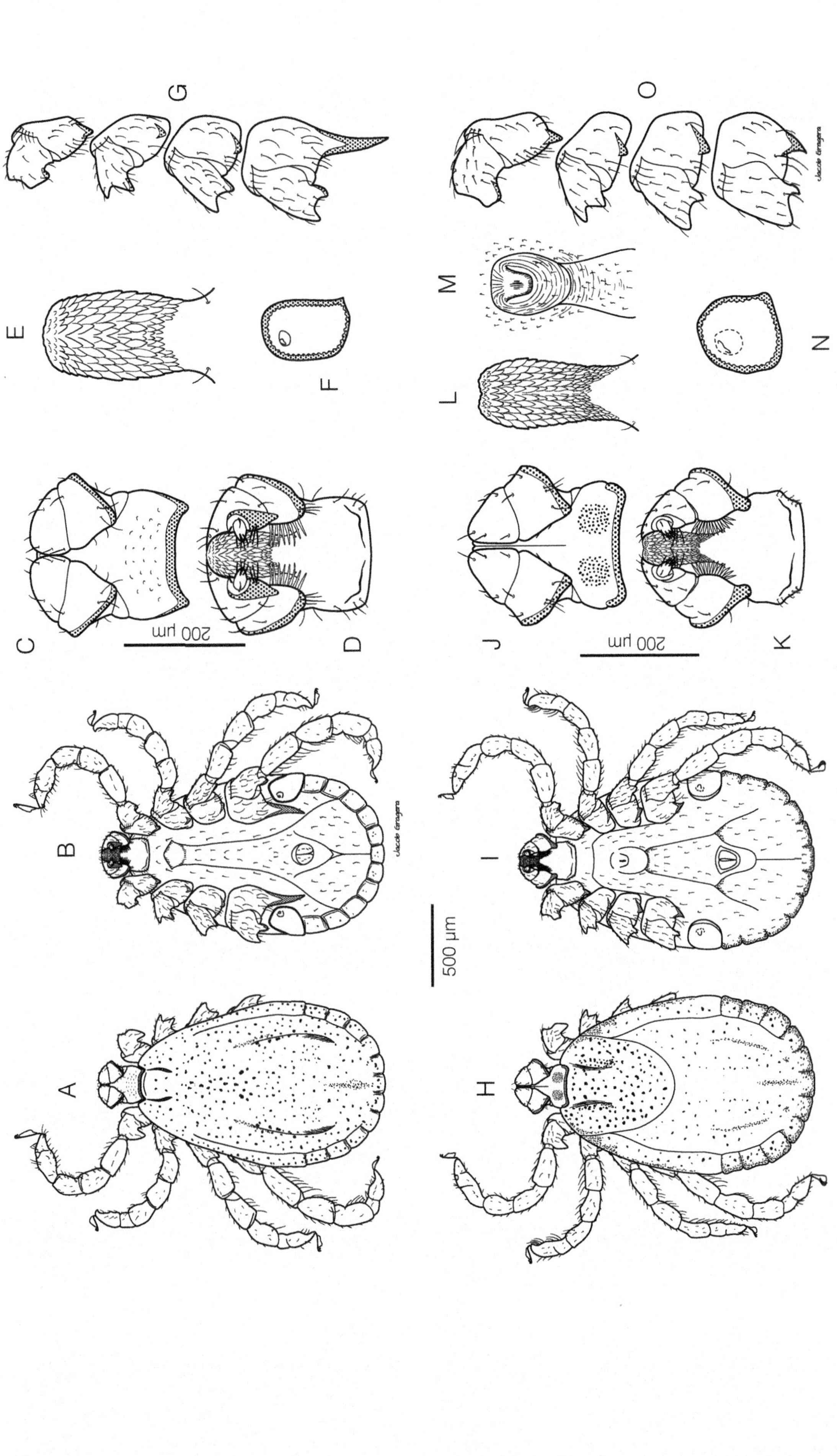

Fig. 89 **A–G** The male of *Hae. punctata*. **A** Dorsal, **B** ventral, **C** capitulum, dorsal, **D** capitulum, ventral, **E** hypostome, **F** spiracular plate, **G** coxae and trochanters I–IV. **H–O** The female of *Hae. punctata*. **H** Dorsal, **I** ventral, **J** capitulum, dorsal, **K** capitulum, ventral, **L** hypostome, **M** genital opening, **N** spiracular plate, **O** coxae and trochanters I–IV. Illustrations from specimens collected in Spain, Portugal, France, Netherlands, Germany, Italy and Turkey

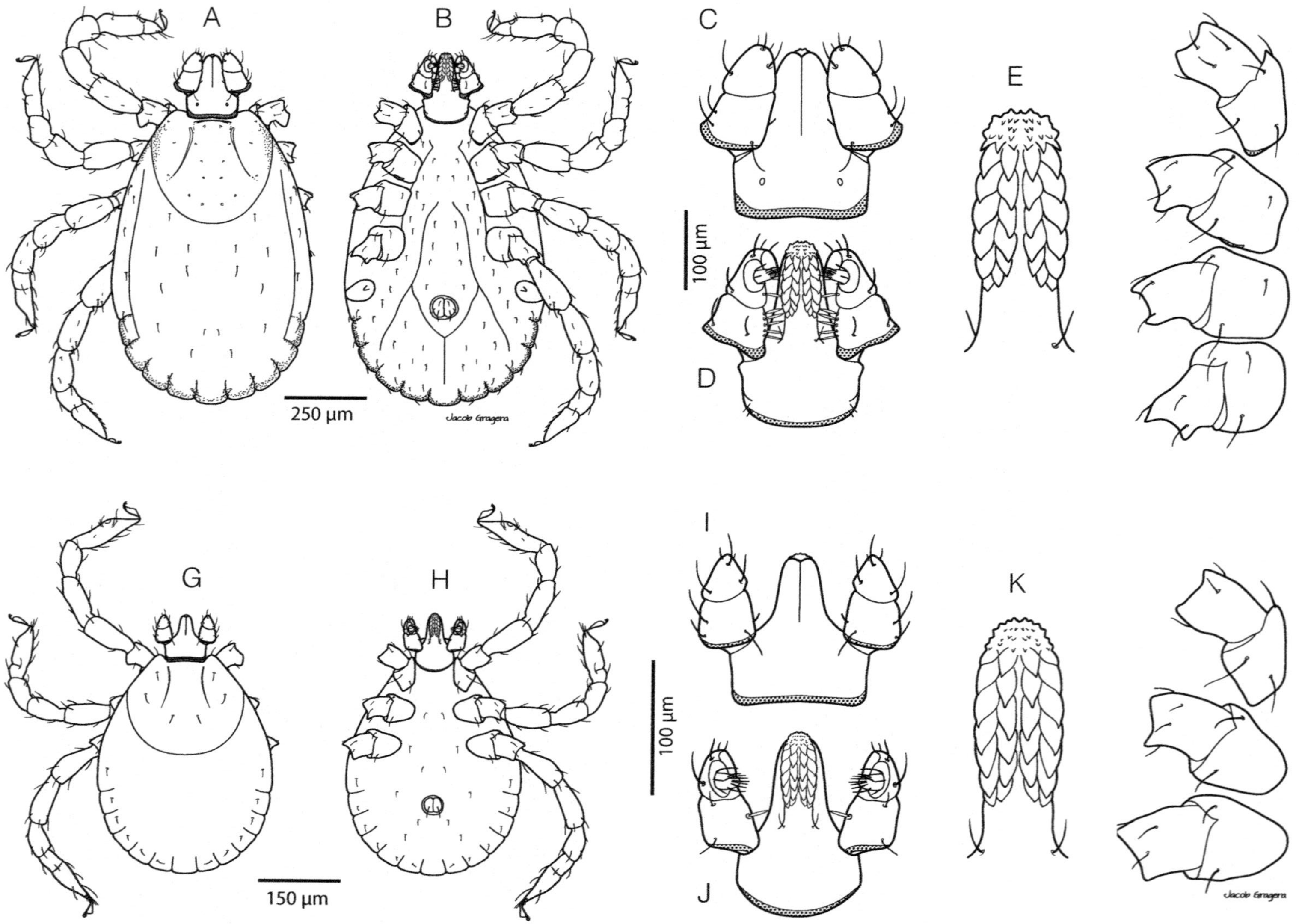

Fig. 90 **A–F** The nymph of *Hae. punctata*. **A** Dorsal, **B** ventral, **C** capitulum, dorsal, **D** capitulum, ventral, **E** hypostome, **F** coxae and trochanters I–IV. **G–L** The larva of *Hae. punctata*. **G** Dorsal, **H** ventral, **I** capitulum, dorsal, **J** capitulum, ventral, **K** hypostome, **L** coxae and trochanters I–III. Illustrations from specimens collected in Spain, France, Netherlands, Germany, Italy and Turkey

Morocco, Algeria, Tunisia, Libya and Egypt (Arthur 1963; Pavlov 1963; Nosek 1971; Brocklesby and Barnett 1972; Liebisch et al. 1976; Garben et al. 1982; Walter 1982; Estrada-Peña et al. 1990; Jaenson et al. 1994; Hillyard 1996; Papadopoulos et al. 1996; Tälleklint 1996; Punda-Polic et al. 2002; Razmi et al. 2002; Estrada-Peña et al. 2004; Bursali et al. 2010; Radulović et al. 2011; Chochlakis et al. 2011; Santos-Silva et al. 2011). Mertins and Schlater (1991) reported the introduction of *Hae. punctata* to the USA on imported ostriches (*Struthio camelus*) from Europe and Africa; however, there is no evidence that it has become established there.

Vectorial Capacity and Pathogen Burden

Haemaphysalis punctata has been associated with Bhanja, Palma, Tribec, tick-borne encephalitis and Crimean-Congo haemorrhagic fever viruses, and is able to transmit louping-ill virus (Nosek 1971; Labuda and Nuttall 2004; De la Fuente et al. 2008; Farkas et al. 2012). *Rickettsia sibirica, R. helvetica, R. massiliae*, *Anaplasma bovis*, *A. centrale* and *Coxiella burnetii* have been detected in *Hae. punctata* (Stoker and Marmion 1955; Nosek 1971; Barandika et al. 2008; Hornok et al. 2010; Tijsse-Klasen et al. 2013; Palomar et al. 2015). In south-eastern Spain, *Hae. punctata* was found to be infected with a *Rickettsia* sp. near *R. hoogstraalii* and it also appeared to be infected with an *R. monacensis*-like rickettsia (Márquez 2008). Furthermore, it is a vector of various *Babesia* spp., such as *Babesia bigemina*, *Babesia major* and *Babesia motasi* (Nosek 1971; Liebisch et al. 1976; Morzaria et al. 1977; Alani and Herbert 1988a; Curioni et al. 2004; García-Sanmartín et al. 2008; Farkas et al. 2012). It is associated with various *Theileria* spp. (Nosek 1971; Brocklesby and Barnett 1972; Alani and Herbert 1988b; L'Hostis and Seegers 2002; García-Sanmartín et al. 2008), but the taxonomic status of these should be revised. Tälleklint (1996) showed that nymphs of *Hae. punctata* from three Swedish Baltic islands were infected with *Borrelia burgdorferi* s.l. Other pathogens associated with *Hae. punctata* are *Anaplasma phagocytophilum* and *Francisella tularensis* (Olsufyev and Petrov 1968; Nosek 1971; Barandika et al. 2008; Farkas et al. 2012). *Hae. punctata* has been confirmed as a cause of tick paralysis in goats and sheep (Mans et al. 2008).

References

Alani AJ, Herbert IV (1988a) The morphometrics of *Babesia motasi* (Wales) and its transmission by *Haemaphysalis punctata* (Canestrini and Fanzago 1877) to sheep. Vet Parasitol 30:87–95. https://dx.doi.org/10.1016/0304-4017(88)90155-0

Alani AJ, Herbert IV (1988b) Pathogenesis of infection with *Theileria recondita* (Wales) isolated from *Haemaphysalis punctata* from North Wales. Vet Parasitol 28:293–301. https://dx.doi.org/10.1016/0304-4017(88)90076-3

Arthur DR (1963) British ticks. p 213. Butterworths, London

Barandika JF, Hurtado A, García-Sanmartín J, Juste RA, Anda P, Garcia-Perez AL (2008) Prevalence of tick-borne zoonotic bacteria in questing adult ticks from northern Spain. Vector Borne Zoonot Dis 8:829–835. https://dx.doi.org/10.1089/vbz.2008.0023

Brocklesby D, Barnett S (1972) The tick *Haemaphysalis punctata*, shown to be a vector of *Theileria mutans* in Britain. Vet Rec 90:512–513. https://dx.doi.org/10.1136/vr.90.18.512

Bursali A, Tekin S, Orhan M, Keskin A, Ozkan M (2010) Ixodid ticks (Acari: Ixodidae) infesting humans in Tokat Province of Turkey: species diversity and seasonal activity. J Vector Ecol 35:180–186. https://dx.doi.org/10.1111/j.1948-7134.2010.00075.x

Chochlakis D, Ioannou I, Sandalakis V, Dimitriou T, Kassinis N, Papadopoulos B, Tselentis Y, Psaroulaki A (2011) Spotted fever goup rickettsiae in ticks in Cyprus. Microb Ecol 63:314–323. https://dx.doi.org/10.1007/s00248-011-9926-4

Curioni V, Cerquetella S, Scuppa P, Pasqualini L, Beninati T, Favia G (2004) Lyme disease and babesiosis: preliminary findings on the transmission risk in highly frequented areas of the Monti Sibillini National Park (Central Italy). Vector-Borne Zoonot Dis 4:214–220. https://dx.doi.org/10.1089/vbz.2004.4.214

de la Fuente J, Estrada-Peña A, Venzal JM, Kocan KM, Sonenshine DE (2008) Overview: ticks as vectors of pathogens that cause disease in humans and animals. Front Bio sci 13:6938–6946

Estrada-Peña A, Bouattour A, Camicas J, Walker A (2004) Ticks of domestic animals in the Mediterranean region: a guide to identification of species. University of Zaragoza, Zaragoza

Estrada-Peña A, Dehesa V, Sánchez C (1990) The seasonal dynamics of *Haemaphysalis punctata*, *Rhipicephalus bursa* and *Dermacentor marginatus* (Acari: Ixodidae) on sheep of Pais Vasco (Spain). Acarologia 31:17–24

Farkas R, Estrada-Peña A, Jaenson TGT, Pascucci I, Madder M (2012) Basic biology and geographical distribution of tick species involved in the transmission of animal pathogens, including zoonoses. In: Salman M, Tarrés-Call J (eds) Ticks and tick-borne diseases: geographical distribution and control strategies in the Euro-Asian region. CABI, Boston, pp 6–26

Garben AFM, Vos H, van Bronswijk JEMH (1982) *Haemaphysalis punctata* Canestrini and Fanzago 1877, a tick of pastured sea dunes on the island of Texel (The Netherlands). Acarologia 23:19–25

García-Sanmartín J, Barandika JF, Juste RA, Garcìa-Pérez AL, Hurtado A (2008) Distribution and molecular detection of *Theileria* and *Babesia* in questing ticks from northern Spain. Med Vet Entomol 22:318–325. https://dx.doi.org/10.1111/j.1365-2915.2008.00748.x

Gilot B (1985) Repartition et écologie d'*Haemaphysalis punctata* (Canestrini et Fanzago, 1877) (Acarina, Ixodoidea) dans les Alpes françaises et leur avant-pays. Rev Iber Parasitol 45:25–40

Guglielmone AA, Robbins RG, Apanaskevich DA, Petney TN, Estrada-Peña A, Horak IG (2014) The hard ticks of the world. (Acari: Ixodida: Ixodidae). Springer, Dordrecht, p 738

Hillyard PD (1996) Ticks of North-West Europe. Field Studies Council, Shrewsbury, UK, p 147

Hoogstraal H (1959) Biological observations on certain Turkish *Haemaphysalis* ticks (Ixodoidea, Ixodidae). J Parasitol 45:227–232. https://dx.doi.org/10.2307/3286538

Hornok S, Meli ML, Perreten A, Farkas R, Willi B, Beugnet F, Lutz H, Hofmann-Lehmann R (2010) Molecular investigation of hard ticks (Acari: Ixodidae) and fleas (Siphonaptera: Pulicidae) as potential vectors of rickettsial and mycoplasmal agents. Vet Microbiol 140:98–104. https://dx.doi.org/10.1016/j.vetmic.2009.07.013

Jaenson TGT, Tälleklint L, Lundqvist L, Olsen B, Chirico J, Mejlon H (1994) Geographical distribution, host associations, and vector roles of ticks (Acari: Ixodidae, Argasidae) in Sweden. J Med Entomol 31:240–256. https://dx.doi.org/10.1093/jmedent/31.2.240

L'Hostis M, Seegers H (2002) Tick-borne parasitic diseases in cattle: current knowledge and prospective risk analysis related to the ongoing evolution in French cattle farming systems. Vet Res 33:599–611. https://dx.doi.org/10.1051/vetres:2002041

Labuda M, Nuttall PA (2004) Tick-borne viruses. Parasitology 129: S221–S245. https://dx.doi.org/10.1017/S0031182004005220

Liebisch A, Melfsen J, Rahman MS (1976) Zum Vorkommen der Zecke *Haemaphysalis punctata* (Can. et Fanz., 1877) und *Babesia major* beim Rind in Norddeutschland. Berliner und Münchener Tierärztliche Wochenschrift 89:477–480

Mans BJ, Gothe R, Neitz AWH (2008) Tick toxins: perspectives on paralysis and other forms of toxicoses caused by ticks. In: Bowman AS, Nuttall PA (eds) Ticks: biology, disease and control. Cambridge University Press, pp 108–126

Márquez FJ (2008) Spotted fever group *Rickettsia* in ticks from southeastern Spain natural parks. Exp Appl Acarol 45:185–194. https://dx.doi.org/10.1007/s10493-008-9181-7

Mertins JW, Schlater JL (1991) Exotic ectoparasites of ostriches recently imported into the United States. J Wildl Dis 27:180–182. https://dx.doi.org/10.7589/0090-3558-27.1.180

Morzaria SP, Brocklesby DW, Harradine DL (1977) Experimental transmission of *Babesia major* by *Haemaphysalis punctata*. Res Vet Sci 23:261–262

Nosek J (1971) The ecology, bionomics, and behaviour of *Haemaphysalis* (*Aboimisalis*) *punctata* tick in Central Europe. Z Parasitenk 37:198–210. https://dx.doi.org/10.1007/BF00259499

Nosek J, Lichard M, Sztankay M (1967) The ecology of ticks in the Tribec and Hronský Inovec Mountains. Bull World Health Organ 36(Suppl):49–59

Olsufyev NG, Petrov VG (1968) Discovery of *Haemaphysalis concinna* Koch ticks naturally infected by the tularemia agent. Tr Inst Zool Akad Kazakh SSR 12:54–56

Palomar AM, Portillo A, Santibáñez P, Mazuelas D, Roncero L, García-Álvarez L, Santibáñez S, Gutiérrez Ó, Oteo JA (2015) Detection of tick-borne *Anaplasma bovis*, *Anaplasma phagocytophilum* and *Anaplasma centrale* in Spain. Med Vet Entomol 29:349–353. https://dx.doi.org/10.1111/mve.12124

Papadopoulos B, Morel PC, Aeschlimann A (1996) Ticks of domestic animals in the Macedonia region of Greece. Vet Parasitol 63:25–40. https://dx.doi.org/10.1016/0304-4017(95)00877-2

Pavlov P (1963) Research on "tick paralysis" observed in chickens in Bulgaria and caused by nymphs of *Haemaphysalis punctata*. Can. and Franz. Ann Parasitol Hum Comp 38:459–461

Pomerantzev BI (1950) Ixodid ticks (Ixodidae). Fauna SSSR, Paukoobraznye, n. s., (41), 4, p 224 (English translation by Elbl A, edited by Anastos G. The American Institute of Biological Sciences, Washington)

Punda-Polic V, Petrovec M, Trilar T, Duh D, Bradaric N, Klismanic Z, Avsic-Zupanc T (2002) Detection and identification of spotted fever group rickettsiae in ticks collected in Southern Croatia. Exp Appl Acarol 28:169–176. https://dx.doi.org/10.1023/A:1025334113190

Radulovic Z, Chochlakis D, Tomanovic S, Milutinovic M, Tselentis Y, Psaroulaki A (2011) First detection of spotted fever group rickettsiae in ticks in Serbia. Vector Borne Zoonot Dis 11:111–115. https://dx.doi.org/10.1089/vbz.2009.0254

Razmi G, Naghibi A, Aslani M, Fathivand M, Dastjerdi K (2002) An epidemiological study on ovine babesiosis in the Mashhad suburb area, province of Khorasan. Iran Vet Parasitol 108:109–115. https://dx.doi.org/10.1016/S0304-4017(02)00203-0

Santos-Silva MM, Beati L, Santos AS, De Sousa R, Núncio MS, Melo P, Santos-Reis M, Fonseca C, Formosinho P, Vilela C, Bacellar F (2011) The hard-tick fauna of mainland Portugal (Acari: Ixodidae): an update on geographical distribution and known associations with hosts and pathogens. Exp Appl Acarol 55:85–121. https://dx.doi.org/10.1007/s10493-011-9440-x

Stoker MGP, Marmion BP (1955) Q Fever in Britain: Isolation of *Rickettsia burneti* from the tick *Haemaphysalis punctata*. J Hyg (Lond.) 53:322–327. https://dx.doi.org/10.1017/S0022172400000802

Tälleklint L (1996) Lyme borreliosis spirochetes in *Ixodes ricinus* and *Haemaphysalis punctata* ticks (Acari: Ixodidae) on three islands in the Baltic Sea. Exp Appl Acarol 20:467–476. https://dx.doi.org/10.1007/BF00053310

Tijsse-Klasen E, Hansford KM, Jahfari S, Phipps P, Sprong H, Medlock JM (2013) Spotted fever group rickettsiae in *Dermacentor reticulatus* and *Haemaphysalis punctata* ticks in the UK. Parasit Vectors 6:212. https://dx.doi.org/10.1186/1756-3305-6-212

Walter G (1982) Transmission of *Babesia microti* by nymphs of *Dermacentor marginatus*, *D. reticulatus*, *Haemaphysalis punctata*, *Rhipicephalus sanguineus* and *Ixodes hexagonus*. Z Parasitenkd 66:353–354

Haemaphysalis sulcata Canestrini and Fanzago, 1877 (Figs. 91–93)

Z. Vatansever

Life Cycle and Host Preferences

Haemaphysalis sulcata is three-host exophilous tick. Adults mostly feed on domestic and wild ungulates, while immature ticks have a special preference for reptiles. They are also rarely found on ground-nesting birds and there are also a few records from small mammals (Pomerantzev 1950; Filippova 1997; Keskin et al. 2013). Under laboratory conditions, where a part of the population may act as two-host ticks, female ticks lay about 2000–3000 eggs 10–30 days after detaching. Embryonic development takes about 18–40 days. Nymphs moult to adults in 23–52 days (Pomerantzev 1950). In nature, female ticks lay eggs during March–April and the larvae hatch during May–June. Moulting of fed larvae and nymphs takes place in rock crevices, under stones and in the burrows of small, wild mammals. Some of the moulted nymphs feed during the same season so that they may overwinter either in the fed or unfed stage. The whole life cycle takes about 1–2 years (Pomerantzev 1950; Grebenyuk 1966; Filippova 1997).

Ecology

This species occurs mostly in steppe with a semi-desert character and is usually not present in areas with high humidity. It is widespread mostly in wormwood foothills, mountain steppe, dry steppe and semi-desert habitats. It has been recorded in mountain valleys up to 2200–2500 m altitude (Pomerantzev 1950; Grebenyuk 1966; Berdyev 1980; Filippova 1997). Adult infestations on animals have two peaks in the cold months of the year: March–May and October–November. Immatures are active in the hot months: larvae between April and August, nymphs between April and October (Pomerantzev 1950; Grebenyuk 1966).

Distribution

Haemaphysalis sulcata has been recorded from southern Russia, Dagestan, Azerbaijan, Armenia, Turkey, Iran, the Middle East, Central Asia, the Balkans, Italy, Mediterranean islands and North Africa (Pomerantzev 1950; Merdivenci 1969; Filippova 1997; Bouattour et al. 1999; Estrada-Peña et al. 2004; Bursali et al. 2012; Keskin et al. 2013).

Vectorial Capacity and Pathogen Burden

Haemaphysalis sulcata is known to transmit *Brucella abortus* under laboratory conditions (Grebenyuk 1966). It has been reported to transmit *Anaplasma ovis* and *Coxiella brunetti* (Berdyev 1980). Massive infections of *Hae. sulcata* with *Rickettsia felis*-like bacteria were reported from Croatia (Duh et al. 2006). This rickettsia was recently classified as *Rickettsia hoogstraalii* (Duh et al. 2010).

Z. Vatansever (✉)
Department of Parasitology, Faculty of Veterinary Medicine, Kafkas University, Kars, Turkey
e-mail: zativet@gmail.com

A. Estrada-Peña et al. (eds.), *Ticks of Europe and North Africa*,
https://doi.org/10.1007/978-3-319-63760-0_47

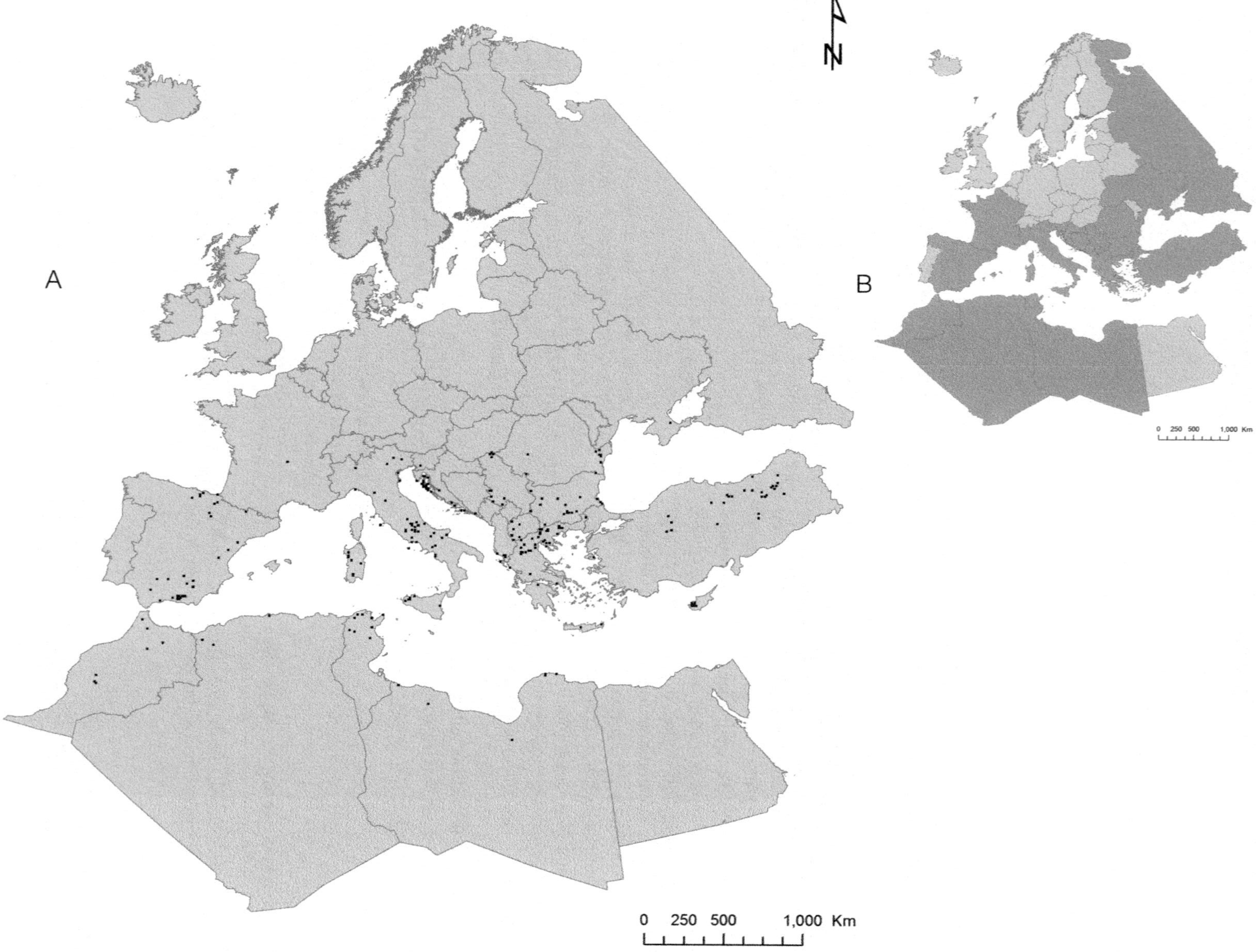

Fig. 91 **A** Distribution of *Hae. sulcata* in Europe and Northern Africa (10 × 10 km grid presence with black dots). **B** Countries where the species has been reported are marked in dark grey

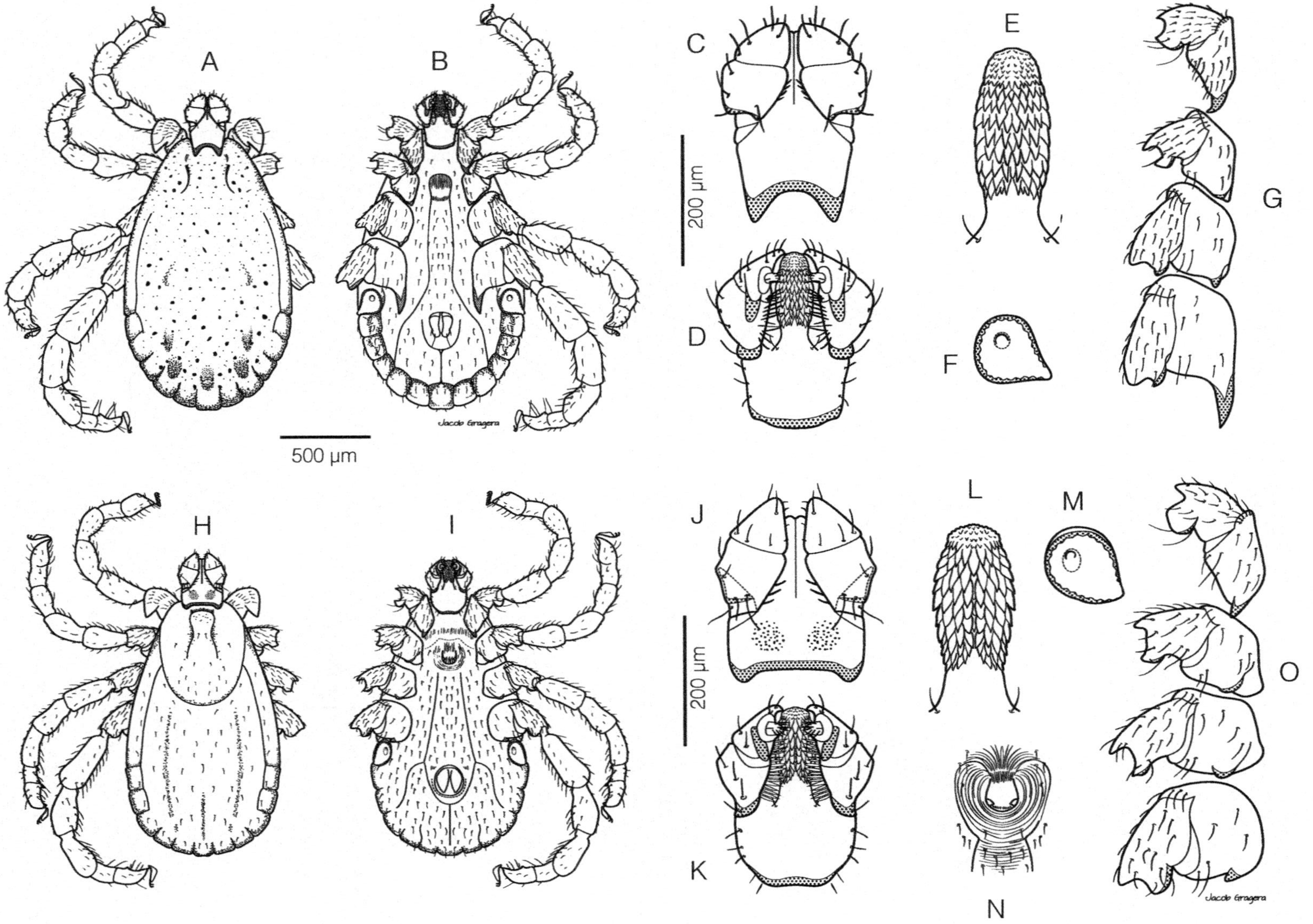

Fig. 92 **A–G** The male of *Hae. sulcata*. **A** Dorsal, **B** ventral, **C** capitulum, dorsal, **D** capitulum, ventral, **E** hypostome, **F** spiracular plate, **G** coxae and trochanters I–IV. **H–O** The female of *Hae. sulcata*. **H** Dorsal, **I** ventral, **J** capitulum, dorsal, **K** capitulum, ventral, **L** hypostome, **M** genital opening, **N** spiracular plate, **O** coxae and trochanters I–IV. Illustrations from specimens collected in Spain, France, Italy and Greece

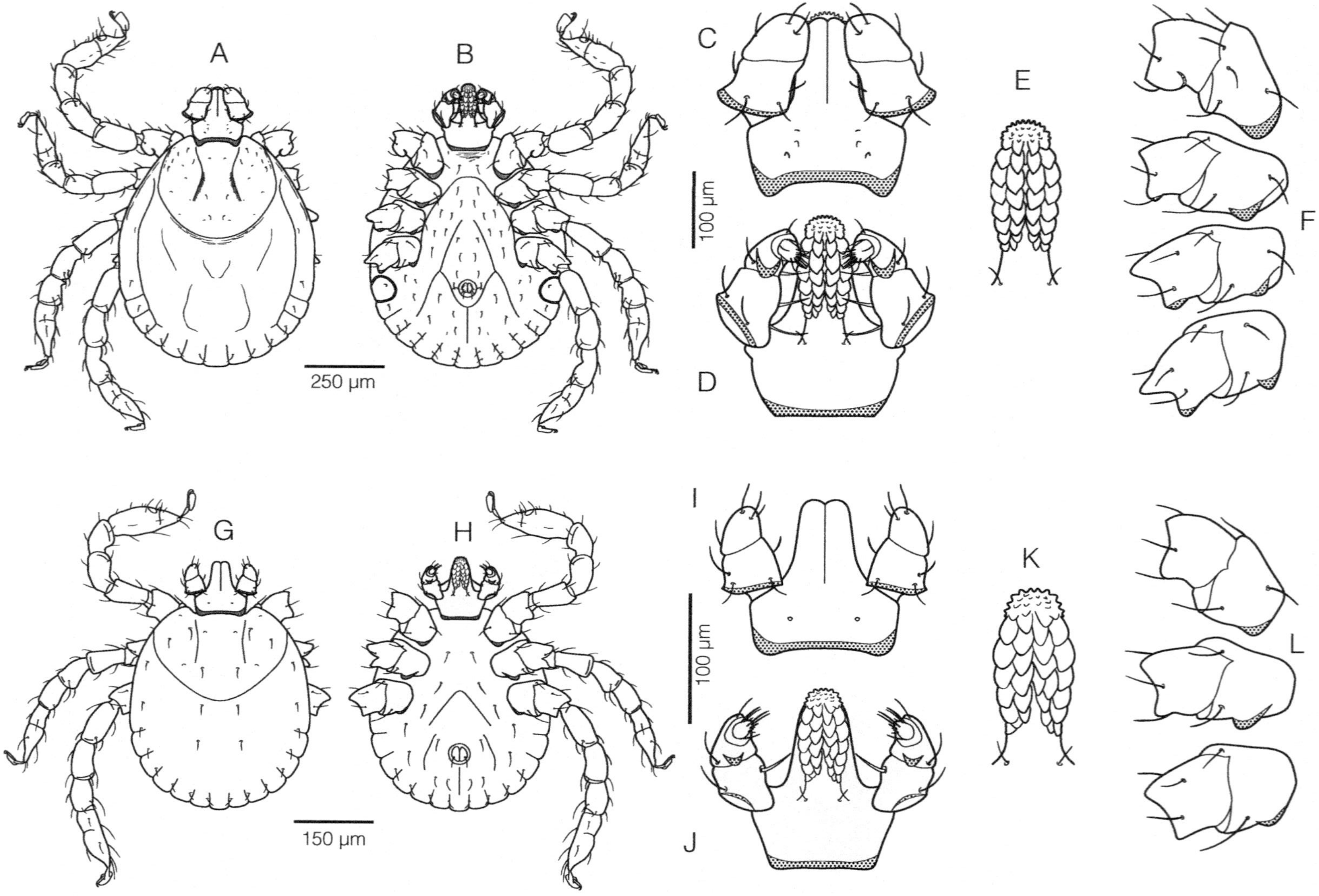

Fig. 93 **A–F** The nymph of *Hae. sulcata*. **A** Dorsal, **B** ventral, **C** capitulum, dorsal, **D** capitulum, ventral, **E** hypostome, **F** coxae and trochanters I–IV. **G–L** The larva of *Hae. sulcata*. **G** Dorsal, **H** ventral, **I** capitulum, dorsal, **J** capitulum, ventral, **K** hypostome, **L** coxae and trochanters I–III. Illustrations from specimens collected in Spain, France, Italy and Greece

References

Berdyev A (1980) Ecology of ixodid ticks of Turkmenistan and their importance in natural focal disease epizootiology. Ylym, Ashgabat (in Russian)

Bouattour A, Darghouth MA, Daoud A (1999) Distribution and ecology of ticks (Acari: Ixodidae) infesting livestock in Tunisia: an overview of eighth years field collections. Parassitologia 41(Suppl 1):5–10

Bursali A, Keskin A, Tekin S (2012) A review of the ticks (Acari: Ixodida) of Turkey: species diversity, hosts and geographical distribution. Exp Appl Acarol 57:91–104

Duh D, Punda-Polic V, Avsic-Zupanc T, Bouyer D, Walker DH, Popov VL, Jelovsek M, Gracner M, Trilar T, Bradaric N, Kurtti TJ, Strus J (2010) *Rickettsia hoogstraalii* sp. nov., isolated from hard- and soft-bodied ticks. Int J Syst Evol Microbiol 60:977–984

Duh D, Punda-PoliĆ V, Trilar T, Petrovec M, BradariĆ N, AvŠIČ-ŽUpanc T (2006) Molecular identification of *Rickettsia felis*-like bacteria in *Haemaphysalis sulcata* ticks collected from domestic animals in southern Croatia. Ann New York Acad Sci 1078: 347–351

Estrada-Peña A, Bouattour A, Camicas JL, Walker AR (2004) Ticks of domestic animals in the Mediterranean region. A guide to identification of species. University of Zaragoza, Zaragoza

Filippova NA (1997) Ixodid ticks of subfamily Amblyomminae. In the fauna of Russia and neighboring countries. New Series no. 145: Arachnoidea, vol 4(5), St. Petersburg, Nauka (in Russian)

Grebenyuk RV (1966) Ixodid ticks (Parasitiformes, Ixodidae) of Kirgizia. Institute of Biology, Academy of Science of Kirgiz Republic, Frunze (in Russian)

Keskin A, Bursali A, Kumlutas Y, Ilgaz C, Tekin S (2013) Parasitism of immature stages of *Haemaphysalis sulcata* (Acari: Ixodidae) on some reptiles in Turkey. J Parasitol 99:752–755

Merdivenci A (1969) Türkiye keneleri üzerine araştırmalar. İstanbul Üniversitesi, Cerrahpaşa Tıp Fakültesi, İstanbul

Pomerantzev BI (1950) Ixodid ticks (Ixodidae). In Fauna of the USSR, New series 41: Arachnoidea, vol 4(2), Leningrad, Nauk (in Russian)

Haemaphysalis caucasica Olenev, 1928 (Figs. 94–96)

T. N. Petney and M. P. Pfäffle

Life-Cycle and Host Preferences

Very little is known about *Hae. caucasica.* Hosts include hares (Leporidae), which are listed as the most common host of all life history stages, bear, jackal and fox (Pomerantzev 1950; Guglielmone et al. 2014). Nymphs have been found on lizards (Guglielmone et al. 2014). It has been recorded from humans (Filippova 1997).

Ecology

This is a rare species. It occurs in steppe areas, either on the plains but more commonly in foothills and mountainous regions where it is most commonly found from March to July (Pomerantzev 1950).

Distribution

Haemaphysalis caucasica has a relatively wide range including Ukraine, southern Russia (including Crimea), Azerbaijan, the Republic of Tajikistan, Kyrgyzstan and Iran (Pomerantzev 1950; Reznik 1966; Filippova et al. 1976).

Vectorial Capacity and Pathogen Burden

Nothing is known about the vector ability or the pathogens associated with *Hae. caucasica.*

References

Filippova NA (1997) Ixodid ticks of subfamily Amblyomminae. In the fauna of Russia and neighboring countries. New Series no. 145: Arachnoidea, vol 4(5), St. Petersburg, Nauka (in Russian)

Filippova NA, Neronov VM, Farhang-Azad A (1976) On the hard ticks (Acarina, Ixodidae) of small mammals in Iran. Entomol Rev 55:467–478

Guglielmone AA, Robbins RG, Apanaskevich DA, Petney TN, Estrada-Peña A, Horak IG (2014) The hard ticks of the world (Acari: Ixodida: Ixodidae). Springer, Dordrecht, p 738

Pomerantzev BI (1950) Fauna of USSR Arachnida vol 4, no 2, Ixodid ticks (Ixodidae). Moscow, Academy of Sciences USSR. English translation, Institute of Acarology, University of Marylamd

Reznik PA (1966) On distribution area features of ticks belonging to the genus *Haemaphysalis* in the Soviet fauna. Tezisy Dokl 1 Akarol Soveshch 171–172

T. N. Petney (✉)
State Museum of Natural History, Erbprinzenstr. 13, 76133 Karlsruhe, Germany
e-mail: trevor.petney@kit.edu

M. P. Pfäffle
Department of Ecology and Parasitology, Institute of Zoology, Karlsruhe Institute of Technology, Karlsruhe, Germany
e-mail: miripfaeffle@web.de

A. Estrada-Peña et al. (eds.), *Ticks of Europe and North Africa*,
https://doi.org/10.1007/978-3-319-63760-0_48

Fig. 94 **A** Distribution of *Hae. caucasica* in Europe and Northern Africa (10 × 10 km grid presence with black dots). **B** Countries where the species has been reported are marked in dark grey

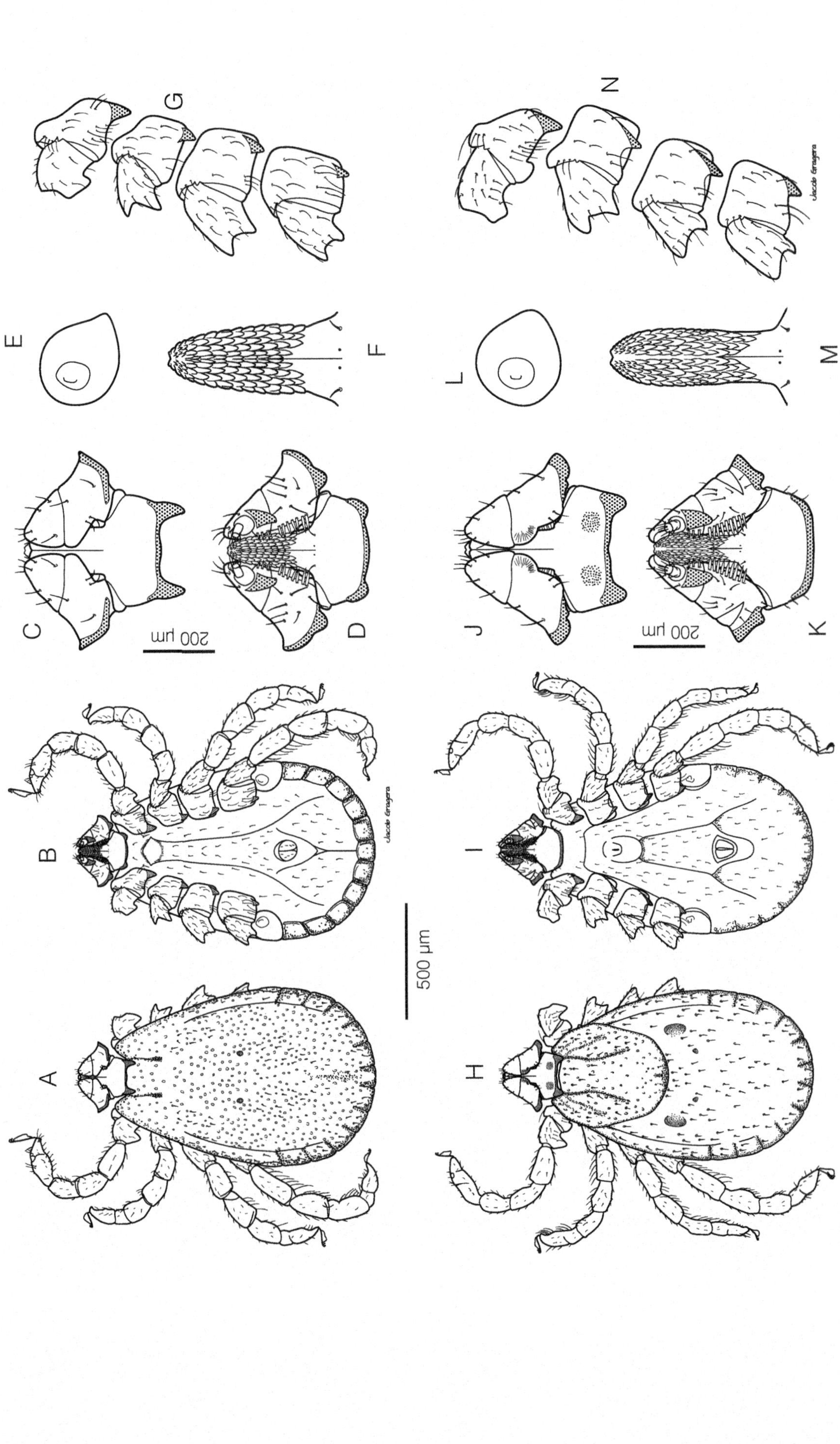

Fig. 95 **A–G** The male of *Hae. caucasica.* **A** Dorsal, **B** ventral, **C** capitulum, dorsal, **D** capitulum, ventral, **E** spiracular plate, **F** hypostome, **G** coxae and trochanters I–IV. **H–N** The female of *Hae. caucasica.* **H** Dorsal, **I** ventral, **J** capitulum, dorsal, **K** capitulum, ventral, **L** spiracular plate, **M** hypostome, **N** coxae and trochanters I–IV. Illustrations from specimens collected in Turkey and complemented from illustrations by Filippova (1997)

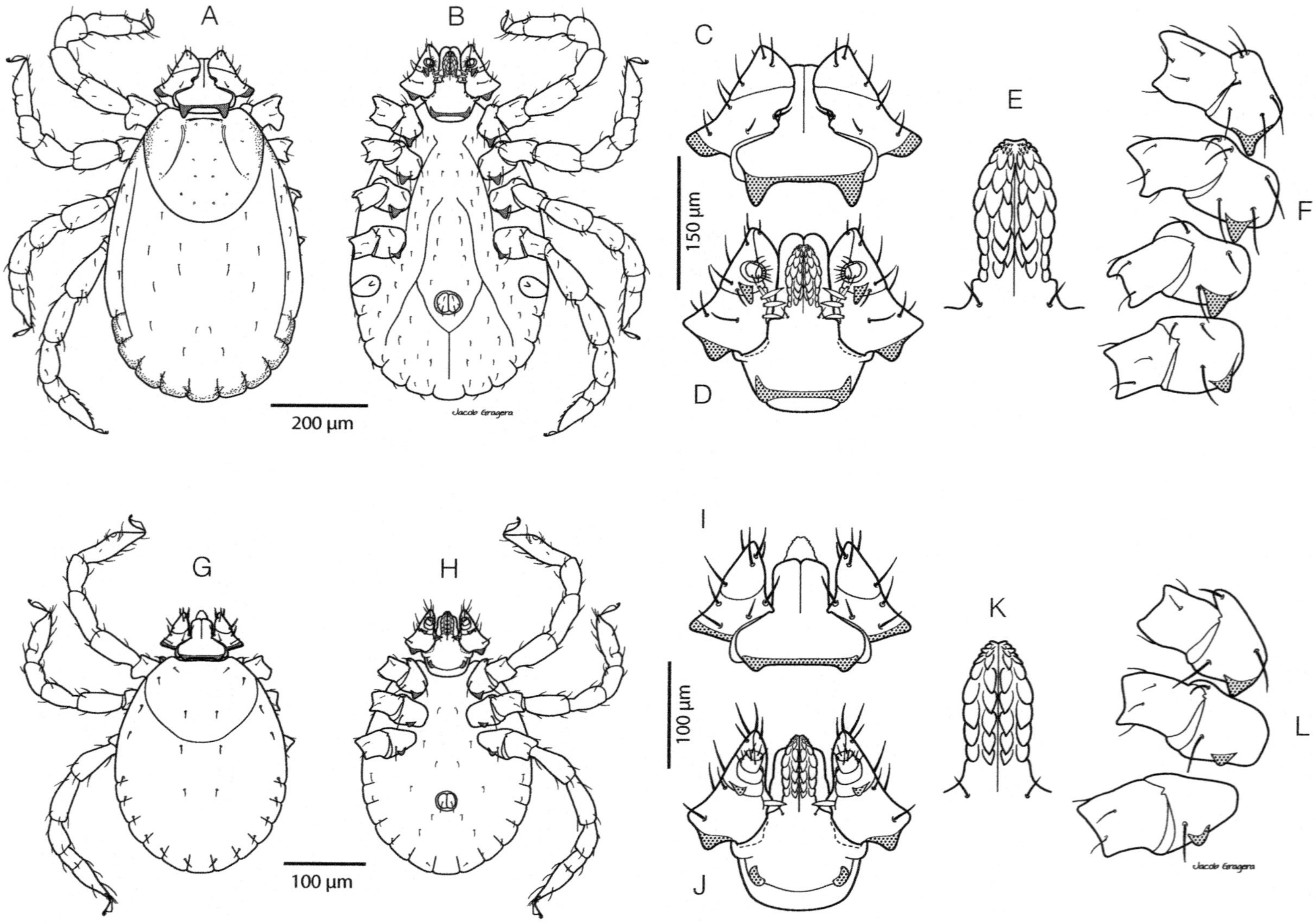

Fig. 96 **A–F** The nymph of *Hae. caucasica*. **A** Dorsal, **B** ventral, **C** capitulum, dorsal, **D** capitulum, ventral, **E** hypostome, **F** coxae and trochanters I–IV. **G–L** The larva of *Hae. caucasica*. **G** Dorsal, **H** ventral, **I** capitulum, dorsal, **J** capitulum, ventral, **K** hypostome, **L** coxae and trochanters I–III. Illustrations from specimens collected in Turkey and complemented from illustrations by Filippova (1997)

Haemaphysalis concinna Koch, 1844 (Figs. 97–99)

M. P. Pfäffle, Z. Vatansever, and T. N. Petney

Life Cycle and Host Preferences

Haemaphysalis concinna is a three-host tick the life cycle of which can be completed within 3 years under natural conditions (Nosek 1971). Under laboratory conditions, with a relative humidity between 90 and 95% and temperatures between 20–24 °C, the developmental cycle can be completed within 194–252 days (Nosek et al. 1967; Nosek 1971). The immatures feed on wide variety of small and medium-sized mammals and birds but rarely on reptiles (Pomerantzev 1950; Nosek et al. 1967; Olsufyev and Petrov 1968; Nosek 1971; Filippova 1997; Hornok et al. 2014). Adults feed on wild and domestic ungulates, carnivores, such as foxes and dogs, and medium-sized insectivores, such as hedgehogs (Pomerantzev 1950; Nosek et al. 1967; Olsufyev and Petrov 1968; Nosek 1971; Filippova 1997; Földvári and Farkas 2005; Sréter-Lancz et al. 2006).

Ecology

Haemaphysalis concinna can be active throughout the year (Barandika et al. 2008), but peak activity and duration of activity can vary depending on the region and the year. In general, most adults are found from around April to October/November with a peak activity in late spring or summer (Pomerantzev 1950; Nosek et al. 1967; Nosek 1971; Földvári and Farkas 2005; Széll et al. 2006; Hornok 2009; Coipan et al. 2012; Duscher et al. 2013). The larvae are active from the end of May until November and the nymphs from the middle of April until October (Nosek et al. 1967; Nosek 1971). Host-seeking activity seems to be related to soil temperature (Hubálek et al. 2003). *Haemaphysalis concinna* can occur in a variety of different habitats, including deciduous forests, mixed-forests, mixed hornbeam-oak forests with bush-undergrowth, forest clearings and the margin of oak forests, lake coasts, river basins, and in shoreline vegetation (Pomerantzev 1950; Nosek et al. 1967; Nosek 1971; Kozuch and Nosek 1980; Filippova 1997). It is a thermophilic species (Hubálek et al. 2003) and prefers humid habitats (Pomerantzev 1950; Olsufyev and Petrov 1968; Bauch 1990). In the Czech Republic, its distribution is restricted to areas with annual temperatures of 8 °C and an annual precipitation of 600–700 mm (Nosek 1971).

Distribution

This is a Eurasian species found between 38°N and 56°N latitudes (Filippova 1997). It has been reported from Austria, Belarus, Bulgaria, China, Croatia, the Czech Republic, France, Germany, Hungary, Iran, Italy, Japan, Poland, Romania, Russia, Serbia, Slovakia, Spain, and Turkey (Olsufyev and Petrov 1968; Nosek 1971; Bauch 1990; Kahl et al. 1992; Jaenson et al. 1994; Shpynov et al. 2004a, b; Siuda et al. 2006; Sun et al. 2006; Széll et al. 2006; Barandika et al. 2006, 2008; Blaschitz et al. 2008; Uspensky et al. 2008; Ando et al. 2010; Briciu et al. 2011; Radulović et al. 2011; Krčmar 2012; Shemshad et al. 2012; Tekin et al. 2012; Dantas-Torres and Otranto 2013; Švehlová et al. 2014) and from Cyprus on migrating birds (Kaiser et al. 1974).

M. P. Pfäffle (✉)
Department of Ecology and Parasitology, Institute of Zoology, Karlsruhe Institute of Technology, Karlsruhe, Germany
e-mail: miripfaeffle@web.de

Z. Vatansever
Department of Parasitology, Faculty of Veterinary Medicine, Kafkas University, Kars, Turkey
e-mail: zativet@gmail.com

T. N. Petney
State Museum of Natural History, Erbprinzenstr. 13, 76133 Karlsruhe, Germany
e-mail: trevor.petney@kit.edu

A. Estrada-Peña et al. (eds.), *Ticks of Europe and North Africa*,
https://doi.org/10.1007/978-3-319-63760-0_49

Fig. 97 **A** Distribution of *Hae. concinna* in Europe and Northern Africa (10 × 10 km grid presence with black dots). **B** Countries where the species has been reported are marked in dark grey

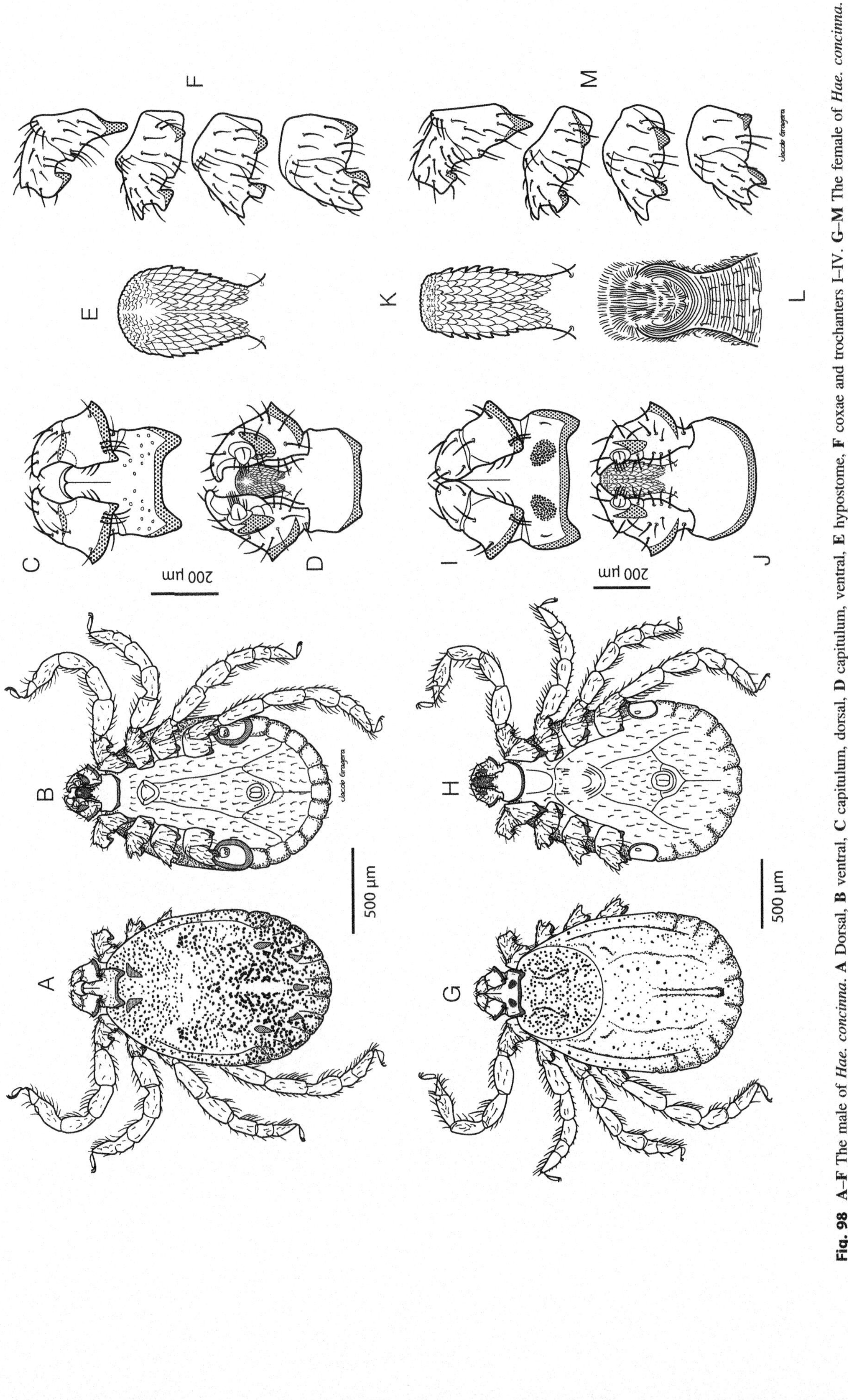

Fig. 98 **A–F** The male of *Hae. concinna*. **A** Dorsal, **B** ventral, **C** capitulum, dorsal, **D** capitulum, ventral, **E** hypostome, **F** coxae and trochanters I–IV. **G–M** The female of *Hae. concinna*. **G** Dorsal, **H** ventral, **I** capitulum, dorsal, **J** capitulum, ventral, **K** hypostome, **L** genital opening, **M** coxae and trochanters I–IV. Illustrations from specimens collected in Spain and France, and complemented from illustrations by Filippova (1997)

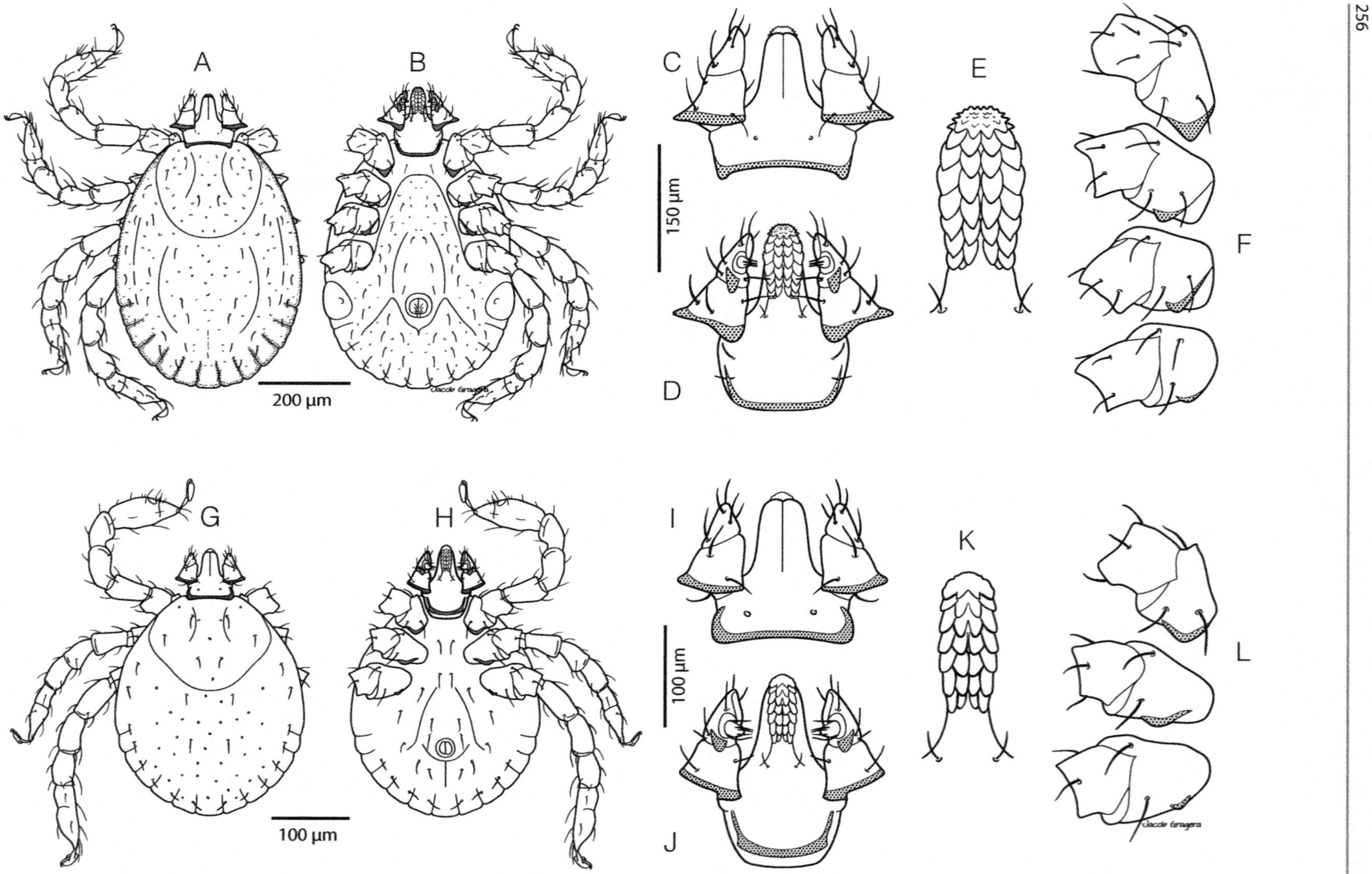

Fig. 99 **A–F** The nymph of *Hae. concinna*. **A** Dorsal, **B** ventral, **C** capitulum, dorsal, **D** capitulum, ventral, **E** hypostome, **F** coxae and trochanters I–IV. **G–L** The larva of *Hae. concinna*. **G** Dorsal, **H** ventral, **I** capitulum, dorsal, **J** capitulum, ventral, **K** hypostome, **L** coxae and trochanters I–III. Illustrations from specimens collected in Spain and France, and complemented from illustrations by Filippova (1997)

Vectorial Capacity and Pathogen Burden

Haemaphysalis concinna is the vector of a variety of *Rickettsia* spp., including *Rickettsia heilongjiangensis*, *Rickettsia sibirica* (Jongejan and Uilenberg 2005; Shpynov et al. 2006; Dantas-Torres et al. 2012), *Rickettsia helvetica*, *Candidatus* "Rickettsia rara" and *Candidatus* "Rickettsia kotlanii" (Raoult et al. 2005; Sréter-Lancz et al. 2006; Hornok et al. 2010). It is also a vector of *Anaplasma phagocytophilum*, *A. bovis*, *Coxiella burnetii* and *Francisella tularensis* (Olsufyev and Petrov 1968; Řeháček et al. 1991; Guryčová et al. 1995; Khazova and Iastrebov 2001; Spitalska and Kocianova 2003; Shpynov et al. 2004a, 2004b; Barandika et al. 2008; Dantas-Torres et al. 2012; Kreizinger et al. 2013). *Haemaphysalis concinna* is able to acquire infection with *Borrelia garinii* by feeding on infected mice, however, it is not able to maintain the infection transstadially (Sun and Xu 2003). *Borrelia burgdorferi* s.l. has been detected in this tick (Rigó et al. 2011), however, it is unclear whether it is suitable as a vector, as spirochete prevalence is normally very low if present (see e.g., Kahl et al. 1992). It is a known vector of the tick-borne encephalitis virus (Kozuch and Nosek 1980; Khazova and Iastrebov 2001; Dantas-Torres et al. 2012) and has been associated with Crimean-Congo haemorrhagic fever virus (Tekin et al. 2012), Omsk haemorrhagic fever virus and Tamdy virus (Hubalek et al. 1989).

References

Ando S, Kurosawa M, Sakata A, Fujita H, Sakai K, Sekine M, Katsumi M, Saitou W, Yano Y, Takada N, Takano A, Kawabata H, Hanaoka N, Watanabe H, Kurane I, Kishimoto T (2010) Human *Rickettsia heilongjiangensis* infection, Japan. Emerg Infect Dis 16:1306–1308. https://dx.doi.org/10.3201/eid1608.100049

Barandika JF, Berriatua E, Barral M, Juste RA, Anda P, Garciá-Pérez A (2006) Risk factors associated with ixodid tick species distributions in the Basque region in Spain. Med Vet Entomol 20:177–188. https://dx.doi.org/10.1111/j.1365-2915.2006.00619.x

Barandika JF, Hurtado A, García-Sanmartín J, Juste RA, Anda P, Garciá-Pérez A (2008) Prevalence of tick-borne zoonotic bacteria in questing adult ticks from northern Spain. Vector Borne Zoonot Dis 8:829–835. https://dx.doi.org/10.1089/vbz.2008.0023

Bauch R (1990) *Ixodes ricinus*, *Haemaphysalis concinna* and *Dermacentor reticulatus* (Ixodida, Ixodidae) in the East German district of Leipzig. Angew Parasitol 31:57–64

Blaschitz M, Narodoslavsky-Gföller M, Kanzler M, Walochnik J, Stanek G (2008) *Borrelia burgdorferi* sensu lato genospecies in questing *Ixodes ricinus* ticks in Austria. Int J Med Microbiol 298:168–176. https://dx.doi.org/10.1016/j.ijmm.2007.10.001

Briciu VT, Titilincu A, Tăţulescu DF, Carstina D, Lefkaditis M, Mihalca AD (2011) First survey on hard ticks (Ixodidae) collected from humans in Romania: possible risks for tick-borne diseases. Exp Appl Acarol 54:199–204. https://dx.doi.org/10.1007/s10493-010-9418-0

Coipan EC, Vladimirescu AF, Ciolpan O, Teodorescu I (2012) Tick species (Acari: Ixodoidea) distribution, seasonality and host associations in Romania. Trav Mus Natl d'Hist Nat "Grigore Antipa" 54:301–317. https://dx.doi.org/10.2478/v10191-011-0018-y

Dantas-Torres F, Otranto D (2013) Species diversity and abundance of ticks in three habitats in southern Italy. Ticks Tick Borne Dis 4:251–255. https://dx.doi.org/10.1016/j.ttbdis.2012.11.004

Dantas-Torres F, Chomel BB, Otranto D (2012) Ticks and tick-borne diseases: a one health perspective. Trends Parasitol 28:437–446. https://dx.doi.org/10.1016/j.pt.2012.07.003

Duscher GG, Feiler A, Leschnik M, Joachim A (2013) Seasonal and spatial distribution of ixodid tick species feeding on naturally infested dogs from Eastern Austria and the influence of acaricides/repellents on these parameters. Parasit Vectors 6:76. https://dx.doi.org/10.1186/1756-3305-6-76

Filippova NA (1997) Ixodid ticks of subfamily Amblyomminae. In the fauna of Russia and neighboring countries. New Series no. 145: Arachnoidea, vol 4(5), St. Petersburg, Nauka (in Russian)

Földvári G, Farkas R (2005) Ixodid tick species attaching to dogs in Hungary. Vet Parasitol 129:125–131. https://dx.doi.org/10.1016/j.vetpar.2004.11.032

Guryčová D, Kocianová E, Výrosteková V, Řeháček J (1995) Prevalence of ticks infected with *Francisella tularensis* in natural foci of tularemia in western Slovakia. Eur J Epidemiol 11:469–474. https://dx.doi.org/10.1007/BF01721235

Hornok S (2009) Allochronic seasonal peak activities of *Dermacentor* and *Haemaphysalis* spp. under continental climate in Hungary. Vet Parasitol 163:366–369. https://dx.doi.org/10.1016/j.vetpar.2009.03.048

Hornok S, Meli ML, Perreten A, Farkas R, Willi B, Beugnet F, Hofmann-Lehmann R (2010) Molecular investigation of hard ticks (Acari: Ixodidae) and fleas (Siphonaptera: Pulicidae) as potential vectors of rickettsial and mycoplasmal agents. Vet Microbiol 140:98–104. https://dx.doi.org/10.1016/j.vetmic.2009.07.013

Hornok S, Kováts D, Csörgő T Meli ML, Gönczi E, Hadnagy Z, Takács N, Farkas R, Hofmann-Lehmann R (2014) Birds as potential reservoirs of tick-borne pathogens: first evidence of bacteraemia with *Rickettsia helvetica*. Parasit Vectors 7:128. https://dx.doi.org/10.1186/1756-3305-7-128

Hubalek Z, Juricova Z, Halouzka J et al (1989) Arboviruses associated with birds in southern Moravia, Czecheslovakia. Prirodoved Pr Ust Ceskoslov Akad Ved Brne 23:1–50

Hubálek Z, Halouzka J, Juricová Z (2003) Host-seeking activity of ixodid ticks in relation to weather variables. J Vector Ecol 28:159–165

Jaenson TGT, Tälleklint L, Lundqvist L, Olsen B, Chirico J, Mejlon H (1994) Geographical distribution, host associations, and vector roles of ticks (Acari: Ixodidae, Argasidae) in Sweden. J Med Entomol 31:240–256. https://dx.doi.org/10.1093/jmedent/31.2.240

Jongejan F, Uilenberg G (2005) The global importance of ticks. Parasitology 129:S3. https://dx.doi.org/10.1017/S0031182004005967

Kahl O, Janetzki C, Gray JS et al (1992) Tick infection rates with *Borrelia*: *Ixodes ricinus* versus *Haemaphysalis concinna* and *Dermacentor reticulatus* in two locations in eastern Germany. Med Vet Entomol 6:363–366. https://dx.doi.org/10.1111/j.1365-2915.1992.tb00634.x

Kaiser MN, Hoogstraal H, Watson GE (1974) Ticks (Ixodoidea) on migrating birds in Cyprus, fall 1967 and spring 1968, and epidemiological considerations. Bull Entomol Res 64:97–110. https://dx.doi.org/10.1017/S0007485300027024

Khazova TG, Iastrebov VK (2001) Combined focus of tick-borne encephalitis, tick-borne rickettsiosis and tularemia in the habitat of *Haemaphysalis concinna* in south central Siberia. Zh Mikrobiol Epidemiol Immunobiol 1:78–80 (in Russian)

Kozuch O, Nosek J (1980) Experimental transmission of tick-borne encephalitis (TBE) virus by *Haemaphysalis concinna* ticks. Acta Virol 24:377

Krčmar S (2012) Hard ticks (Acari, Ixodidae) of Croatia. ZooKeys 234:19–57. https://dx.doi.org/10.3897/zookeys.234.3658

Kreizinger Z, Hornok S, Dán A, Hresko S, Makrai L, Magyar T, Bhide M, Erdélyi K, Hofmann-Lehmann R, Gyuranecz M (2013) Prevalence of *Francisella tularensis* and *Francisella*-like endosymbionts in the tick population of Hungary and the genetic variability of *Francisella*-like agents. Vector Borne Zoonot Dis 13:160–163. https://dx.doi.org/10.1089/vbz.2012.1065

Nosek J (1971) The ecology, bionomics and behaviour of *Haemaphysalis* (*Haemaphysalis*) *concinna* tick. Z Parasitenkd 36:233–241. https://dx.doi.org/10.1007/BF00348561

Nosek J, Lichard M, Sztankay M (1967) The ecology of ticks in the Tribec and Hronský Inovec mountains. Bull World Health Organ 36(Suppl 1):49–59

Olsufyev NG, Petrov VG (1968) Discovery of *Haemaphysalis concinna* Koch ticks naturally infected by the tularemia agent. Tr Inst Zool Akad Kazakh SSR 12:54–56

Pomerantzev BI (1950) Fauna of U.S.S.R. Arachnida, IV(2) Ixodid ticks (Ixodidae). Academy of Sciences USSR, Leningrad

Radulović Z, Chochlakis D, Tomanović S, Mihaljica D, Tselentis Y, Psaroulaki A (2011) First detection of spotted fever group rickettsiae in ticks in Serbia. Vector Borne Zoonot Dis 11:111–115. https://dx.doi.org/10.1089/vbz.2009.0254

Raoult D, Fournier P-E, Eremeeva M, Graves S, Kelly PJ, Oteo JA, Sekeyova Z, Tamura A, Tarasevich I, Zhang L (2005) Naming of rickettsiae and rickettsial diseases. Ann N Y Acad Sci 1063:1–12. https://dx.doi.org/10.1196/annals.1355.002

Řeháček J, Úrvölgyi J, Kocianová E, Sekeyová Z, Vavreková M, Kováčová E (1991) Extensive examination of different tick species for infestation with *Coxiella burnetii* in Slovakia. Eur J Epidemiol 7:299–303. https://dx.doi.org/10.1007/BF00145682

Rigó K, Gyuranecz M, Tóth AG, Földvári G (2011) Detection of *Borrelia burgdorferi* sensu lato and *Anaplasma phagocytophilum* in small mammals and ectoparasites in Hungary. Vector Borne Zoonot Dis 11:1499–1501. https://dx.doi.org/10.1089/vbz.2011.0608

Shemshad M, Shemshad K, Sedaghat MM, Shokri M, Barmaki A, Baniardalani M, Rafinejad J (2012) First survey of hard ticks (Acari: Ixodidae) on cattle, sheep and goats in Boeen Zahra and Takistan counties. Iran. Asian Pac J Trop Biomed 2:489–492. https://dx.doi.org/10.1016/S2221-1691(12)60082-3

Shpynov SN, Rudakov NV, Iastrebov VK, Leonova GN, Khazova TG, Egorova NV, Borisova ON, Preĭder VP, Bezrukov GV, Fedorov EG, Fedianin AP, Sherstneva MB, Turyshev AG, Gavrilov AP, Tankibaev MA, Fournier PE, Raoult D (2004a) New evidence for the detection of *Ehrlichia* and *Anaplasma* in *Ixodes* ticks in Russia and Kazakhstan. Meditsinskaia parazitologiia i parazitarnye bolezni 2:10–14

Shpynov SN, Rudakov NV, Iastrebov VK, Khazova TG, Fournier PE, Raoult D (2004b) Detection of *Rickettsia hulinii* in ticks of the *Haemaphysalis concinna* species in Russia, in Russian. Zh Mikrobiol Epidemiol Immunobiol 26–29

Shpynov S, Fournier P-E, Rudakov N, Tarasevich I, Raoult D (2006) Detection of members of the genera *Rickettsia*, *Anaplasma*, and *Ehrlichia* in ticks collected in the Asiatic part of Russia. Ann N Y Acad Sci 1078:378–383. https://dx.doi.org/10.1196/annals.1374.075

Siuda K, Majszyk A, Nowak M (2006) Ticks (Acari: Ixodida) parasitizing birds (Aves) in Poland. Biol Lett 43:147–151

Spitalska E, Kocianova E (2003) Tick-borne microorganisms in southwestern Slovakia. Ann N Y Acad Sci 990:196–200. https://dx.doi.org/10.1111/j.1749-6632.2003.tb07362.x

Sréter-Lancz Z, Széll Z, Kovács G, Egyed L, Márrialigeti K, Sréter T (2006) rickettsiae of the spotted-fever group in ixodid ticks from Hungary: identification of a new genotype ('*Candidatus* Rickettsia kotlanii'). Ann Trop Med Parasitol 100:229–236. https://dx.doi.org/10.1179/136485906X91468

Sun Y, Xu R (2003) Ability of *Ixodes persulcatus*, *Haemaphysalis concinna* and *Dermacentor silvarum* ticks to acquire and transstadially transmit *Borrelia garinii*. Exp Appl Acarol 31:151–160. https://dx.doi.org/10.1023/B:APPA.0000005119.30172.43

Sun J, Li K, Wei H, Zhong-Wu H (2006) Studies of *Haemaphysalis concinna* outbreaks in Cangxi County of Sichuan Province. Vet Sci China 36:601–605 (in Chinese)

Švehlová A, Berthová L, Sallay B, Boldiš V, Sparagano OA, Špitalská E (2014) Sympatric occurrence of *Ixodes ricinus*, *Dermacentor reticulatus* and *Haemaphysalis concinna* ticks and *Rickettsia* and *Babesia* species in Slovakia. Ticks Tick Borne Dis 5:600–605. https://dx.doi.org/10.1016/j.ttbdis.2014.04.010

Széll Z, Sréter-Lancz Z, Márialigeti K, Sréter T (2006) Temporal distribution of *Ixodes ricinus*, *Dermacentor reticulatus* and *Haemaphysalis concinna* in Hungary. Vet Parasitol 141:377–379. https://dx.doi.org/10.1016/j.vetpar.2006.06.008

Tekin S, Bursali A, Mutluay N, Keskin A, Dundar E (2012) Crimean-Congo hemorrhagic fever virus in various ixodid tick species from a highly endemic area. Vet Parasitol 186:546–552. https://dx.doi.org/10.1016/j.vetpar.2011.11.010

Uspensky I, Robinson WH, Bajomi D (2008) Ticks (Acari: Ixodoidea) as urban pests and vectors with special emphasis on ticks outside their geographical range. In: 6th International Conference on Urban Pests, Budapest, Hungary, 13–16 July 2008. International conference on urban pests (ICUP), pp 333–347

Haemaphysalis parva (Neumann, 1897) (Figs. 100–102)

Z. Vatansever

Life-Cycle and Host Preferences

Haemaphysalis parva is a three-host exophilic tick. Adults infest large and medium sized wild and domestic animals and immatures feed on wide range of medium and small sized wild mammals (hedgehogs, hares, rodents, and carnivores), ground-feeding birds and reptiles (Filippova 1997; Kolonin 1978; Pomerantzev 1950). Human infestations with both adults and immatures have also been reported (Keskin et al. 2015; Pomerantzev 1950). Under laboratory conditions, the entire life-cycle of *Hae. parva* is completed in 50–80 days. Engorged females begin to oviposit in 3–49 days after drop-off. Oviposition takes 12–20 days and larvae hatch in 25–36 days. Larvae feed for 2–7 days and moult to nymphs in 15–28 days. Nymphs feed for 2–8 days and moult to adults in 17–26 days. Female ticks feed for 3–14 days and lay up to 3000 eggs (Pomerantzev 1950).

Ecology

Haemaphysalis parva prefers various types of mountain steppe and lower mountain forest habitat. It cannot survive in steppe climates with low humidity (Pomerantzev 1950), and is mostly observed around places with grazing livestock within altitudes of 0–1, 250 m a.sl. (Filippova 1997). Adults of *Hae. parva* are characterized by autumn-winter-spring activity. Parasitism occurs on livestock from August to May with a peak in October and November. Immatures are active from March to August with a peak in June and July (Filippova 1997; Pomerantzev 1950).

Distribution

Haemaphysalis parva is a Mediterranean species. It is reported from Italy, Romania, Turkey, Israel, Irak, Iran, the southern Federal Districts of Russia (northern Caucasus), Moldova, the Ukraine, Georgia, Armenia, Azerbaijan, Turkmenistan, Tajikistan, Egypt and Libya (Dantas-Torres and Otranto 2013; Hoogstraal 1959; Kolonin 1978; Kurtpinar 1954; Mihalca et al. 2012; Pomerantzev 1950).

Vectorial Capacity and Pathogen Burden

Haemaphysalis parva can transmit *Babesia ovis* experimentally and is associated with ovine babesiosis cases in the cold months in Russia (Filippova 1997; Pomerantzev 1950). It has been cited to harbour *Coxiella burnetii* (the causative agent of Q fever) and *Franciesella tularensis* (the causative agent of tularemia) (Filippova 1997). *Rickettsia hoogstraali* was demonstrated in ticks infesting humans (Orkun et al. 2014) and Candidatus '*Rickettsia goldwasserii*' in ticks feeding on golden jackal (Keysary et al. 2011).

Z. Vatansever (✉)
Department of Parasitology, Faculty of Veterinary Medicine, Kafkas University, Kars, Turkey
e-mail: zativet@gmail.com

A. Estrada-Peña et al. (eds.), *Ticks of Europe and North Africa*,
https://doi.org/10.1007/978-3-319-63760-0_50

Fig. 100 **A** Distribution of *Hae. parva* in Europe and Northern Africa (10 × 10 km grid presence with black dots). **B** Countries where the species have been reported are marked in dark grey

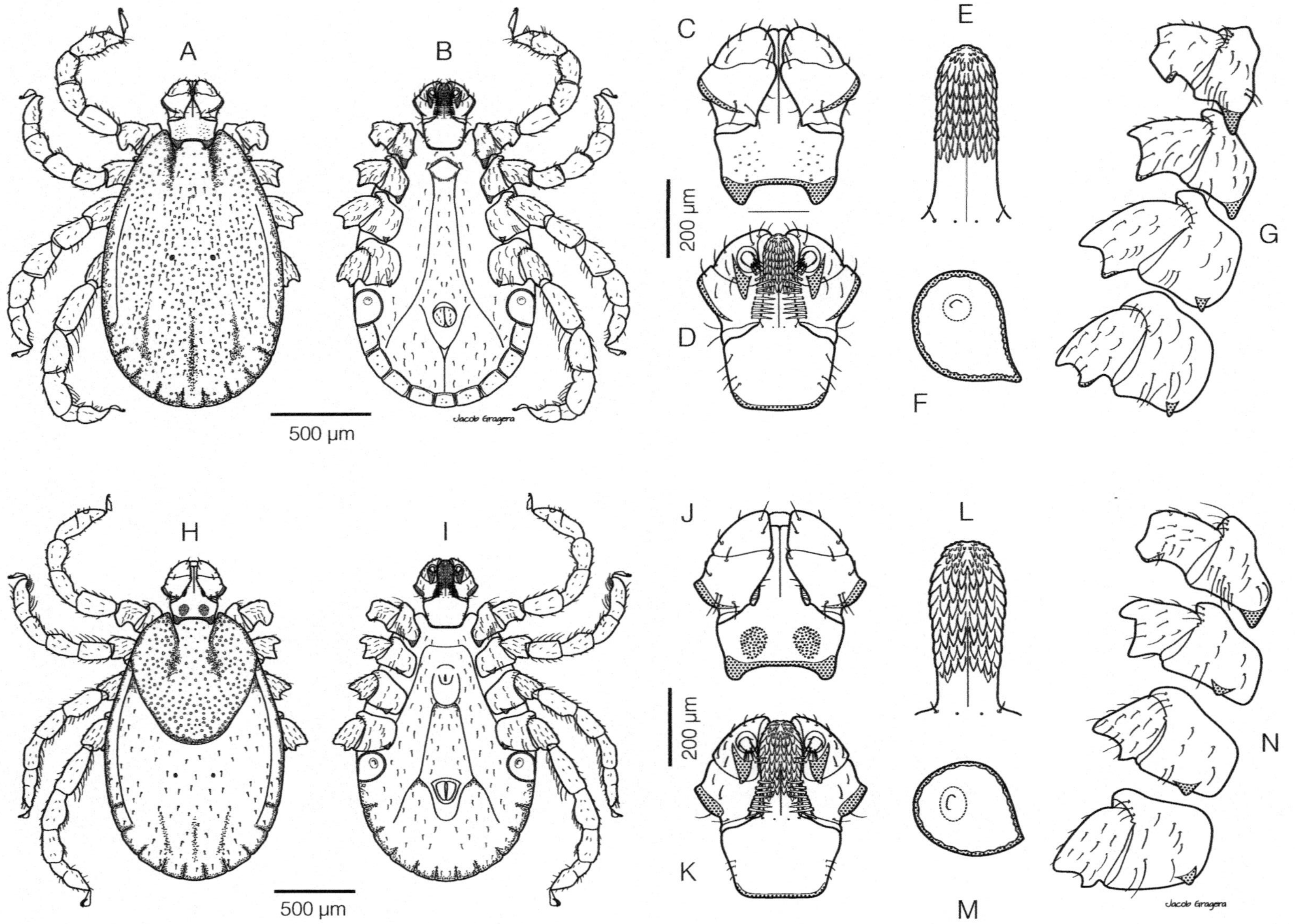

Fig. 101 **A–G** The male of *Hae. parva*. **A** Dorsal, **B** ventral, **C** capitulum, dorsal, **D** capitulum, ventral, **E** hypostome, **F** spiracular plate, **G** coxae and trochanters I–IV. **H–N** The female of *Hae. parva*. **H** Dorsal, **I** ventral, **J** capitulum, dorsal, **K** capitulum, ventral, **L** hypostome, **M** spiracular plate, **N** coxae and trochanters I–IV. Illustrations from specimens collected in Turkey

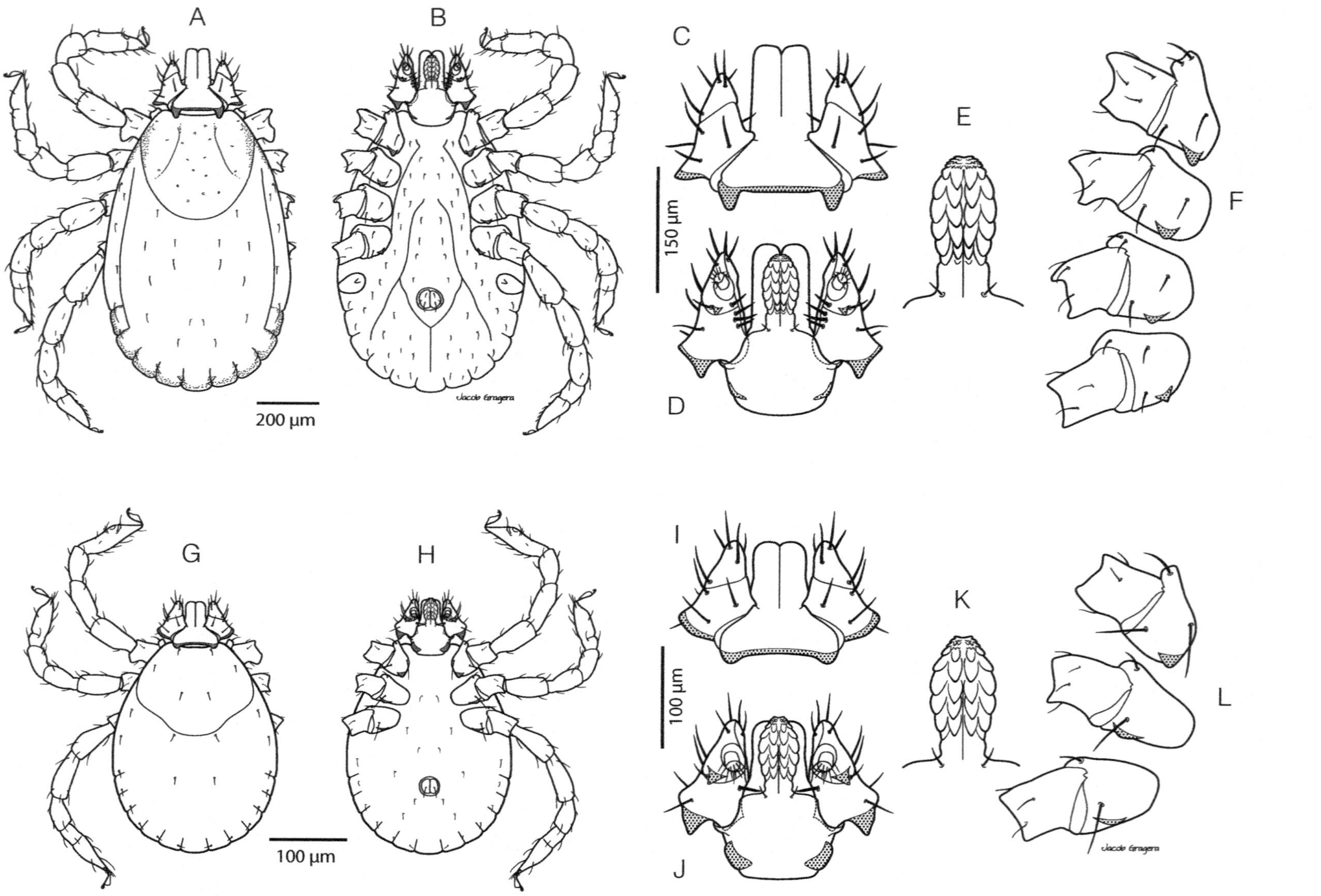

Fig. 102 **A–F** The nymph of *Hae. parva*. **A** Dorsal, **B** ventral, **C** capitulum, dorsal, **D** capitulum, ventral, **E** hypostome, **F** coxae and trochanters I–IV. **G–L** The larva of *Hae. parva*. **G** dorsal, **H** ventral, **I** capitulum, dorsal, **J** capitulum, ventral, **K** hypostome, **L** coxae and trochanters I–III. Illustrations from specimens collected in Turkey

References

Dantas-Torres F, Otranto D (2013) Species diversity and abundance of ticks in three habitats in southern Italy. Ticks Tick Borne Dis 4:251–255

Filippova NA (1997) Ixodid ticks of subfamily Amblyomminae. In Fauna of Russia and neighboring countries: New Series No: 145: Arachnoidea, (St. Petersburg, Nauka) 4(5):483 (in Russian)

Hoogstraal H (1959) Biological observations on certain Turkish *Haemaphysalis* ticks (Ixodoidea, Ixodidae). J Parasitol 45:227–232

Keskin A, Keskin A, Bursali A, Tekin S (2015) Ticks (Acari: Ixodida) parasitizing humans in Corum and Yozgat provinces, Turkey. Exp Appl Acarol 67:607–616

Keysary A, Eremeeva ME, Leitner M, Din AB, Wikswo ME, Mumcuoglu KY, Inbar M, Wallach AD, Shanas U, King R, Waner T (2011) Spotted fever group rickettsiae in ticks collected from wild animals in Israel. Am J Trop Med Hyg 85:919–923

Kolonin GV (1978) World distribution of ixodid ticks. Nauka, Moscow, p 74 (in Russian)

Kurtpinar H (1954) Ticks of Turkey. Güven Matbaasi, Ankara, p 113 (in Turkish)

Mihalca AD, Dumitrache MO, Magdas C, Gherman CM, Domsa C, Mircean V, Ghira IV, Pocora V, Ionescu DT, Siko BS, Cozma V, Sandor AD (2012) Synopsis of the hard ticks (Acari: Ixodidae) of Romania with update on host associations and geographical distribution. Exp Appl Acarol 58:183–206

Orkun O, Karaer Z, Cakmak A, Nalbantoglu S (2014) Spotted fever group rickettsiae in ticks in Turkey. Ticks Tick Borne Dis 5:213–218

Pomerantzev BI (1950) Ixodid ticks (Ixodidae). In Fauna of the USSR, New series 41: Arachnoidea, (Leningrad, Nauk), 4(2):224 (in Russian)

Haemaphysalis erinacei (Pavesi, 1894) (Figs. 103–105)

T. N. Petney, Z. Vatansever, and M. P. Pfäffle

This is considered to be a polytypic species containing a number of morphologically definable subspecies: *Hae. erinacei erinacei* Pavesi, 1844, *Hae. erinacei taurica* Pospelova-Schtom 1940 and *Hae. erinacei turanica* Pospelova-Schtom 1940 (Hoogstraal 1959; Filippova et al. 1993; Guglielmone et al. 2014).

Life-Cycle and Host Preferences

Haemaphysalis erinacei is a three-host species. Life-cycle data for Egyptian specimens can be found in Hoogstraal (1955) and for all three subspecies in Filippova et al. (1993). This species uses various small- and medium-sized mammals as hosts, mainly hedgehogs and their predators (Kolonin 2007). Rodents can be used as the main hosts of both larvae and nymphs (Hoogstraal 1955; Filippova et al. 1976; Bursali et al. 2015). The subspecies *taurica* and *turanica* show more adaptability to non-insectivore hosts than the subspecies *erinacei*, however, their host range is limited and they are seldom found on domestic animals (Hoogstraal 1959). Hedgehog hosts include the North African hedgehog (*Atelerix algirus*), the European hedgehog (*Erinaceus europaeus*), the southern white-breasted hedgehog (*Erinaceus concolor*), long-eared hedgehogs (*Hemiechinus auritus*) and desert hegdehogs (*Paraechinus aethiopicus*) (Pomerantzev 1950; Hoogstraal 1955, 1959; Filippova et al. 1993; Lorusso et al. 2011; Khaldi et al. 2012; Waner et al. 2014). Additionally, this tick is often found on beech marten (*Martes foina*), least weasel (*Mustela nivalis*), and red fox (*Vulpes vulpes*) (Hoogstraal 1959; Abbassian-Lintzen 1960; Otranto et al. 2007; Lorusso et al. 2011). Tovornik and Cerný (1974) report this species from birds.

Ecology

Haemaphysalis erinacei is an endophilic tick, living in the nests or dens of their hosts (Pomerantzev 1950; Lorusso et al. 2011). It is active throughout the year, larvae in April and May, and nymphs predominantly in March (Hoogstraal 1955). It is found in different biotope types such as Mediterranean forests, woodlands and scrub (Guglielmone et al. 2014). Filippova et al. (1993) report that *Hae. e. taurica* occupies humid foothills and middle-altitude mountains while *Hae. e. turanica* is found on arid plains and foothills. In Iran, it was collected in mountainous areas (Ramezani et al. 2014), in Iraq in desert habitats (up to 350 m a.s.l.) (Shamsuddin and Mohammad 1988).

Distribution

Haemaphysalis erinacei is found throughout North Africa as well as the Balkan area and southern Italy (Hoogstraal et al. 1981; Lorusso et al. 2011). *Haemaphysalis e. turanica* comes from Asia Minor and Middle Asia, ranging from Pakistan and Afghanistan across to Kazakhstan and Saudi Arabia, while *Hae. e. taurica* comes from northern Saudi Arabia ranging through Turkey to the Crimea, the Caucasus and parts of Middle and Central Asia (Hoogstraal et al. 1981; Filippova et al. 1993; Bursali et al. 2015). Additional reports of *Hae. erinacei*, mainly without further description of the subspecies, come from Algeria, Bulgaria, Croatia, Jordan, Iran, Iraq, Israel, Kazakhstan, Kyrgyzstan and Uzbekistan (Reznik 1966; Shamsuddin and Mohammad 1988; Fedorova 2005; Rasulov 2007; Gemel and Hörweg 2011; Krčmar 2012;

T. N. Petney (✉)
State Museum of Natural History, Erbprinzenstr. 13, 76133 Karlsruhe, Germany
e-mail: trevor.petney@kit.edu

Z. Vatansever
Department of Parasitology, Faculty of Veterinary Medicine, Kafkas University, Kars, Turkey
e-mail: zativet@gmail.com

M. P. Pfäffle
Department of Ecology and Parasitology, Institute of Zoology, Karlsruhe Institute of Technology, Karlsruhe, Germany
e-mail: miripfaeffle@web.de

A. Estrada-Peña et al. (eds.), *Ticks of Europe and North Africa*,
https://doi.org/10.1007/978-3-319-63760-0_51

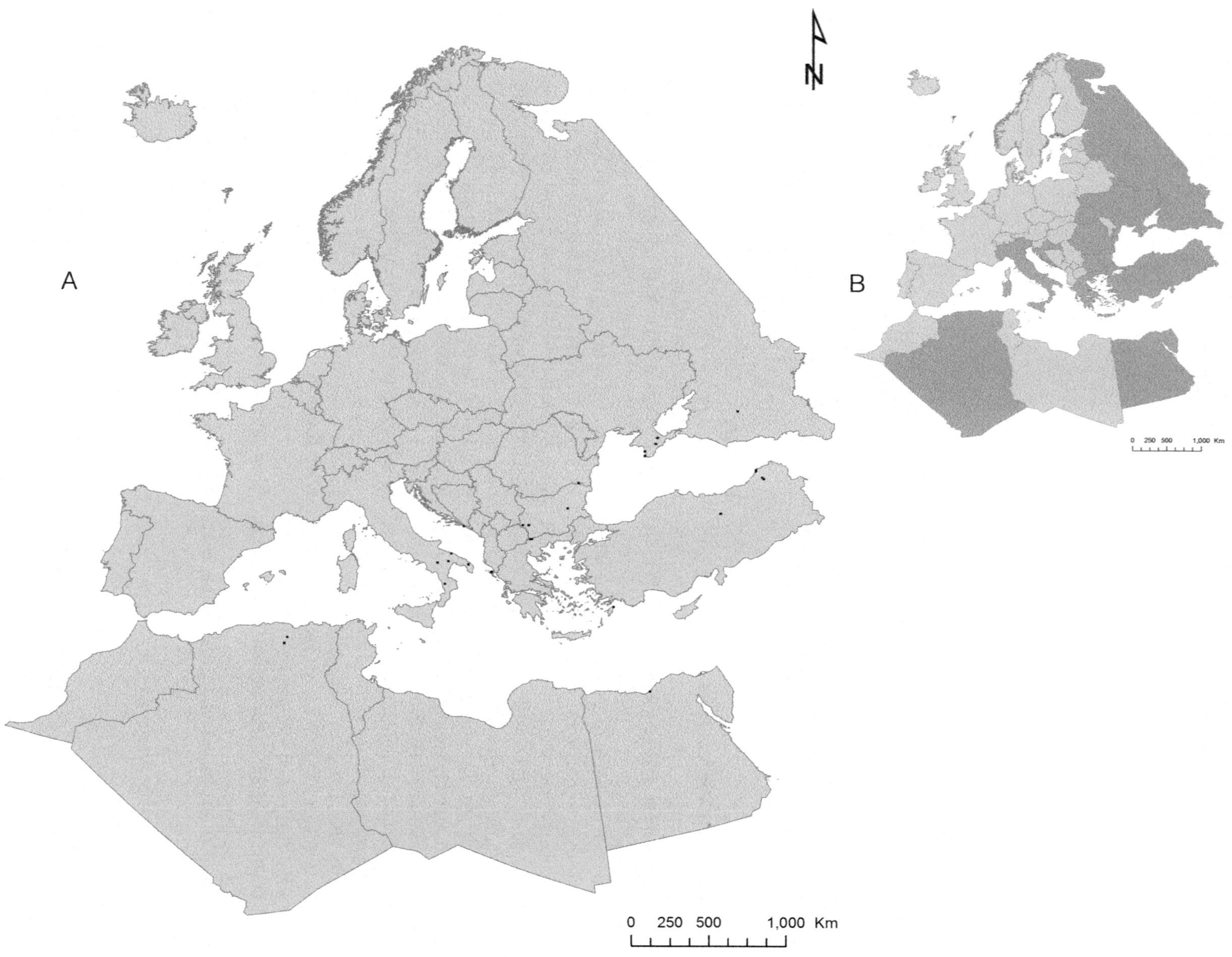

Fig. 103 **A** Distribution of *Hae. erinacei* in Europe and Northern Africa (10 × 10 km grid presence with black dots). **B** countries where the species has been reported are marked in dark grey

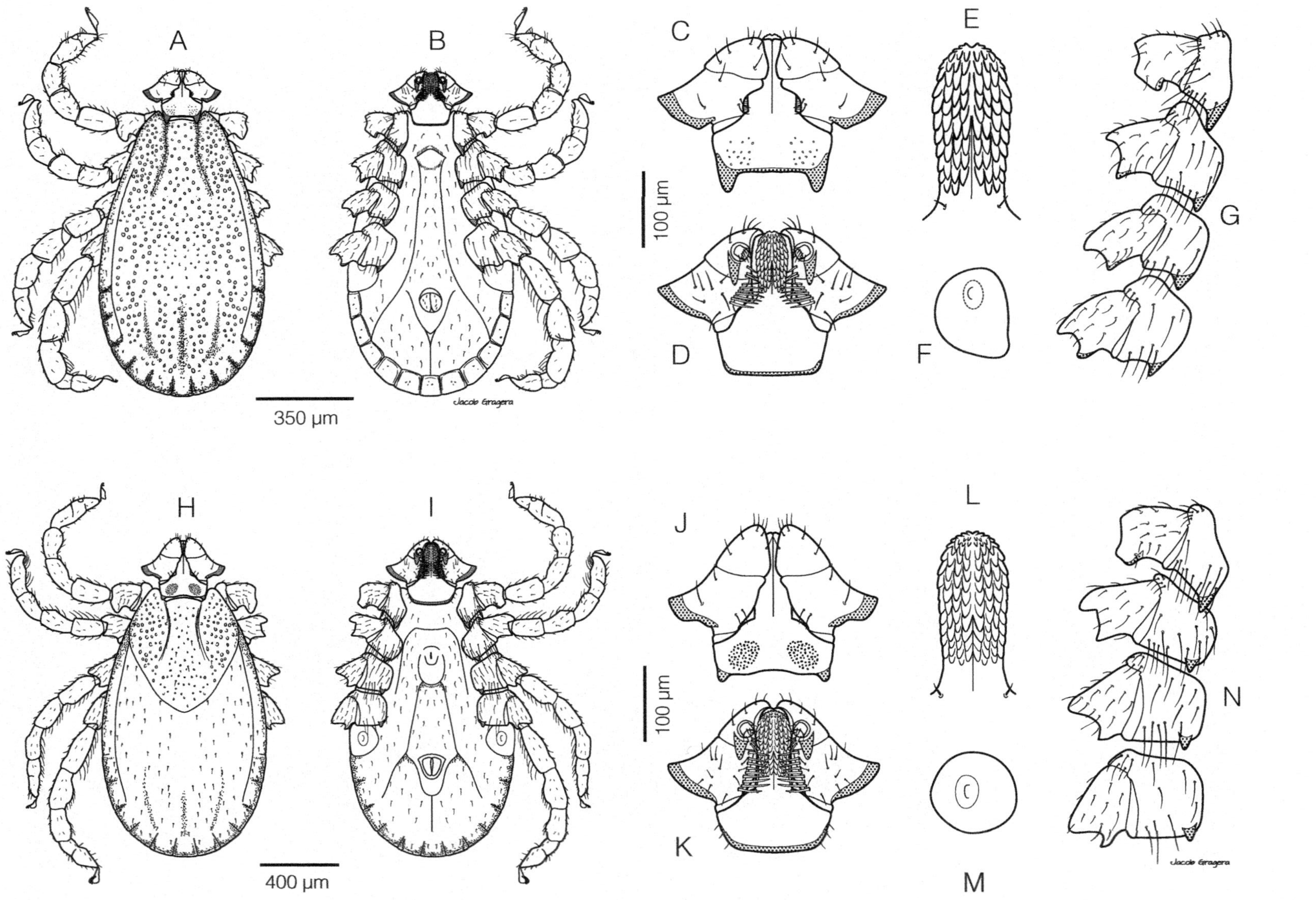

Fig. 104 **A–G** The male of *Hae. erinacei*. **A** Dorsal, **B** ventral, **C** capitulum, dorsal, **D** capitulum, ventral, **E** hypostome, **F** spiracular plate, **G** coxae and trochanters I–IV. **H–N** The female of *Hae. erinacei*. **H** Dorsal, **I** ventral, **J** capitulum, dorsal, **K** capitulum, ventral, **L** hypostome, **M** spiracular plate, **N** coxae and trochanters I–IV. Illustrations from specimens collected in Turkey and Tunisia

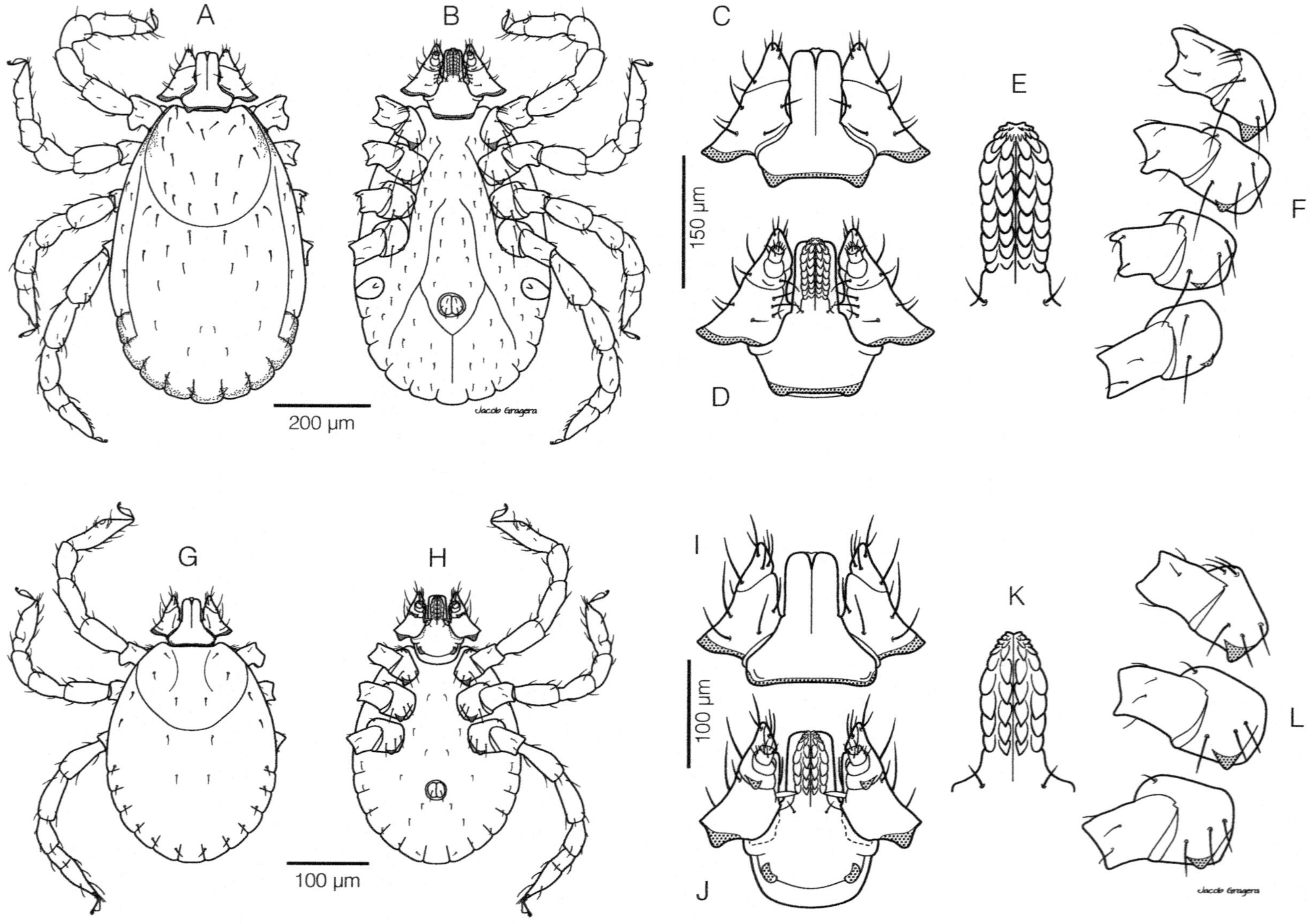

Fig. 105 **A–F** The nymph of *Hae. erinacei*. **A** Dorsal, **B** ventral, **C** capitulum, dorsal, **D** capitulum, ventral, **E** hypostome, **F** coxae and trochanters I–IV. **G–L** The larva of *Hae. erinacei*. **G** Dorsal, **H** ventral, **I** capitulum, dorsal, **J** capitulum, ventral, **K** hypostome, **L** coxae and trochanters I–III. Illustrations from specimens collected in Turkey and Tunisia

Khaldi et al. 2012; Qablan et al. 2012; Waner et al. 2014; Hosseini-Chegeni et al. 2014).

Vectorial Capacity and Pathogen Burden

Haemaphysalis erinacei is a possible reservoir of *Yersinia pestis* (plague) amongst rodents (Pomerantzev 1950). Khaldi et al. (2012) found a novel spotted fever group *Rickettsia* species in 77% of *Hae. erinacei* collected from hedgehogs (*Atelerix algirus* and *Paraechinus aethiopicus*) in Algeria. Waner et al. (2014) report *Rickettsia massiliae* from *Hae. erinacei* in Israel.

References

Abbassian-Lintzen R (1960) A preliminary list of ticks (Acarina: Ixodoidea) occurring in Iran and their distributional data. Acarologia 2:43–61

Bursali A, Keskin A, Şimşek E et al (2015) A survey of ticks (Acari: Ixodida) infesting some wild animals from Sivas, Turkey. Exp Appl Acarol 66:293–299. https://dx.doi.org/10.1007/s10493-015-9898-z

Fedorova SZ (2005) Ixodidae ticks in Bishkek. Med Parazitol (Mosk) 4:34–37

Filippova NA, Neronov VM, Farhang-Azad A (1976) On the hard ticks (Acarina, Ixodidae) of small mammals in Iran. Entomol Rev 55:467–478

Filippova NA, Panova IV, Musatov SA (1993) The taxonomic pattern of the polytypic species *Haemaphysalis erinacei* (Ixodidae). Parazitologia 27:193–215

Gemel R, Hörweg C (2011) Zum Befall der Maurischen Landschildkröte *Testudo graeca* Linnaeus, 1758 durch Zecken, und deren Bedeutung als Vektoren. Ein Literaturüberblick samt eigenen Beobachtungen. Herpetozoa 23:21–30

Guglielmone AA, Robbins RG, Apanaskevich DA, Petney TN, Estrada-Peña A, Horak IG (2014) The hard ticks of the world (Acari: Ixodida: Ixodidae). Springer, Dordrecht, p 738

Hoogstraal H (1955) Notes on African *Haemaphysalis* ticks. I. The Mediterranean littoral hedgegog parasite *H. erinacei* Pavesi, 1884 (Ixodoidea, Ixodidae). J Parasitol 41:221–233

Hoogstraal H (1959) Biological observations on certain Turkish *Haemaphysalis* ticks (Ixodoidea, Ixodidae). J Parasitol 45:227–232. https://dx.doi.org/10.2307/3286538

Hoogstraal H, Wassef HY, Buttiker W (1981) Ticks (Acarina) of Saudi Arabia. Fam. Argasidae, Ixodidae. Fauna Saudi Arabia 3:25–110

Hosseini-Chegeni A, Telmadarraiy Z, Salimi M et al (2014) A record of *Haemaphysalis erinacei* (Acari: Ixodidae) collected from hedgehog and an identification key for the species of *Haemaphysalis* occurring in Iran. Persian J Acarol 3:203–215

Khaldi M, Socolovschi C, Benyettou M et al (2012) rickettsiae in arthropods collected from the Northern African Hedgehog (*Atelerix algirus*) and the desert hedgehog (*Paraechinus aethiopicus*) in Algeria. Comp Immunol Microbiol Infect Dis 35:117–122. https://dx.doi.org/10.1016/j.cimid.2011.11.007

Kolonin GV (2007) Mammals as hosts of Ixodid ticks (Acarina, Ixodidae). Entomol Rev 87:401–412. https://dx.doi.org/10.1134/S0013873807040033

Krčmar S (2012) Hard ticks (Acari, Ixodidae) of Croatia. ZooKeys 234:19–57. https://dx.doi.org/10.3897/zookeys.234.3658

Lorusso V, Lia RP, Dantas-Torres F et al (2011) Ixodid ticks of road-killed wildlife species in southern Italy: new tick-host associations and locality records. Exp Appl Acarol 55:293–300. https://dx.doi.org/10.1007/s10493-011-9470-4

Otranto D, Lia RP, Cantacessi C et al (2007) *Filaria martis* Gmelin 1790 (Spirurida, Filariidae) affecting beech marten (*Martes foina*): morphological description and molecular characterisation of the cytochrome oxidase c subunit I. Parasitol Res 101:877–883. https://dx.doi.org/10.1007/s00436-007-0554-3

Pomerantzev BI (1950) Fauna of U.S.S.R. 4(2). Arachnida: ixodid ticks (Ixodidae). Zoological Institute of the Academy of Science, Washington

Qablan MA, Kubelová M, Siroký P et al (2012) Stray dogs of northern Jordan as reservoirs of ticks and tick-borne hemopathogens. Parasitol Res 111:301–307. https://dx.doi.org/10.1007/s00436-012-2839-4

Ramezani Z, Chavshin AR, Telmadarraiy Z et al (2014) Ticks (Acari: Ixodidae) of livestock and their seasonal activities, northwest of Iran. Asian Pacific J Trop Dis 4:S754–S757. https://dx.doi.org/10.1016/S2222-1808(14)60721-0

Rasulov I (2007) Ticks status in Central Asia with a special emphasis on Uzbekistan. Parasitol Res 101:183–186. https://dx.doi.org/10.1007/s00436-007-0691-8

Reznik PA (1966) On distribution area features of ticks belonging to the genus *Haemaphysalis* in the Soviet fauna. Tezisy Dokl 1 Akarol Soveshch 171–172

Shamsuddin M, Mohammad MK (1988) Incidence, distribution, and host relationships of some ticks (Ixodoidea) in Iraq. J Univ Kuwait 15:321–329

Tovornik D, Cerný V (1974) Finding of *Haemaphysalis erinacei erinacei* Pavesi, 1884 on birds in Yugoslavia. Folia Parasitol (Praha) 21:282

Waner T, Keysary A, Eremeeva ME et al (2014) *Rickettsia africae* and *Candidatus* Rickettsia barbariae in ticks in Israel. Am J Trop Med Hyg 90:920–922. https://dx.doi.org/10.4269/ajtmh.13-0697

Haemaphysalis hispanica Gil Collado, 1938 (Figs. 106–108)

M. M. Santos-Silva

Life Cycle and Host Preferences

Haemaphysalis hispanica is a three-host tick (Hoogstraal and Morel 1970) with a natural life cycle of one year (Marquez 1992). There are no available reports about its life cycle under laboratory conditions. All stages feed predominantly on lagomorphs, particularly the European rabbit, *Oryctolagus cuniculus*. Members of the avian Strigiformes and Falconiformes are exceptionally rare hosts (Estrada-Peña et al. 1985; Guglielmone et al. 2014). There is no difference in host preference between the adult and immature stages (Peréz-Eid 2007). *Haemaphysalis hispanica* has not been reported from humans so far.

Ecology

Haemaphysalis hispanica is restricted to the south-western area of the Palaearctic region. Typical habitats include cold, sub-humid or dry areas, with scarce oak forests, mixed shrub and natural grass (Dias 1994). All stages have their peak activity during spring and summer. Larval and nymphal activity peaks from May to July, and August to September, respectively. Adult activity peaks occur earlier in March and April (Dias 1994). *Haemaphysalis hispanica* presents an endophilic and monotropic behaviour (Gilot et al. 1985; Dias 1994), similar to a nidicolous tick species, with specificity to a single host. The host-seeking behaviour is that of an endophilic tick species that can only be collected inside of host burrows or on the hosts (Estrada-Peña et al. 1992).

Distribution

Haemaphysalis hispanica has been reported in a narrow strip in the south-western Mediterranean region, including southern France, mainland Portugal and mainland Spain (Hoogstraal and Morel 1970; Estrada-Peña 1990; Dias 1994). On the Iberian Peninsula, *Hae. hispanica* is considered a scarce species (Barandika et al. 2006; Toledo et al. 2009; Santos-Silva et al. 2011). Further studies are required to determine its complete range.

Vectorial Capacity and Pathogen Burden

No association with infectious agents is known so far.

M. M. Santos-Silva (✉)
Centro de Estudos de Vectores e Doenças Infecciosas Dr. Francisco Cambournac (CEVDI), Instituto Nacional de Saúde Doutor Ricardo Jorge, Águas de Moura, Portugal
e-mail: m.santos.silva@insa.min-saude.pt

A. Estrada-Peña et al. (eds.), *Ticks of Europe and North Africa*,
https://doi.org/10.1007/978-3-319-63760-0_52

Fig. 106 **A** Distribution of *Hae. hispanica* in Europe and Northern Africa (10 × 10 km grid presence with black dots). **B** Countries where the species has been reported are marked in dark grey

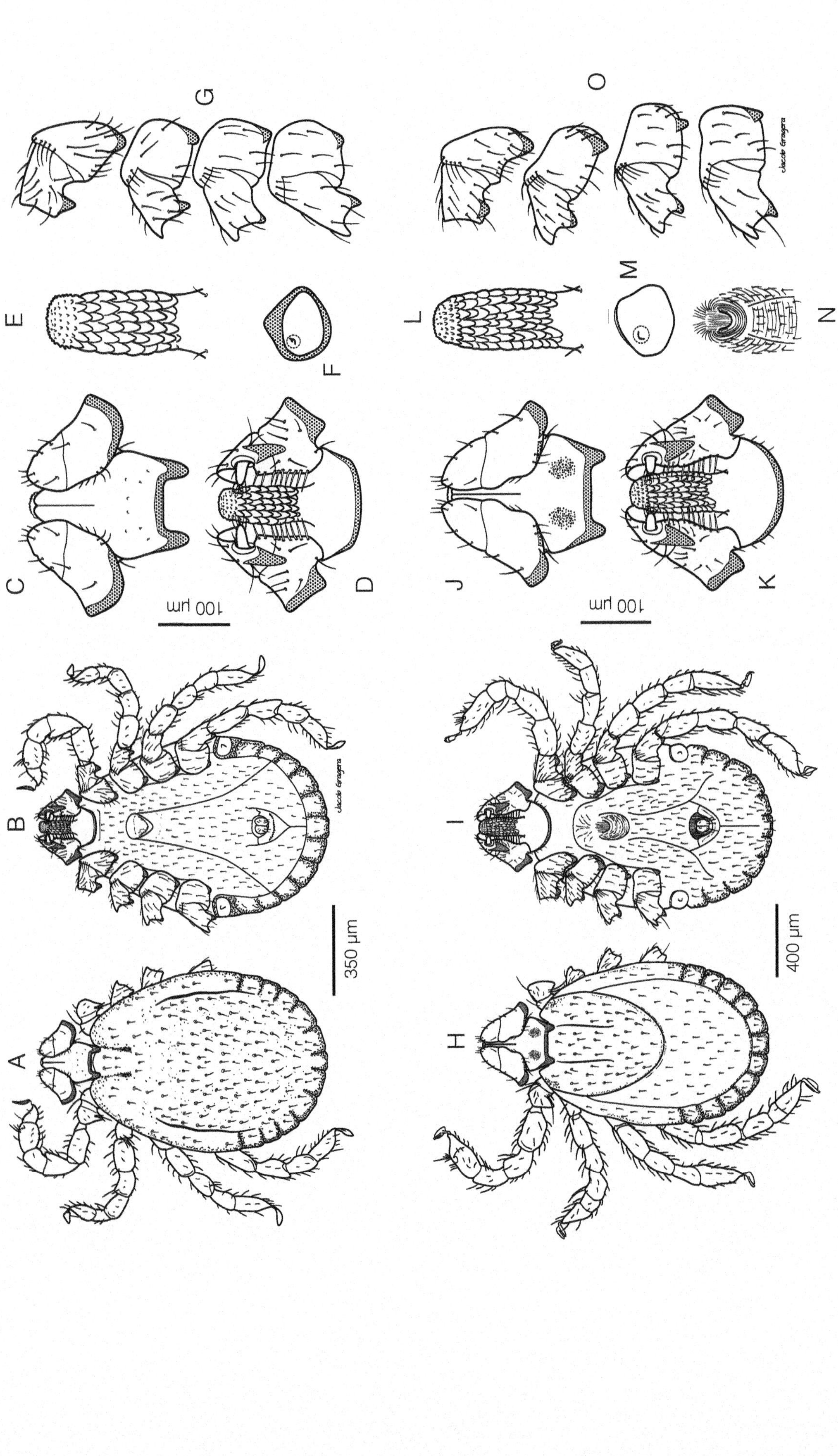

Fig. 107 **A–G** The male of *Hae. hispanica*. **A** Dorsal, **B** ventral, **C** capitulum, dorsal, **D** capitulum, ventral, **E** hypostome, **F** spiracular plate, **G** coxae and trochanters I–IV. **H–O** The female of *Hae. hispanica*. **H** Dorsal, **I** ventral, **J** capitulum, dorsal, **K** capitulum, ventral, **L** hypostome, **M** spiracular plate, **N** genital opening, **O** coxae and trochanters I–IV. Illustrations from specimens collected in Spain

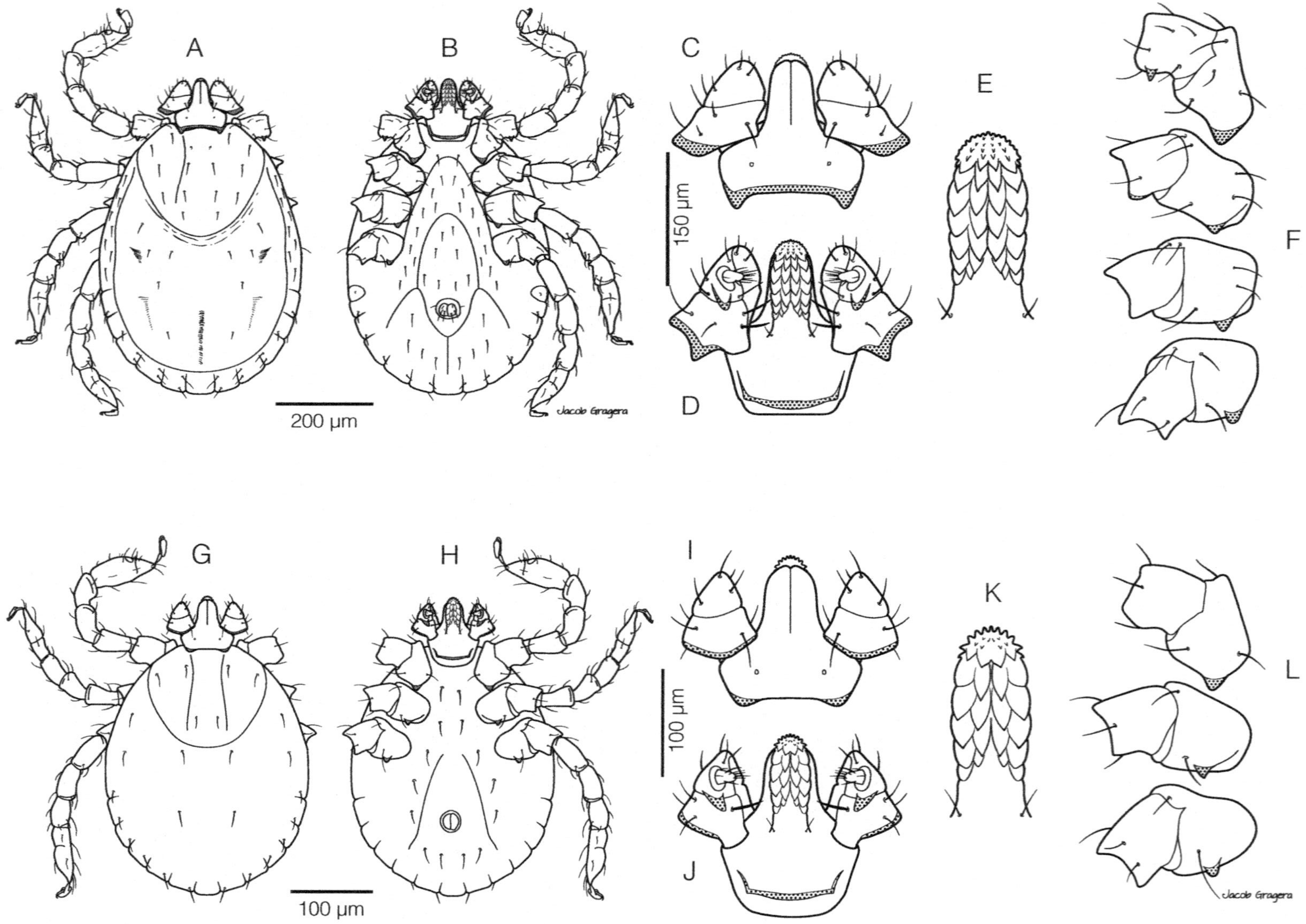

Fig. 108 **A–F** The nymph of *Hae. hispanica*. **A** Dorsal, **B** ventral, **C** capitulum, dorsal, **D** capitulum, ventral, **E** hypostome, **F** coxae and trochanters I–IV. **G–L** The larva of *Hae. hispanica*. **G** Dorsal, **H** ventral, **I** capitulum, dorsal, **J** capitulum, ventral, **K** hypostome, **L** coxae and trochanters I–III. Illustrations from specimens collected in Spain

References

Barandika JF, Berriatua E, Barral M, Juste RA, Anda P, Garcia-Perez AL (2006) Risk factors associated with ixodid tick species distributions in the Basque region in Spain. Med Vet Entomol 20:177–188

Dias JATS (1994) As carraças (Acarina-Ixodoidea) da Península Ibérica. Algumas considerações sobre a sua biogeografia e relacionamento com a ixodofauna Afropaleárctica e Afrotropical. Instituto de Investigação Cientifica Tropical, Estudos, Ensaios e Documentos 158, Lisboa

Estrada-Peña A (1990) Taxonomic relationships of the subgenus *Rhipistoma* (Acari: Ixodidae: *Haemaphysalis*). J Med Entomol 27:399–405

Estrada-Peña A, Lucientis C, Sánchez A, Galindo JG, Meléndez BO, Femenias MG, Hernández JC (1985) Parasitismo accidental de *Haemaphysalis* (*Rhipistoma*) *hispanica* Gil Collado (Acari: Ixodoidea) sobre *Aquila chrysaetos*. Rev Iber Parasitol 45:379–380

Estrada-Peña J, Ósacar J, Gortázar C, Calvete C, Lucientes J (1992) An account of the ticks of the northeastern of Spain (Acarina: Ixodidae). Ann Parasitol Hum Comp 67:42–49

Gilot B, Rogers P, Lachet B (1985) Données biologiques et écologiques sur les tiques de lagomorphes (et plus spécialement celles du lapin de garenne, *Oryctolagus cuniculus* L.) dans les alpes françaises et leur avant-pays. Acarologia 26:335–354

Guglielmone AA, Robbins RG, Apanaskevich DA, Petney TN, Estrada-Peña A, Horak IG (2014) The hard ticks of the world (Acari: Ixodida: Ixodidae). Springer, Dordrecht, p 738

Hoogstraal H, Morel PC (1970) *Haemaphysalis* (*Rhipistoma*) *hispanica* Gil Collado, a parasite of the European rabbit, redescription of adults, and description of immature stages (Ixodoidea: Ixodidae). J Parasitol 56:813–822

Marquez FJ (1992) Dynamique de la population de *Haemaphysalis hispanica* (Acarina, Ixodidae) dans le Sud-Est de l'Espagne. Vie milieu 42:41–49

Pérez-Eid C (2007) Les tiques. Identification, biologie, importance médicale et vétérinaire. Editions Tec & Doc, Paris, p 314

Santos-Silva MM, Beati L, Santos AS, De Sousa R, Núncio MS, Melo P, Santos-Reis M, Fonseca C, Formosinho P, Vilela C, Bacellar F (2011) The hard-tick fauna of mainland Portugal (Acari: Ixodidae): an update on geographical distribution and known associations with hosts and pathogens. Exp Appl Acarol 55:85–121. https://dx.doi.org/10.1007/s10493-011-9440-x

Toledo A, Olmeda S, Escudero R, Jado I, Valcarcel F, Casado-Nistal M, Rodriguez-Vargas M, Gil H, Anda P (2009) Tick-borne zoonotic bacteria in ticks collected from Central Spain. Am J Trop Med Hyg 81:67–74

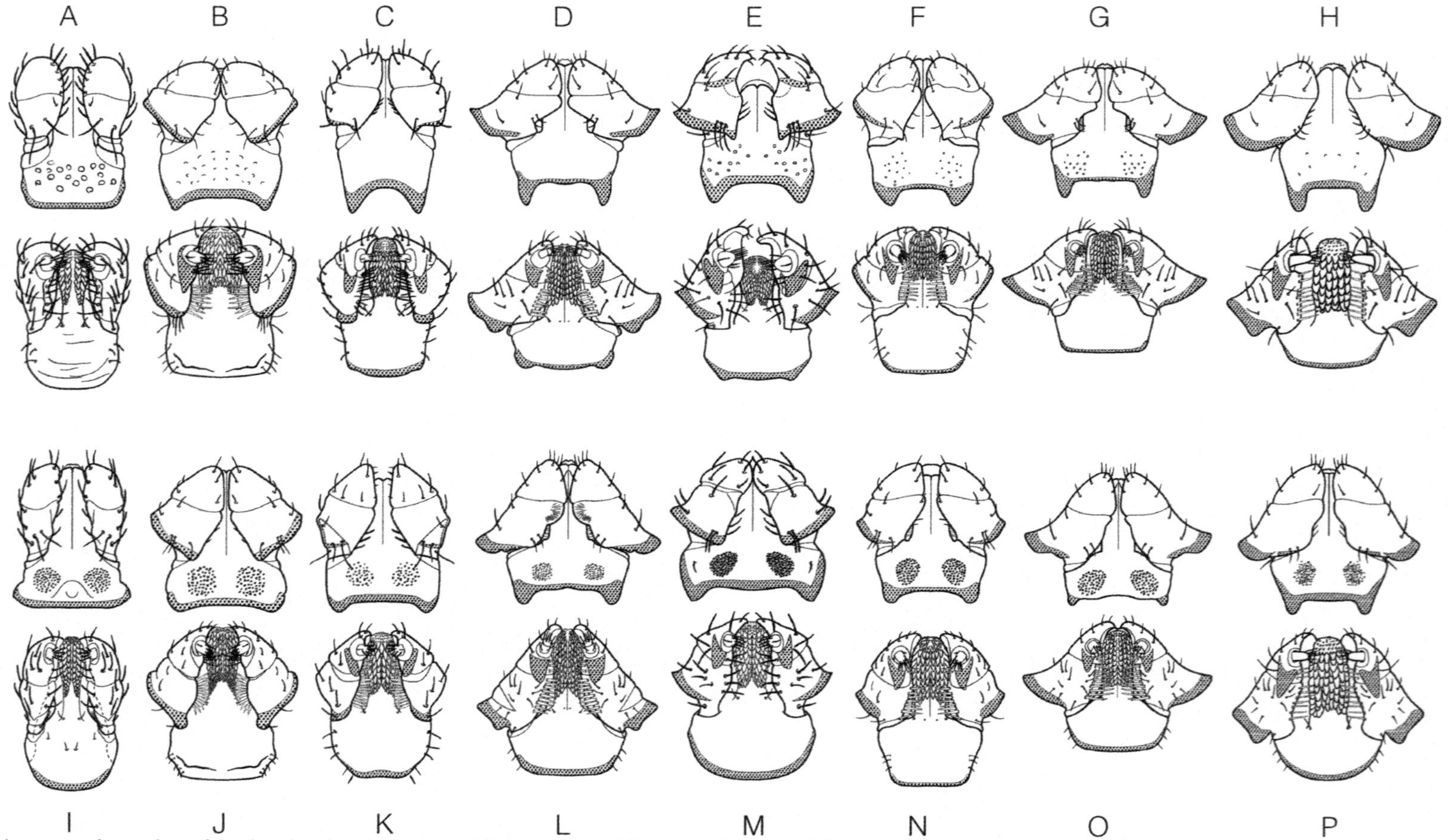

Fig. 109 A compared overview of the dorsal and ventral aspect of the capitulum of the adults of species of the genus *Haemaphysalis* recorded in the western Palaearctic. Males, **A** *Hae. inermis*, **B** *Hae. punctata*, **C** *Hae. sulcata*, **D** *Hae. caucasica*, **E** *Hae. concinna*, **F** *Hae. parva*, **G** *Hae. erinacei*, **H** *Hae. hispanica*. Females, **I** *Hae. inermis*, **J** *Hae. punctata*, **K** *Hae. sulcata*, **L** *Hae. caucasica*, **M** *Hae. concinna*, **N** *Hae. parva*, **O** *Hae. erinacei*, **P** *Hae. hispanica*

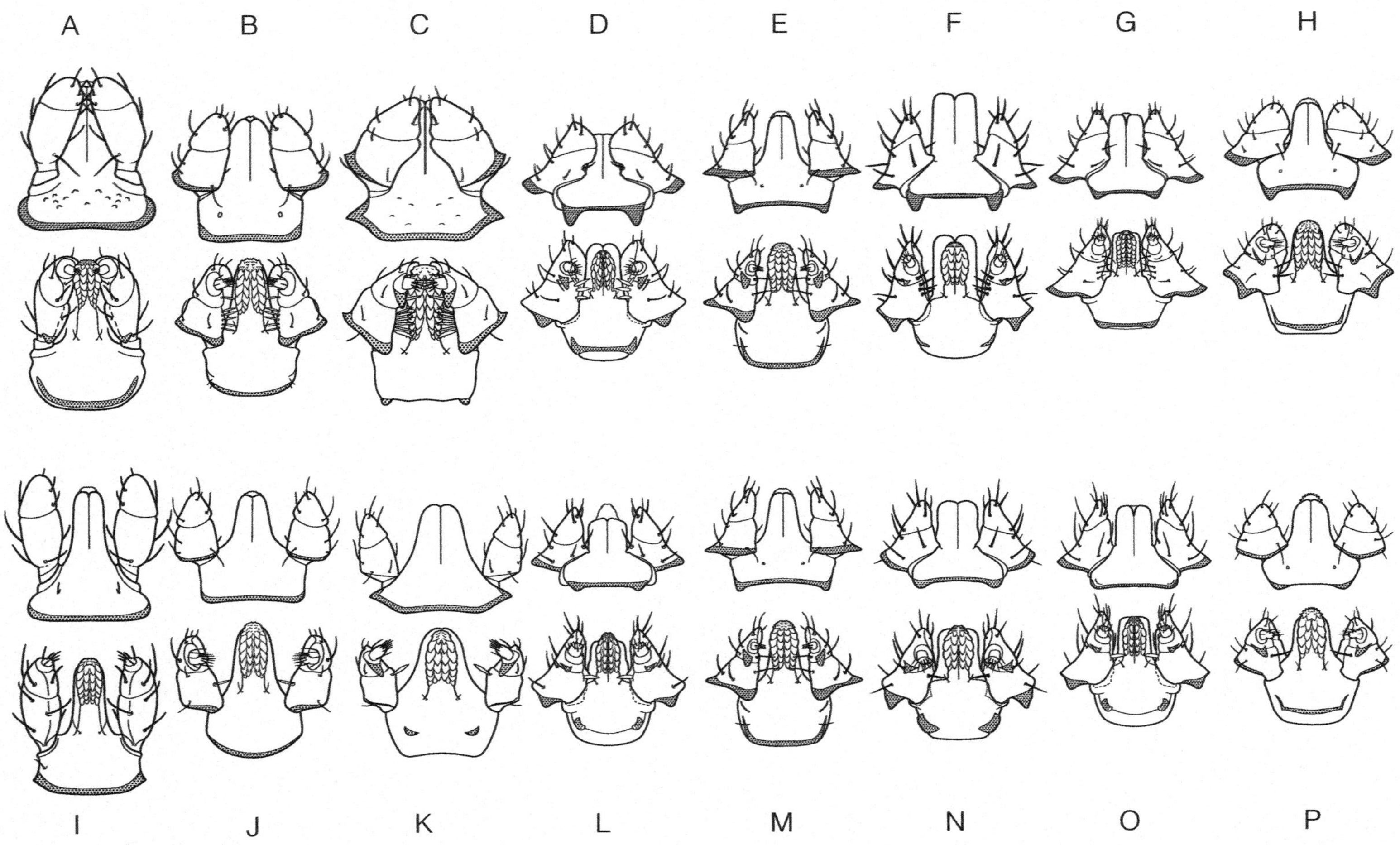

Fig. 110 A compared overview of the dorsal and ventral aspect of the capitulum of the immatures of species of the genus *Haemaphysalis* recorded in the western Palaearctic. Nymphs, **A** *Hae. inermis*, **B** *Hae. punctata*, **C** *Hae. sulcata*, **D** *Hae. caucasica*, **E** *Hae. concinna*, **F** *Hae. parva*, **G** *Hae. erinacei*, **H** *Hae. hispanica*. Larvae, **I** *Hae. inermis*, **J** *Hae. punctata*, **K** *Hae. sulcata*, **L** *Hae. caucasica*, **M** *Hae. concinna*, **N** *Hae. parva*, **O** *Hae. erinacei*, **P** *Hae. hispanica*

Genus *Dermacentor* Koch, 1844

A. Estrada-Peña, M. P. Pfäffle, and T. N. Petney

The genus *Dermacentor* is a representative of the Nearctic, Palaearctic, Asian and Afrotropical tick fauna. Around 14 species of the genus are found in the Palaearctic region, but only two are present in the Western Palaearctic. General works on the genus include Filippova and Panova (1986, 1989) and Filippova et al. (1985). The work by Filippova and Panova (1989) reviews the genus in the former Soviet Union and includes information on *D. reticulatus* (Fabricius, 1794), *D. pavlosvkyi* Olenev, 1927, *D. raskemensis* Pomerantzev, 1946, *D. niveus* Neumann, 1897 (but see below under *D. marginatus*), *D. ushakovae* Filippova and Panova, 1987, *D. marginatus* (Sulzer, 1776), *D. silvarum* Olenev, 1931, *D. pomerantzevi* Serdjukova, 1951, *D. nuttalli* Olenev, 1929 and *D. montanus* Filippova and Panova, 1974. The work by Filippova (1997) is a treatise on the ticks of the subfamily Amblyomminae and also includes the above species. The descriptions available in Filippova and Panova (1986) provide a diagnosis of the subgenera of *Dermacentor* for the immatures.

Dermacentor marginatus is a widely distributed species, with records from North Africa, Spain, France, Italy, Switzerland and the warmer parts of Central Europe (Petney et al. 2012; Rubel et al. 2015). There are adequate redescriptions, illustrations and keys in Filippova (1997), Filippova and Panova (1986, 1989), Feider (1965) and Estrada-Peña and Estrada-Peña (1991). SEM illustrations of adults are available in Cringoli et al. (2005). Zahler et al. (1995) provided a molecular comparison (and separation) between *D. marginatus* and *D. reticulatus*, however, no information was provided on the morphology of the specimens involved, and there is no way to associate the molecular data with the determinations made by the authors. Sixl (1975) provided a key for the separation of the larvae of both species.

Dermacentor daghestanicus Olenev 1929 was originally named *Dermacentor niveus daghestanicus*. The name *Dermacentor daghestanicus*, used first in Schulze (1933), is a synonym of *Dermacentor marginatus*. Kolonin (2009) considered *Dermacentor daghestanicus* valid and *Dermacentor niveus* its synonym, while Camicas et al. (1998) and Tekin et al. (2012) regarded both names as valid. *Dermacentor daghestanicus* is a synonym of *Dermacentor niveus* sensu Filippova (1997). It has also been used as *D. marginatus daghestanicus* by Lotozky (1956). *Dermacentor niveus* Neumann, 1897 was originally named *Dermacentor reticulatus niveus* and given its current status by Yakimov (1923). The name *D. niveus* in Olenev (1927) and Sénevet (1937) is partly a synonym of *Dermacentor marginatus* and a synonym of *D. silvarum* in Filippova (1997). Estrada-Peña and Estrada-Peña (1991) and Moshaverinia et al. (2009) consider *D. niveus* to be a synonym of *D. marginatus*, while it is a synonym of *D. daghestanicus* in Kolonin (2009). The name *D. niveus* is considered valid by Filippova (1997). We consider that empirical evidence exists to consider both *D. daghestanicus* and *D. niveus* synonyms of *D. marginatus* (e.g. examination of the syntypes of *D. niveus* and *D. daghestanicus* by Estrada-Peña and Estrada-Peña 1991). A recent update of the distribution of *D. marginatus* (and *D. reticulatus*) was published by Rubel et al. (2015).

Dermacentor reticulatus is a species that has been reported as invasive in several parts of the Western Palaearctic (summarized by Petney et al. 2012). One of the main synonyms of the species is *D. pictus* (Hermann 1894) (as found in, e.g. Feider 1965) according to Camicas et al. (1998) and Petney et al. (2012) among others. This species is well known in the Western Palaearctic and there are

A. Estrada-Peña (✉)
Department of Animal Health, Faculty of Veterinary Medicine, Zaragoza, Spain
e-mail: aestrada@unizar.es

M. P. Pfäffle
Department of Ecology and Parasitology, Institute of Zoology, Karlsruhe Institute of Technology, Karlsruhe, Germany
e-mail: aestrada@unizar.es

T. N. Petney
State Museum of Natural History, Erbprinzenstr. 13, 76133 Karlsruhe, Germany
e-mail: trevor.petney@kit.edu

A. Estrada-Peña et al. (eds.), *Ticks of Europe and North Africa*,
https://doi.org/10.1007/978-3-319-63760-0_53

adequate redescriptions in Filippova (1997) and Cringoli et al. (2005) who also included SEM pictures and adequate comparisons of *D. marginatus* and *D. reticulatus*. Keys, illustrations and details on distribution are available in Immler (1973), Siuda (1993) and Pérez-Eid (2007).

Key to the Species of Genus *Dermacentor* in Europe and Northern Africa

Females

1a. With a clear spur backwards directed on dorsal palpal article II. Cornua rounded and evident...***Dermacentor reticulatus*** (Fig. 115).
1b. Dorsal spur on palpal dorsal article II absent. Cornua very small...***Dermacentor marginatus*** (Fig. 112).

Males

1a. Palpi wider than basis capituli. With a clear spur directed backwards on dorsal palpal article II...***Dermacentor reticulatus*** (Fig. 115).
1b. Basis capituli wider or at least of the same width as the palpi. Dorsal spur on palpal dorsal article II absent...***Dermacentor marginatus*** (Fig. 112).

Nymphs

1a. Scutal setae at least twice as long as the conscutal setae. Ventral auriculae large, well marked, sharply pointed... ***Dermacentor reticulatus*** (Fig. 116).
1b. Scutal setae of the same length as the conscutal setae. Ventral auriculae small...***Dermacentor marginatus*** (Fig. 113).

Larvae

1a. Ventral auriculae well marked, rounded...***Dermacentor reticulatus*** (Fig. 116).
1b. Ventral auriculae very small or absent, commonly slightly visible as cuticular ridges...***Dermacentor marginatus*** (Fig. 113).

References

Cringoli G, Iori A, Rinaldi L, Veneziano V, Genchi C (2005) Mappe parassitologiche - Zecche. Series Edit. Giuseppe Cringoli, Napoli

Camicas JL, Hervy JP, Adam F, Morel PC (1998) Les tiques du monde. Nomenclature, stades décrits, hôtes, répartition (Acarida, Ixodida). Orstom, Paris

Estrada-Peña A, Estrada-Peña R (1991) Notes on *Dermacentor* ticks: redescription of *D. marginatus* with the synonymies of *D. niveus* and *D. daghestanicus* (Acari: Ixodidae). J Med Entomol 28:1–15

Feider Z (1965) Fauna Republicii Populare Române. Arachnida, vol 5, fasc. 2. Acaromorpha, Suprafamilie Ixodoidea (Capuse) Editura Academiei Republicii Populare Române, Bucuresti, 404 pp

Filippova NA, Panova IV (1986) Evaluation of the apron as a taxonomic character in ticks of the genus *Dermacentor* Koch (Ixodoidea, Ixodidae). Parazitologiya 20:337–346 (in Russian)

Filippova NA, Panova IV (1989) Revision of the genus *Dermacentor* Koch of the fauna of the USSR and adjoining territories (Ixodoidea, Ixodidae). Parazitolog Sb 35:49–55 (in Russian, NIH Library translation by T. Crump)

Filippova NA, Panova IV, Grebenjuk RV (1985) Diagnostics of species of the genus *Dermacentor* Koch of the USSR on their larval phases (Ixodidae). Parazitologiya 20:89–98 (in Russian)

Filippova NA (1997) Ixodid ticks of subfamily Amblyomminae. In the fauna of Russia and neighboring countries. New Series no. 145: Arachnoidea, vol 4(5), St. Petersburg, Nauka (in Russian)

Immler RM (1973) Untersuchungen zur Biologie und Ökologie der Zecke *Dermacentor reticulatus* (Fabricius, 1794) (Ixodidae) in einem endemischen Vorkommensgebiet. Ph.D. dissertation, University of Basel

Kolonin GV (2009) Fauna of ixodid ticks of the world. http://www.kolonin.org/. Accessed 15 May 2012. (Please note this page is not longer working)

Lotozky BV (1956) Identity of the species *Dermacentor marginatus* Sulz. 1776 and *Dermacentor daghestanicus* Ol., 1929 (Ixodidae). Izvestiya Akademy Nauk Tadzhiksoy SSR 15:95–98 (in Russian)

Moshaverinia A, Shayan P, Nabian S, Rahbari S (2009) Genetic evidence for conspecificity between *Dermacentor marginatus* and *Dermacentor niveus*. Parasitol Res 105:1125–1132

Olenev NO (1927) Sur le classification et la distribution géographique des ixodides. Doklady Akademy Nauk SSSR, Series A 1:219–224 (in Russian, with French summary)

Pérez-Eid C (2007) Les tiques. Identification, biologie, importance médicale et vétérinaire. Editions Tec Doc, Paris, p 314

Petney TN, Pfaeffle MP, Skuballa JD (2012) An annotated checklist of the ticks (Acari: Ixodida) of Germany. Syst Appl Acarol 17:115–170

Rubel F, Brugger K, Pfeffer M, Chitimia-Dobler L, Didyk YM, Leverenz S, Kahl O (2015) Geographical distribution of *Dermacentor marginatus* and *Dermacentor reticulatus* in Europe. Ticks Tick-Borne Dis 7:224–233

Sénevet G (1937) Ixodidés. Faune de France, 32, 100 pp

Siuda K (1993) Kleszcze Polski. Monografie Parazytologiczne, Polskie Towarzystwo Parazytologicne (in Polish)

Sixl W (1975) Contribution to the morphology of *Dermacentor reticulatus* and *D. marginatus* larvae (Arach., Acari, Ixodidae). In: Mitteilungen der Abteilung für Zoologie am Landesmuseum Joanneum Graz 04, 11–19

Tekin S, Bursali A, Mutluay N, Keskin A, Dundar E (2012) Crimean-Congo hemorrhagic fever virus in various ixodid tick species from a highly endemic area

Yakimov VL (1923) A propos des ixodidés de Russie. Parasitology 15:253–255

Zahler M, Gothe R, Rinder H (1995) Genetic evidence against a morphologically suggestive conspecificity of *Dermacentor reticulatus* and *D. marginatus* (Acari: Ixodidae). Int J Parasitol 25:1413–1419

Dermacentor marginatus (Sulzer, 1776) (Figs. 111–113)

S. Hornok

Life-Cycle and Host Preferences

Dermacentor marginatus is a three-host tick species. The life-cycle takes 75–163 days under laboratory conditions and the natural developmental cycle is normally completed in one year (Nosek 1972). Larvae emerge from the eggs after 6–12 days and under laboratory conditions feed on white mice for 2–3 days. Preferred hosts of this stage are rodents and small to medium size insectivores (shrews, moles and hedgehogs), as well as lagomorphs and carnivores, including dogs (Nosek 1972). Nymphs feed on white mice or rabbits for 3 and 4 days, respectively. The most important hosts of this stage, in addition to those of larvae, are artiodactyls (Nosek 1972). Adult females feed for 8–11 days and can be found on ungulates, carnivores (but rarely on pet dogs, Földvári and Farkas 2005) and medium-sized insectivores (Nosek 1972). Adults also feed on humans (Estrada-Peña and Jongejan 1999), mainly in the scalp region (Parola et al. 2009). *Dermacentor marginatus* developmental stages have seldom been reported from bats (Filippova et al. 1976) or birds (Feider 1965).

Ecology

Dermacentor marginatus is a typical open country tick species, preferring meadows and pastures, where it may be sympatric with *D. reticulatus* (Hornok and Farkas 2009). In Central Europe, the habitats of *D. marginatus* can be characterized by xerophilic plant communities (Nosek 1972). It can occur up to 1000 m a.s.l. (Hornok and Farkas 2009). The cold tolerance (e.g. host-seeking activity of adults occurs above 0 °C) is similar to that of *D. reticulatus* (Dörr and Gothe 2001). Larvae and nymphs are nidicolous, living in the burrows and nests of small mammals (Meyer-König et al. 2001). The seasonal activity of adult *D. marginatus* shows a higher spring and a lower autumn peak (opposed to that of *D. reticulatus*, Hornok 2009). In Central Europe, adult questing starts in late August and can last until May/June of the next year, including the winter months during days of mild weather (Nosek 1972; Hornok 2009). Larvae and nymphs are active during the summer (Nosek 1972).

Distribution

Dermacentor marginatus has been reported from France, Portugal, Spain, Switzerland, southern Germany and Poland, Austria, the Czech Republic, Slovakia, Hungary, Italy, Romania, the Balkans, Greece, Turkey, Ukraine, Belarus and Russia.

Vectorial Capacity and Pathogen Burden

Dermacentor marginatus is a competent vector of tick-borne encephalitis virus, Crimean-Congo haemorrhagic fever virus and (to a lesser extent than *D. reticulatus*) Omsk haemorrhagic fever virus, *Rickettsia sibirica*, *R. slovaca* (the more frequent causative agent of TIBOLA or DEBONEL in humans: Parola et al. 2009), *R. conorii*, *Babesia caballi* and *Theileria equi* (Nosek 1972). In addition, the following pathogens have been found in this tick species: West Nile virus (Bakonyi et al. 2005), *Coxiella burnetii* (Rehácek et al. 1991), *Rickettsia raoultii* (Spitalská et al. 2012), *R. massiliae*, *Ehrlichia canis*, and *Borrelia afzelii* (Hornok et al. 2013).

S. Hornok (✉)
Department of Parasitology and Zoology, University of Veterinary Medicine, Budapest, Hungary
e-mail: hornok.sandor@univet.hu

A. Estrada-Peña et al. (eds.), *Ticks of Europe and North Africa*,
https://doi.org/10.1007/978-3-319-63760-0_54

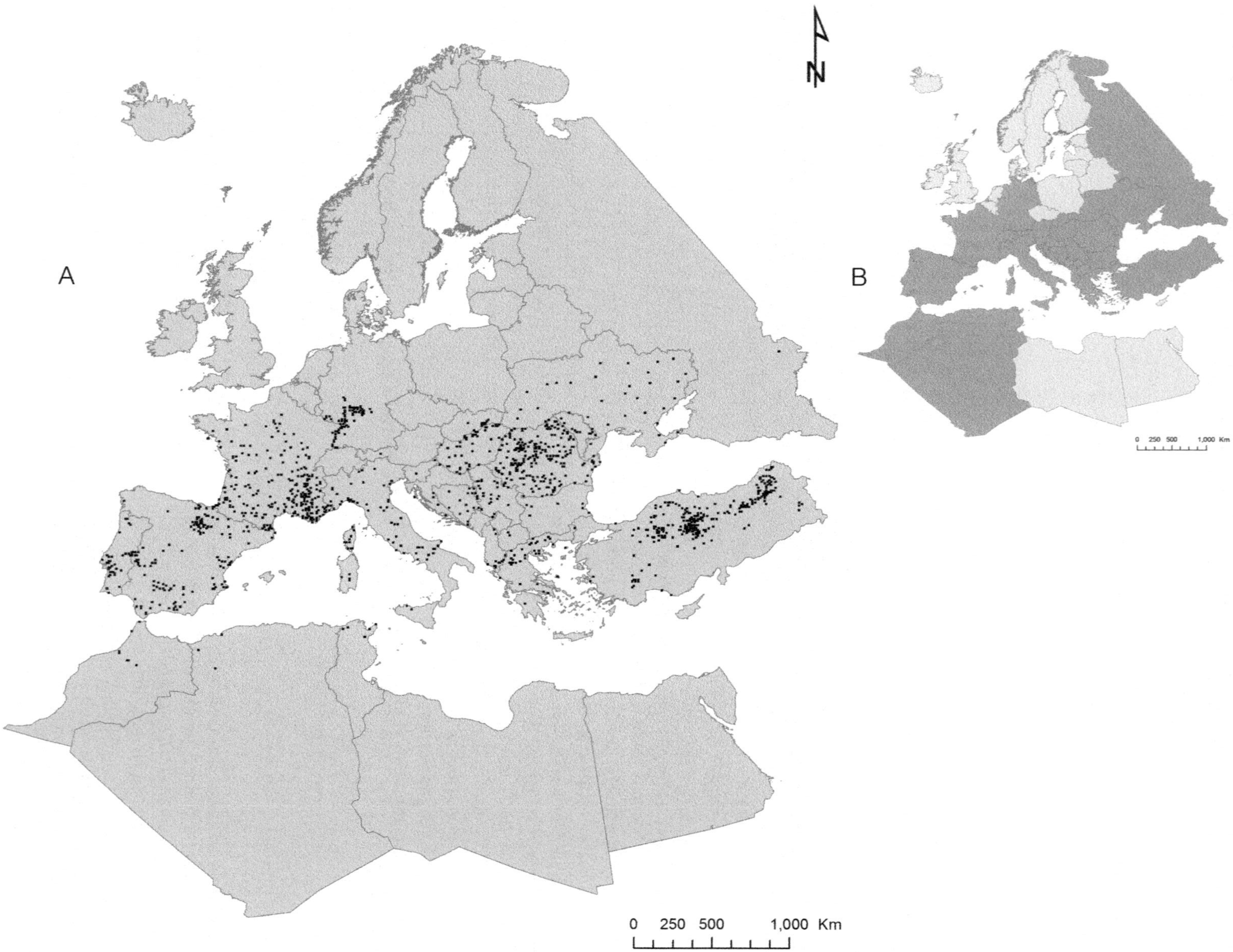

Fig. 111 **A** Distribution of *D. marginatus* in Europe and Northern Africa (10 × 10 km grid presence with black dots). **B** Countries where the species have been reported are marked in dark grey

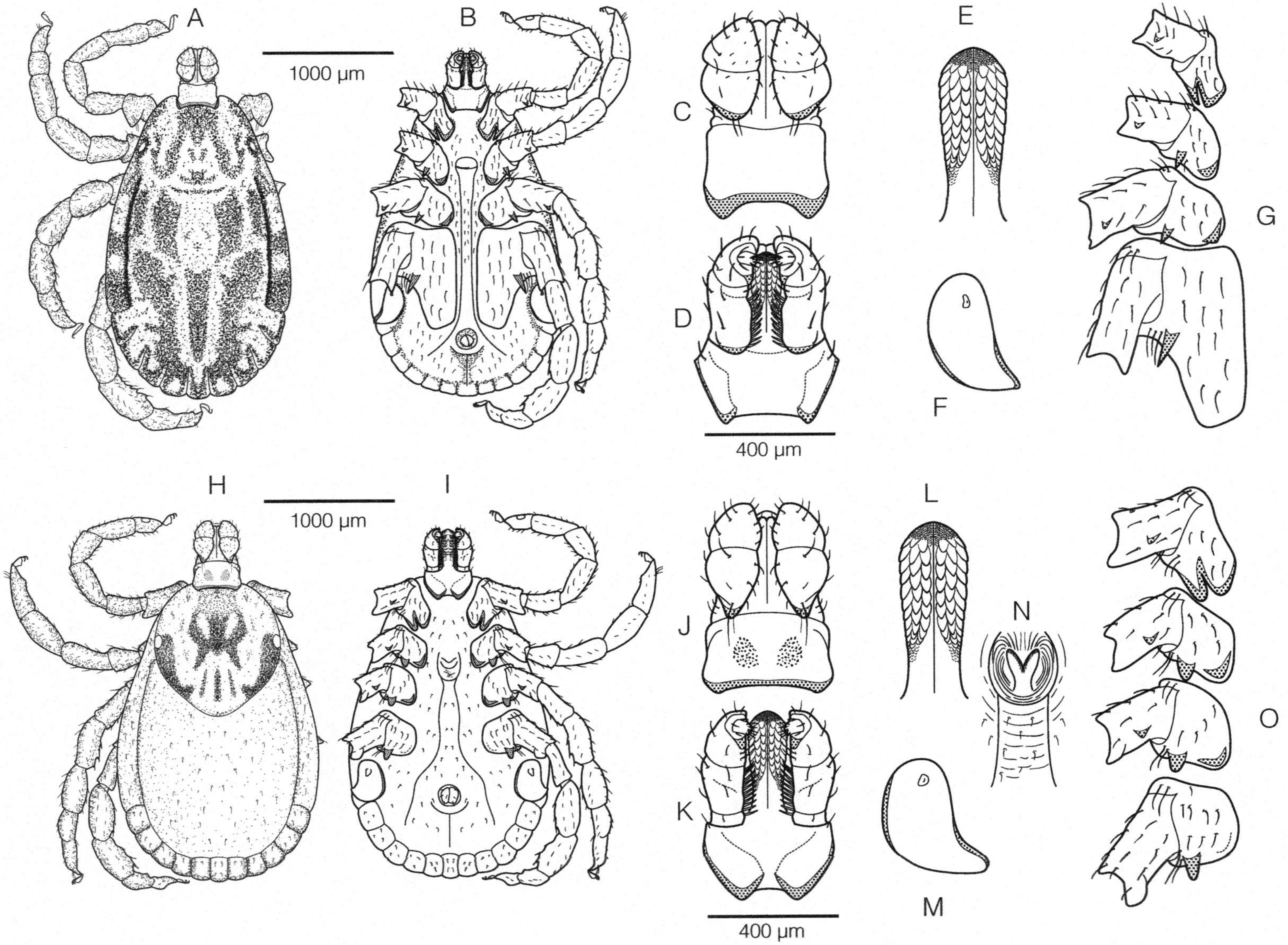

Fig. 112 **A–G** The male of *D. marginatus*. **A** dorsal, **B** ventral, **C** capitulum, dorsal, **D** capitulum, ventral, **E** hypostome, **F** spiracular plate, **G** coxae and trochanters I–IV. **H–O** the female of *D. marginatus*. **H** dorsal, **I** ventral, **J** capitulum, dorsal, **K** capitulum, ventral, **L** hypostome, **M** spiracular plate, **N** genital opening, **O** coxae and trochanters I–IV. Illustrations from specimens collected in Spain, France and Italy

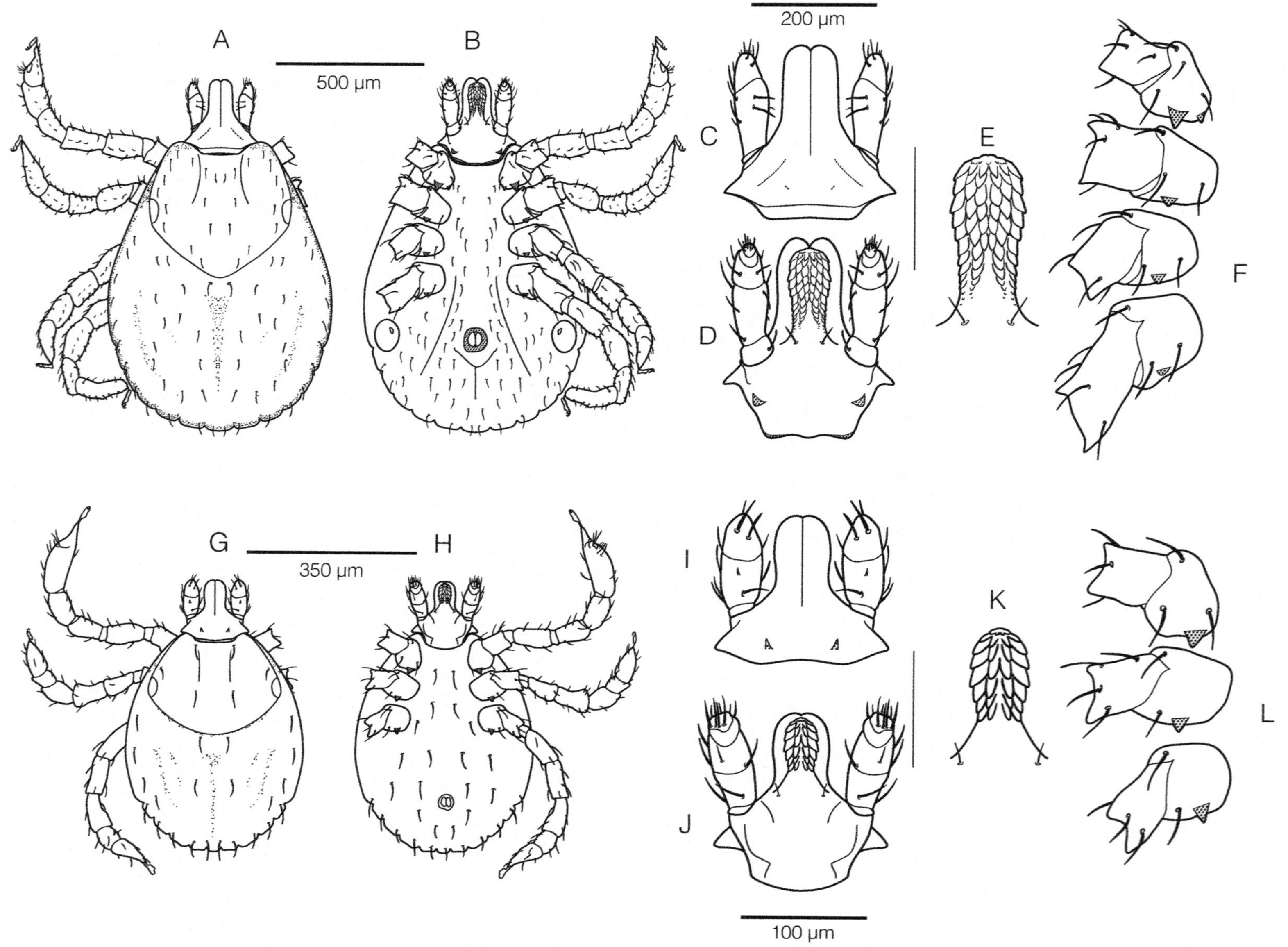

Fig. 113 **A–F** The nymph of *D. marginatus*. **A** dorsal, **B** ventral, **C** capitulum, dorsal, **D** capitulum, ventral, **E** hypostome, **F** coxae and trochanters I–IV. **G–L** the larva of *D. marginatus*. **G** dorsal, **H** ventral, **I** capitulum, dorsal, **J** capitulum, ventral, **K** hypostome, **L** coxae and trochanters I–III. Illustrations from specimens collected in Spain, France and Italy

References

Bakonyi T, Hubálek Z, Rudolf I, Nowotny N (2005) Novel flavivirus or new lineage of West Nile virus, central Europe. Emerg Infect Dis 11:225–231

Dörr B, Gothe R (2001) Cold-hardiness of *Dermacentor marginatus* (Acari: Ixodidae). Exp Appl Acarol 25:151–169

Estrada-Peña A, Jongejan F (1999) Ticks feeding on humans: a review of records on human-biting Ixodoidea with special reference to pathogen transmission. Exp Appl Acarol 23:685–715

Feider Z (1965) Fauna of the popular Republic of Romania, vol 5/2. Acaromorpha, Suprafamily Ixodoidea (in Romanian). Editura Academiei Republicii Populare Romane, Bucharest

Filippova NA, Neronov VM, Farhang-Azad A (1976) Data on the ixodid fauna (Acarina, Ixodidae) of small mammals in Iran (in Russian). Entomole Obozren 55:467–479

Földvári G, Farkas R (2005) Ixodid tick species attaching to dogs in Hungary. Vet Parasitol 129:125–131

Hornok S (2009) Allochronic seasonal peak activities of *Dermacentor* and *Haemaphysalis* spp. under continental climate in Hungary. Vet Parasitol 163:366–369

Hornok S, Farkas R (2009) Influence of biotope on the distribution and peak activity of questing ixodid ticks in Hungary. Med Vet Entomol 23:41–46

Hornok S, Fuente J, Horváth G, Fernández de Mera IG, Wijnveld M, Tánczos B, Farkas R, Jongejan F (2013) Molecular evidence of *Ehrlichia canis* and *Rickettsia massiliae* in ixodid ticks of carnivores from South Hungary. Acta Vet Hung 61:42–50

Meyer-König A, Zahler M, Gothe R (2001) Studies on the critical water mass and the rehydration potential of unfed adult *Dermacentor marginatus* and *D. reticulatus* ticks (Acari: Ixodidae). Exp Appl Acarol 25:505–516

Nosek J (1972) The ecology and public health importance of *Dermacentor marginatus* and *D. reticulatus* ticks in Central Europe. Folia Parasitol 19:93–102

Parola P, Rovery C, Rolain JM, Brouqui P, Davoust B, Raoult D (2009) *Rickettsia slovaca* and *R. raoultii* in tick-borne rickettsioses. Emerg Infect Dis 15:1105–1108

Rehácek J, Urvölgyi J, Kocianová E, Sekeyová Z, Vavreková M, Kovácová E (1991) Extensive examination of different tick species for infestation with *Coxiella burnetii* in Slovakia. Eur J Epidemiol 7:299–303

Spitalská E, Stefanidesová K, Kocianová E, Boldiš V (2012) *Rickettsia slovaca* and *Rickettsia raoultii* in *Dermacentor marginatus* and *Dermacentor reticulatus* ticks from Slovak Republic. Exp Appl Acarol 57:189–197

Dermacentor reticulatus (Fabricius, 1794) (Figs. 114–116)

S. Hornok

Life-Cycle and Host Preferences

Dermacentor reticulatus is a three-host tick species. The life-cycle can be completed in 80–125 days under laboratory conditions, but in natural biotopes it usually lasts for two years, because most adults hibernate twice before feeding (Nosek 1972). Larvae emerge from the eggs after 12–19 days and under laboratory conditions feed on white mice for 3 days. Preferred hosts of this stage include rodents and small to medium size insectivores (shrews, moles, hedgehogs), as well as lagomorphs (Nosek 1972). Nymphs may feed on white mice for 3–4 days, and their most important hosts, in addition to those of the larvae, are artiodactyls and carnivores (Nosek 1972), including dogs (Hornok et al. 2013). Adult females feed for 7–11 days and can be found on ungulates, carnivores (especially dogs) and medium-sized insectivores and lagomorphs (Nosek 1972). Adults also feed on humans (Estrada-Peña and Jongejan 1999), mainly in the scalp region (Parola et al. 2009). *Dermacentor reticulatus* developmental stages have rarely been reported from bats (Neumann 1911), birds (Nosek 1972; Akimov and Nebogatkin 2011) and reptiles (Nowak-Chmura and Siuda 2012).

Ecology

Dermacentor reticulatus is a typical open country tick species, preferring meadows and pastures, where it may be sympatric with *D. marginatus* (Hornok and Farkas 2009). In Central Europe, the habitats of *D. reticulatus* usually have a partly dry vegetation covering (Hornok and Farkas 2009), and are frequently found in river basins or along lake shores (Nosek 1972; Kiewra and Czulowska 2013). *Dermacentor reticulatus* occurs up to 1000 m altitude a.s.l. (Hornok and Farkas 2009). This tick species is cold-tolerant, and host-seeking adults can be found on the vegetation at temperatures above 0 °C (Hubálek et al. 2003). On the other hand, larvae and nymphs are nidicolous and live in the burrows and nests of small mammals (Meyer-König et al. 2001). The seasonal activity of adult *D. reticulatus* is characterized by a higher autumn and a lower spring peak, i.e. in Central Europe questing can be observed from late August until May/June of the next year, including winter months with days of mild weather (Nosek 1972; Hornok 2009). Larvae and nymphs are active during the summer (Nosek 1972).

Distribution

Dermacentor reticulatus has been reported from the United Kingdom, France, Portugal and Spain (excluding Mediterranean regions), the Netherlands, Belgium, Switzerland, Germany, Poland, Austria, the Czech Republic, Slovakia, Hungary, Slovenia, northern Italy and the northern Balkans (Croatia, Serbia, Romania, Bulgaria and Moldova), Ukraine, Lithuania, Latvia, Belarus and Russia. Expansion of its geographical range has been observed in central, then towards both north-western and north-eastern Europe over the past decades (Sréter et al. 2005; Dautel et al. 2006; Bullová et al. 2009; Siroký et al. 2011; Karbowiak 2014; Jongejan et al. 2015; Paulauskas et al. 2015; Rubel et al. 2016). This species frequently occurs in urban biotopes (Hornok et al. 2014).

Vectorial Capacity and Pathogen Burden

Dermacentor reticulatus is a competent vector of tick-borne encephalitis and Omsk haemorrhagic fever viruses, *Rickettsia sibirica*, *R. raoultii* (causing TIBOLA or DEBONEL in humans), *R. conorii*, *Francisella tularensis*, *Anaplasma marginale*, *Babesia canis*, *B. caballi* and *Theileria equi*

S. Hornok (✉)
Department of Parasitology and Zoology, University of Veterinary Medicine, Budapest, Hungary
e-mail: hornok.sandor@univet.hu

A. Estrada-Peña et al. (eds.), *Ticks of Europe and North Africa*,
https://doi.org/10.1007/978-3-319-63760-0_55

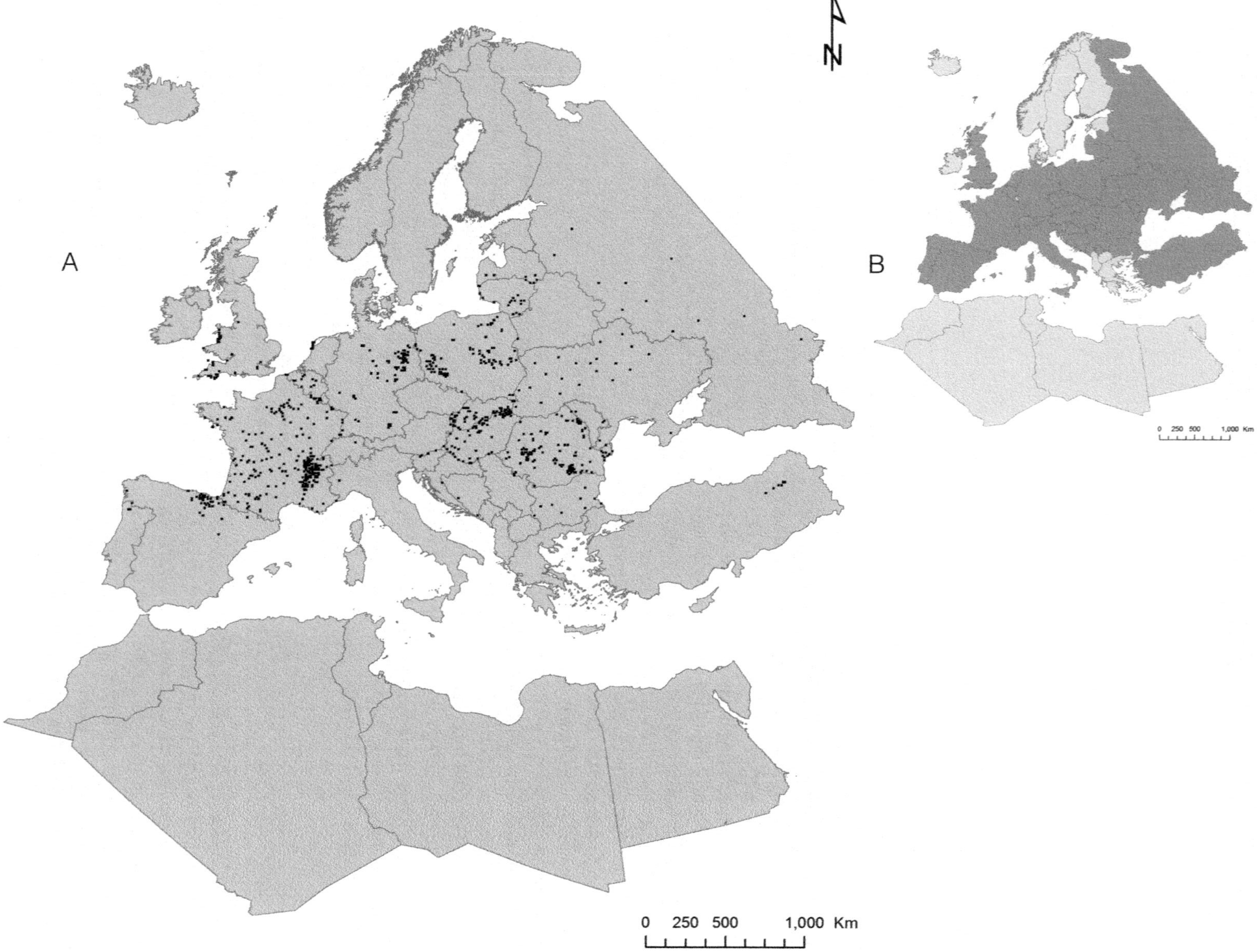

Fig. 114 **A** Distribution of *D. reticulatus* in Europe and Northern Africa (10 × 10 km grid presence with black dots). **B** Countries where the species have been reported are marked in dark grey

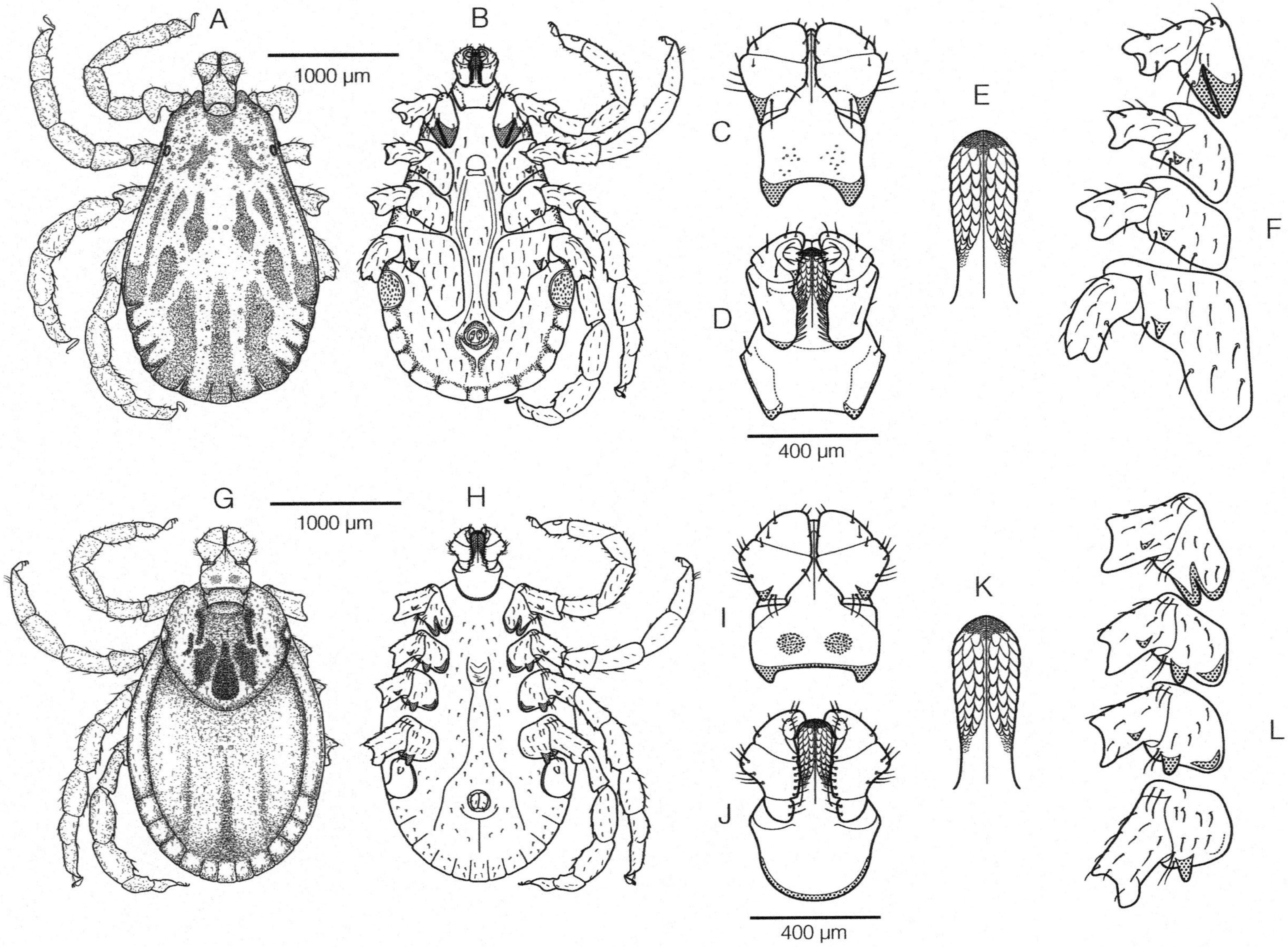

Fig. 115 **A–F** The male of *D. reticulatus*. **A** Dorsal, **B** ventral, **C** capitulum, dorsal, **D** capitulum, ventral, **E** hypostome, **F** coxae and trochanters I–IV. **G–L** The female of *D. reticulatus*. **G** Dorsal, **H** ventral, **I** capitulum, dorsal, **J** capitulum, ventral, **K** hypostome, **L** coxae and trochanters I–IV. Illustrations from specimens collected in Spain, Poland, Czech Republic and Germany

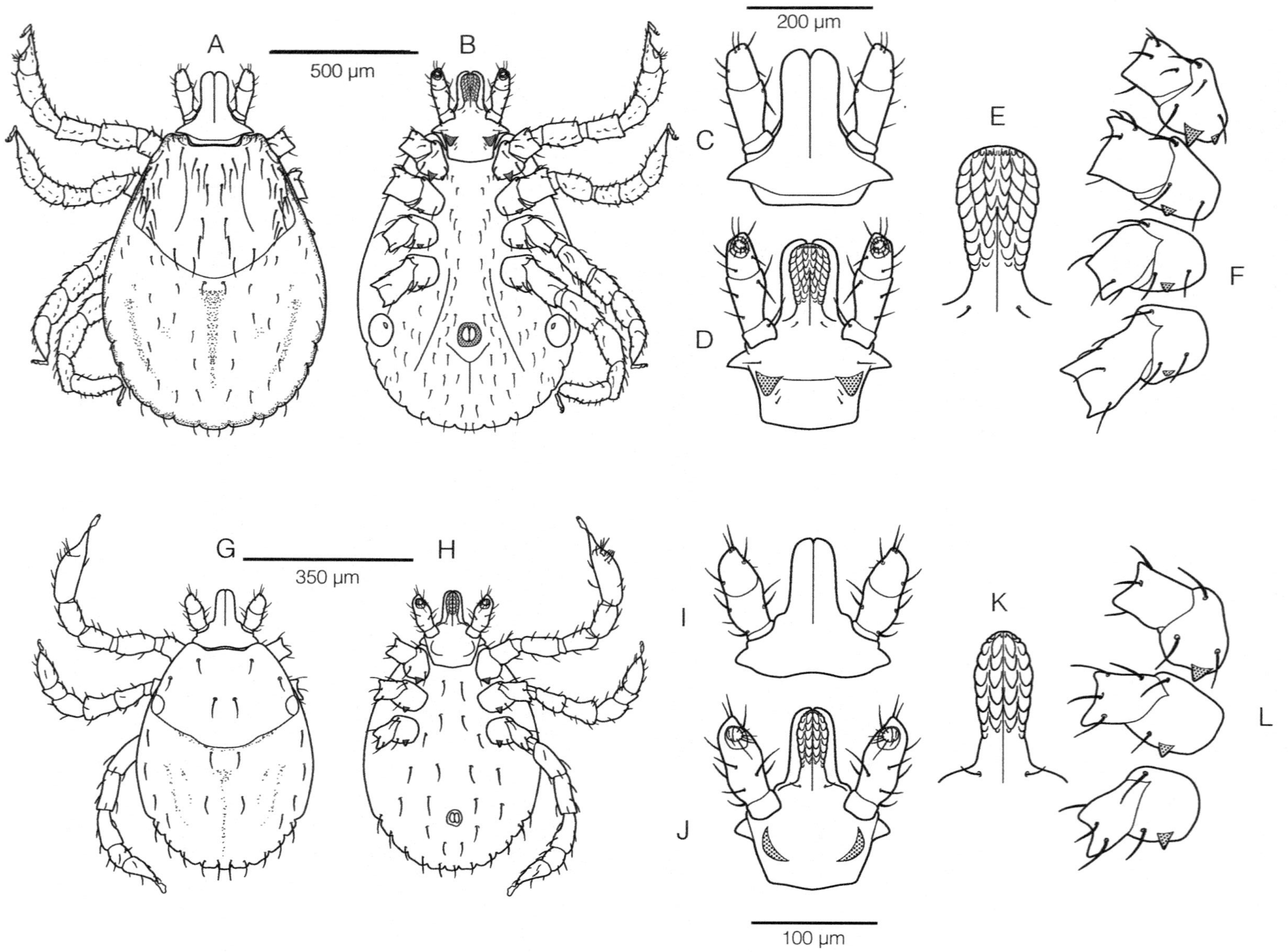

Fig. 116 **A–F** The nymph of *D. reticulatus*. **A** Dorsal, **B** ventral, **C** capitulum, dorsal, **D** capitulum, ventral, **E** hypostome, **F** coxae and trochanters I–IV. **G–L** The larva of *D. reticulatus*. **G** Dorsal, **H** ventral, **I** capitulum, dorsal, **J** capitulum, ventral, **K** hypostome, **L** coxae and trochanters I–III. Illustrations from specimens collected in Spain, Poland, Czech Republic and Germany

(Nosek 1972; Jongejan et al. 2015). In addition, the following pathogens have been found in this tick species: *Coxiella burnetii* (Rehácek et al. 1991), *Bartonella* spp. (Rar et al. 2005), *Rickettsia helvetica* (Dobec et al. 2009), *R. slovaca* (Spitalská et al. 2012), *Borrelia burgdorferi* s.l. (Kahl et al. 1992), *Anaplasma phagocytophilum* and *Babesia microti* (Jongejan et al. 2015).

References

Akimov IA, Nebogatkin IV (2011) Distribution of ticks of the genus *Dermacentor* (Acari, Ixodidae) in Ukraine. Vestnik Zoologii 45:35–40

Bullová E, Lukán M, Stanko M, Petko B (2009) Spatial distribution of *Dermacentor reticulatus* tick in Slovakia in the beginning of the 21st century. Vet Parasitol 165:357–360

Dautel H, Dippel C, Oehme R, Hartelt K, Schettler E (2006) Evidence for an increased geographical distribution of *Dermacentor reticulatus* in Germany and detection of *Rickettsia* sp. RpA4. Int J Med Microbiol 296(Suppl 40):149–156

Dobec M, Golubic D, Punda-Polic V, Kaeppeli F, Sievers M (2009) *Rickettsia helvetica* in *Dermacentor reticulatus* ticks. Emerg Infect Dis 15:98–100

Estrada-Peña A, Jongejan F (1999) Ticks feeding on humans: a review of records on human-biting Ixodoidea with special reference to pathogen transmission. Exp Appl Acarol 23:685–715

Hornok S (2009) Allochronic seasonal peak activities of *Dermacentor* and *Haemaphysalis* spp. under continental climate in Hungary. Vet Parasitol 163:366–369

Hornok S, Farkas R (2009) Influence of biotope on the distribution and peak activity of questing ixodid ticks in Hungary. Med Vet Entomol 23:41–46

Hornok S, Fuente J, Horváth G, Fernández de Mera IG, Wijnveld M, Tánczos B, Farkas R, Jongejan F (2013) Molecular evidence of *Ehrlichia canis* and *Rickettsia massiliae* in ixodid ticks of carnivores from South Hungary. Acta Vet Hung 61:42–50

Hornok S, Meli ML, Gönczi E, Halász E, Takács N, Farkas R, Hofmann-Lehmann R (2014) Occurrence of ticks and prevalence of *Anaplasma phagocytophilum* and *Borrelia burgdorferi* s.l. in three types of urban biotopes: forests, parks and cemeteries. Ticks Tick Borne Dis 5:785–789

Hubálek Z, Halouzka J, Juricová Z (2003) Host-seeking activity of ixodid ticks in relation to weather variables. J Vector Ecol 28: 159–165

Jongejan F, Ringenier M, Putting M, Berger L, Burgers S, Kortekaas R, Lenssen J, van Roessel M, Wijnveld M, Madder M (2015) Novel foci of *Dermacentor reticulatus* ticks infected with *Babesia canis* and *Babesia caballi* in the Netherlands and in Belgium. Parasit Vectors 8:232

Kahl O, Janetzki C, Gray JS, Stein J, Bauch RJ (1992) Tick infection rates with Borrelia: *Ixodes ricinus* versus *Haemaphysalis concinna* and *Dermacentor reticulatus* in two locations in eastern Germany. Med Vet Entomol 6:363–366

Karbowiak G (2014) The occurrence of the *Dermacentor reticulatus* tick-its expansion to new areas and possible causes. Ann Parasitol 60:37–47

Kiewra D, Czulowska A (2013) Evidence for an increased distribution range of *Dermacentor reticulatus* in south-west Poland. Exp Appl Acarol 59:501–506

Meyer-König A, Zahler M, Gothe R (2001) Studies on the critical water mass and the rehydration potential of unfed adult *Dermacentor marginatus* and *D. reticulatus* ticks (Acari: Ixodidae). Exp Appl Acarol 25:505–516

Neumann LG (1911) Ixodidae: *Dermacentor reticulatus*. Das Tierreich 26:98–100

Nosek J (1972) The ecology and public health importance of *Dermacentor marginatus* and *D. reticulatus* ticks in Central Europe. Folia Parasitol 19:93–102

Nowak-Chmura M, Siuda K (2012) Ticks of Poland. Review of contemporary issues and latest research. Ann Parasitol 58:125–155

Parola P, Rovery C, Rolain JM, Brouqui P, Davoust B, Raoult D (2009) *Rickettsia slovaca* and *R. raoultii* in tick-borne Rickettsioses. Emerg Infect Dis 15:1105–1108

Paulauskas A, Radzijevskaja J, Mardosaitė-Busaitienė D, Aleksandravičienė A, Galdikas M, Krikštolaitis R (2015) New localities of *Dermacentor reticulatus* ticks in the Baltic countries. Ticks Tick Borne Dis 6:630–635

Rar VA, Fomenko NV, Dobrotvorsky AK, Livanova NN, Rudakova SA, Fedorov EG, Astanin VB, Morozova OV (2005) Tickborne pathogen detection, Western Siberia, Russia. Emerg Infect Dis 11:1708–1715

Rehácek J, Urvölgyi J, Kocianová E, Sekeyová Z, Vavreková M, Kovácová E (1991) Extensive examination of different tick species for infestation with *Coxiella burnetii* in Slovakia. Eur J Epidemiol 7:299–303

Rubel F, Brugger K, Pfeffer M, Chitimia-Dobler L, Didyk YM, Leverenz S et al (2016) Geographical distribution of *Dermacentor marginatus* and *Dermacentor reticulatus* in Europe. Ticks Tick-Borne Dis 7(1):224–233

Siroký P, Kubelová M, Bednář M, Modrý D, Hubálek Z, Tkadlec E (2011) The distribution and spreading pattern of *Dermacentor reticulatus* over its threshold area in the Czech Republic-how much is range of this vector expanding? Vet Parasitol 183:130–135

Spitalská E, Stefanidesová K, Kocianová E, Boldiš V (2012) *Rickettsia slovaca* and *Rickettsia raoultii* in *Dermacentor marginatus* and *Dermacentor reticulatus* ticks from Slovak Republic. Exp Appl Acarol 57:189–197

Sréter T, Széll Z, Varga I (2005) Spatial distribution of *Dermacentor reticulatus* and *Ixodes ricinus* in Hungary: evidence for change? Vet Parasitol 128:347–351

Genus *Rhipicephalus* Koch, 1844

A. Estrada-Peña, M. P. Pfäffle, and T. N. Petney

To date, there are eight species of the genus *Rhipicephalus* known from the Western Palaearctic, namely, *R. guilhoni* Morel and Vassiliades 1963 (so far only from the Canary Islands, probably introduced); *R. camicasi* Morel et al. 1976 (so far only in Egypt); *R. sanguineus* s.l.; *R. turanicus* Pomerantzev, 1940; *R. pusillus* Gil Collado 1936; *R. rossicus* Yakimov and Kol-Yakimova, 1911; and *R. bursa* Canestrini and Fanzago, 1878. The eighth species is *R. annulatus* (Say, 1821), which was formerly included in the genus *Boophilus*.

This, however, may not represent a stable situation because of the intense debate regarding the identity of certain species (one or several) included in *R. sanguineus* s.l. and the actual range of *R. turanicus* (Dantas-Torres et al. 2013; Levin et al. 2012; Nava et al. 2012, 2014). Early molecular studies highlighted these problems by showing that identical taxa had been misdiagnosed as different species (Pegram et al. 1987a, b; Ribeiro et al. 1996; Walker et al. 2000; Beati and Keirans 2001) or different species had been identified as conspecifics (Szabó et al. 2005; Burlini et al. 2010; Moraes-Filho et al. 2011; Levin et al. 2012; Nava et al. 2012). Most notably, much confusion exists between different "populations" of *R. sanguineus* s.l. and *R. turanicus*, with early molecular studies showing that many samples identified as *R. turanicus* were in fact *R. sanguineus* s.l. (Santos-Silva et al. 2011). In the case of *R. camicasi*, which is a tick typically distributed south of the Sahara, some specimens have been collected in Egypt (Morel 1980). We have not had the opportunity to examine these specimens and verify their identity, but we include the species in this list of Western Palaearctic ticks. This species has also been found in Sudan and in some areas of the Nile Valley, suggesting a population extending from the Ethiopian plains to the Mediterranean (Jongejan et al. 1987).

The conclusion of these long-lasting debates is that the name "*R. sanguineus* s.s." should not be used to name a tick, because the type is lost and there is no reliable description of the tick. It is necessary to collect specimens from the same area where the type was collected, redescribing and naming a neotype, and, most importantly, rising a colony through several generations to demonstrate that no hybrid forms exist in the collected specimens. An account of the morphological variability of what has been classically considered as *R. sanguineus* was published by Dantas-Torres et al. (2013). For the sake of simplicity, we will adhere here to the morphological concepts by Filippova (1997) when discussing *R. sanguineus* s.l. or *R. turanicus*, but the reader must be aware that such a consensus does not exist. The readers should also be aware that, as a consequence of the debate on the taxonomic status of *R. sanguineus* s.l., the data on sequences available in GenBank are also likely to be "contaminated" by outdated identifications, and therefore, we cannot recommend their use. Importantly, not every *Rhipicephalus* collected on carnivores in the Mediterranean region should be immediately considered to be *R. sanguineus* (s.s. or s.l.).

With these recommendations in mind, every stage of four of these species, namely, *R. bursa*, *R. sanguineus* (given the above reservations), *R. turanicus*, and *R. rossicus*, is adequately described, keyed, compared, and illustrated by Filippova (1997) in her revision of the subfamily Amblyomminae. Additional information on the larvae of these species can be found in Filippova (1989), with valuable information about body measurements and size range. Additional keys for the larvae and nymphs of some species of Palaearctic *Rhipicephalus* can be obtained from Shatas

A. Estrada-Peña (✉)
Department of Animal Health, Faculty of Veterinary Medicine, Zaragoza, Spain
e-mail: aestrada@unizar.es

M. P. Pfäffle
Department of Ecology and Parasitology, Institute of Zoology, Karlsruhe Institute of Technology, Karlsruhe, Germany
e-mail: miripfaeffle@web.de

T. N. Petney
State Museum of Natural History, Erbprinzenstr. 13, 76133 Karlsruhe, Germany
e-mail: trevor.petney@kit.edu

A. Estrada-Peña et al. (eds.), *Ticks of Europe and North Africa*,
https://doi.org/10.1007/978-3-319-63760-0_56

(1956). SEM pictures of these four species, together with *R. pusillus*, can be found in Cringoli et al. (2005).

Rhipicephalus pusillus is a poorly known species. It is monoxenous on the Mediterranean wild rabbit (*Oryctolagus cuniculus*). There are reports of this tick on wild ruminants and livestock, which we believe are misidentifications of small specimens of *R. bursa*, because the adults have the same punctuation pattern on the scutum and the males have wide adanal plates, similar to those of the males of *R. bursa*. This species is awaiting a complete redescription, but some information can be found in Walker et al. (2000). Keys and SEM pictures can be found in Cringoli et al. (2005). These keys include only *R. pusillus*, *R. bursa*, *R. turanicus*, and *R. sanguineus* s.l. (as *R. sanguineus* s.s.). The keys available from Manilla (1998) refer only to these four species, with explicit mention of the difficulty of finding reliable details for the separation of immatures, which are, on the other hand, easily separated by Filippova (1997).

Both *R. camicasi* and *R. guilhoni* are tropical species that seem to be invasive in a few parts of the Western Palaearctic. Information and keys for the adults of these species can be found in Walker et al. (2000) and Walker et al. (2003). The former reference includes some sketches of the capitulum of the immatures, without further illustrations or keys; the second is exclusively devoted to the adults. Additional information on these two species can be found in their original descriptions, namely, Morel et al. (1976) for *R. camicasi* and Morel and Vassiliades (1963) for *R. guilhoni*. A source of information for both species is Pegram et al. (1987b) in which SEM pictures of the two species, together with illustrations of "*R. turanicus*" and "*R. sanguineus*", are included. It must be noted that the "*R. sanguineus*" found in Africa has proven to be different from the European populations (Nava et al. 2014), and that the specimens of *R. turanicus* included in this study were collected far from its type locality (Nava et al. 2014). Therefore, the issue of the identity of the specimen pictures as either *sanguineus* or *turanicus* in this paper still persists.

Key to the Species of Genus *Rhipicephalus* in Europe and Northern Africa

Females

Note: We were unable to examine material of *R. camicasi* and *R. guilhoni*. The keys provided are tentative. No illustrations are provided for these two species because of the unavailability of reliable material.

1a. Without festoons...***Rhipicephalus annulatus*** (Fig. 134)
1b. With festoons...**2**
2a. All of the pores of the spiracular plate are equal in size. They are scarce in the basal region and very dense on the dorsal prolongation of the spiracular plate, which does not lie at an obtuse angle. Ventromedial setae of palpal segments I and II are rounded in section throughout their length, becoming narrower (cone-like) toward the end...***Rhipicephalus bursa*** (Fig. 118)
2b. Pores of the spiracular plate with different sizes according to their position. In the basal area of the spiracular plate, all sizes of pores can be found. The dorsal prolongation of the spiracular plate at an obtuse angle, with a minimum of 12 pores. Ventromedial setae of palpal segments I and II flattened and thickened from the base to their mid-length...**3**
3a. Spiracular plate with a relatively narrow dorsal prolongation. The dorsal portion of the spiracular plate from its medial longitudinal line to the end, pointed and straight. Marginal pores on the spiracular plate present or absent. Ventromedial setae of palpal segments I and II plumose, each seta with more than two rows of longitudinal barbs. The vestibular segment of the vagina is large and has a funnel shape...**4**
3b. Spiracular plate with a wide dorsal prolongation. The dorsal portion of the spiracular plate from its medial longitudinal line to the end, rounded. Marginal pores on the spiracular plate present on all of the surface. Ventromedial setae of palpal segments I and II have one row of teeth on the caudal face (medially). The vestibular segment of the vagina is narrow...**7**
4a. Scutum moderately punctate...**5**
4b. All areas of the scutum quite lightly punctate...**6**
5a. Marginal pores in the anteromedial sector of the spiracular plate absent. Ventromedial hairs of palpal segments I and II plumose, with three longitudinal barbs each. Cervical grooves wide and superficial. Scutum as wide as long...***Rhipicephalus sanguineus*** s.l. (Fig. 128)
5b. Marginal pores in the anteromedial sector of the spiracular plate present. Ventromedial hairs of the palp segments I and II plumose, with six to eight longitudinal barbs each. Cervical grooves narrow, short, and deep. Scutum clearly longer than wide...***Rhipicephalus pusillus*** (Fig. 121)
6a. Genital aperture with distinct hyaline flaps...***Rhipicephalus guilhoni*** (a tropical species, so far reported only from the Canary Islands within the Western Palaearctic)
6b. Genital aperture without distinct hyaline flaps...***Rhipicephalus camicasi*** (a tropical species, so far reported only from Egypt within the Western Palaearctic)
7a. Third palpal segment almost triangle shaped. Spiracular plate oval and elongated. Posterior part of the scutal margin pointed...***Rhipicephalus turanicus*** (Fig. 131)

7b. Third palpal segment trapeze-like. Spiracular plate very broad. Posterior part of the scutal margin widely rounded... ***Rhipicephalus rossicus*** (Fig. 124)

Males

Note: We were unable to examine material of *R. camicasi* and *R. guilhoni*. The keys provided are tentative. No illustrations are provided for these two species because of the unavailability of reliable material.

1a. Without festoons...***Rhipicephalus annulatus*** (Fig. 134)
2b. With festoons...**2**
2a. Spiracular plates with all pores with the same size and shape. Ventromedial setae of palpal segments I and II rounded in cross section over their entire length and conical, inserted at a distance which approximately equals the diameter of one seta. Adanal plates very wide, almost as right angle triangle...***Rhipicephalus bursa*** (Fig. 118)
2b. Spiracular plates with pores of different size and shape according to the regions of the plate. Ventromedial setae of palpal segments I and II flattened and broadening toward the middle of their length, located close to each other. Adanal plates variously shaped but never as wide as long...**3**
3a. Adanal plates approximately twice as long as wide... ***Rhipicephalus rossicus*** (Fig. 124)
3b. Adanal plates more than twice as long as wide...**4**
4a. Adanal plates without a pointed prolongation internally. Spiracular plate comma shaped, with a long dorsal prolongation. Marginal pores absent in the anteromedial sector ...**5**
4b. Adanal plates with a pointed and small prolongation on internally, posteriorly to a medial concavity...**6**
5a. Lateral grooves long and well visible. Cervical grooves short, narrow, and deep. Dorsal punctations smooth and scarce. Dorsal prolongation of the spiracle about one-half the breadth of the adjacent festoons...***Rhipicephalus sanguineus*** s.l. (Fig. 128)
5b. Lateral grooves short, only slightly marked by the presence of larger scutal punctations. Cervical grooves relatively wide and very deep. Dorsal punctations deep, abundant...***Rhipicephalus pusillus*** (Fig. 121)
6a. Cervical fields very large, with dense punctations scattered over the scutal surface. Spiracular plate elongated, with a relatively broad ventral part progressively narrowing toward the end, having a short dorsal prolongation. Pores of the spiracular plate completely covering the plate...***Rhipicephalus turanicus*** (Fig. 131)
6b. Cervical fields small, punctations on scutal surface diffuse. Spiracular plate wide or narrow, with pores not completely covering the plate...**7**
7a. Marginal grooves shallow, quite indistinct, and lightly punctate. Spiracular plate very narrow, with pores large and sparse...***Rhipicephalus camicasi*** (a tropical species, so far reported only from Egypt within the Western Palaearctic)
7b. Marginal grooves deep, distinct, and heavily punctate. Spiracular plate relatively wide, with very small pores... ***Rhipicephalus guilhoni*** (a tropical species, so far recorded from the Canary Islands within the Western Palaearctic)

Nymphs

Note: We were unable to examine material of *R. camicasi* and *R. guilhoni.* The keys provided are tentative. No illustrations are provided for these two species because of the unavailability of reliable material.

1a. Without festoons...***Rhipicephalus annulatus*** (Fig. 135)
1b. With festoons...**2**
2a. Coxae IV with one internal spur of variable size and shape...**3**
2b. Coxae IV without spurs...**7**
3a. Capitulum essentially quadrangular, lateral margins of the basis capituli not protruding into acute angles...***Rhipicephalus bursa*** (Fig. 119)
3b. Capitulum not as above, with lateral margins of the basis capituli clearly protruding into more or less well developed acute angles...**4**
4a. Setae in the medial part of the scutum clearly longer than the lateral ones...***Rhipicephalus guilhoni*** (a tropical species, so far recorded from the Canary Islands within the western Palaearctic)
4b. Setae on the medial part of the scutum of the same length than the lateral ones...**5**
5a. Auriculae very well marked. Lateral prolongations of the basis capituli wide and rounded. With a small spur on the ventral side of palpal segment I. Internal spur of coxa I of the same size or even larger than the external one...***Rhipicephalus pusillus*** (Fig. 122)
5b. Internal spur of coxa I smaller than the external one ...**6**
6a. Auriculae long, acute. Scutal setae of about the same length than alloscutal setae...***Rhipicephalus turanicus*** (Fig. 132)
6b. Auriculae short, rounded. Scutal setae about half the length of the alloscutal setae...***Rhipicephalus sanguineus*** s.l. (Fig. 129)
7a. Coxae II and II with mere indications of spurs, as curricular ridges...***Rhipicephalus camicasi*** (a tropical species, so far reported only from Egypt within the Western Palaearctic)
7b. Coxae II and III with internal spurs well visible... ***Rhipicephalus rossicus*** (Fig. 125)

Larvae

Note: We were unable to examine material of *R. camicasi* and *R. guilhoni*. We are not aware of specific morphological details adequate for the separation of these species from the rest of the species. The keys provided do not include the separation of the larvae of these two species.

1a. Without festoons…***Rhipicephalus annulatus*** (Fig. 135)
1b. With festoons…**2**
2a. Capitulum quadrangular or subrectangular. Lateral sides of basis capituli not protruding laterally. Without a small spur on the ventral side of palpal segment I…***Rhipicephalus bursa*** (Fig. 119)
2b. Capitulum triangular or with palpi narrowing at their tips. Lateral sides of the basis capituli protruding laterally at relatively acute angles, more or less pronounced. With or without a small spur on the ventral side of palpal segment I…**3**
3a. Very small size, not longer than 250 μm. With a very small, backward directed spur on the ventral side of the palpal segment I…***Rhipicephalus pusillus*** (Fig. 122)
3b. Body size above 300 μm. Without spurs on the ventral side of the palpal segment I…**4**
4a. Palpi trapezoidal, apex blunt with bluntly rounded margin. Lateral margins of palpi straight, gradually rounded. Lateral prolongations of basis capituli sharp, but slightly protruding, situated anterior to the midlength of the basis, not visible ventrally. Median festoon of a width approximately equal to those adjoining it…***Rhipicephalus sanguineus*** s.l. (Fig. 129)
4b. Palpi triangular. Apex of palpi triangular or conical. Lateral protrusions of the basis capituli protruding far in both dorsal and ventral views. Median festoon of a width half that of those adjoining it. Ventral spur on segment I of palpi small…**5**
5a. Palpi triangular, long, with a well developed and narrow conical apex. Lateral margins of palpi convex from the narrow apex to the basis, slightly concave near the apex. Lateral expansions of the basis capituli triangular, protruding acutely and laterally from both dorsal and ventral views. Second and third coxae of similar size. Posthypostomal setae long, reaching the posterior denticles of the hypostome. Internal palpal setae wide and feathery. Posterior margin of scutum obliquely rounded…***Rhipicephalus rossicus*** (Fig. 125)
5b. Palpi short, apex in the form of a short and wide triangle. Lateral margins of the palpi oblique from the narrow apex to the wide base, which is slightly enlarged near the base. Lateral prolongations of the basis capituli rounded, protruding moderately and obliquely from the midlength of the basis, and visible from both dorsal and ventral views. Posthypostomal setae short, not reaching the posterior denticles of the hypostome. Internal setae of the palpi narrow, feathered. Posterior margin of the scutum convex…***Rhipicephalus turanicus*** (Fig. 132)

References

Beati L, Keirans JE (2001) Analysis of the systematic relationships among ticks of the genera *Rhipicephalus* and *Boophilus* (Acari: Ixodidae) based on mitochondrial 12S ribosomal DNA gene sequences and morphological characters. J Parasitol 87:32–48

Burlini L, Teixeira KS, Szabó MJ, Famadas K (2010) Molecular dissimilarities of *Rhipicephalus sanguineus* (Acari: Ixodidae) in Brazil and its relation with samples throughout the world: is there a geographical pattern? Exp Appl Acarol 50:361–374

Cringoli G, Iori A, Rinaldi L, Veneziano V, Genchi C (2005) Mappe parassitologiche—Zecche. Series Edit. Giuseppe Cringoli, Napoli

Dantas-Torres F, Latrofa MF, Annoscia G, Giannelli A, Parisi A, Otranto D (2013) Morphological and genetic diversity of *Rhipicephalus sanguineus* s.l. from the New and Old Worlds. Parasites Vectors 6:213

Filippova NA (1989) The identification of species of the genus *Rhipicephalus* (Ixodidae) from the fauna of USSR on the larval phase. Parazitologiya 23:104–117 (in Russian)

Filippova NA (1997) Ixodid ticks of subfamily Amblyomminae. In the fauna of Russia and neighboring countries. New Series no. 145: Arachnoidea, vol 4(5), St. Petersburg, Nauka (in Russian)

Jongejan F, Zivkovic D, Pegram RG, Tatchell RJ, Fison T, Latif AA, Paine G (1987) Ticks (Acari: Ixodidae) of the Blue and White Nile ecosystems in the Sudan with particular reference to the *Rhipicephalus sanguineus* group. Exp Appl Acarol 3:331–346

Levin ML, Studer E, Killmaster LF, Zemtsova GE, Mumcuoglu KY (2012) Crossbreeding between different geographical populations of the brown dog tick *Rhipicephalus sanguineus* (Acari: Ixodidae). Exp Appl Acarol 58:51–68

Manilla G (1998) Fauna d'Italia. Acari, Ixodida. Edizioni Calderini, Bologna, p 280

Moraes-Filho J, Marcili A, Nieri-Bastos FA, Richtzenhain LJ, Labruna MB (2011) Genetic analysis of ticks belonging to the *Rhipicephalus sanguineus* group in Latin America. Acta Trop 117:51–55

Morel PC (1980) Study on Etiophian Ticks (Acarida, Ixodida). Institute d'Élevage et de Médecine Vétérinaire des Pays Tropicaux. Mimeographed, 332 pp

Morel PC, Vassiliades G (1963) Les *Rhipicephalus* du groupe *sanguineus*: espèces africaines (Acarines: Ixodoidea). Rev Elev Méd vet Pays Trop 15:343–386

Morel PC, Mouchet J, Rodhain F (1976) Description de *Rhipicephalus camicasi* n. sp. (Acarines: Ixodida) des steppes subdesertiques de la plain afar. Rev Elev Méd Vet Pays Trop 29:337–340

Nava S, Mastropaolo M, Venzal JM, Mangold AJ, Guglielmone AA (2012) Mitochondrial DNA analysis of *Rhipicephalus sanguineus* s. l. (Acari: Ixodidae) in the Southern Cone of South America. Vet Parasitol 190:547–555

Nava S, Estrada-Peña A, Petney T, Beati L, Labruna ML, Szabó MPJ, Venzal JM, Mastropaolo M, Mangold AJ, Guglielmone AA (2014) The taxonomic status of *Rhipicephalus sanguineus* (Latreille, 1806). Vet Parasitol 208:2–8

Pegram RG, Clifford CM, Walker J, Keirans JE (1987a) Clarification of the *Rhipicephalus sanguineus* group (Acari, Ixodoidea, Ixodidae). I. *R. sulcatus* Neumann, 1908 and *R. turanicus* Pomerantsev, 1936. Syst Parasitol 10:3–26

Pegram RG, Keirans JE, Clifford CM, Walker JB (1987b) Clarification of the *Rhipicephalus sanguineus* group (Acari, Ixodoidea, Ixodidae). II. *Rh. sanguineus* (Latreille, 1806) and related species. Syst Parasitol 10:27–44

Ribeiro AL, Faccini JLH, Daemon E (1996) Estudo das variacões morfológicas de *Rhipicephalus sanguineus* (Latreille, 1806) (Acari: Ixodidae) no Brasil. Rev Univ Rural Sér Ciênc Vida 18:25–33

Santos-Silva MM, Beati L, Santos AS, De Sousa R, Núncio MS, Melo P, Bacellar F (2011) The hard-tick fauna of mainland Portugal (Acari: Ixodidae): an update on geographical distribution and known associations with hosts and pathogens. Exp Appl Acarol 55:85–121

Shatas JF (1956) Larvae and nymphs of some species of the genus *Rhipicephalus* Koch (Acarina, Ixodidae). Ent Rev 35:944–955 (in Russian, English NAMRU3 translation T79)

Szabó MPJ, Mangold AJ, Joao CF, Bechara GH, Guglielmone AA (2005) Biological and DNA evidence of two dissimilar populations of the *Rhipicephalus sanguineus* tick group (Acari, Ixodidae) in South America. Vet Parasitol 130:131–140

Walker JB, Keirans JE, Horak IG (2000) The genus *Rhipicephalus* (Acari, Ixodidae): a guide to the brown ticks of the world. Cambridge University Press

Walker AR, Bouattour A, Camicas JL, Estrada-Peña A, Horak IG, Latif AA, Pegram RG, Preston PM (2003) Ticks of domestic animals in Africa: a guide to identification of species. Bioscience reports, Edinburgh, pp 3–210

Rhipicephalus bursa Canestrini and Fanzago, 1878 (Figs. 117–119)

Z. Vatansever

Life Cycle and Host Preferences

Rhipicephalus bursa is a two-host tick with a 1-year life cycle. All developmental stages feed on sheep, goats, cattle, horses, donkeys and rarely wild ungulates, although the preferred hosts are sheep and goats. Occasionally, it can be found on dogs and hares (Pomerantzev 1950; Balashov 1998). Human infestations have also been reported (Bursali et al. 2012; Kar et al. 2013). Under laboratory conditions, the life cycle is completed in 99–254 (mean 142) days (Yeruham et al. 2000). Female ticks engorge in 7–12 days (Göksu 1969), but this can be as short as 3 days (Yeruham et al. 2000). Preoviposition, oviposition, larval pre-hatching and hatching periods take 5.1, 16.5, 31.2 and 7 days on average, respectively (Yeruham et al. 2000). Larval/nymphal feeding on lambs is completed in 14–18 days. Engorged nymphs moult to adults in 13–25 days (Göksu 1969; Yeruham et al. 2000).

Ecology

Rhipicephalus bursa is closely associated with sheep breeding and is distributed mostly in areas with a humid winter and long dry summer. It can be found at altitudes up to 1950 m a.s.l. (Filippova 1997; Göksu 1969; Yeruham et al. 1998). *Rhipicephalus bursa* has a very marked seasonal activity pattern; adults are active in midsummer and immatures in autumn–winter (Göksu 1969; Yeruham et al. 1996). In Israel, adults are active from April to July with a peak in May. Larvae and nymphs infest animals between October and February, with a peak in December (Yeruham et al. 1996). In Turkey, adults are active between May and August, with a peak in June. Larvae and nymphs infest animals from the end of September until January, with a peak in October and November (Göksu 1969). In northern Spain, adults are active between June and September (Estrada-Pena et al. 1990). Under field conditions, adults of *R. bursa* become active when the average day and night temperatures reach 18 and 12 °C, respectively (Yeruham et al. 1998). Below temperatures of 13 °C, engorged females may oviposit but larval development is inhibited (Göksu 1969). Unfed larvae cluster together and stay in an inactive form for 5–6 months during the dry and hot summer. They become active when temperatures drop and humidity increases at the end of September (Göksu 1969; Yeruham et al. 1998). Engorged nymphs then overwinter and moult to adults in the following summer (Filippova 1997; Göksu 1969; Yeruham et al. 1998).

Distribution

The distribution of *R. bursa* covers the whole Mediterranean Region (southern Europe, North Africa and the Middle East), the Ukraine, southern Russia, Georgia, Azerbaijan, Iran, Kazakhstan, Turkmenistan and Uzbekistan (Pomerantzev 1950; Yeruham et al. 1996; Filippova 1997).

Vectorial Capacity and Pathogen Burden

Rhipicephalus bursa is the main biological vector of *Babesia ovis* (ovine babesiosis) and plays an important role in the transmission of *Anaplasma ovis*, *A. marginale* and *A. centrale* (Yeruham et al. 1998; Koenen et al. 2013). Several *Rickettsia* species (*R. aeshlimannii, R. conorii, R. felis, R. massiliae* and *R. sibirica*) (Gargili et al. 2012; Parola et al. 2013) and *Theileria ovis* (Aktas et al. 2006) have been found in *R. bursa*. The AP92 strain of the Crimean–Congo haemorrhagic fever virus (Papa et al. 2014) was isolated from *R. bursa* collected from goats, but the vectorial capacity of this tick still needs to be proven.

Z. Vatansever (✉)
Department of Parasitology, Faculty of Veterinary Medicine, Kafkas University, Kars, Turkey
e-mail: zativet@gmail.com

A. Estrada-Peña et al. (eds.), *Ticks of Europe and North Africa*,
https://doi.org/10.1007/978-3-319-63760-0_57

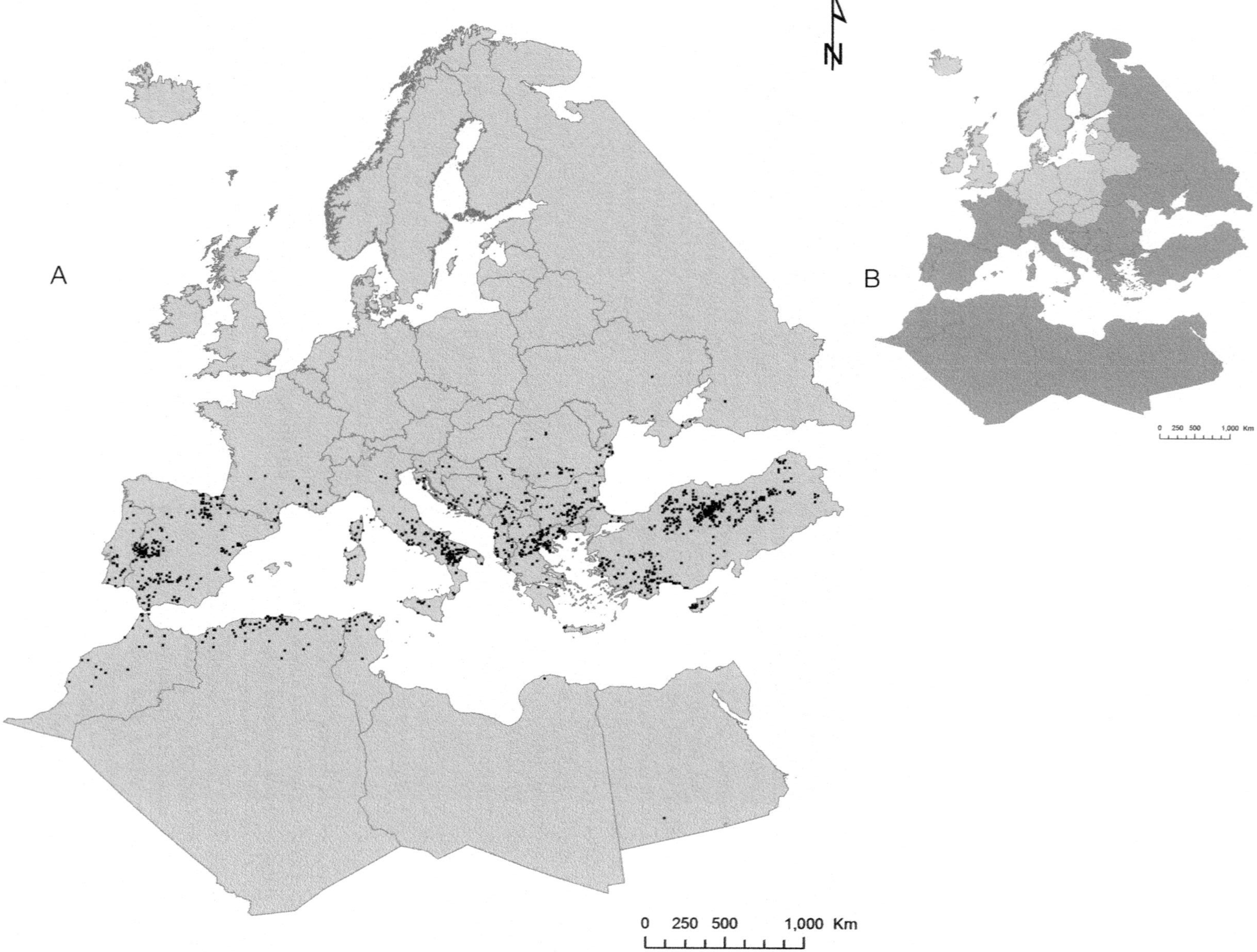

Fig. 117 **A** Distribution of *R. bursa* in Europe and Northern Africa (10 × 10 km grid presence with black dots). **B** Countries where the species has been reported are marked in dark grey

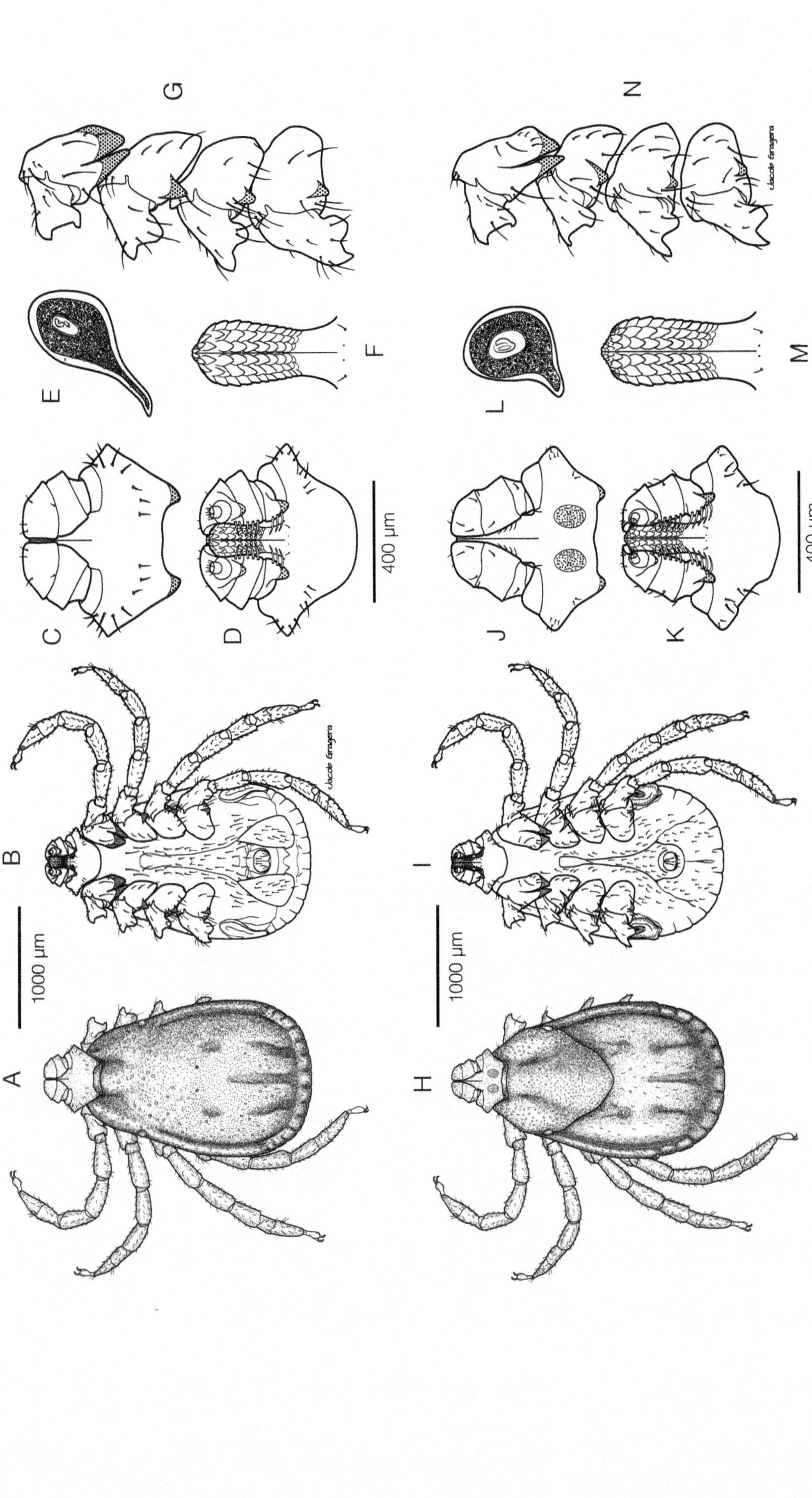

Fig. 118 **A–G** The male of *R. bursa*. **A** Dorsal, **B** ventral, **C** capitulum, dorsal, **D** capitulum, ventral, **E** spiracular plate, **F** hypostome, **G** coxae and trochanters I–IV. **H–N** The female of *R. bursa*. **H** Dorsal, **I** ventral, **J** capitulum, dorsal, **K** capitulum, ventral, **L** spiracular plate, **M** hypostome, **N** coxae and trochanters I–IV. Illustrations from specimens collected in Spain, Portugal, France, Italy and Morocco

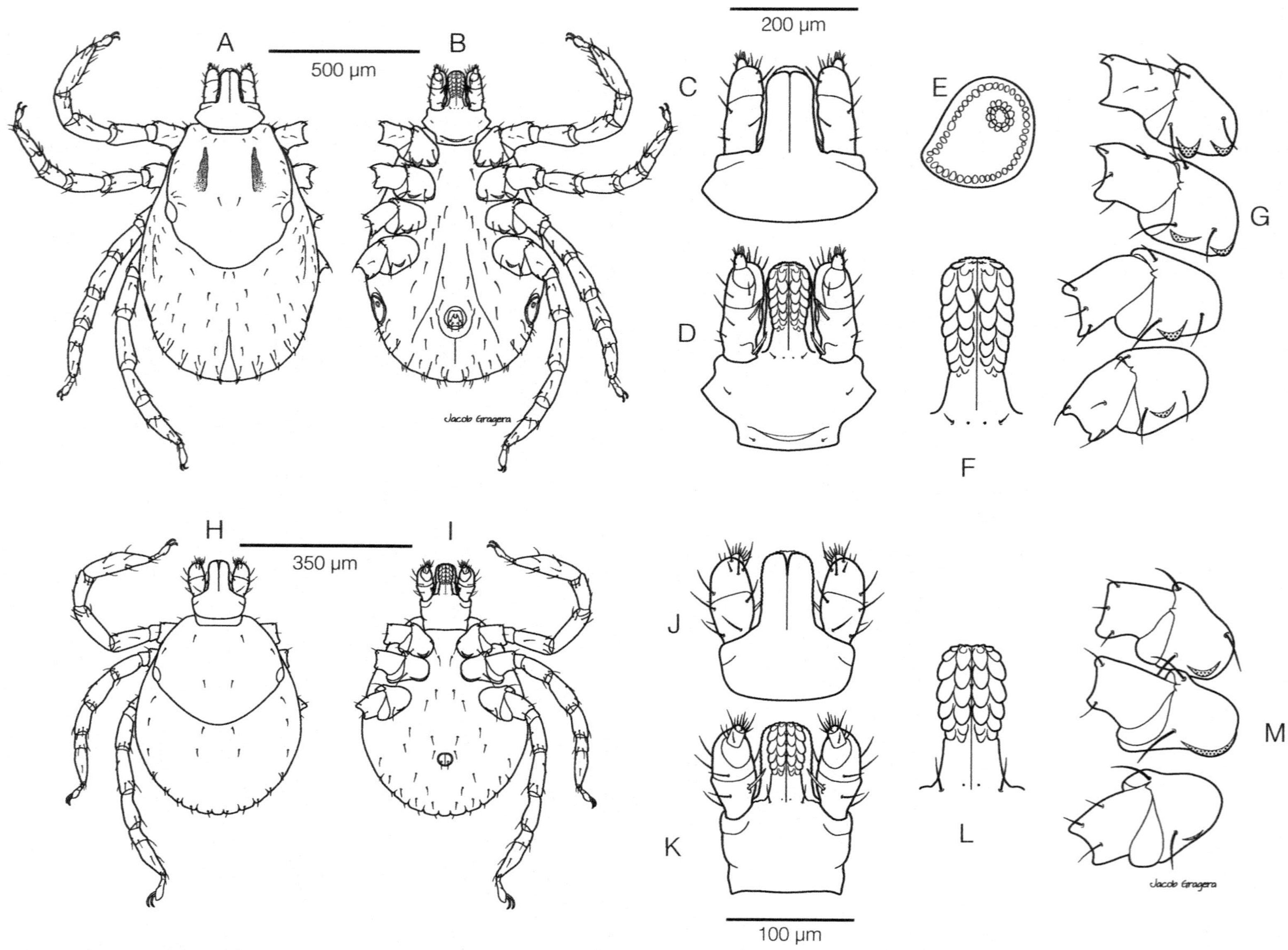

Fig. 119 **A–G** The nymph of *R. bursa*. **A** Dorsal, **B** ventral, **C** capitulum, dorsal, **D** capitulum, ventral, **E** spiracular plate; **F** hypostome, **G** coxae and trochanters I–IV. **H–M** The larva of *R. bursa*. **H** Dorsal, **I** ventral, **J** capitulum, dorsal, **K** capitulum, ventral, **L** hypostome, **M** coxae and trochanters I–III. Illustrations from material moulted from engorged specimens collected in Spain, Portugal and Morocco, or hatching from engorged females collected in Italy and Portugal

References

Aktas M, Altay K, Dumanli N (2006) PCR-based detection of *Theileria ovis* in *Rhipicephalus bursa* adult ticks. Vet Parasitol 140:259–263

Balashov YS (1998) Ixodid ticks - parasites and vectors of diseases. Nauka, Sankt Petersburg, p 287 (in Russian)

Bursali A, Keskin A, Tekin S (2012) A review of the ticks (Acari: Ixodida) of Turkey: species diversity, hosts and geographical distribution. Exp Appl Acarol 57:91–104

Estrada-Pena A, Dehesa V, Sanchez C (1990) The seasonal dynamics of *Haemaphysalis punctata*, *Rhipicephalus bursa* and *Dermacentor marginatus* (Acari: Ixodidae) on sheep of Pais Vasco (Spain). Acarologia 31:17–24

Filippova NA (1997) Ixodid ticks of subfamily Amblyomminae. In: Fauna of Russia and Neighboring Countries: New Series No: 145: Arachnoidea, vol 4(5). St. Petersburg, Nauka, p 483 (in Russian)

Gargili A, Palomar AM, Midilli K, Portillo A, Kar S, Oteo JA (2012) *Rickettsia* species in ticks removed from humans in Istanbul, Turkey. Vector Borne Zoonotic Dis 12:938–941

Göksu K (1969) Bio-ecological studies of *Rhipicephalus bursa* Canestrini and Fanzago, 1877 (sic) (Acarina: Ixodoidea) under field and laboratory conditions. Ankara Üniversitesi Veteriner Fakültesi Dergisi 16:295–312 (in Turkish)

Kar S, Dervis E, Akin A, Ergonul O, Gargili A (2013) Preferences of different tick species for human hosts in Turkey. Exp Appl Acarol 61:349–355

Koenen F, Pascucci I, Jaenson TGT, Madder M, Sousa Rd, Estrada-Peña A, Farkas R, Salman M (2013) Tick-borne infections (including zoonoses) in Europe and the Mediterranean basin. CABI, Wallingford, pp 33–75

Papa A, Chaligiannis I, Kontana N, Sourba T, Tsioka K, Tsatsaris A, Sotiraki S (2014) A novel AP92-like Crimean-Congo hemorrhagic fever virus strain, Greece. Ticks Tick Borne Dis 5:590–593

Parola P, Paddock CD, Socolovschi C, Labruna MB, Mediannikov O, Kernif T, Abdad MY, Stenos J, Bitam I, Fournier PE, Raoult D (2013) Update on tick-borne rickettsioses around the world: a geographic approach. Clin Microbiol Rev 26:657–702

Pomerantzev BI (1950) Ixodid ticks (Ixodidae). In: Fauna of the USSR, New Series 41: Arachnoidea, vol 4(2). Leningrad, Nauk, p 224 (in Russian)

Yeruham I, Hadani A, Galker F (1998) Some epizootiological and clinical aspects of ovine babesiosis caused by *Babesia ovis*–a review. Vet Parasitol 74:153–163

Yeruham I, Hadani A, Galker F (2000) The life cycle of *Rhipicephalus bursa* Canestrini and Fanzago, 1877 (Acarina: Ixodidae) under laboratory conditions. Vet Parasitol 89:109–116

Yeruham I, Hadani A, Galker F, Rosen S (1996) The seasonal occurrence of ticks (Acari: Ixodidae) on sheep and in the field in the Judean area of Israel. Exp Appl Acarol 20:47–56

Rhipicephalus pusillus Gil Collado, 1936 (Figs. 120–122)

M. M. Santos-Silva

Life-Cycle and Host Preferences

Rhipicephalus pusillus is a three host tick with a presumed natural life-cycle of one year (Walker et al. 2000). Under laboratory conditions, the life cycle is completed in less than 165 days (Santos-Silva and Filipe 1998). All stages primary feed on lagomorphs, particularly *Oryctolagus cuniculus*. Other wild and domestic hosts from several mammalian orders, including carnivores, ungulates, insectivores and rodents are recognized hosts for this species (Walker et al. 2000; Santos-Silva et al. 2011). Aves are considered exceptionally hosts (Guglielmone et al. 2014). It has been reported from humans (Santos-Silva et al. 2011).

Ecology

Rhipicephalus pusillus is found in the Palaearctic region, mainly in Mediterranean climatic areas. The typical habitats include Iberian conifer forests, south-west Iberian Mediterranean sclerophyllous and mixed forests (Guglielmone et al. 2014), but it also occurs in other habitats inhabited by the European rabbit (Dias 1994). Larvae and nymphs are active during summer from July to September. Adults are active all year, peaking between February and June (Gilot 1985; Dias 1994). *Rhipicephalus pusillus* usually shows an endophilic and monotropic behaviour (Dias 1994; Sobrinho et al. 2012). On some occasions it can also act exophilically and be found outside of rabbit burrows or parasitizing other hosts (Gilot et al. 1985; Ruiz-Fons et al. 2006; Santos-Silva et al. 2011). It can establish stable populations independently of rabbits (Sobrinho et al. 2012). The host-seeking behaviour of this tick includes two different strategies: one endophilic, nidicolous in which the tick waits for their host, and the other as an active seeker.

Distribution

Rhipicephalus pusillus has been reported from southern France, Italy, Morocco, Portugal, Spain, and Tunisia (Morel and Vassiliades 1962; Walker et al. 2000).

Vectorial Capacity and Pathogen Burden

This species is known to carry and may transmit *Rickettsia sibirica mongolitimonae* to humans (Sousa et al. 2006). *Rickettsia massiliae* has also been associated with this tick (Márquez 2008). The vectorial ability of *R. pusillus* is poorly known and further studies are required.

M. M. Santos-Silva (✉)
Centro de Estudos de Vectores e Doenças Infecciosas Dr. Francisco Cambournac (CEVDI), Instituto Nacional de Saúde Doutor Ricardo Jorge, Águas de Moura, Portugal
e-mail: m.santos.silva@insa.min-saude.pt

A. Estrada-Peña et al. (eds.), *Ticks of Europe and North Africa*,
https://doi.org/10.1007/978-3-319-63760-0_58

Fig. 120 **A** Distribution of *R. pusillus* in Europe and Northern Africa (10 × 10 km grid presence with black dots). **B** Countries where the species has been reported are marked in dark grey

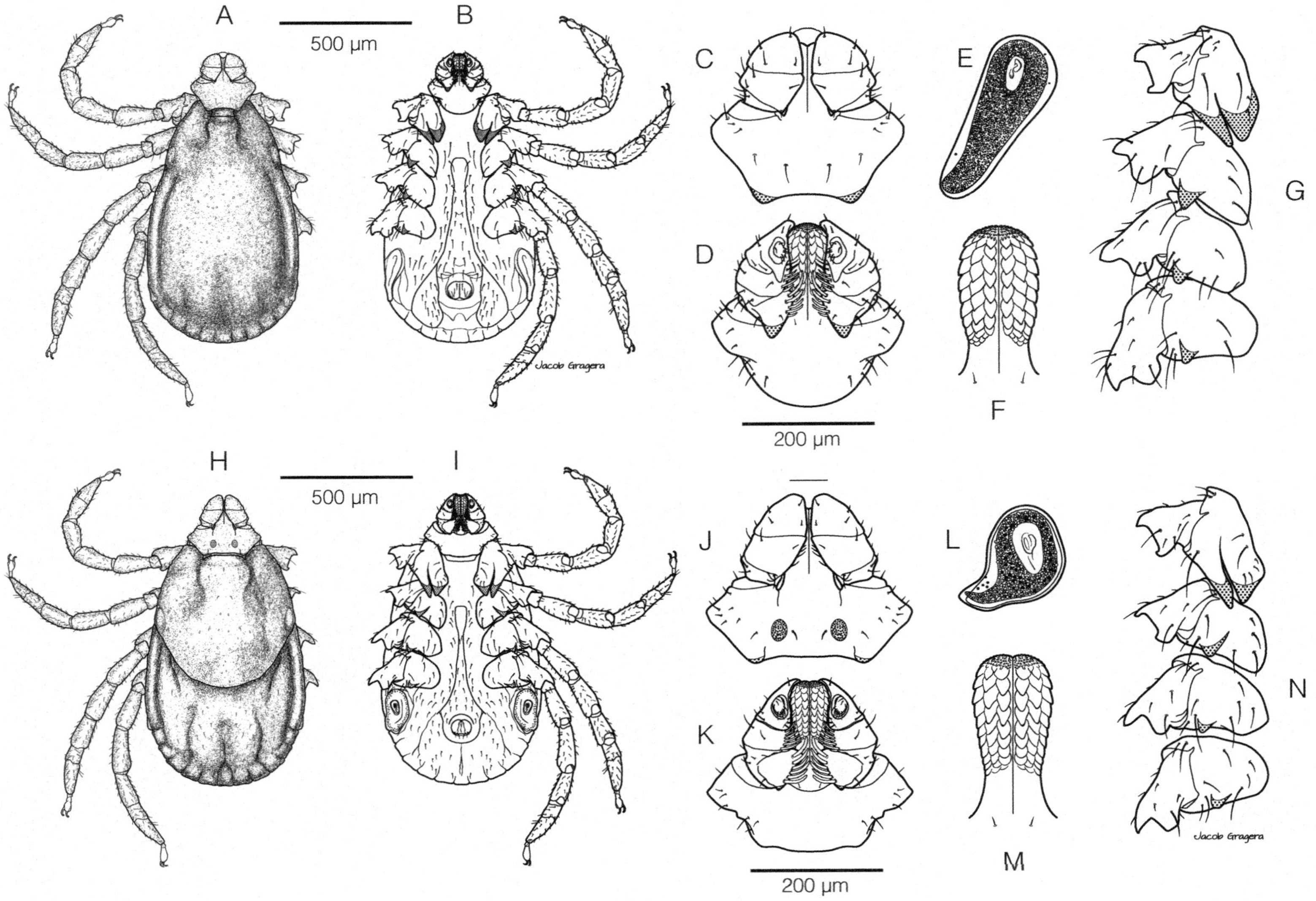

Fig. 121 **A–G** The male of *R. pusillus*. **A** Dorsal, **B** ventral, **C** capitulum, dorsal, **D** capitulum, ventral, **E** spiracular plate; **F** hypostome, **G** coxae and trochanters I–IV. **H–N** The female of *R. pusillus*. **H** Dorsal, **I** ventral, **J** capitulum, dorsal, **K** capitulum, ventral, **L** spiracular plate; **M** hypostome, **N** coxae and trochanters I–IV. Illustrations from specimens collected in Spain

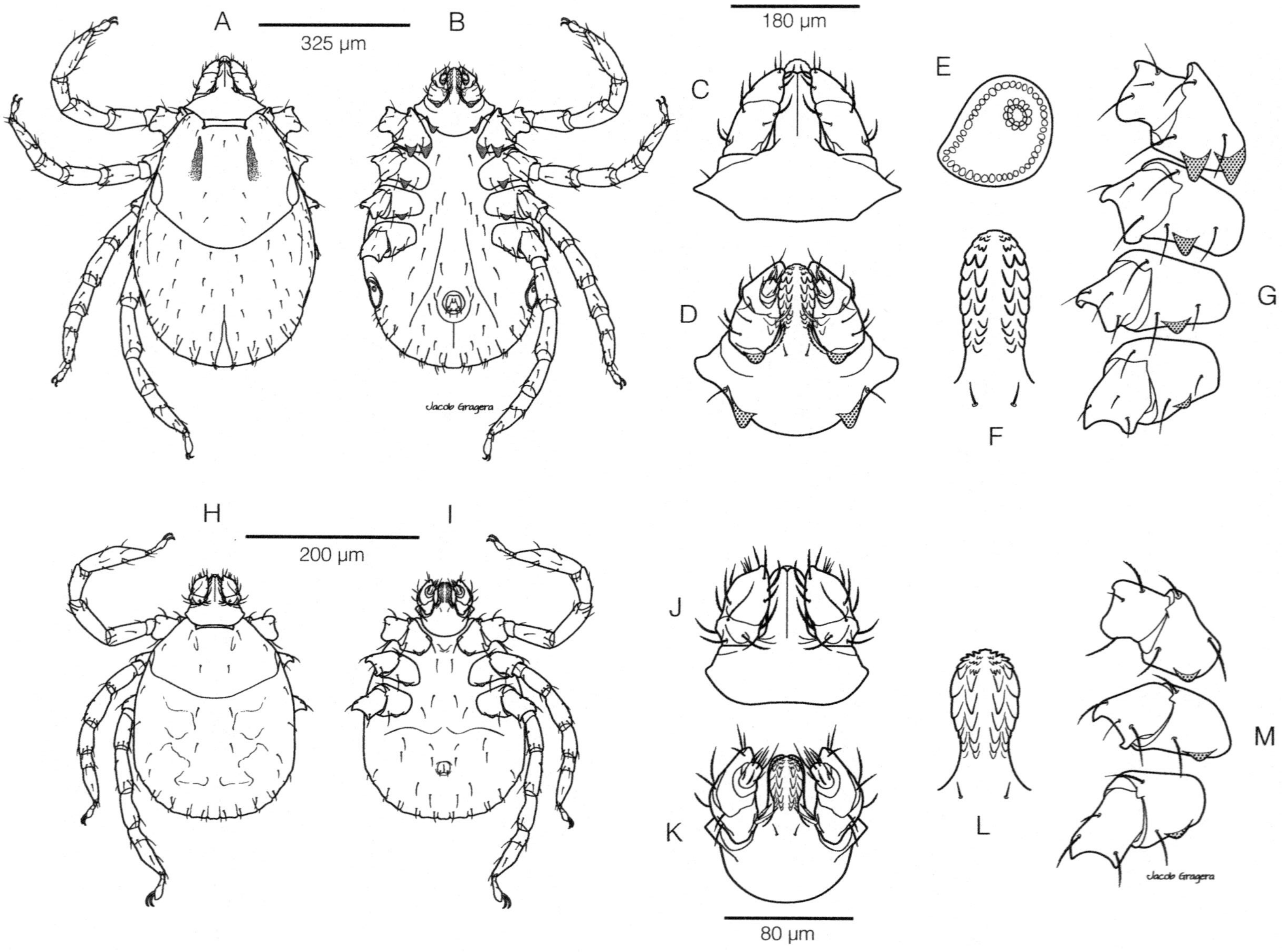

Fig. 122 **A–F** The nymph of *R. pusillus*. **A** Dorsal, **B** ventral, **C** capitulum, dorsal, **D** capitulum, ventral, **E** spiracular plate, **F** hypostome, **G** coxae and trochanters I–IV. **H–M** The larva of *R. pusillus*. **H** Dorsal, **I** ventral, **J** capitulum, dorsal, **K** capitulum, ventral, **L** hypostome, **M** coxae and trochanters I–III. Illustrations from specimens collected in Spain

References

Dias JATS (1994) As carraças (Acarina-Ixodoidea) da Península Ibérica. Algumas considerações sobre a sua biogeografia e relacionamento com a ixodofauna Afropaleárctica e Afrotropical. Instituto de Investigação Cientifica Tropical, Estudos, Ensaios e Documentos 158, Lisboa

Gilot B, Rogers P, Lachet B (1985) Données biologiques et écologiques sur les tiques de lagomorphes (et plus spécialement celles du lapin de garenne, *Oryctolagus cuniculus* L.) dans les alpes françaises et leur avant-pays. Acarologia 26(84):335–354

Guglielmone AA, Robbins RG, Apanaskevich DA, Petney TN, Estrada-Peña A, Horak IG (2014) The hard ticks of the world (Acari: Ixodida: Ixodidae). Springer, Dordrecht, p 738

Márquez FJ (2008) Spotted fever group *Rickettsia* in ticks from southeastern Spain natural parks. Exp Appl Acarol 2008(45):185–194. https://dx.doi.org/10.1007/s10493-008-9181-7

Morel PC, Vassiliades G (1962) Les *Rhipicephalus* du group *sanguineus*: Espèces africaines. Rev d'Elev Med Veterin Pays Trop 15:343–386

Ruiz-Fons F, Fernández-de-Mera IG, Acevedo P, Höfle U, Vicente J, de la Fuente J, Gortazár C (2006) Ixodid ticks parasitizing Iberian red deer (*Cervus elaphus hispanicus*) and European wild boar (*Sus scrofa*) from Spain: geographical and temporal distribution. Vet Parasitol 140:133–142

Santos-Silva MM, Filipe AR (1998) Ciclos biológicos de ixodídeos (Ixodoidea: Ixodidae) em condições de laboratório. Rev Port Cienc Vet 527:143–148

Santos-Silva MM, Beati L, Santos AS, De Sousa R, Núncio MS, Melo P, Santos-Reis M, Fonseca C, Formosinho P, Vilela C, Bacellar F (2011) The hard-tick fauna of mainland Portugal (Acari: Ixodidae): an update on geographical distribution and known associations with hosts and pathogens. Exp Appl Acarol 55:85–121. https://dx.doi.org/10.1007/s10493-011-9440-x

Sobrino R, Millán J, Oleaga A, Gortázar C, de la Fuente J, Ruiz-Fons F (2012) Ecological preferences of exophilic and endophilic ticks (Acari: Ixodidae) parasitizing wild carnivores in the Iberian Peninsula. Vet Parasitol 184:248–257

Sousa R, Barata C, Vitorino L, Santos-Silva M, Carrapato C, Torgal J, Walker D, Bacellar F (2006) *Rickettsia sibirica* isolation from patient and detection in ticks, Portugal. Emerg Infect Dis 12:1103–1108

Walker JB, Keirans JE, Horak IG (2000) The Genus *Rhipicephalus* (Acari, Ixodidae): a guide to the brown ticks of the world. Cambridge University Press, Cambridge

Rhipicephalus rossicus Yakimov and Kol-Yakimova, 1911 (Figs. 123–125)

A. D. Mihalca, M. O. Dumitrache, and G. D'Amico

Life Cycle and Host Preferences

This is a three-host tick with a natural life cycle of 2–3 years (Emchuk 1967; Walker et al. 2000). Under laboratory conditions, the life cycle is completed in 77–108 days (Kolomietz 1936). All stages feed on a large variety of hosts. At least nine families of birds and 17 families of mammals have been reported as hosts (Mihalca et al. 2015). Amphibians and reptiles are exceptionally rare hosts (Filippova 1997; Akimov and Nebogatkin 2013). There seems to be no differences in host preference between adults and the immature stages (Mihalca et al. 2015). This tick has been reported on humans several times (Mihalca et al. 2015).

Ecology

Rhipicephalus rossicus is one of the dominant tick species found in the Eurasian steppe. The typical habitats include mostly river basin valleys but also dry forests and xeric shrublands from steppic regions (Guglielmone et al. 2014; Mihalca et al. 2015). Habitat preference is attributed mainly to host abundance and availability rather than to abiotic factors (Shatas and Bystrova 1954). The tick has been found at various altitudes, ranging from 0 to 1500 m a.s.l. (Mihalca et al. 2015). There is an evident seasonal dynamics throughout all the distribution range. Larvae and nymphs are active from as early as March to as late as November. Adults are active all year. All stages have the peak activity between April and July (Pomerantzev et al. 1940; Emchuk 1960; Dumitrache et al. 2014; Mihalca et al. 2015). *R. rossicus* is a non-nidicolous tick. The host-seeking behaviour is unknown, but it is probably similar to other *Rhipicephalus* ticks which actively move and 'hunt' their host.

Distribution

It has been reported in Armenia, Azerbaijan, Bulgaria, China, Dagestan, Egypt, Georgia, Iran, Israel, Kazakhstan, Tajikistan, Moldavia, Poland, Romania, Russia, Turkey, the Ukraine and Uzbekistan. The limits of its distribution are between 31.3–51.3°N and 23.8–88.1°E (Mihalca et al. 2015).

Vectorial Ability and Pathogen Burden

Experimental proof for the vectorial capacity of *R. rossicus* is available for *Francisella tularensis*, Crimean-Congo hemorrhagic fever virus and West Nile virus (for a complete review, see Mihalca et al. 2015). Other pathogens have been detected in *R. rossicus* by various laboratory methods, but there is no experimental evidence for its vectorial ability. These include: *Theileria equi*, *Babesia bigemina* and *Coxiella burnetii* (see Mihalca et al. 2015).

A. D. Mihalca · M. O. Dumitrache · G. D'Amico (✉)
Department of Parasitology and Parasitic Diseases, University of Agricultural Sciences and Veterinary Medicine Cluj-Napoca, Cluj-Napoca, Romania
e-mail: gianluca.damico@usamvcluj.ro

A. D. Mihalca
e-mail: amihalca@usamvcluj.ro

M. O. Dumitrache
e-mail: mirabela.dumitrache@usamvcluj.ro

A. Estrada-Peña et al. (eds.), *Ticks of Europe and North Africa*,
https://doi.org/10.1007/978-3-319-63760-0_59

Fig. 123 **A** Distribution of *R. rossicus* in Europe and Northern Africa (10 × 10 km grid presence with black dots). **B** Countries where the species has been reported are marked in dark grey

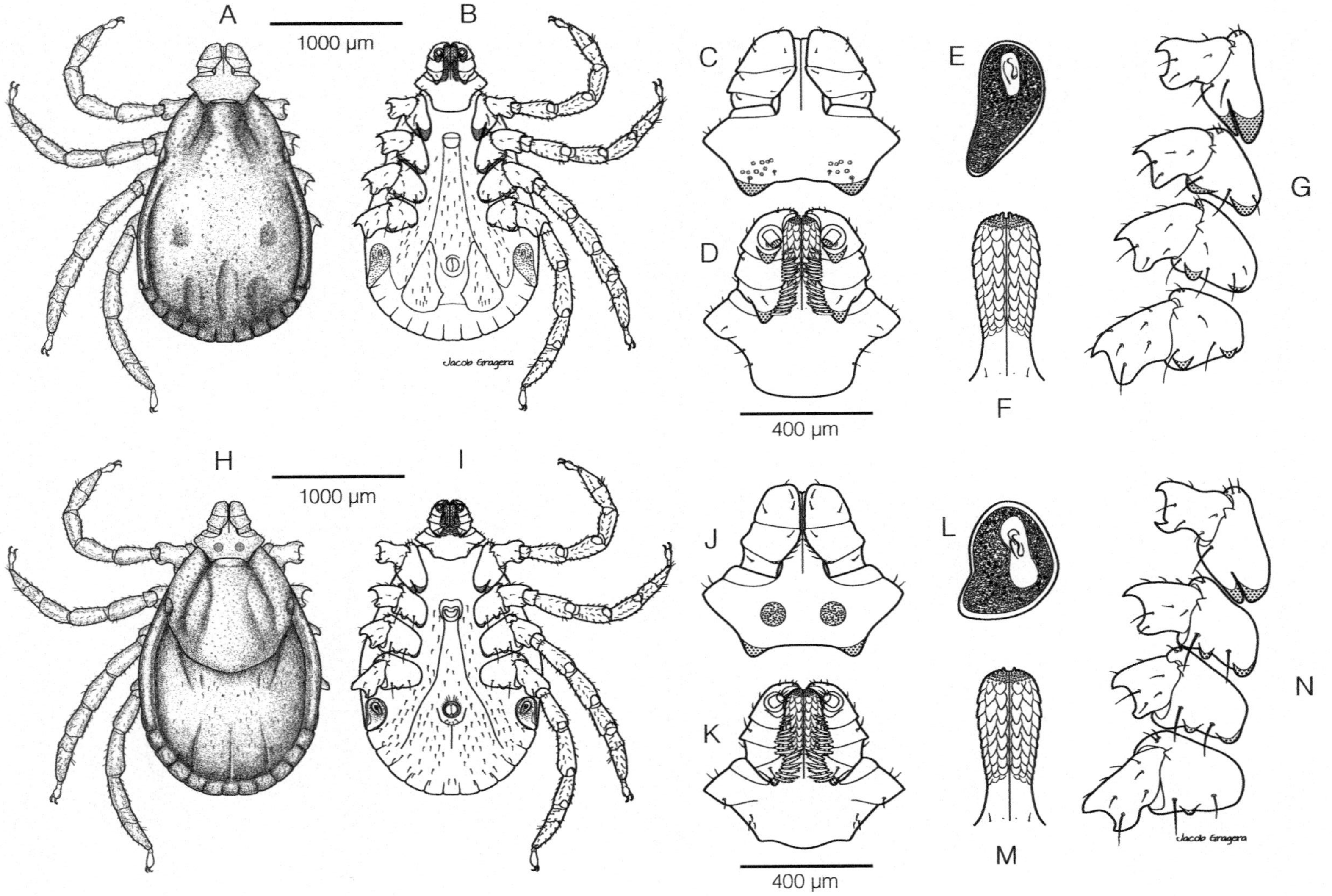

Fig. 124 **A–G** The male of *R. rossicus*. **A** Dorsal, **B** ventral, **C** capitulum, dorsal, **D** capitulum, ventral, **E** spiracular plate; **F** hypostome, **G** coxae and trochanters I–IV. **H–N**—The female of *R. rossicus*. **H** Dorsal, **I** ventral, **J** capitulum, dorsal, **K** capitulum, ventral, **L** spiracular plate; **M** hypostome, **N** coxae and trochanters I–IV. Illustrations from specimens collected in Romania

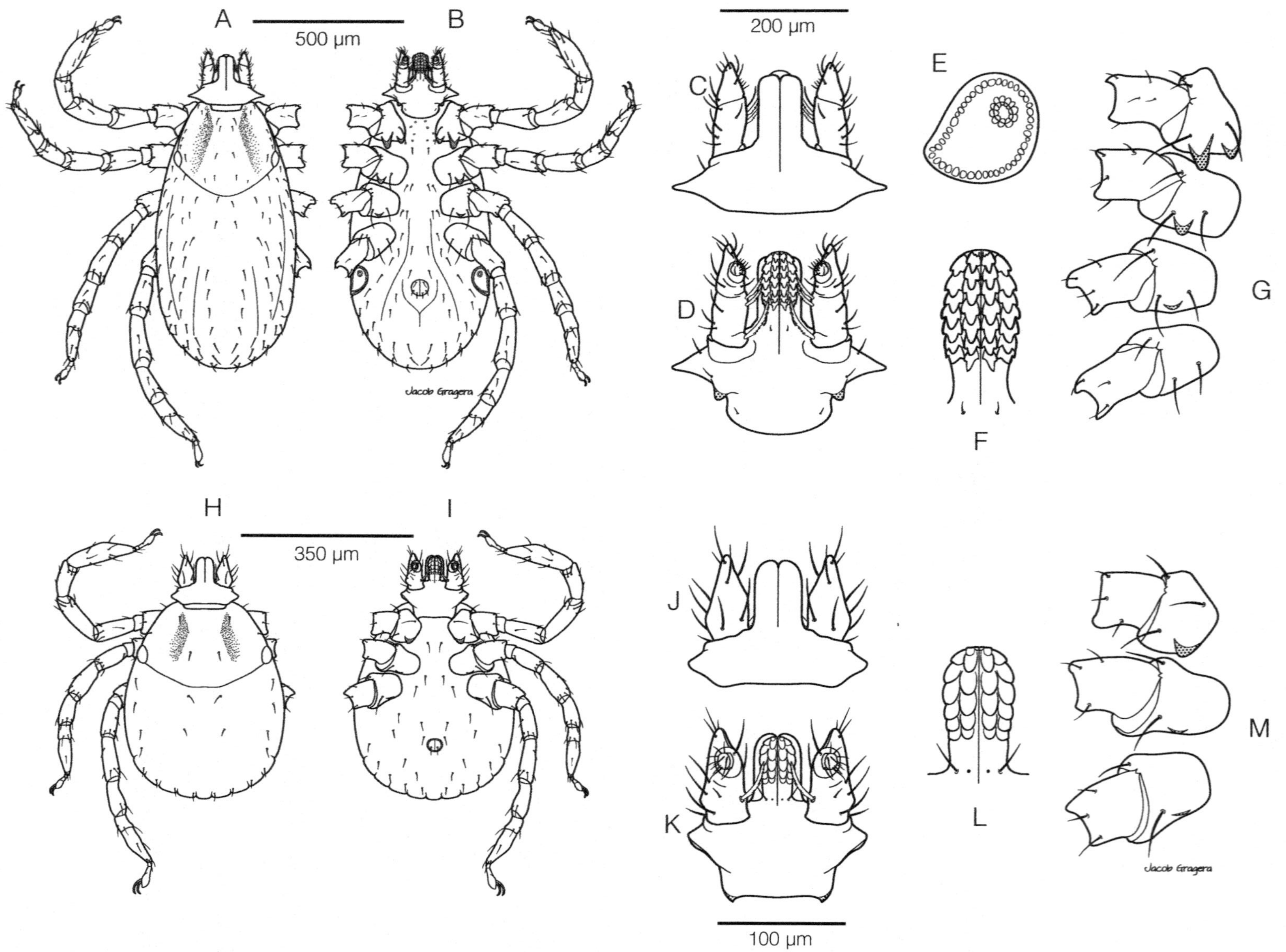

Fig. 125 **A–F** The nymph of *R. rossicus*. **A** Dorsal, **B** ventral, **C** capitulum, dorsal, **D** capitulum, ventral, **E** spiracular plate, **F** hypostome, **G** coxae and trochanters I–IV. H–M—The larva of *R. rossicus*. **H** Dorsal, **I** ventral, **J** capitulum, dorsal, **K** capitulum, ventral, **L** hypostome, **M** coxae and trochanters I–III. Illustrations from specimens collected in Romania

References

Akimov IA, Nebogatkin IV (2013) Ticks of the genus *Rhipicephalus* (Acari, Ixodidae) and their distribution in Ukraine. Vestnik Zoologii 47(3):e28–e34

Dumitrache MO, Kiss B, Dantas-Torres F, Latrofa MS, D'Amico G, Sándor AD, Mihalca AD (2014) Seasonal dynamics of *Rhipicephalus rossicus* attacking domestic dogs from the steppic region of southeastern Romania. Parasit Vectors 7:97

Emchuk EM (1960) [Fauna of Ukraine, Volume 25/1]. Akademii Nauk Ukrainskoi SSR, Kiev (in Ukrainian)

Emchuk EM (1967) Landscape distribution of Ixodid ticks and natural focal diseases in Ukraine. Tezisy Dokl Nauch Konf Ukrain Respub Nauch Obshch Parasit, Izd "Naukova Dumka", Kiev, pp 51–54

Filippova NA (1997) Ixodid ticks of subfamily Amblyomminae. In the fauna of Russia and neighboring countries. New Series no. 145: Arachnoidea, vol 4(5), St. Petersburg, Nauka (in Russian)

Guglielmone AA, Robbins RG, Apanaskevich DA, Petney TN, Estrada-Peña A, Horak IG (2014) The hard ticks of the world (Acari: Ixodida: Ixodidae). Springer, Dordrecht, p 738

Kolomietz JS (1936) Zur Frage der Verbreitung und Lebensweise der Zecke *Rhipicephalus rossicus* Yak. Und Kol-Yak. In der Ukraine. Z Parasitenk 8(5):538–541

Mihalca AD, Kalmár Z, Dumitrache MO (2015) *Rhipicephalus rossicus*, a neglected tick at the margin of Europe: a review of its distribution, ecology and medical importance. Med Vet Entomol 29:215–224

Pomerantzev BI, Matkashvili NV, Lotoskii BV (1940) An ecological and faustic outline of Ixodidae ticks occurring in Transcaucasia. Parasitol Sborn Zool Inst Acad Nauk SSSR (Leningrad) 7:100–133

Shatas YF, Bystrova NA (1954) Role of ixodid ticks in maintenance of natural foci of tularemia. Zh Mikrobiol Epidemiol Immunobiol 6:55–61

Walker JB, Keirans JE, Horak IG (2000) The genus *Rhipicephalus* (Acari, Ixodidae): a guide to the brown ticks of the world. Cambridge University Press, Cambridge

Rhipicephalus camicasi Morel, Mouchet and Rodhain, 1976 (Fig. 126)

A. Estrada-Peña

Life Cycle and Host Preferences

Rhipicephalus camicasi is a recently described species about which relatively little is known. It is a three-host species (Pegram 1984). To date, domestic animals, particularly cattle, sheep, goats and camels, are the most commonly reported hosts (Pegram et al. 1987). In Yemen, it is the most common rhipicephalid on domesticated animals. This species has also been collected from zebras and Cape hares (Morel 1980). The hosts of the immature stages are unknown, but most probably are 'mouse-like' rodents (Morel 1980). In both Ethiopia and Somalia, *R. camicasi* appears to be most active during the dry season (Walker et al. 2000).

Ecology

Rhipicephalus camicasi is found in arid and semi-arid habitats generally receiving less than 250 l of rainfall per square metre annually. It appears to be common in deciduous bushland and Acacia thickets, semi-desert grassland and shrub-land. (Pegram et al. 1987).

Distribution

Rhipicephalus camicasi was originally collected in the steppe and desert climatic regions of north-eastern Africa, concentrated in Sudan, Ethiopia, Somalia and Kenya, as well as outlying populations in Yemen, Saudi Arabia, Jordan and Lebanon. It has been reported from Egypt (Morel at el. 1976; Walker et al. 2003).

Vectorial Capacity and Pathogen Burden

Matei et al. (2016) suggested the possible role of this tick in the transmission of *Anaplasma platys* in dogs from Lamu Island, Kenya.

A. Estrada-Peña (✉)
Department of Animal Health, Faculty of Veterinary Medicine, Zaragoza, Spain
e-mail: aestrada@unizar.es

A. Estrada-Peña et al. (eds.), *Ticks of Europe and North Africa*,
https://doi.org/10.1007/978-3-319-63760-0_60

Fig. 126 **A** Distribution of *R. camicasi* in Europe and Northern Africa (10 × 10 km grid presence with black dots). **B** Countries where the species has been reported are marked in dark grey

References

Matei IA, D'Amico G, Yao PK, Ionică AM, Kanyari PW, Daskalaki AA, Dumitrache MO, Sándor AD, Gherman CM, Qablan M5, Modrý, Mihalca AD (2016) Molecular detection of *Anaplasma platys* infection in free-roaming dogs and ticks from Kenya and Ivory Coast. Parasit Vectors 9:157

Morel PC (1980) Study on Ethiopian ticks (Acarida, Ixodida). Institut d'Élevage et de Médecine Vétérinaire des Pays Tropicaux, Maisons-Alfort, Paris

Morel PC, Mouchet J, Rodhain F (1976) Description de *Rhipicephalus camicasi* n.sp. (Acariens, Ixodida) des steppes subdésertiques de la plaine afar. Rev Elevage Med Vet Payx Trop 29:337–340

Pegram RG (1984) Biosystematic studies on the genus *Rhipicephalus*: the *R. sanguineus* and the *R. simus* groups (Ixodoidea, Ixodidae), 160 pp + 90 plates. Ph.D. Thesis, Brunel University, London

Pegram RG, Keirans JE, Clifford CM, Walker JB (1987) Clarification of the *Rhipicephalus sanguineus* group (Acari, Ixodoidea, Ixodidae). II. *R. sanguineus* (Latreille, 1806) and related species. Syst Parasitol 10:27–44

Walker JB, Keirans JE, Horak IG (2000) The genus *Rhipicephalus* (Acari, Ixodidae), a guide to the brown ticks of the world. Cambridge University Press, Cambridge, UK

Walker A, Bouattour A, Camicas JL, Estrada-Peña A, Hirak IG, Latif AA, Pegram RG, Preston PM (2003) Ticks of domestic animals in Africa. A guide to identification of species. Bioscience Reports, Edinburgh, UK

Rhipicephalus guilhoni Morel and Vassiliades, 1963

T. N. Petney, G. D'Amico, and M. P. Pfäffle

Life Cycle and Host Preferences

Rhipicephalus guilhoni is a three-host tick; the life cycle of which takes a year to complete (Walker et al. 2003). It has been recorded from a variety of hosts including domestic animals, mostly cattle and sheep, but also dogs and cats, and wild animals including carnivores, insectivores, artiodactyls, perissodactyls, lagomorphs, rodents and non-passerine birds (Jongejan et al. 1987; Walker et al. 2000, 2003). This species has also been recorded from humans (Camicas et al. 1990).

Ecology

The immature stages of *R. guilhoni* are nidicolous, while adults are exophilic (Gray et al. 2013). Limited data from northern Nigeria suggest that the adults start their activity during the rainy season (Walker et al. 2000). In Senegal, a sharp peak of adults and nymphs also occurs in September–October at the end of the rainy season (Camicas et al. 1990). A similar pattern is found in Sudan (Mohammed and Hassan 2007). Overall, the adults are more abundant from May to June in the rainy season until December and January in the cooler season (Walker et al. 2003). In general, it is found in areas with between 250 and 500 mm rainfall/year reaching 750 mm in the Sudanian savanna, grass scrub zone, swamp vegetation, as well as Sahel *Acacia* savanna, deciduous bushland and desert and xeric shrublands (Jongejan et al. 1987; Walker et al. 2000; Guglielmone et al. 2014). Populations of *R. guilhoni* appear to be sensitive to drought, deforestation and cultivation (Walker et al. 2000).

Distribution

Rhipicephalus guilhoni occurs in a band across Africa bordering the southern Sahara (Senegal and Mauritania in the west to Ethiopia in the east, occurring as far southern as South Sudan (Jongejan et al. 1987; Walker et al. 2000, 2003). It has also been found in Tenerife, Canary Islands, on hedgehogs (*Atelerix* spp., Petney et al. unpublished data).

Vectorial Capacity and Pathogen Burden

Very few data are available on the pathogens associated with this species. In Senegal, *R. guilhoni* has been found naturally infected with Crimean-Congo haemorrhagic fever virus (Camicas et al. 1990; Wilson et al. 1990), *Rickettsia massiliae* (Mediannikov et al. 2010a) and *Coxiella burnetii* (Mediannikov et al. 2010b). Wad Medani virus (Main et al. 1980) and Ngoye virus (Grard et al. 2006) have also been recorded from this species.

T. N. Petney (✉)
State Museum of Natural History, Erbprinzenstr. 13, 76133 Karlsruhe, Germany
e-mail: trevor.petney@kit.edu

G. D'Amico
Department of Parasitology and Parasitic Diseases, University of Agricultural Sciences and Veterinary Medicine Cluj-Napoca, Cluj-Napoca, Romania
e-mail: gianluca.damico@usamvcluj.ro

M. P. Pfäffle
Department of Ecology and Parasitology, Institute of Zoology, Karlsruhe Institute of Technology, Karlsruhe, Germany
e-mail: miripfaeffle@web.de

A. Estrada-Peña et al. (eds.), *Ticks of Europe and North Africa*,
https://doi.org/10.1007/978-3-319-63760-0_61

References

Camicas J-L, Wilson ML, Cornet J-P et al (1990) Ecology of ticks as potential vectors of Crimean-Congo hemorrhagic fever virus in Senegal: epidemiological implications. Arch Virol Suppl 1:303–322. https://dx.doi.org/10.1007/978-3-7091-9091-3

Grard G, Lemasson J-J, Sylla M et al. (2006) Ngoye virus: a novel evolutionary lineage within the genus *Flavivirus*. J Gen Virol 87:3273–3277. https://dx.doi.org/10.1099/vir.0.82071-0

Gray J, Dantas-Torres F, Estrada-Peña A, Levin M (2013) Systematics and ecology of the brown dog tick, *Rhipicephalus sanguineus*. Ticks Tick Borne Dis 4:171–180. https://dx.doi.org/10.1016/j.ttbdis.2012.12.003

Guglielmone AA, Robbins RG, Apanaskevich DA, Petney TN, Estrada-Peña A, Horak IG (2014) The hard ticks of the world (Acari: Ixodida: Ixodidae). Springer, Dordrecht, p 738

Jongejan F, Zivkovic D, Pegram RG et al. (1987) Ticks (Acari: Ixodidae) of the Blue and White Nile ecosystems in the Sudan with particular reference to the *Rhipicephalus sanguineus* group. Exp Appl Acarol 3:331–346. https://dx.doi.org/10.1007/BF01193169

Main AJ, Kloter KO, Camicas J-L et al. (1980) Wad medani and soldado viruses from ticks (Ixodoidea) in West Africa. J Med Entomol 17:380–382. https://dx.doi.org/10.1093/jmedent/17.4.380

Mediannikov O, Diatta G, Fenollar F et al. (2010a) Tick-borne rickettsioses, neglected emerging diseases in rural Senegal. PLoS Negl Trop Dis 4:e821. https://dx.doi.org/10.1371/journal.pntd.0000821

Mediannikov O, Fenollar F, Socolovschi C et al. (2010b) *Coxiella burnetii* in humans and ticks in rural Senegal. PLoS Negl Trop Dis 4:e654. https://dx.doi.org/10.1371/journal.pntd.0000654

Mohammed MS, Hassan SM (2007) Distribution and population dynamics of ticks (Acari: Ixodidae) infesting sheep in Sennar State, Sudan. Onderstepoort J Vet Res 74:301–306

Walker AR, Bouattour A, Camicas J-L et al. (2003) Ticks of domestic animal in Africa: a guide to identification of species. Bioscience Reports., Edinburgh

Walker JB, Keirans JE, Horak IG (2000) The genus *Rhipicephalus* (Acari, Ixodidae): a guide to the brown ticks of the world. Cambridge University Press, Cambridge

Wilson ML, Gonzalez J-P, LeGuenno B et al (1990) Epidemiology of Crimean-Congo hemorrhagic fever in Senegal: temporal and spatial patterns. Arch Virol Suppl 1:323–340. https://dx.doi.org/10.1007/978-3-7091-9091-3_35

Rhipicephalus sanguineus s.l. (Latreille, 1806) (Figs. 127–129)

F. Dantas-Torres and D. Otranto

The taxonomy of the brown dog tick, *R. sanguineus* sensu lato (s.l.), is currently under discussion as there is no type and no proper description for this species (for recent detailed reviews, see Dantas-Torres and Otranto 2015; Nava et al. 2015). Genetic and cross-breeding studies revealed that ticks morphologically identified as '*R. sanguineus*' belong to at least two well-defined lineages, designated as temperate and tropical lineages (Burlini et al. 2010; Moraes-Filho et al. 2011; Nava et al. 2012), with additional operative taxonomic units being recently molecularly and morphologically identified (Dantas-Torres et al. 2013). As the type locality was defined as 'habitat in Gallia' (Latreille 1806), the temperate lineage probably represents the actual *R. sanguineus* s.l. However, until a proper redescription of the species with the definition of the neotype is available, all ticks morphologically compatible with available descriptions of '*R. sanguineus*' (e.g. Walker et al. 2000) should be referred to as *R. sanguineus* s.l (Guglielmone et al. 2014). In this chapter, we will provide general information available for *R. sanguineus* s.l.

Life Cycle and Host Preferences

Rhipicephalus sanguineus s.l. is a three-host tick. Under laboratory conditions, the life cycle may be completed in 2–4 months, depending on factors such as temperature and host availability (Dantas-Torres et al. 2011). Unfed adult ticks can survive for more than 1 year without taking a blood meal (Dantas-Torres et al. 2012). All developmental stages feed primarily on dogs, but eventually on other hosts, including rodents, birds and humans (for a review, see Dantas-Torres 2010).

Ecology

Rhipicephalus sanguineus s.l. is the most common tick found on dogs in urban areas around the world. This tick may also be highly prevalent in rural areas, but even in such areas, it is typically associated with dog shelters and human houses. Indeed, *R. sanguineus* s.l. is essentially a nidicolous tick found in close association with dogs. This tick is typically found in shelters and kennels where confined dogs may be highly infested. In houses with tick-infested dogs, *R. sanguineus* s.l. are usually found in the places where the dogs stay during the night, which may be a doghouse in the backyard or even inside the house. Engorged females are often found on the wall where they usually hide themselves in cracks and crevices, i.e. places where they are more protected from unsuitable weather conditions (Dantas-Torres 2010). Nonetheless, *R. sanguineus* s.l. ticks are well adapted to a range of climatic conditions. In particular, the so-called tropical lineage is usually found in regions with average mean temperatures >20 °C, whereas the temperate lineage is present in regions with average annual temperatures <20 °C (Zemtsova et al. 2016). In tropical and subtropical regions, *R. sanguineus* s.l. ticks can complete more than one generation per year (Silveira et al. 2009), whereas in temperate countries, they typically produce a single generation per year, occurring mainly from spring to autumn (Lorusso et al. 2010) and overwintering in southern Europe as adults (Ramos et al. 2014).

F. Dantas-Torres (✉)
Centro de Pesquisas Aggeu Magalhães, Fundação Oswaldo Cruz, Recife, Brazil
e-mail: filipe.vet@globo.com

D. Otranto
Dipartimento di Medicina Veterinaria, Università degli Studi di Bari, Bari, Italy
e-mail: domenico.otranto@uniba.it

A. Estrada-Peña et al. (eds.), *Ticks of Europe and North Africa*,
https://doi.org/10.1007/978-3-319-63760-0_62

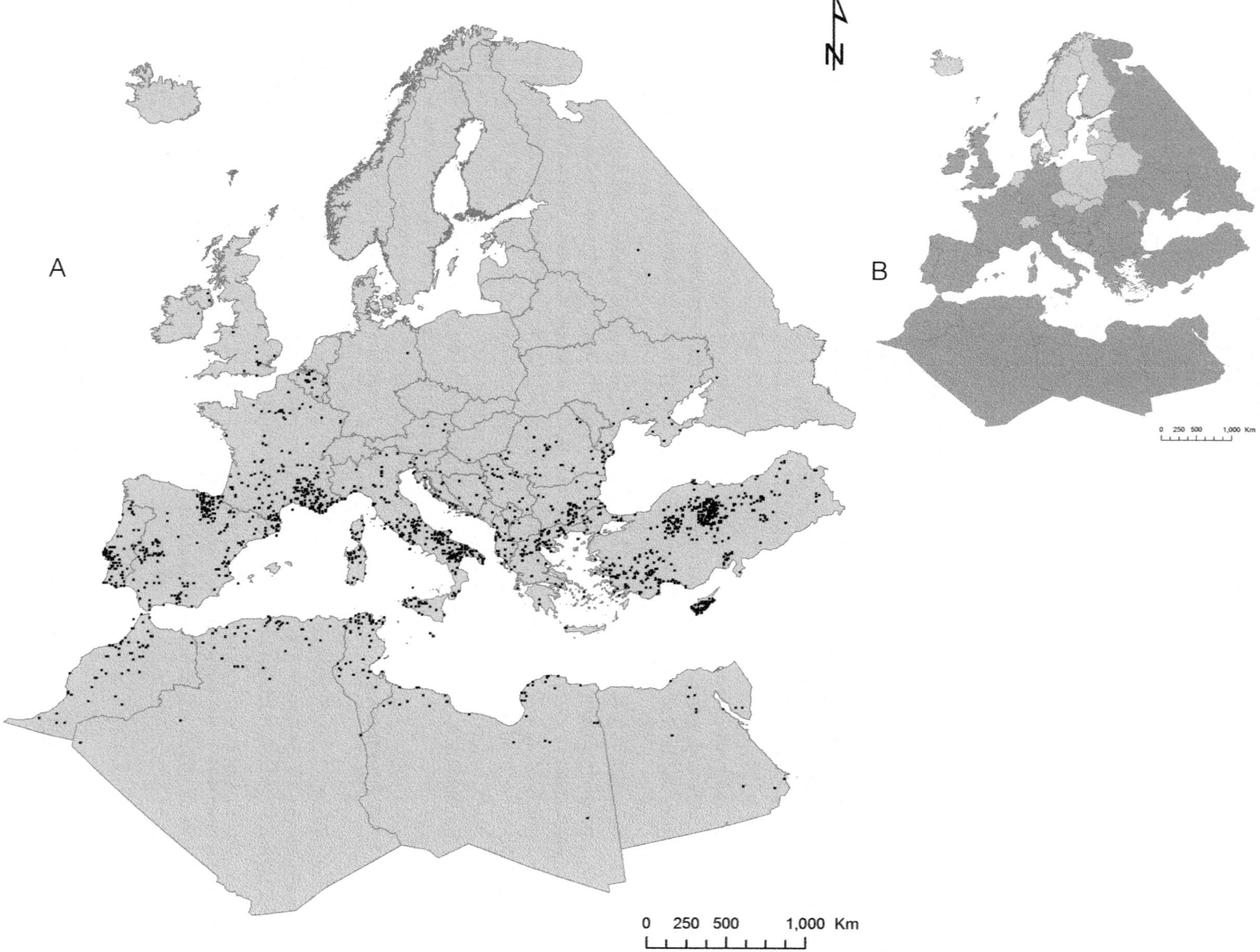

Fig. 127 **A** Distribution of *R. sanguineus* s.l. in Europe and Northern Africa (10 × 10 km grid presence with black dots). **B** Countries where the species has been reported are marked in dark grey

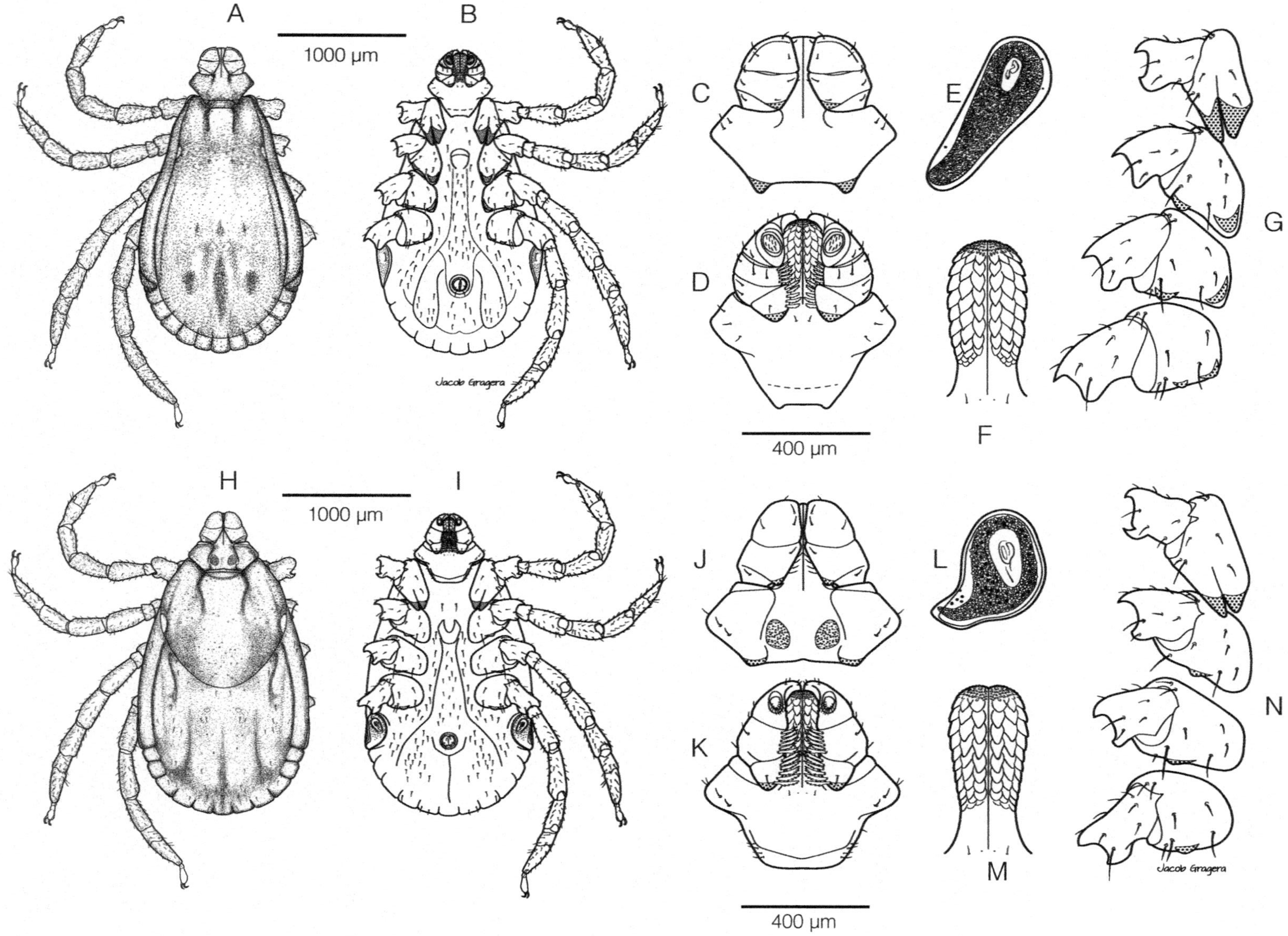

Fig. 128 **A–G** The male of *R. sanguineus* s.l. **A** Dorsal, **B** ventral, **C** capitulum, dorsal, **D** capitulum, ventral, **E** spiracular plate; **F** hypostome, **G** coxae and trochanters I–IV. **H–N** The female of *R. sanguineus* s.l. **H** Dorsal, **I** ventral, **J** capitulum, dorsal, **K** capitulum, ventral, **L** spiracular plate; **M** hypostome, **N** coxae and trochanters I–IV. Illustrations from specimens collected in kennels in Portugal, Spain, France and Italy

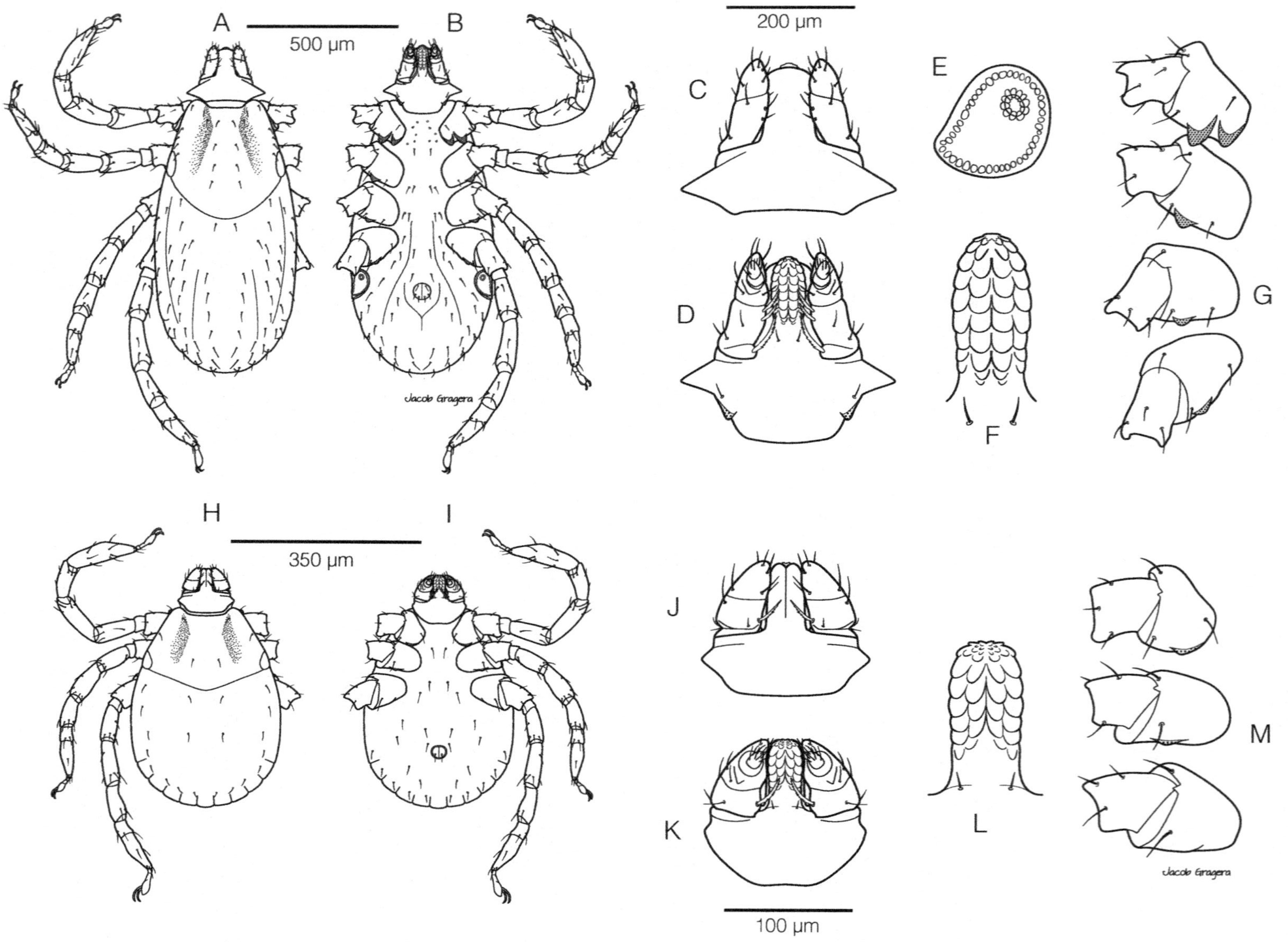

Fig. 129 **A–F** The nymph of *R. sanguineus* s.l. **A** Dorsal, **B** ventral, **C** capitulum, dorsal, **D** capitulum, ventral, **E** spiracular plate, **F** hypostome, **G** coxae and trochanters I–IV. **H–M** The larva of *R. sanguineus* s.l. **H** Dorsal, **I** ventral, **J** capitulum, dorsal, **K** capitulum, ventral, **L** hypostome, **M** coxae and trochanters I–III. Illustrations from a colony of ticks collected in southern France

Distribution

Rhipicephalus sanguineus s.l. has a cosmopolitan distribution, occurring mainly in the tropical and subtropical zones but also in temperate regions. Sporadic findings of brown dog ticks in northern countries (e.g. the UK, Hansford et al. 2014) have been published, but there is limited evidence indicating that these ticks could establish themselves permanently in such cold regions. The limits of its distribution range are thought to be from latitude 20°N to below 30°S for the temperate lineage (Burlini et al. 2010).

Vectorial Capacity and Pathogen Burden

Laboratory and field studies have implicated *R. sanguineus* s.l. in the transmission of numerous pathogens, including *Babesia vogeli*, *Ehrlichia canis*, *Hepatozoon canis*, *Rickettsia conorii* and *Rickettsia rickettsii*, among others (for complete review, see Dantas-Torres 2008). Different lineages of *R. sanguineus* s.l. appear to have different vector competences for certain pathogens, including *E. canis* (Moraes-Filho et al. 2015).

References

Burlini L, Teixeira KR, Szabó MP, Famadas KM (2010) Molecular dissimilarities of *Rhipicephalus sanguineus* (Acari: Ixodidae) in Brazil and its relation with samples throughout the world: is there a geographical pattern? Exp Appl Acarol 50:361–374

Dantas-Torres F (2008) The brown dog tick, *Rhipicephalus sanguineus* (Latreille, 1806) (Acari: Ixodidae): from taxonomy to control. Vet Parasitol 152:173–185

Dantas-Torres F (2010) Biology and ecology of the brown dog tick, *Rhipicephalus sanguineus*. Parasit Vectors 2010(3):26

Dantas-Torres F, Figueredo LA, Otranto D (2011) Seasonal variation in the effect of climate on the biology of *Rhipicephalus sanguineus* in southern Europe. Parasitology 138:527–536

Dantas-Torres F, Giannelli A, Otranto D (2012) Starvation and overwinter do not affect the reproductive fitness of *Rhipicephalus sanguineus*. Vet Parasitol 185:260–264

Dantas-Torres F, Latrofa MS, Annoscia G, Giannelli A, Parisi A, Otranto D (2013) Morphological and genetic diversity of *Rhipicephalus sanguineus* sensu lato from the new and old worlds. Parasit Vectors 6:213

Dantas-Torres F, Otranto D (2015) Further thoughts on the taxonomy and vector role of *Rhipicephalus sanguineus* group ticks. Vet Parasitol 208:9–13

Guglielmone AA, Robbins RG, Apanaskevich DA, Petney TN, Estrada-Peña A, Horak IG (2014) The hard ticks of the world (Acari: Ixodida: Ixodidae). Springer, Dordrecht, p 738

Hansford KM, Pietzsch ME, Cull B, Medlock JM (2014) Importation of *R. sanguineus* into the UK via dogs: tickborne diseases. Vet Rec 175:385–386

Latreille PA (1806) Genera crustaceorum et insectorum secundumordinem naturalem in familia disposita, iconibus exemplisqueplurimis explicata. Paris Argentorati 1:302

Lorusso V, Dantas-Torres F, Lia RP, Tarallo VD, Mencke N, Capelli G, Otranto D (2010) Seasonal dynamics of the brown dog tick, *Rhipicephalus sanguineus*, on a confined dog population in Italy. Med Vet Entomol 24:309–315

Moraes-Filho J, Krawczak FS, Costa FB, Soares JF, Labruna MB (2015) Comparative evaluation of the vector competence of four South American populations of the *Rhipicephalus sanguineus* group for the bacterium *Ehrlichia canis*, the agent of canine monocytic ehrlichiosis. PLoS ONE 10:e0139386

Moraes-Filho J, Marcili A, Nieri-Bastos FA, Richtzenhain LJ, Labruna MB (2011) Genetic analysis of ticks belonging to the *Rhipicephalus sanguineus* group in Latin America. Acta Trop 117:51–55

Nava S, Estrada-Peña A, Petney T, Beati L, Labruna MB, Szabó MP, Venzal JM, Mastropaolo M, Mangold AJ, Guglielmone AA (2015) The taxonomic status of *Rhipicephalus sanguineus* (Latreille, 1806). Vet Parasitol 208:2–8

Nava S, Mastropaolo M, Venzal JM, Mangold AJ, Guglielmone AA (2012) Mitochondrial DNA analysis of *Rhipicephalus sanguineus* sensu lato (Acari: Ixodidae) in the Southern Cone of South America. Vet Parasitol 190:547–555

Ramos RA, Giannelli A, Carbone D, Baneth G, Dantas-Torres F, Otranto D (2014) Occurrence of *Hepatozoon canis* and *Cercopithifilaria bainae* in an off-host population of *Rhipicephalus sanguineus* sensu lato ticks. Ticks Tick Borne Dis 5:311–314

Silveira JA, Passos LM, Ribeiro MF (2009) Population dynamics of *Rhipicephalus sanguineus* (Latrielle, 1806) in Belo Horizonte, Minas Gerais state, Brazil. Vet Parasitol 161:270–275

Walker JB, Keirans JE, Horak IG (2000) The genus *Rhipicephalus* (Acari, Ixodidae): a guide to the brown ticks of the world. Cambridge University Press, Cambridge

Zemtsova GE, Apanaskevich DA, Reeves WK, Hahn M, Snellgrove A, Levin ML (2016) Phylogeography of *Rhipicephalus sanguineus* sensu lato and its relationships with climatic factors. Exp Appl Acarol 69:191–203

Rhipicephalus turanicus Pomerantzev, 1940 (Figs. 130–132)

F. Dantas-Torres, D. Otranto, M. M. Santos-Silva, and Z. Vatansever

Life Cycle and Host Preferences

Rhipicephalus turanicus is a three-host tick that lives in lowland desert and semi-desert, steppe and open woodlands (Filippova 1997). This species is under discussion due to the morphological, biological and molecular variability of ticks identified as *R. turanicus* in different countries. For this reason, many records in the literature are currently considered speculative pending confirmation (Guglielmone et al. 2014). Indeed, genetic data indicate that ticks identified as *R. turanicus* in some regions (e.g. in some African countries) may actually represent different species (Beati and Keirans 2001; Dantas-Torres et al. 2013a). This species feeds on mammals (e.g. rodents, lagomorphs, canids, felids and mustelids), birds and lizards (Filippova 1997; Dantas-Torres et al. 2011, 2013a). Larvae and nymphs have also been found on birds, while the adult stage also occurs on lizards (Filippova 1997). Human infestation has been reported (Filippova 1997).

F. Dantas-Torres (✉)
Centro de Pesquisas Aggeu Magalhães, Fundação Oswaldo Cruz, Recife, Brazil
e-mail: filipe.vet@globo.com

D. Otranto
Dipartimento di Medicina Veterinaria, Università degli Studi di Bari, Bari, Italy
e-mail: domenico.otranto@uniba.it

M. M. Santos-Silva
Centro de Estudos de Vectores e Doenças Infecciosas Dr. Francisco Cambournac (CEVDI), Instituto Nacional de Saúde Doutor Ricardo Jorge, Águas de Moura, Portugal
e-mail: m.santos.silva@insa.min-saude.pt

Z. Vatansever
Department of Parasitology, Faculty of Veterinary Medicine, Kafkas University, Kars, Turkey
e-mail: zativet@gmail.com

A. Estrada-Peña et al. (eds.), *Ticks of Europe and North Africa*,
https://doi.org/10.1007/978-3-319-63760-0_63

Ecology

Adults of *R. turanicus* have been found questing during the spring and summer months. In southern Italy, this species is mostly found in grasslands and meadow habitats (Dantas-Torres et al. 2013b). In the same area, two males and a nymph of *R. turanicus* were found on Corsican hares in early autumn (Dantas-Torres et al. 2011). Considering the uncertainties about the identity of ticks identified as *R. turanicus*, data on the ecology of this tick is uncertain.

Distribution

Rhipicephalus turanicus is a Palaearctic species whose precise geographical distribution is unknown, considering that many distribution records are currently speculative (Filippova 1997; Dantas-Torres et al. 2013a; Guglielmone et al. 2014). It is heavily suspected that what has been classically described as '*R. turanicus*' in Europe west to Turkey might, in fact, represent another entity. It is important to notice that molecular sequences available in GenBank are expected to have the same degree of unreliability as mentioned for the records or the knowledge of the ecology of this species. There are currently only two valid sequences of *R. turanicus* in GenBank, acknowledged as coming from this species from specimens determined by N.A. Filippova.

Vectorial Capacity and Pathogen Burden

Since the taxonomic status of *R. turanicus* is under discussion, its role as a vector of human pathogens, including *Rickettsia massiliae*, *R. conorii* and other microorganisms molecularly detected in ticks designated as this species, needs confirmation. The reliability of tick identification in some old studies on the vector role of *R. turanicus* for

Fig. 130 **A** Distribution of *R. turanicus* in Europe and Northern Africa (10 × 10 km grid presence with black dots). **B** Countries where the species has been reported are marked in dark grey

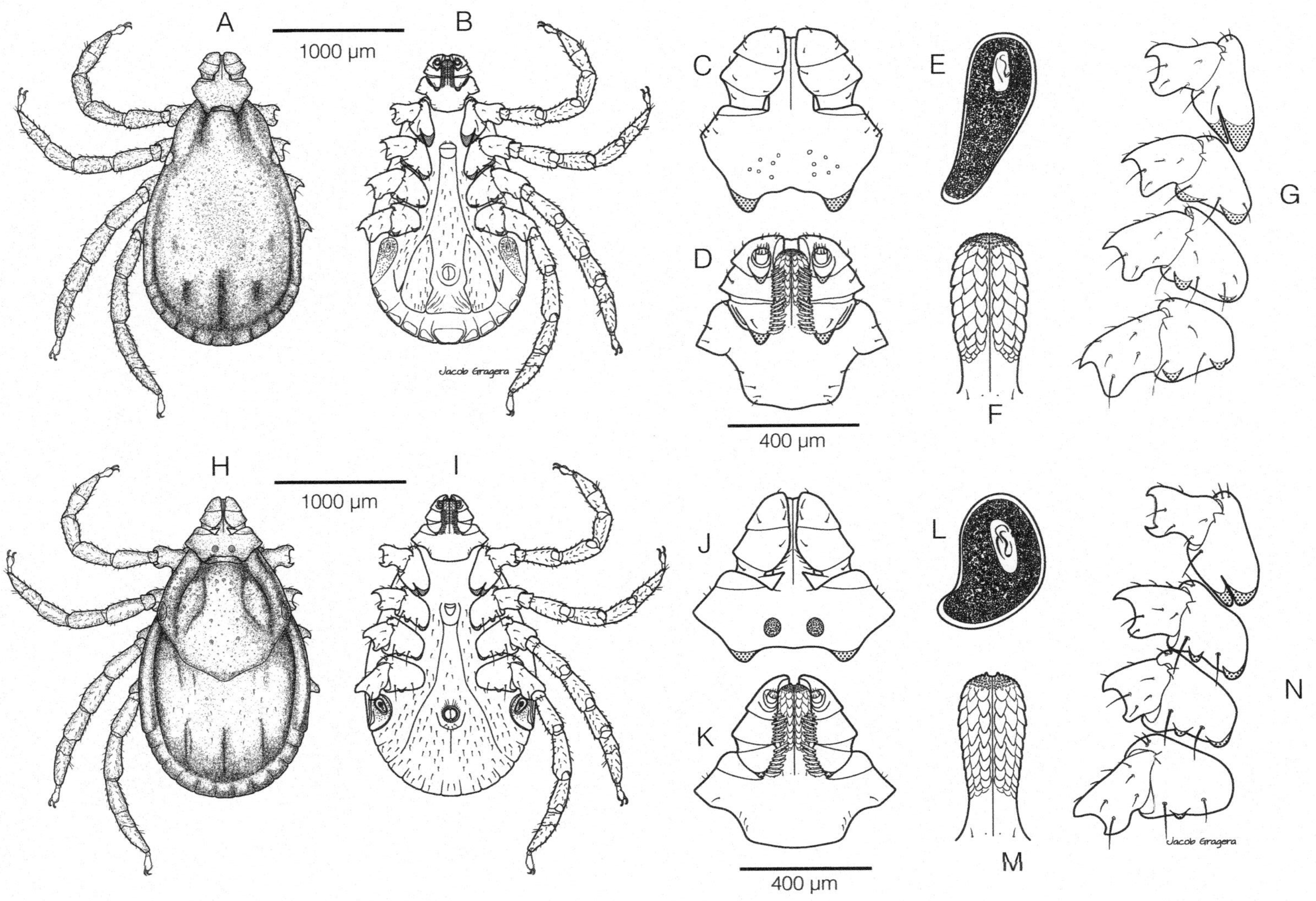

Fig. 131 **A–G** The male of *R. turanicus*. **A** Dorsal, **B** ventral, **C** capitulum, dorsal, **D** capitulum, ventral, **E** spiracular plate, **F** hypostome, **G** coxae and trochanters I–IV. **H–N** The female of *R. turanicus*. **H** Dorsal, **I** ventral, **J** capitulum, dorsal, **K** capitulum, ventral, **L** spiracular plate, **M** hypostome, **N** coxae and trochanters I–IV. Illustrations redrawn from Filippova (1997)

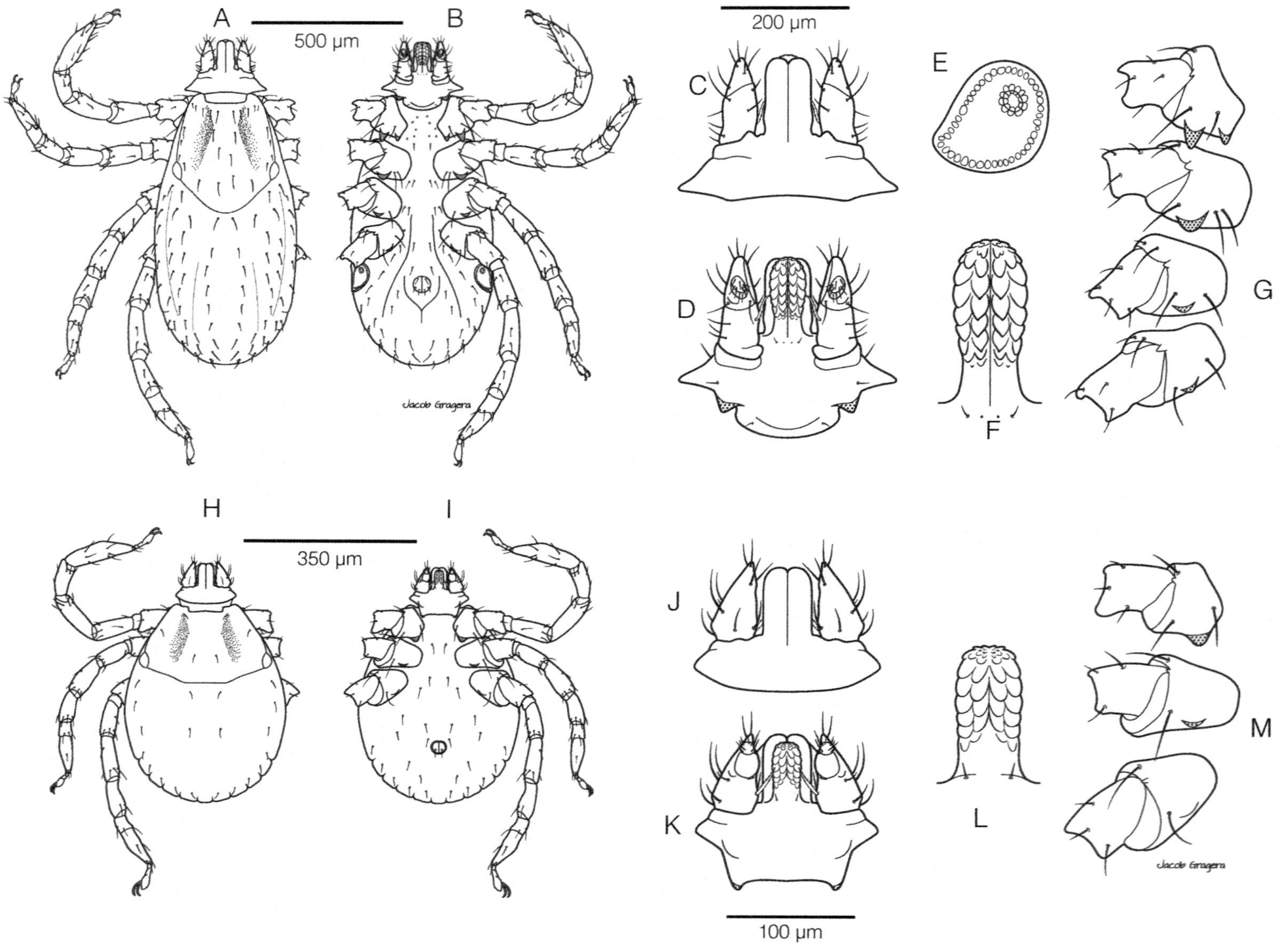

Fig. 132 **A–G** The nymph of *R. turanicus*. **A** Dorsal, **B** ventral, **C** capitulum, dorsal, **D** capitulum, ventral, **E** spiracular plate, **F** hypostome, **G** coxae and trochanters I–IV. **H–M** The larva of *R. turanicus*. **H** Dorsal, **I** ventral, **J** capitulum, dorsal, **K** capitulum, ventral, **L** hypostome, **M** coxae and trochanters I–III. Illustrations redrawn from Filippova (1997)

different pathogens may be argued. Further studies are needed to determine its participation in transmission.

References

Achutan HN, Mahadevan S, Lalitha M (1980) Studies on the developmental forms of *Babesia bigemina* and *Babesia canis* in ixodid ticks. Indian Vet J 57:181–184

Balashov YS, Daiter AB (1973) Bloodsucking arthropods and rickettsiae. Nauka, Leningrad (in Russian)

Beati L, Keirans JE (2001) Analysis of the systematic relationships among ticks of the genera *Rhipicephalus* and *Boophilus* (Acari: Ixodidae) based on mitochondrial 12S ribosomal DNA gene sequences and morphological characters. J Parasitol 87:32–48

Berdyev A (1980) Ecology of ixodid ticks of Turkmenistan and their importance in epizootiology of natural-foci diseases. Ylym, Ashkhabad, Turkmenistan (in Russian)

Dantas-Torres F, Testini G, DiGeronimo PM, Lorusso V, Mallia E, Otranto D (2011) Ticks infesting the endangered Italian hare (*Lepus corsicanus*) and their habitat in an ecological park in southern Italy. Exp Appl Acarol 53:95–102

Dantas-Torres F, Latrofa MS, Annoscia G, Giannelli A, Parisi A, Otranto D (2013a) Morphological and genetic diversity of *Rhipicephalus sanguineus* sensu lato from the new and old worlds. Parasit Vectors 6:213

Dantas-Torres F, Lia RP, Capelli G, Otranto D (2013b) Efficiency of flagging and dragging for tick collection. Exp Appl Acarol 61:119–127

Filippova NA (1997) Ixodid ticks of subfamily Amblyomminae. In the fauna of Russia and neighboring countries. New Series no. 145: Arachnoidea, vol 4(5), St. Petersburg, Nauka (in Russian)

Guglielmone AA, Robbins RG, Apanaskevich DA, Petney TN, Estrada-Peña A, Horak IG (2014) The hard ticks of the world (Acari: Ixodida: Ixodidae). Springer, Dordrecht, p 738

Keysary A, Massung RF, Inbar M, Wallach AD, Shanas U, Mumcuoglu KY, Waner T (2007) Molecular evidence for *Anaplasma phagocytophilum* in Israel. Emerg Infect Dis 13:1411–1412

Mitchell CJ, Spillett JJ (1968) Ecological notes on *Rhipicephalus turanicus* Pomerantzev in West Bengal, India (Acarina: Ixodidae). J Med Entomol 5:5–8

Walker JB, Keirans JE, Horak IG (2000) Genus *Rhipicephalus* (Acari, Ixodidae). A guide to the brown ticks of the world. Cambridge University Press, Cambridge

Rhipicephalus annulatus (Say, 1821) (Figs. 133–135)

G. D'Amico, A. D. Mihalca, and A. Estrada-Peña

Life Cycle and Host Preferences

Rhipicephalus annulatus formerly belonged to the genus *Boophilus* (as *B. annulatus*), and most records prior to 2001 list it under this name. Molecular studies led to *Boophilus* becoming a subgenus of the genus *Rhipicephalus* (Beati and Keirans 2001; Guglielmone et al. 2014). *Rhipicephalus annulatus* is a one-host tick with all stages feeding on the same animal. The life cycle can take less than 2 months to complete. Two generations per year can occur under favourable conditions. This fast life cycle can lead to massive infestations on animals resulting in huge economic losses (Spickler and Rovid 2007). The larvae quest on vegetation looking for a host (Emmanuel et al. 2011). Cattle represent the main hosts of *R. annulatus*, but sheep, goats and wild ungulates are also occasionally infested. The source of infestation in other hosts seems to be maintained exclusively by cattle. The preferred feeding sites are legs, belly, neck and pectoral area (Walker et al. 2003). It has occasionally been reported on humans and dogs.

Ecology

The habitat of *R. annulatus* is characterized by Mediterranean and savanna vegetation, but it is also found in humid localities within steppe areas that have hot and dry seasons. During the summer, this species can survive for more than 3–4 months without feeding (Emmanuel et al. 2011). In cooler seasons, the larvae may live without food for up to 6 months. In northern Africa, the ticks are active from September to January with a peak in October. Its ecology is similar in many aspects to the other species of the subgenus *Boophilus* (Walker et al. 2003).

Distribution

The origin of this species is still controversial. Morel (1969) designated the Mediterranean area as its origin. However, there is some agreement that the subgenus *Boophilus* could have originated in parts of Southeast Asia. In Africa, *R. annulatus* is found along two strips, with the Sahara Desert separating both areas: the Palaearctic and the Afrotropical. This suggests that there was originally either only one population that became separated by the Sahara Desert or two different colonization events. This tick is widely distributed in parts of Africa with Mediterranean and savanna climate: Algeria, Egypt, Morocco, Tunisia and Libya. It is considered to be mainly a northern and western African tick, but it has also been reported from the Sudan, the Central African Republic and the Democratic Republic of Congo. *Rhipicephalus annulatus* has been reported from southern countries of the former U.S.S.R. such as the Ukraine, in Mediterranean regions and Asia (Spickler and Rovid 2007). In Europe, this species is known from Romania, Bulgaria, Albania, Croatia, Serbia, Montenegro, Kosovo, Cyprus, Greece, Turkey, Italy, France (Corsica), Portugal and Spain.

Vectorial Capacity and Pathogen Burden

Rhipicephalus annulatus is well-known for its role in the transmission of babesiosis (*Babesia bigemina* and *B. bovis*) to cattle (Walker et al. 2003). *Rhipicephalus annulatus* is also a competent vector for *Anaplasma marginale* to cattle,

G. D'Amico (✉) · A. D. Mihalca
Department of Parasitology and Parasitic Diseases, University of Agricultural Sciences and Veterinary Medicine Cluj-Napoca, Cluj-Napoca, Romania
e-mail: gianluca.damico@usamvcluj.ro

A. D. Mihalca
e-mail: amihalca@usamvcluj.ro

A. Estrada-Peña
Department of Animal Health, Faculty of Veterinary Medicine, Zaragoza, Spain
e-mail: aestrada@unizar.es

A. Estrada-Peña et al. (eds.), *Ticks of Europe and North Africa*,
https://doi.org/10.1007/978-3-319-63760-0_64

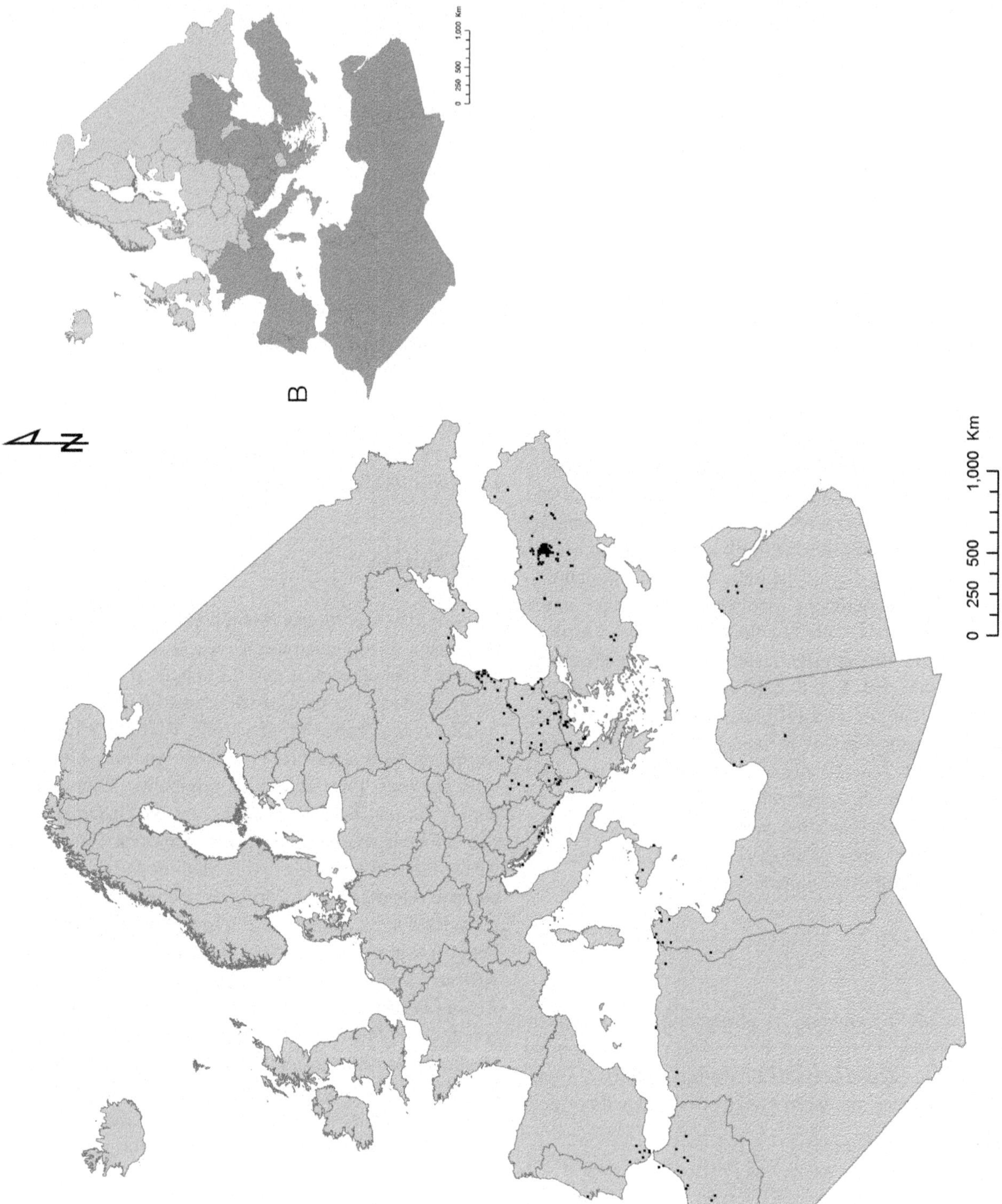

Fig. 133 **A** Distribution of *R. annulatus* in Europe and Northern Africa (10 × 10 km grid presence with black dots). **B** Countries where the species has been reported are marked in dark grey

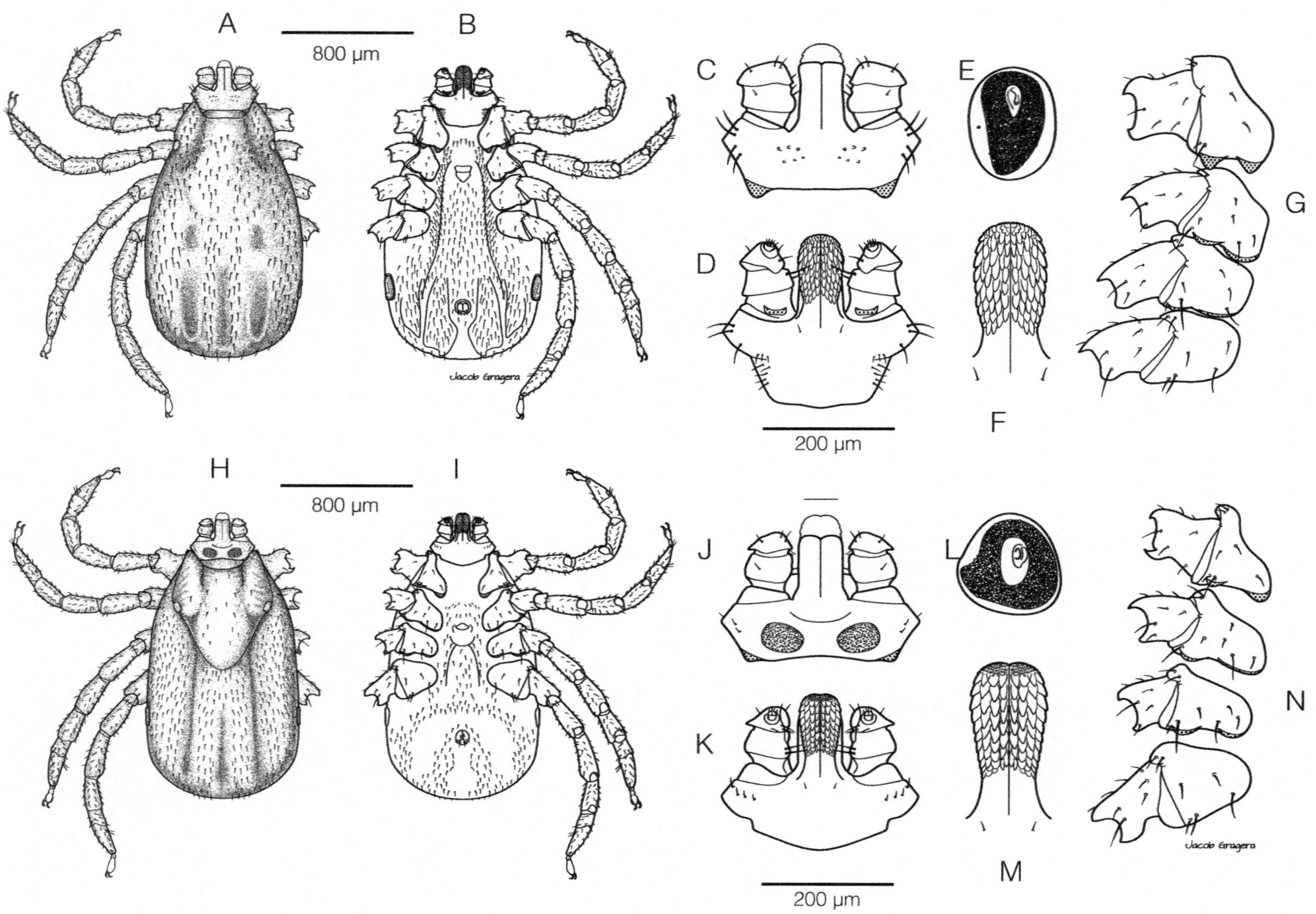

Fig. 134 **A–G** The male of *R. annulatus*. **A** Dorsal, **B** ventral, **C** capitulum, dorsal, **D** capitulum, ventral, **E** spiracular plate, **F** hypostome, **G** coxae and trochanters I–IV. **H–N** The female of *R. annulatus*. **H** Dorsal, **I** ventral, **J** capitulum, dorsal, **K** capitulum, ventral, **L** spiracular plate, **M** hypostome, **N** coxae and trochanters I–IV. Illustrations form specimens collected in Spain, Morocco and Tunisia

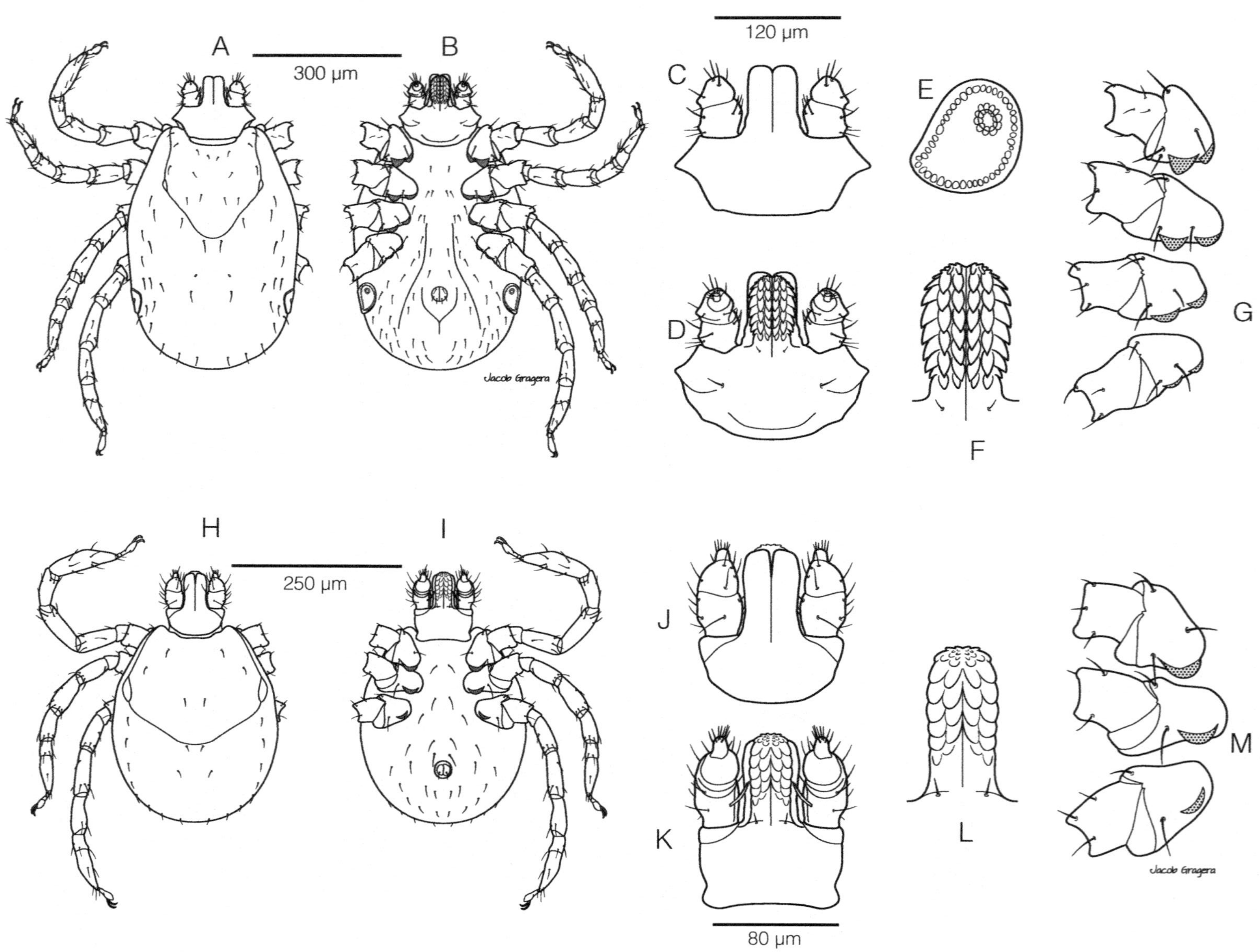

Fig. 135 **A–G** The nymph of *R. annulatus*. **A** Dorsal, **B** ventral, **C** capitulum, dorsal, **D** capitulum, ventral, **E** spiracular plate, **F** hypostome, **G** coxae and trochanters I–IV. **H–M** The larva of *R. annulatus*. **H** Dorsal, **I** ventral, **J** capitulum, dorsal, **K** capitulum, ventral, **L** hypostome, **M** coxae and trochanters I–III. Illustrations from specimens collected in Spain, Morocco and Tunisia

causing bovine anaplasmosis (Walker et al. 2003). Recent studies present molecular evidence for the presence of *Ehrlichia* sp. in *R. annulatus* from Tahiti (Laroche et al. 2016). *Rickettsia africae* has been detected in this species in Senegal, Guinea and Nigeria (Mediannikov et al. 2010, 2012; Reye et al. 2012), but its vectorial role remains unclear.

References

Beati L, Keirans JE (2001) Analysis of the systematic relationships among ticks of the genera *Rhipicephalus* and *Boophilus* (Acari: Ixodidae) based on mitochondrial 12S ribosomal DNA gene sequences and morphological characters. J Parasitol 87:32–48

Emmanuel N, Julius B, Amafu-Dey H, Richard OS (2011) Some aspects of the reproductive biology of female *Rhipicephalus annulatus* (Say, 1821) in the laboratory. Int J Anim Vet Adv 3(6):419–421

Guglielmone AA, Robbins RG, Apanaskevich DA, Petney TN, Estrada-Peña A, Horak IG (2014) The hard ticks of the world (Acari: Ixodida: Ixodidae). Springer, Dordrecht, p 738

Laroche M, Marie J, Mediannikov O, Almeras L, Berenger JM, Musso D, Raoult D, Parola P (2016) A novel ehrlichial agent detected in tick in French Polynesia. Ticks Tick-Borne Dis 7(6):1203–1208

Mediannikov O, Diatta G, Fenollar F, Sokhna C, Trape JF, Raoult D (2010) Tick-borne rickettsioses, neglected emerging diseases in rural Senegal. PLoS Negl Trop Dis 4(9):pii: e821

Mediannikov O, Davoust B, Socolovschi C, Tshilolo L, Raoult D, Parola P (2012) Spotted fever group rickettsiae in ticks and fleas from the Democratic Republic of the Congo. Ticks Tick Borne Dis 3:371–373

Morel PC (1969) Contribution à la connaissance de la distribution des tiques (Acariens, Ixodidae et Amblyommidae) en Ethiopie continentale. Ph.D. thesis, p 388, (in French)

Reye AL, Arinola OG, Hubschen JM, Muller CP (2012) Pathogen prevalence in ticks collected from the vegetation and livestock in Nigeria. Appl Environ Microbiol 78(8):2562–2568

Spickler C, Rovid A (2007) *Rhipicephalus (Boophilus) annulatus*, 2007. At http://www.cfsph.iastate.edu/DiseaseInfo/factsheets.php

Walker AR (2003) Ticks of domestic animals in Africa: a guide to identification of species. Bioscience reports, Edinburgh, pp 3–210

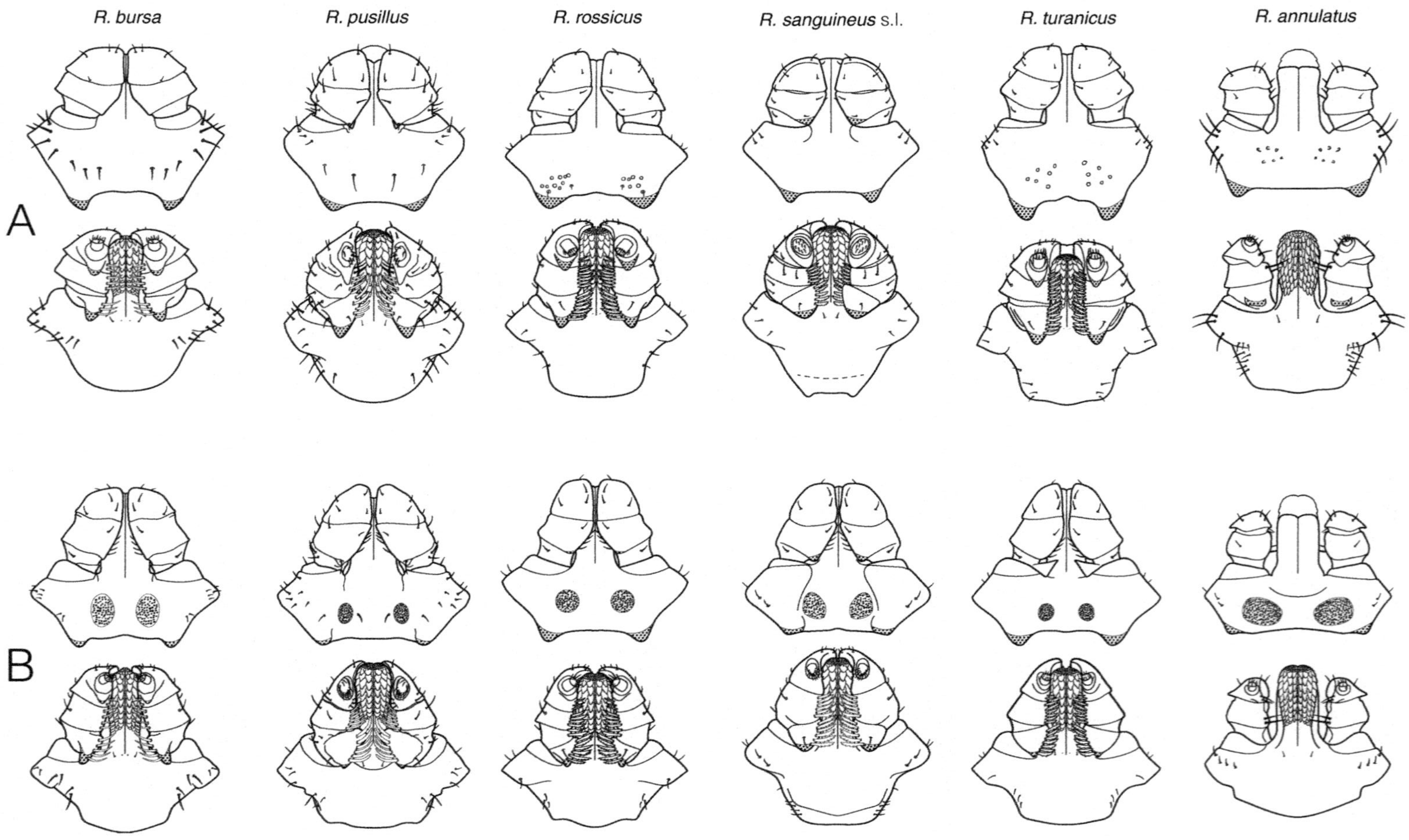

Fig. 136 Composite view of the capitulum of the adults of the species of *Rhipicephalus* reported in Europe and northern Africa. **A** Males (dorsal and ventral); **B** females (dorsal and ventral)

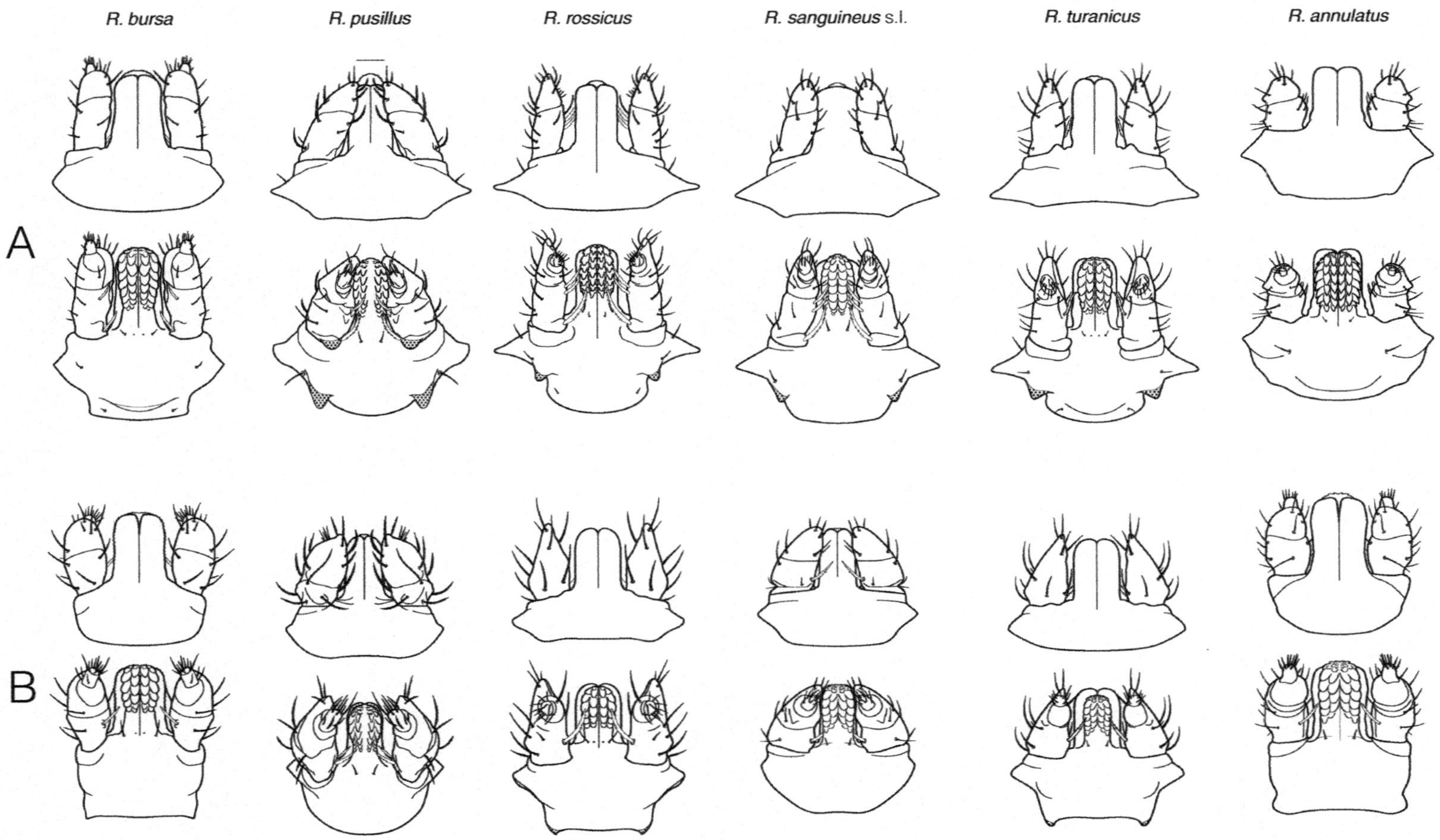

Fig. 137 Composite view of the capitulum of the immatures of the species of *Rhipicephalus* reported in Europe and northern Africa. **A** Nymphs (dorsal and ventral); **B** larvae (dorsal and ventral)

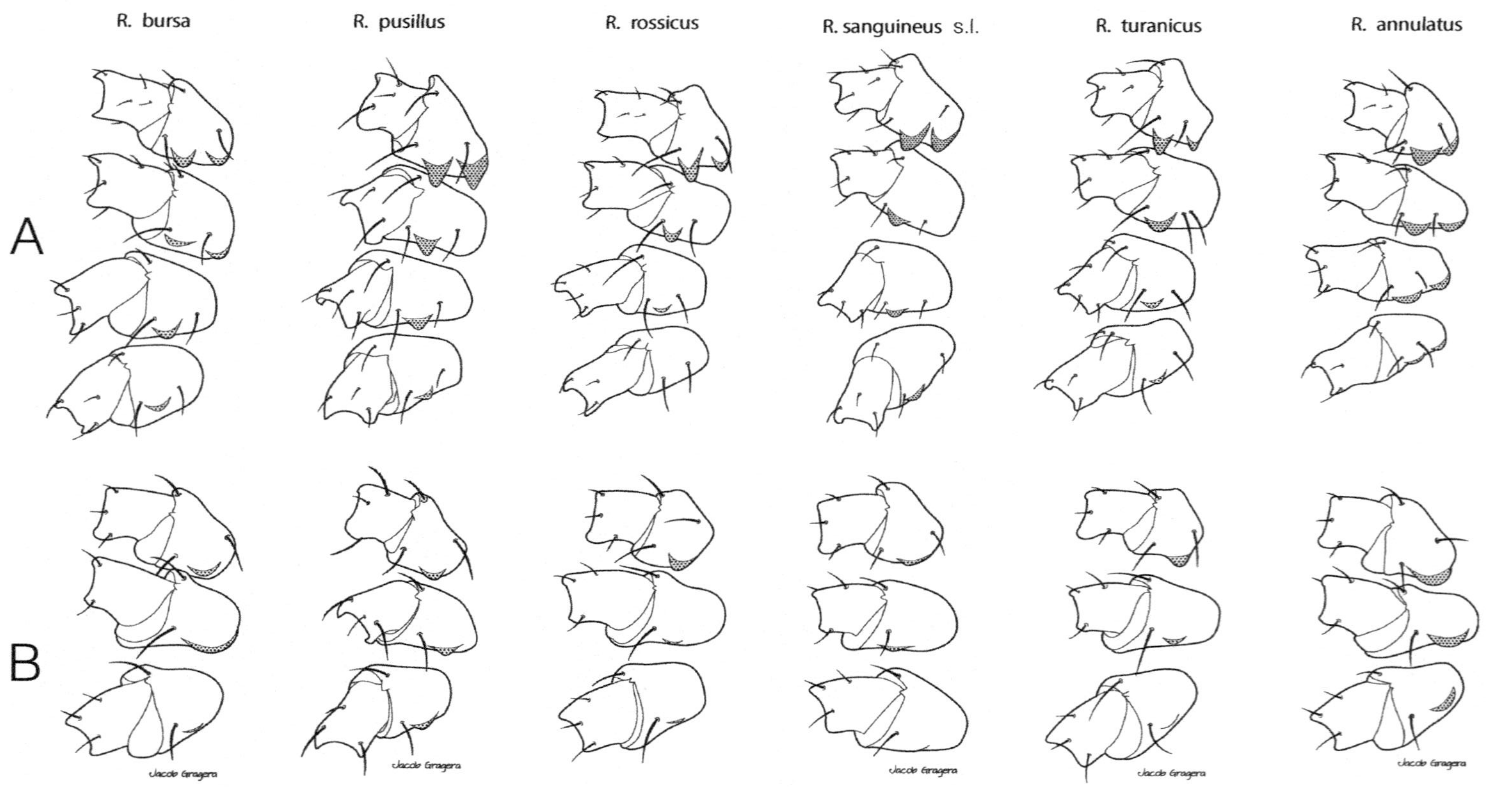

Fig. 138 Composite view of the coxal and first trochanter of the immatures of the species of *Rhipicephalus* reported in Europe and northern Africa. **A** Nymphs; **B** larvae

Genus *Hyalomma* Koch, 1844

A. Estrada-Peña, M. P. Pfäffle, and T. N. Petney

The genus *Hyalomma* contains species of ticks involved in the transmission of a number of significant pathogens in the Mediterranean region, Africa and Asia. All species in the genus are restricted to the Palaearctic, Indo-Malaysian and Afrotropical regions. The immatures of these species pose considerable difficulties in identification and we strongly advise researchers not to attempt to identify the engorged nymphs or larvae of *Hyalomma* to species level. It is of extreme importance to keep the engorged nymphs alive and allow them to moult to obtain adults. Adequate keys for the determination of the adults were recently published in a series of studies by Apanaskevich and co-workers (see below, separately for each species). It is important to realize that classic studies on the determination of these ticks contain substantial mistakes, at least with respect to several species. Our recommendation is to follow the most recent criteria for the identification of the species in this genus, and thus not generate more confusion in reporting the geographical distributions of ticks from the genus *Hyalomma*. Because of the confusion regarding the distribution of the species of this genus, we will include comments about the countries in which they have been collected to date as an additional guide for researchers.

Several aspects of the systematics of *Hyalomma* are unclear, and many of these difficulties relate to the identification of species within the subgenus *Euhyalomma* Filippova, 1984 (Apanaskevich and Horak 2008). One of the most difficult groups from the taxonomic perspective is the *Hy. marginatum* species complex. This was considered to consist of a single extremely polymorphic species, *Hy. marginatum* Koch, 1844, of which the overall distribution very nearly coincides with that of the entire genus, which is present in southern Europe, Asia Minor, Arabia, central, southern and, to a limited extent, continental South-east Asia and Africa. It was generally accepted that this species contains four subspecies, with a further fifth taxon belonging to this group recently re-established as a full species in South Africa. The former group of subspecies included in *Hy. marginatum* is now considered to be formed by three species in the Palaearctic region, namely *Hy. marginatum*, *Hy. rufipes* Koch, 1844 and *Hy. turanicum* Pomerantzev 1946. *Hyalomma turanicum* could possibly occur in Israel, Lebanon and Turkey but is found predominantly in sub-Saharan Africa. Apanaskevich and Horak (2008) were unable to confirm its presence in Cyprus, Libya and Tunisia, and specimens previously identified there as *Hy. marginatum turanicum* are actually *Hy. marginatum*. Keys and illustrations for the adults of *Hy. marginatum*, *Hy. rufipes* and *Hy. turanicum* are present in Walker et al. (2003). The most authoritative source of illustrations of the adults and immatures of these three species is Apanaskevich and Horak (2008).

Hyalomma rufipes is a native Afrotropical species, however, adults species have also been recorded in Russia (Astrakhan region), the Ukraine, Malta, Tunisia and Libya, as well as other countries in the Eastern Palaearctic (according to the compilation by Apanaskevich and Horak 2008). Therefore, whether introduced or not, it seems to be established in the Palaearctic region, and additional efforts to separate it from other species are necessary. *Hyalomma rufipes* has been confused with *Hy. impressum* Koch, 1844. Some authors consider it a subspecies or variety of *Hy. impressum* Koch, 1844, while others have used the name *Hy. impressum* for *Hy. rufipes*. This confusion was resolved by Feldman-Muhsam (1954). Subsequently, most Western workers considered *Hy. rufipes* as a separate species, but in Russia, many scientists continued to use the name *Hy.*

A. Estrada-Peña (✉)
Department of Animal Health, Faculty of Veterinary Medicine, Zaragoza, Spain
e-mail: aestrada@unizar.es

M. P. Pfäffle
Department of Ecology and Parasitology, Institute of Zoology, Karlsruhe Institute of Technology, Karlsruhe, Germany
e-mail: miripfaeffle@web.de

T. N. Petney
State Museum of Natural History, Erbprinzenstr. 13, 76133 Karlsruhe, Germany
e-mail: trevor.petney@kit.edu

A. Estrada-Peña et al. (eds.), *Ticks of Europe and North Africa*,
https://doi.org/10.1007/978-3-319-63760-0_65

impressum for *Hy. rufipes* (*Hy. marginatum impressum* or *Hy. plumbeum impressum* of Pomerantzev 1946, 1950). Some years later, Hoogstraal and Kaiser (1960) concluded that *Hy. rufipes* should be a subspecies of *Hy. marginatum*, namely *Hy. marginatum rufipes*. Nevertheless, some researchers continued to use the name *Hy. rufipes* as a specific epithet for this taxon.

Pomerantzev (1946) originally described *Hy. turanicum* as a subspecies of *Hy. marginatum*. However, Hoogstraal (1956), after examining specimens of *Hy. marginatum turanicum* mainly from South Africa (which are actually *Hy. glabrum* Delpy, 1949), stated that *Hy. marginatum turanicum* is a separate species. Hoogstraal and Kaiser (1960) later concluded that this taxon is indeed a subspecies of *Hy. marginatum*. Conversely, Russian and Eastern European workers have considered *Hy. turanicum* to be a subspecies of *Hy. plumbeum* since 1950. Other than the study by Apanaskevich and Horak (2008), keys to separate adults of some *Hyalomma* spp. are available in Cringoli et al. (2005). It must be explicitly noted, however, that while Apanaskevich and Horak (2008) determined several characters of the adults that make it possible to identify the species in the group at this life stage without hesitation, they could not find reliable characters for distinguishing the immature stages of some species. Although some characters of the adults, such as the density of punctations on the conscutum and scutum, might be fairly variable, a combination of all the proposed diagnostic features provides reliable adult identification.

Other species of *Hyalomma* found in the Western Palaearctic are: *Hy. aegyptium* (Linnaeus, 1758) *Hy. dromedarii* Koch, 1844, *Hy. franchinii* Tonelli-Rondelli, 1932, *Hy. impeltatum* Schulze and Schlottke, 1930, *Hy. scupense* Schulze, 1919, *Hy. lusitanicum* Koch, 1844, *Hy. anatolicum* Koch, 1844 and *Hy. excavatum* Koch, 1844. *Hyalomma aegyptium* is primarily a tick of tortoises, but several reports mention it as a parasite of humans (i.e. Široký et al. 2006). Other studies reported the importance of *Hy. aegyptium* as a reservoir of *Rickettsia aeschlimannii* (Bitam et al. 2009), and its involvement in the transmission of *Borrelia turcica*, a species not belonging to the *Borrelia burgdorferi* group (Güner et al. 2004; Takano et al. 2010). *Hyalomma aegyptium* has been adequately described and illustrated by Apanaskevich (2003a). Keys and SEM figures are available in Manilla and Giannetto (1996). Adults are described and illustrated in Pomerantzev (1950), Hoogstraal (1956) and Kaiser and Hoogstraal (1963). The immatures of this species were described and illustrated in Apanaskevich (2003a) together with a list of records and hosts.

The Mediterranean *Hy. lusitanicum* Koch, 1844, was redescribed by Apanaskevich et al. (2008a), providing features to morphologically separate it from the closely related *Hy. franchinii* and *Hy. excavatum*. The immatures of *Hy. lusitanicum* were also described and illustrated by Pérez-Eid and Cabrita (2003). The distribution of *Hy. lusitanicum* is restricted to the western part of the Mediterranean subregion countries of the Palaearctic zoogeographic region: France, Italy, Portugal, Spain (including the Canary Islands), Algeria and Morocco (Starkoff 1958; Morel 1959; Tendeiro 1962; Manilla 1998). The known distribution of *Hy. franchinii* is restricted to the eastern countries of the Mediterranean subregion of the Palaearctic zoogeographic region: Egypt, Libya, Tunisia and Israel (Hoogstraal 1956; Hoogstraal and Kaiser 1958a, b, 1960; Cwilich and Hadani 1962; Bouattour et al. 1999; Apanaskevich et al. 2008a). *Hyalomma franchinii* can be found sympatrically with *Hy. aegyptium, Hy. anatolicum, Hy. dromedarii, Hy. excavatum, Hy. impeltatum, Hy. marginatum, Hy. rufipes, Hy. turanicum* and *Hy. scupense*.

Hyalomma impeltatum Schulze and Schlottke, 1930 is widely distributed in North Africa, the northern parts of West, Central and East Africa, Arabia, Asia Minor, the Middle East and the south-western parts of Central and Southern Asia. In the Western Palaearctic, it is present in Algeria, Egypt, Libya, Mauritania, Mali, Morocco, Tunisia, the western Sahara, Israel, Jordan and Turkey (for a list of references, see the compilation by Apanaskevich and Horak 2009). This species has not yet been reported on the European side of the Mediterranean basin. Descriptions of *Hy. impeltatum* are present in Hoogstraal (1956) and Hoogstraal et al. (1981). Descriptions, keys and illustrations for the adults are in Walker et al. (2003). All of the stages were described by Kaiser and Hoogstraal (1968) as *Hy. erythraeum* Tonelli-Rondelli, 1932, which is a synonym of *Hy. impeltatum*. The larva and nymph were schematically described and illustrated as *Hy. brumpti* by Rousselot (1948). The larva was redescribed by Camicas (1970). All of the stages were redescribed and illustrated by Apanaskevich and Horak (2009).

Hyalomma scupense Schulze, 1919, was originally described as an independent species. In the same year, but in a later issue of the same journal, Schulze described a similar species that he called *Hy. detritum* Schulze, 1919. Since then, both have been considered as proper species, although differentiation between them was practically impossible because it was based on a few variable characters. The main criterion for distinguishing between the species was of an ecological nature, namely that *Hy. scupense* is a one-host species while *Hy. detritum* is a two-host species. For a detailed list of the historical changes in the determination of both *Hy. detritum* and *Hy. scupense*, the reader is referred to Apanaskevich et al. (2010). The current situation is thus confused as to the names and their status: Traditionally some workers consider both *Hy. scupense* and *Hy. detritum* as full species; others believe *Hy. scupense* is a subspecies of *Hy. detritum* and a third group accepts *Hy. detritum* as a junior synonym of *Hy. scupense*. Apanaskevich et al. (2010)

studied, redescribed and illustrated all parasitic stages and thereby authenticated only one species, *Hy. scupense*. Descriptions and illustrations of the adults as *Hy. scupense* are available in several publications, but the most useful are those by Pomerantzev (1950) and Hoogstraal (1956). The nymph and larva are schematically illustrated and described as *Hy. mauritanicum* in Senevet (1924, 1928), as *Hy. scupense* and as *Hy. detritum* in Ogandzhanyan (1953) and Dzhaparidze (1960) and as *Hy. scupense* in Feider (1965). A complete list of synonymies was provided by Apanaskevich et al. (2010). This species is known from the following countries in the Western Palaearctic: Albania, Algeria, Bosnia and Herzegovina, Bulgaria, Croatia, Egypt, France, Greece, Italy, Israel, Jordan, Lebanon, Libya, the Republic of Macedonia, Moldova, Montenegro, Morocco, Romania, Russia (south of the European part and north Caucasus), Serbia, Spain, Sudan, Tunisia, Turkey and the Ukraine (Hoogstraal 1956; Kolonin 1983; Apanaskevich et al. 2010).

Hyalomma dromedarii Koch, 1844 was originally described as an independent species Apanaskevich et al. 2008b. Later, however, Neumann (1901) considered it a variety of the 'catch-all' species *Hy. aegyptium*. Complete redescriptions and illustrations of all the parasitic stages can be found in Apanaskevich et al. (2008a). In the Western Palaearctic, *Hy. dromedarii* is known from Algeria, Egypt, Libya, Mauritania, Morocco, Spain (the Canary Islands) and Tunisia (data compiled by Apanaskevich et al. 2008a).

Koch further described two distinct species, *Hy. anatolicum* Koch, 1844 and *Hy. excavatum* Koch, 1844. Pomerantzev (1946, 1950) considered *Hy. anatolicum* and *Hy. excavatum* as subspecies of the polymorphic species *Hy. anatolicum*. Based on the examination of a large number of ticks, as well as perusal of the type descriptions, Hoogstraal and Kaiser (1959) concluded that the names *Hy. anatolicum excavatum* and *Hy. anatolicum anatolicum* proposed by Pomerantzev (1946) for the larger and smaller subspecies, respectively, were correct, and that the reverse determination, despite its common usage in most non-Russian studies on ticks, was incorrect. Apanaskevich and Horak (2005) redescribed and illustrated all of the parasitic stages of these two important species, which are now regarded as independent species without subspecific status. They also provided numerous details for the separation of all stages of both species. Both species are present in northern Africa and it is uncertain as to exactly which of the two occur in southern European countries such as Bulgaria, Italy and the countries that constituted the former Yugoslavia. Apanaskevich (2003b) stated that *Hy. excavatum* is present in Greece and Cyprus, but this does not exclude the possibility that *Hy. anatolicum* might also occur there. SEM images of *Hy. excavatum* nymphs (as *Hy. anatolicum excavatum*) and a comparison with nymphs of *Hy. marginatum* and *Hy. dromedarii* are available in Abdel-Shafy (2008).

Other than the publications mentioned above, there are keys and illustrations for the larvae of *Hy. marginatum*, *Hy. rufipes*, *Hy. turanicum*, *Hy. aegyptium*, *Hy. scupense*, *Hy. anatolicum*, *Hy. excavatum*, *Hy. asiaticum* and *Hy. dromedarii* by Apanaskevich and Filippova (2007). Please note that some of the species included in that publication, such as the group of species formerly included in the *Hy. marginatum* complex, still have the old designation as subspecies.

Key to the Species of Genus *Hyalomma* in Europe and Northern Africa

Females.

We were unable to examine adequate material of *Hy. aegyptium*, *Hy. franchinii* and *Hy. asiaticum*. The keys provided are tentative. No illustrations are provided for these species because of the unavailability of reliable material for every stage.

1a. Coxa I with two short spurs, separate and almost equal in length…***Hyalomma aegyptium***

1b. Coxa I deeply incised with two contiguous, unequal spurs (the internal one wide and flat, the external one conical and pointed)…**2**

2a. Scutum with marbled white spots that converge irregularly and extend to the basis capituli, palpi and legs. Genital opening with a convex anterior margin and lateral margins converging to a point with their outline neither prominent nor depressed…**3**

2b. Scutum uniformly coloured, without marbled colouration. Legs uniformly coloured or with rings of pale enamelling. Genital opening with lateral margins not converging to a point and with depressed or prominent outlines…**4**

3a. Lateral spur of coxa I subequal to internal spur…**10**

3b. Lateral spur of coxa I longer than internal spur…***Hyalomma excavatum*** (Fig. 162).

4a. Scutum smooth and shiny, with sparse punctations. Cervical grooves superficial, not extending as far as the posterolateral margins. Legs uniformly coloured, without marbled periarticular rings. Genital opening triangular, almost equilateral, with rounded angles and outline gradually depressed posteriorly…***Hyalomma scupense*** (Fig. 153).

4b. Scutum rugose with dense and large punctations. Cervical grooves deep, extending to the posterolateral margins. Ivory-coloured enamelling on the leg segments indistinct. Genital opening variously shaped…5

5a. Punctations on the cervical grooves commonly confluent, deep, giving the aspect of large ovoid holes on the cervical fields near the eyes…***Hyalomma anatolicum*** (Fig. 159).

5b. Punctations on the cervical grooves not confluent but deep, giving the aspect of large ovoid depressions on the cervical fields...**6**
6a. Circumspiracular setae sparse. Dorsal ivory-coloured strip on segments of leg IV complete. Punctations on the scutum usually sparse. Dorsal prolongation of the spiracular plate broad...***Hyalomma marginatum*** (Fig. 140).
6b. Circumspiracular setae sparse or very dense. The segments of leg IV have a small to medium-sized ivory-coloured spot...**7**
7a. Punctations on the scutum and circumspiracular setae very dense. Scutum dark-brown. Dorsal prolongation of the spiracular plate very narrow...***Hyalomma rufipes*** (Fig. 143).
7b. Punctations on the scutum sparse or dense. Dorsal prolongation of the spiracular plate which can be variously shaped. Circumspiracular setae never very dense...**8**
8a. Moderately deep cervical grooves. Preatrial fold of the genital opening bulging. Internal spur on coxa I relatively narrow, with tapering apex...***Hyalomma impeltatum*** (Fig. 150).
8b. Cervical grooves very deep...**9**
9a. Preatrial fold of genital operculum flat. Internal spur on coxa I broad and blunt...***Hyalomma dromedarii*** (Fig. 147).
9b. Preatrial fold of genital operculum bulging. Internal spur on coxa I relatively narrow...***Hyalomma asiaticum***
10a. Scutum very smooth with few punctations...***Hyalomma franchinii***
10b. Scutum with relatively dense large and medium punctations...***Hyalomma lusitanicum*** (Fig. 156).

Males.

We were unable to examine material of *Hy. aegypitium*, *Hy. franchinii* and *Hy. asiaticum*. The keys provided are tentative. No illustrations are provided for these three species because of the unavailability of reliable material.

1a. Coxa I with two short spurs, separate and almost equal in length...***Hyalomma aegyptium***
1b. Coxa I deeply incised, with two contiguous, unequal spurs (internal one wider and flat, external one conical and pointed)...**2**
2a. Lateral grooves superficial. Conscutum with whitish spots...**10**
2b. Lateral grooves deep. Conscutum without whitish spots. Adanal plates with posterior margins shorter than inner margins. Legs uniformly coloured or with stripes or depigmented rings...**3**
3a. Lateral grooves very long, extending from the festoons almost to the eyes. Conscutum smooth with very few large punctations. Posteromedian and paramedian grooves well defined. Festoons fused, parma present. Adanal plates long and narrow, with concave inner margins and anteromedial angles almost meeting behind the anus. Accessory adanal plates crescent-shaped, small, only exceptionally projecting from posterior margin of the idiosome. Legs uniformly coloured and without stripes or depigmented rings...***Hyalomma scupense*** (Fig. 153).
3b. Lateral grooves short or long. Conscutum rough, with small and/or large punctations; cervical grooves extending beyond the level of the eyes. Posteromedian and paramedian grooves more or less visible. Festoons may or may not be fused. Parma present or absent. Adanal plates generally extending beyond the posterior margin of the idiosome even in unfed specimens. Legs with areas of enamelling...**4**
4a. With more than six setae on the medial side of palpal segment I. Cervical grooves long and deep. Adanal plates distinctly curved medially. Distal leg segments with enamel. Posteromedian groove reaching parma...***Hyalomma dromedarii*** (Fig. 147).
4b. With less than six setae on the medial side of the palpal segment I. Lateral grooves short or long, deep or shallow...**5**
5a. Cervical grooves shallow, about one-third of the length of the conscutum. Lateral grooves long, almost reaching the eyes. Parma absent...**6**
5b. Cervical grooves deep. Lateral grooves short...**7**
6a. Large punctations rare, small punctations especially dense on the caudal field. A complete longitudinal dorsal, ivory strip of enamelling on each leg segment, particularly distinct on the hind legs. Circumspiracular setae sparse. Perforated portion of the dorsal prolongation of the spiracular plate broad...***Hyalomma marginatum*** (Fig. 140).
6b. Entire conscutum covered with very dense medium-sized punctations. Perforated portion of the dorsal prolongation of the spiracular plate extremely narrow. Circumspiracular setae very dense...***Hyalomma rufipes*** (Fig. 143).
7a. Ivory marbling often present on the conscutum. Posteromedian groove separated from the parma by a distinct ridge connected to the paraparmal festoons. Marginal grooves short...***Hyalomma excavantum*** (Fig. 162).
7b. Ivory marbling usually absent from the conscutum...**8**
8a. Posteromedian groove separated from the parma by contiguous punctations...***Hyalomma anatolicum*** (Fig. 159).
8b. Posteromedian groove reaching the parma...**9**
9a. Cervical grooves deep and long...***Hyalomma asiaticum***
9b. Cervical grooves deep and short...***Hyalomma impeltatum*** (Fig. 150).
10a. Conscutum with relatively dense large, medium and small punctations. Adanal plates relatively broad with a clear concavity in their posterior margin. Posterolateral spur of coxa I longer or subequal to posteromedian spur...***Hyalomma lusitanicum*** (Fig. 156).
10b. Conscutum very smooth with few punctations. Adanal plates relatively broad with a very small concavity on their posterior margin, posterolateral spur of coxa I distinctly shorter than posteromedian spur...***Hyalomma franchinii***

Nymphs and Larvae

Note: According to the literature on the identification of immatures of the genus *Hyalomma*, most of the species cannot be adequately separated in the absence of large collections of adequate material. We do not include keys for these stages and strongly recommend that the engorged stages be allowed to moult in order to obtain flat adults for accurate identification.

References

Abdel-Shafy S (2008) Scanning electron microscopy and comparative morphology of *Hyalomma anatolicum excavatum*, *H. dromedarii* and *H. marginatum marginatum* (Acari: Ixodidae) based on nymphs. Acarologia XLVIII:3–18

Apanaskevich DA (2003a) To diagnostics of *Hyalomma (Hyalomma) aegyptium* (Acari: Ixodidae). Parazitologiya 37:47–59 (in Russian)

Apanaskevich DA (2003b) Differentiation of closely related species *Hyalomma anatolicum* and *H. excavatum* (Acari, Ixodidae) based on a study of all life cycle stages, throughout their entire geographical range. Parazitologiya 37:259–280

Apanaskevich DA, Horak IG (2005) The genus *Hyalomma* Koch, 1844. II. Taxonomic status of *H. (Euhyalomma) anatolicum* Koch, 1844 and *H. (E.) excavatum* Koch, 1844 (Acari, Ixodidae) with redescriptions of all stages. Acarina 13:181–197

Apanaskevich DA, Horak IG (2008) The genus *Hyalomma* Koch, 1844: V. Re-evaluation of the taxonomic rank of taxa comprising the *H. (Euhyalomma) marginatum* Koch complex of species (Acari: Ixodidae) with redescription of all parasitic stages and notes on biology. Int J Acarol 34:13–42

Apanaskevich DA, Horak IG (2009) The genus *Hyalomma* Koch, 1844. IX. Redescription of all parasitic stages of *H. (Euhyalomma) impeltatum* Schulze & Schlottke, 1930 and *H. (E.) somalicum* Tonelli Rondelli, 1935 (Acari: Ixodidae). Syst Parasitol 73:199–218

Apanaskevich DA, Santos-Silva MM, Horak IG (2008a) The genus *Hyalomma* Koch, 1844. IV. Redescription of all parasitic stages of *H. (Euhyalomma) lusitanicum* Koch, 1844 and the adults of *H. (E.) franchinii* Tonelli Rondelli, 1932 (Acari: Ixodidae) with a first description of its immature stages. Folia Parasitol 55:61–74

Apanaskevich DA, Schuster AL, Horak IG (2008b) The genus *Hyalomma*: VII. Redescription of all parasitic stages of *H. (Euhyalomma) dromedarii* and *H. (E.) schulzei* (Acari: Ixodidae). J Parasitol 45:817–831

Apanaskevich DA, Filippova NA, Horak IG (2010) The genus *Hyalomma* Koch, 1844. X. Redescription of all parasitic stages of *H. (Euhyalomma) scupense* Schulze, 1919 (=*H. detritum* Schulze) (Acari: Ixodidae) and notes on its biology. Folia Parasitol 57:69–78

Bitam I, Kernif T, Harrat Z, Parola P, Raoult D (2009) First detection of *Rickettsia aeschlimannii* in *Hyalomma aegyptium* from Algeria. Clin Microbiol Infect 15:253–254

Bouattour A, Darghouth MA, Daoud A (1999) Distribution and ecology of ticks (Acari: Ixodidae) infesting livestock in Tunisia: an overview of eight years field collections. Parassitologia 41:5–10

Camicas JL (1970) Contribution a l'e ́tude des tiques du Sénégal (Acarina, Ixodoidea). I. Les larves d'*Amblyomma* Koch et de *Hyalomma* Koch. Acarologia 12:71–102

Cringoli G, Iori A, Rinaldi L, Veneziano V, Genchi C (2005) Mappe parassitologiche—Zecche. Series Edit. Giuseppe Cringoli, Napoli

Cwilich R, Hadani A (1962) The identification of the tick *Hyalomma franchinii*, Tonelli-Rondelli, 1932. Refu Vet 19:178–180

Dzhaparidze N.I., 1960. Ixodid ticks of Georgia. Izd Akad Nauk Gruzinskoy SSR, Tbilisi, 295 p (in Russian)

Feider Z (1965) Fauna Republicii Populare Române. Arachnida, vol 5, fasc. 2. Acaromorpha, Suprafamilie Ixodoidea (Capuse) Editura Academiei Republicii Populare Române, Bucuresti, 404 p

Feldman-Muhsam B (1954) Revision of the genus *Hyalomma*. I. Description of Koch's types. Bull Res Council Isr 4:150–170

Filippova NA (2007) Type specimens of argasid and ixodid ticks (Ixodoidea: Argasidae, Ixodidae) in the collection of the Zoological Institute, Russian Academy of Sciences (St. Petersburg). Entomol Rev 88:1002–1011 (original version in Russian, Entomologicheskoe Obozrenie, 87, 939–951)

Güner ES, Watanabe M, Hashimoto N, Kadosaka T, Kawamura Y, Ezaki T, Kawabata H, Imai Y, Kaneda K, Masuzawa T (2004) *Borrelia turcica* sp. nov., isolated from the hard tick *Hyalomma aegyptium* in Turkey. Int J Syst Evol Microbiol 54:1649–1652

Hoogstraal H (1956) African Ixodoidea. I. Ticks of the Sudan (with special reference to Equatoria Province and with preliminary reviews of the genera *Boophilus*, *Margaropus* and *Hyalomma*). U. S. Navy, Washington D.C., 1101 p

Hoogstraal H, Kaiser MN (1958a) Observations on Egyptian *Hyalomma* ticks (Ixodoidea, Ixodidae). I. Parasitism of lizards by nymphs. Ann Entomol Soc Am 51:7–12

Hoogstraal H, Kaiser MN (1958b) Observations on Egyptian *Hyalomma* ticks (Ixodoidea, Ixodidae). 4. Identity, distribution, and hosts of *H. franchinii* Tonelli-Rondelli (new combination). Systematic status of *H. tunesiacum* Sc. & Sc. and its subspecies. Ann Entomol Soc Am 51:397–400

Hoogstraal H, Kaiser MN (1959) Observations on Egyptian *Hyalomma* ticks (Ixodoidea, Ixodidae). 5. Biological notes and differences in identity of *H. anatolicum* and its subspecies *anatolicum* Koch and *excavatum* Koch among Russian and other workers. Identity of *H. lusitanicum* Koch. Ann Entomol Soc Am 52:243–261

Hoogstraal H, Kaiser MN (1960) Observations on ticks (Ixodoidea) of Libya. Ann Entomol Soc Am 53:445–457

Hoogstraal H, Wassef HY, Büttiker W (1981) Ticks (Acarina) of Saudi Arabia: fam. Argasidae, Ixodidae. Fauna Saudi Arabia 3:25–110

Kaiser MN, Hoogstraal H (1963) The *Hyalomma* ticks (Ixodoidea, Ixodidae) of Afghanistan. J Parasitol 130–139

Kaiser MN, Hoogstraal H (1968) Re-description of *Hyalomma (H.) erythraeum* Tonelli-Rondelli (resurrected), description of the female and immature stages, and hosts and distribution in Ethiopia and Somali Republic. Ann Entomol Soc Am 61:1228–1235

Kolonin GV (1983) World distribution of ixodid ticks. Genera *Hyalomma*, *Aponomma*, *Amblyomma*. Izd Nauka, Moscow, 120 p (in Russian)

Manilla G (1998) Fauna d'Italia. Acari, Ixodida. Edizioni Calderini, Bologna, p 280

Manilla G, Giannetto S (1996) Scanning electron microscopy study for an analytical key to the Hyalomminae (Ixodidae) species to be found in Italy. Parassitologia 38:537–541

Morel PC (1959) Les *Hyalomma* (Acariens, Ixodidae) de France. Ann Parasitol Hum Comp 34:552–555

Neumann G (1901) Revision de la famille des Ixodideés (4e mémoire). Mém Soc Zool Fr 14:249–372

Ogandzhanyan AM (1953) Larvae and nymphs of ticks of the genus *Hyalomma* Koch of Armenian SSR. Zool Sb Zool Inst Akad Nauk Armyan SSR 8:149–167 (in Russian)

Pérez-Eid C, Cabrita J (2003) La larve et la nymphe de *Hyalomma (Hyalomma) lusitanicum* Koch, 1844 (Acari: Ixodida): description morphologique, habitats, hotes. Acarologia 43:327–335

Pomerantzev BI (1946) Ticks (Ixodidae) of the USSR and neighbouring countries. Opred Faune SSSR Zool Inst Akad Nauk SSSR, vol 26. Izd Akad Nauk SSSR, Moscow, Leningrad, 28 p (in Russian)

Pomerantzev BI (1950) Ixodid ticks (Ixodidae). Fauna SSSR, Paukoobraznye, vol 4, No. 2. Izd Akad Nauk SSSR, Moscow, Leningrad, 224 p (in Russian)

Rousselot, R. (1948). *Hyalomma brumpti* Delpy, 1946. Description de la larve et de la nymphe. Biologie. Annales de Parasitologie Humaine et Comparée, 23:31–34.

Sénevet G (1924) Description de la nymphe de *Hyalomma mauritanicum* Sénevet 1922. Arch Inst Pasteur Algérie 2:233–234

Sénevet G (1928) Clé pour la détermination des larves d'Ixodidés trouvées sur les bovines en Algérie. Arch Inst Pasteur Algérie 6:42–46

Široký P, Petrželková KJ, Kamler M, Mihalca AD, Modrý D (2006) *Hyalomma aegyptium* as dominant tick in tortoises of the genus *Testudo* in Balkan countries, with notes on its host preferences. Exp Appl Acarol 40:279–290

Starkoff O (1958) Ixodoidea d'Italia. Il Pensiero Scientifico, Roma, p 385

Takano A, Goka K, Une Y, Shimada Y, Fujita H, Shiino T, Kawabata H (2010) Isolation and characterization of a novel Borrelia group of tick-borne borreliae from imported reptiles and their associated ticks. Environ Microbiol 12:134–146

Tendeiro J (1962) Revisão sistemática dos Ixodídeos Portugue- ses. Bol Pecu 30:5–131

Walker AR, Bouattour A, Camicas JL, Estrada-Peña A, Horak IG, Latif AA, Pegram RG, Preston PM (2003) Ticks of domestic animals in Africa: a guide to identification of species. Bioscience reports, Edinburgh, pp 3–210

Hyalomma marginatum Koch, 1844 (Figs. 139–141)

M. M. Santos-Silva and Z. Vatansever

Life-Cycle and Host Preferences

Hyalomma marginatum is a two-host tick with one generation per year in nature (Hoogstraal 1979). Under laboratory conditions, the shortest time to complete the life-cycle is variable, ranging from 73 to 97 days, and seems to depend on the host species (Berezin 1971; Ouhelli 1994; Yukari et al. 2011; Gargili et al. 2013). Adults feed on a wide variety of mammals, mainly wild and domestic ungulates, particularly bovines (Pomerantzev 1950; Berezin 1971; Hoogstraal 1979; Apanaskevich 2004). The larvae and nymphs are specific to small mammals (leporids and insectivores) and to ground-dwelling birds of various taxonomic groups (Pomerantzev 1950; Apanaskevich 2004; Apanaskevich and Horak 2008; Kotti et al. 2001). Other hosts referred to in the literature require confirmation (Guglielmone et al. 2014). This tick is often reported as biting humans (Petrova-Piontkovskaya 1947; Karaer et al. 2011; Santos-Silva et al. 2011; Keskin et al. 2015).

Ecology

Hyalomma marginatum has a Palaearctic distribution showing ecological plasticity. It is adapted to several biogeographical regions supporting an extensive range of abiotic conditions (e.g. temperature and humidity), from humid Mediterranean climates to the arid environments of steppe regions (Estrada-Peña and Venzal 2007). All stages are more active during spring and summer and are rarely seen during winter. Larvae are mostly active in June and July and nymphs first appear in July (Tokhov Iu 2009). Adults begin to appear in March, reach their peak activity in May and gradually decrease until August, although some individuals can be observed until October or even November (Hoogstraal 1956; Berezin 1971; Hoogstraal 1979; Estrada-Peña and Venzal 2007; Tokhov Iu 2009; Estrada-Peña et al. 2011). *Hyalomma marginatum* is a non-nidicolous tick. Its host-seeking behaviour relies on the use of an active locomotory hunting strategy (Petrova-Piontkovskaya 1947; Emelianova 2006; Tokhov Iu 2009; Estrada-Peña et al. 2011). Whenever a sensorial, chemical and mechanical stimulus indicates the presence of a host they run rapidly several metres across the ground to attack, similar to other *Hyalomma* species such as *Hy. asiaticum* and *Hy. dromedarii* (Petrova-Piontkovskaya 1947; Sonenshine 2005; Emelianova 2006; Romanenko 2007).

Distribution

The distribution of *Hy. marginatum* includes southern Europe, northern Africa and some parts of Asia. This tick has been reported in Albania, Algeria, Armenia, Azerbaijan, Bosnia and Herzegovina, Bulgaria, Croatia, Cyprus, Egypt, France, Georgia, Greece, Iran, Iraq, Israel, Italy, Kosovo, Libya, the Republic of Macedonia, Moldova, Montenegro, Morocco, Portugal, Romania, Russia, Serbia, Spain, Syria, Tunisia, Turkey, Turkmenistan and the Ukraine (Pomerantzev 1950; Apanaskevich and Horak 2008). Occasional records from Central and Northern Europe, such as Finland, Germany, Hungary, the United Kingdom and from other world areas, including the Afrotropical (Apanaskevich and Horak 2008), Neotropical (Labruna et al. 2001) and Neartic regions (Burridge 2011) are most probably a consequence of its immature stages being transported attached to migratory birds or due to human travel and imported animals, rather than representing well-established populations (Kampen

M. M. Santos-Silva (✉)
Centro de Estudos de Vectores e Doenças Infecciosas Dr. Francisco Cambournac (CEVDI), Instituto Nacional de Saúde Doutor Ricardo Jorge, Águas de Moura, Portugal
e-mail: m.santos.silva@insa.min-saude.pt

Z. Vatansever
Department of Parasitology, Faculty of Veterinary Medicine, Kafkas University, Kars, Turkey
e-mail: zativet@gmail.com

A. Estrada-Peña et al. (eds.), *Ticks of Europe and North Africa*,
https://doi.org/10.1007/978-3-319-63760-0_66

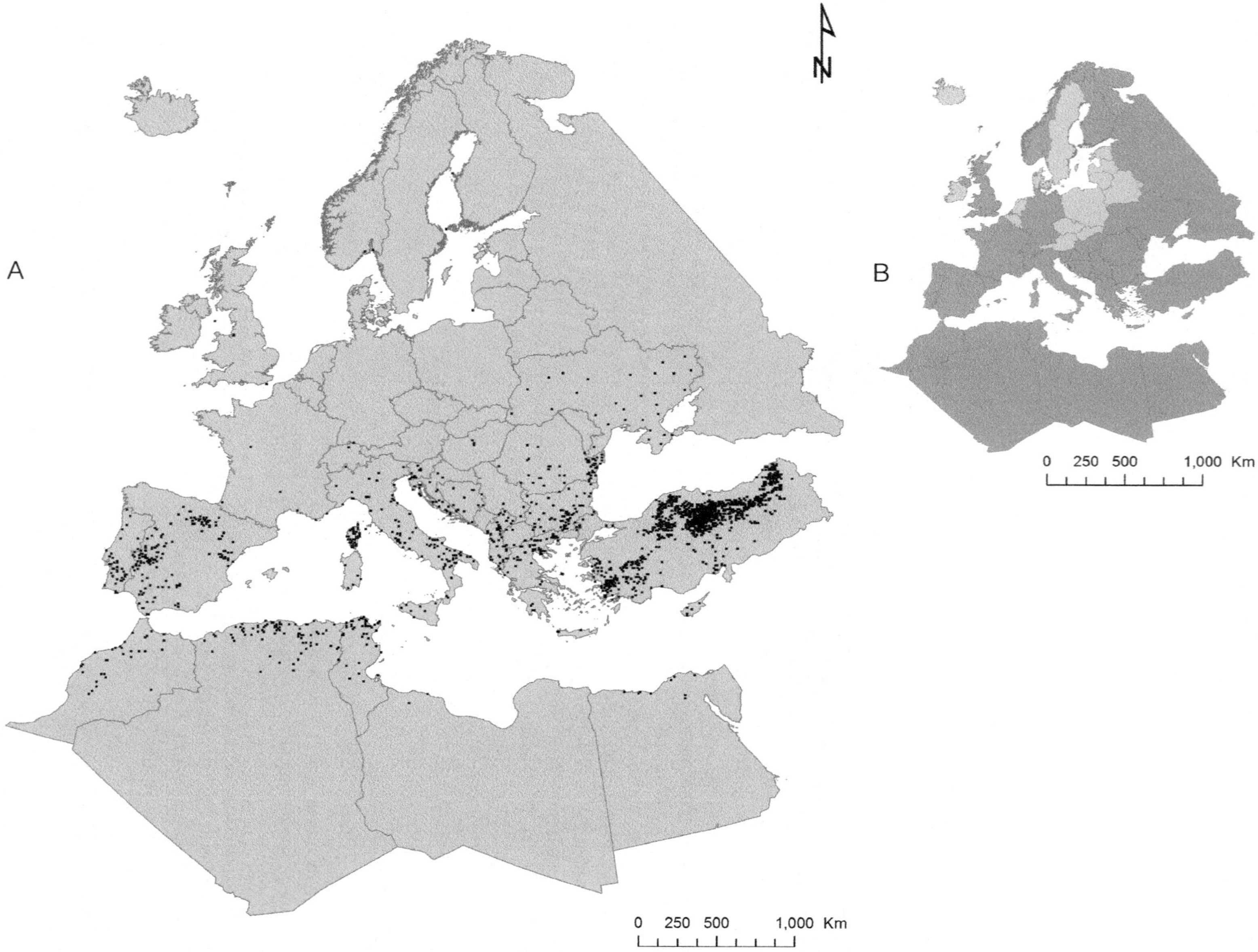

Fig. 139 **A** Distribution of *Hy. marginatum* in Europe and Northern Africa (10 × 10 km grid presence with black dots). **B** Countries where the species has been reported are marked in dark grey

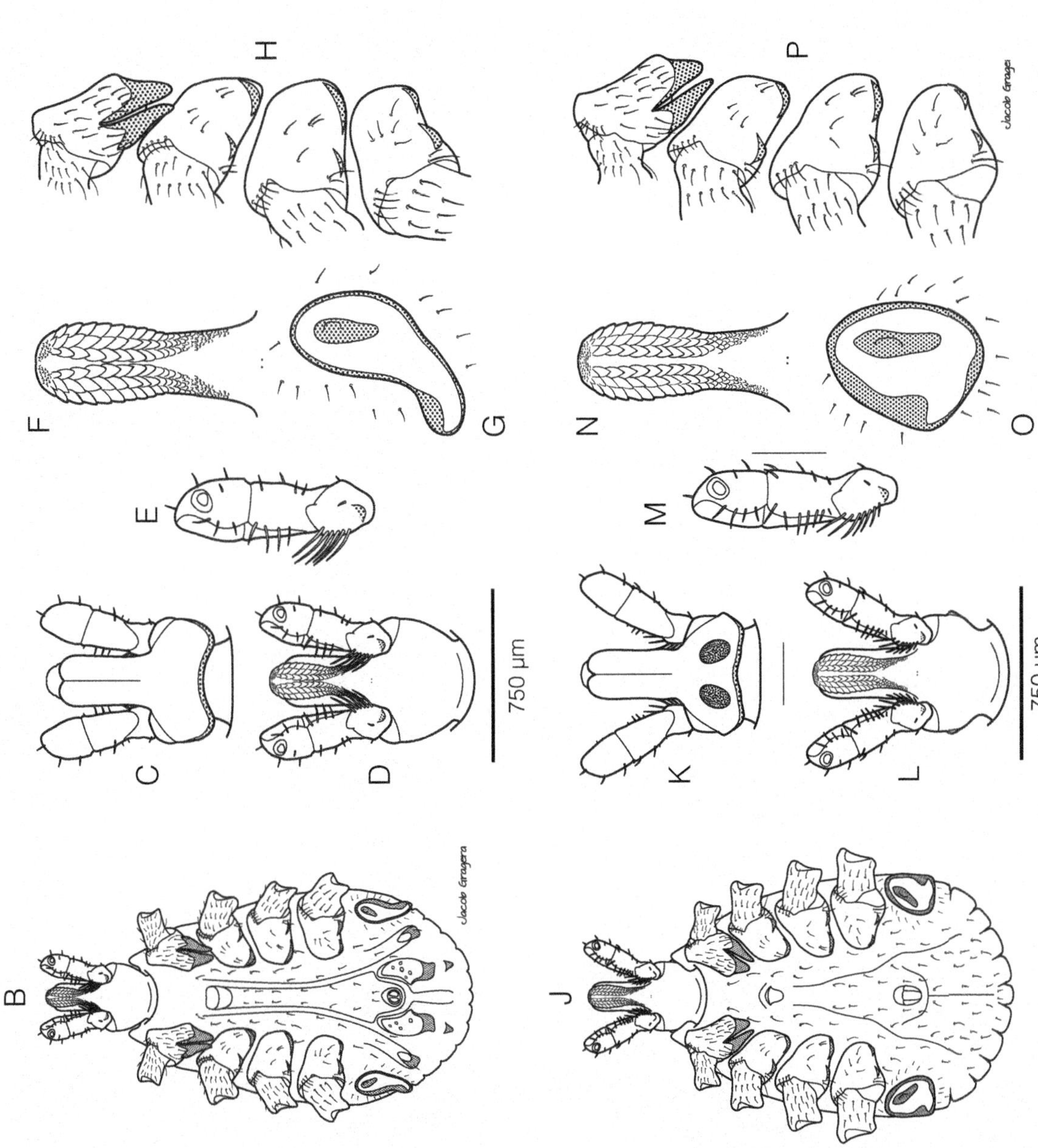

Fig. 140 **A–H** The male of *Hy. marginatum*. **A** Dorsal, **B** ventral, **C** capitulum, dorsal, **D** capitulum, ventral, **E** palpi, lateral view, **F** hypostome, **G** spiracular plate, **H** coxae and trochanters I–IV. **I–P** The female of *Hy. marginatum*. **I** Dorsal, **J** ventral, **K** capitulum, dorsal, **L** capitulum, ventral, **M** palpi, lateral view, **N** hypostome, **O** spiracular plate, **P** coxae and trochanters I–IV. Illustrations from specimens collected in Spain, Italy and Turkey

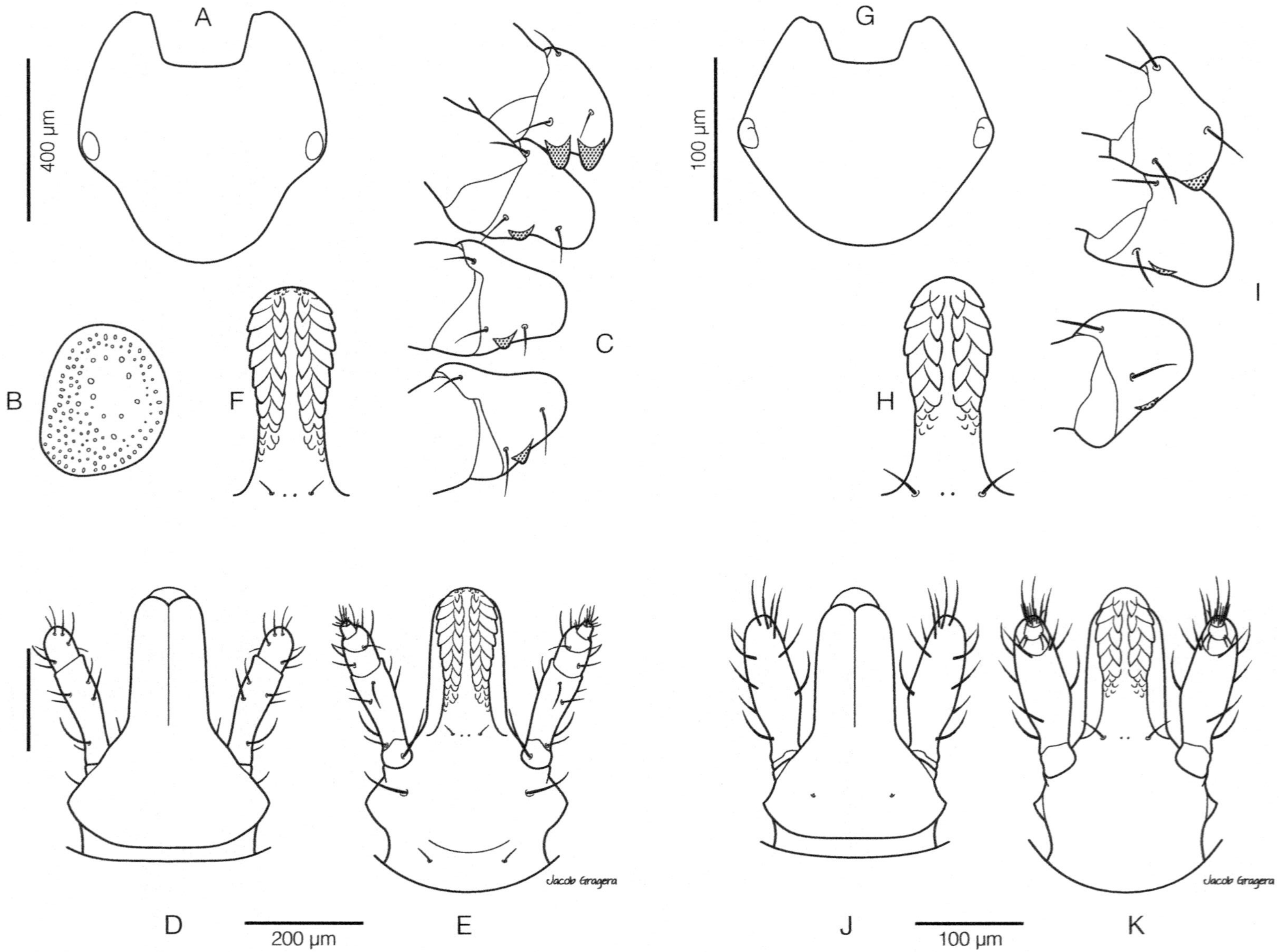

Fig. 141 **A–F** The nymph of *Hy. marginatum*. **A** Dorsal scutum, **B** spiracular plate, **C** coxae I–IV, **D** capitulum, dorsal, **E** capitulum, ventral, **F** hypostome. **G–K** The larva of *Hy. marginatum*. **G** Dorsal plate, **H** hypostome, **I** coxae I–III, **J** capitulum, dorsal, **K** capitulum, ventral. Illustrations from specimens obtained in the laboratory, from engorged females collected in Spain and Tunisia

et al. 2007; Jameson et al. 2012; Hornok et al. 2013; Guglielmone et al. 2014). An expansion of its distribution might be expected in the Mediterranean region (Estrada-Peña and Venzal 2007).

Vectorial Capacity and Pathogen Burden

Hyalomma marginatum is the primary vector to humans of Crimean-Congo haemorrhagic fever virus (Hoogstraal 1979; Ergonul 2006). Experimental transmission attempts with West Nile virus were performed under laboratory conditions. These demonstrated that *Hy. marginatum* could be involved in the natural circulation of this virus in Portugal (Formosinho and Santos-Silva 2006). Other pathogens have been associated with this tick species: *Anaplasma marginale* (De la Fuente et al. 2004), *A. phagocytophylum* (Keysary et al. 2007), *Babesia bigemina* (Iori et al. 2010), *B. bovis* (Iori et al. 2010), *B. caballi* (Estrada-Peña et al. 2004), *B. occultans* (Ioniţă et al. 2013; Aktas et al. 2014), *Borrelia lusitaniae* (De Michelis et al. 2000), *Coxiella burnetii* (Hoogstraal 1956), Dhori virus (Filipe and Casals 1979); *Rickettsia aeschlimannii* (Beati et al. 1997), *Theileria annulata* (Georges et al. 2001), *T. equi* (Iori et al. 2010) and *T. orientalis/sergenti/buffeli*-group (Ioniţă et al. 2013). Further studies using experimental assays are needed to validate the vectorial capacity of *Hy. marginatum* for the majority of these pathogens.

References

Aktas M, Vatansever Z, Ozubek S (2014) Molecular evidence for trans-stadial and transovarial transmission of *Babesia occultans* in *Hyalomma marginatum* and *Rhipicephalus turanicus* in Turkey. Vet Parasitol 204:369–371

Apanaskevich DA (2004) Host-parasite relationships of the genus *Hyalomma* Koch, 1844 (Acari, Ixodidae) and their connection with microevolutionary process. Parazitologiia 38:515–523 (in Russian)

Apanaskevich DA, Horak IG (2008) The genus *Hyalomma* Koch, 1844: v. re-evaluation of the taxonomic rank of taxa comprising the *H.* (*Euhyalomma*) *marginatum* Koch complex of species (Acari: Ixodidae) with redescription of all parasitic stages and notes on biology. Int J Acarol 34:13–42

Beati L, Meskini M, Thiers B, Raoult D (1997) *Rickettsia aeschlimannii* sp. nov., a new spotted fever group rickettsia associated with *Hyalomma marginatum* ticks. Int J Syst Bacteriol 47(2):548–554

Berezin VV (1971) Investigation of the ecology of arboviruses in river deltas of the Caspian and Azov sea basins (Avtoref. Diss. Soisk. Uchen. Step. Dokt. Biol. Nauk). Inst. Polio. Virusn. Entsefalitov, Akad. Med. Nauk SSSR, Moscow. 37 p (in Russian) (in English, NAMRU3-T1160)

Burridge MJ (2011) Non-native and invasive ticks: threats to human and animal health in the United States. University Press of Florida, Gainesville, FL, USA, p 448

De la Fuente J, Vicente J, Höflea U, Ruiz-Fons F, Mera I, Van Den Bussche R, Kocan K, Gortazar C (2004) *Anaplasma* infection in free-ranging Iberian red deer in the region of Castilla-La Mancha, Spain. Vet Microbiol 100:163–173

De Michelis S, Sewell HS, Collares-Pereira M, Santos-Reis M, Schouls LM, Benes V, Holmes EC, Kurtenbach K (2000) Genetic diversity of *Borrelia burgdorferi* sensu lato in ticks from mainland Portugal. J Clin Microbiol 38(6):2128–2133

Emelianova IN (2006) Hyalomma Koch, 1844 (ACARI: IXODIDAE) ticks of Central Precaucasia and surrounding territories (distribution, ecology, role in the natural foci of Crimean-Congo heamorrhagic fever). MSc thesis in biology. State University of Stavropol, Stavropol, 146 p

Ergonul O (2006) Crimean-Congo haemorrhagic fever. Lancet Infect Dis 6:203–214

Estrada-Peña A, Venzal JM (2007) Climate niches of tick species in the Mediterranean region: modeling of occurrence data, distributional constraints, and impact of climate change

Estrada-Peña A, Bouattour A, Camicas J-L, Walker AR (2004) Ticks of domestic animals in the Mediterranean area. A guide to identification of species. University of Zaragoza, Zaragoza, Spain

Estrada-Peña A, Martinez Aviles M, Munoz Reoyo MJ (2011) A population model to describe the distribution and seasonal dynamics of the tick *Hyalomma marginatum* in the Mediterranean Basin. Transbound Emerg Dis 2011

Filipe AR, Casals C (1979) Isolation of Dhori from *Hyalomma marginatum* ticks in Portugal. Intervirol 11:124–127

Formosinho P, Santos-Silva MM (2006) Experimental infection of *Hyalomma marginatum* ticks with West Nile virus. Acta Virol 50:175–180

Gargili A, Thangamani S, Bente D (2013) Influence of laboratory animal hosts on the life cycle of *Hyalomma marginatum* and implications for an in vivo transmission model for Crimean-Congo hemorrhagic fever virus. Front Cell Infect Microbiol 3:39

Georges K, Loria GR, Riili S, Greco A, Caracappa S, Jongejan F, Sparagano O (2001) Detection of haemoparasites in cattle by reverse line blot hybridisation with a note on the distribution of ticks in Sicily. Vet Parasitol 99:273–286

Guglielmone AA, Robbins RG, Apanaskevich DA, Petney TN, Estrada-Peña A, Horak IG (2014) The hard ticks of the world (Acari: Ixodida: Ixodidae). Springer, Dordrecht, p 738

Hoogstraal H (1956) African Ixodoidea. 1. Ticks of the Sudan. [Washington]: United States. Navy Dept. Bureau of Medicine and Surgery, United States. Naval Medical Research Unit No. 3

Hoogstraal H (1979) The epidemiology of tick-borne Crimean-Congo hemorrhagic fever in Asia, Europe, and Africa. J Med Entomol 15:307–417

Hornok S, Csorgo T, de la Fuente J, Gyuranecz M, Privigyei C, Meli ML, Kreizinger Z, Gönczi E, Fernández de Mera IG, Hofmann-Lehmann R (2013) Synanthropic birds associated with high prevalence of tick-borne rickettsiae and with the first detection of *Rickettsia aeschlimannii* in Hungary. Vector Borne Zoonotic Dis 13:77–83

Ioniţă M, Mitrea IL, Pfister K, Hamel D, Silaghi C (2013) Molecular evidence for bacterial and protozoan pathogens in hard ticks from Romania. Vet Parasitol 196:71–76. https://dx.doi.org/10.1016/j.vetpar.2013.01.016

Iori A, Gabrielli S, Calderini P, Moretti A, Pietrobelli M, Tampieri MP, Galuppi R, Cancrini G (2010) Tick reservoirs for piroplasms in central and northern Italy. Vet Parasitol 170:291–296. https://dx.doi.org/10.1016/j.vetpar.2010.02.027

Jameson LJ, Morgan PJ, Medlock JM, Watola G, Vaux AG (2012) Importation of *Hyalomma marginatum*, vector of Crimean-Congo haemorrhagic fever virus, into the United Kingdom by migratory birds. Ticks Tick Borne Dis 3:95–99

Kampen H, Poltz W, Hartelt K, Wolfel R, Faulde M (2007) Detection of a questing *Hyalomma marginatum marginatum* adult female (Acari, Ixodidae) in southern Germany. Exp Appl Acarol 43:227–231

Karaer Z, Guven E, Nalbantoglu S, Kar S, Orkun O, Ekdal K, Kocak A, Akcay A (2011) Ticks on humans in Ankara, Turkey. Exp Appl Acarol 54:85–91

Keskin A, Keskin A, Bursali A, Tekin S (2015) Ticks (Acari: Ixodida) parasitizing humans in Corum and Yozgat provinces, Turkey. Exp Appl Acarol 67:607–616. https://dx.doi.org/10.1007/s10493-015-9966-4

Keysary A, Massung RF, Inbar M, Wallach AD, Shanas U, Mumcuoglu KY, Waner T (2007) Molecular evidence for *Anaplasma phagocytophilum* in Israel. Emerg Infect Dis 13:1411–1412. https://dx.doi.org/10.3201/eid1309.070455

Kotti BK, Shaposhnikova LI, Evchenko Iu M, Levchenko BI, Surkhaev DB, Korzhov PN, Tokhov Iu M (2001) [*Hyalomma marginatum* Koch in Stavropol' region]. Zh Mikrobiol Epidemiol Immunobiol, pp 105–108

Labruna MB, Pinter A, Roncati NV, Pereira MC, Faccini JL (2001) Exotic ticks on horses imported from Portugal to Brazil. Arq Bras Med Vet Zoo 53:179–182

Ouhelli H (1994) Comparative development of *Hyalomma marginatum* (Koch, 1844), *H. detritum* (Schulze, 1919), *H. anatolicum excavatum* (Koch, 1844), *H. lusitanicum* (Koch, 1844) and *H. dromedarii* (Koch, 1844) under laboratory conditions. Acta Parasitol 39:153–157

Petrova-Piontkovskaya SP (1947) Biological and ecological data on the tick *Hyalomma marginatum* Koch in the northwestern Crimean hemorrhagic fever focus. Nov Med 5:21–24 (Translation 864 (T864), Medical Zoology Department, US NAMRU, Cairo, Egypt)

Pomerantzev BI (1950) [Ixodid ticks (Ixodidae)]. Fauna of the USSR, Paukoobraznyye. New Series 41, 4:224

Romanenko VN (2007) Ecological bases of ethology of pasture ticks (Parasitiformes, Ixodidae) in finding and attacking hosts (in Russian). Bull Tomsk State Univ, pp 224–228

Santos-Silva MM, Beati L, Santos AS, De SR, Nuncio MS, Melo P, Santos-Reis M, Fonseca C, Formosinho P, Vilela C et al (2011) The hard-tick fauna of mainland Portugal (Acari: Ixodidae): an update on geographical distribution and known associations with hosts and pathogens. Exp Appl Acarol 55:85–121

Sonenshine DE (2005) The biology of tick vectors of human disease. In: Goodman JL, Dennis DT, Sonenshine DE (eds) Tick-borne diseases of humans. ASM Press, Washington, DC, pp 12–36

Tokhov Iu M (2009) Faunistic complex of Ixodidae in Stavropol Region: (dissemination, epizootic and epidemiologic value, control measures). Anti-plaque Institute, Stavropol (in Russian)

Yukari BA, Nalbantoglu S, Karaer Z, Inci A, Eren H, Sayin F (2011) [Some biological features of *Hyalomma marginatum* in the laboratory]. Turkiye parazitolojii dergisi/Turkiye Parazitoloji Dernegi. Acta Parasitol Turcica/Turkish Soc Parasitol 35:40–42

Hyalomma rufipes Koch, 1844 (Figs. 142–144)

Z. Vatansever

Life Cycle and Host Preferences

Hyalomma rufipes is a two-host exophilic tick with a one-year life cycle. Its host preferences are similar to those of *Hy. marginatum*. Larvae and nymphs feed on a large variety of birds and leporids; adults infest mostly domestic cattle. Sheep, goats, horses, camels, buffaloes, giraffes and rhinoceroses are listed among the other hosts of adults; antelopes and ostriches are less preferred (Hoogstraal 1956; Apanaskevich and Horak 2008). This species is reported to frequently infest humans (Horak et al. 2002). The life cycle of *Hy. rufipes* can be completed in as little as 5 months (Hoogstraal 1956). Under laboratory conditions preoviposition, oviposition and hatching take 7.1, 9 and 29.3 days on average, respectively (Knight et al. 1978). Larvae feed on hares for 8–15 days and moult to nymphs on the host in 4–6 days. Nymphs feed on the same host for 6–15 days. The total larval/nymphal feeding period takes 22.8 days on average (Knight et al. 1978). Female ticks feed on animals for 5–12 days (Hoogstraal 1956).

Ecology

There are very few data available on the ecology of *Hy. rufipes*. It is active during the summer months and is found in desert, semi-desert and arid lands with low rainfall. Its distribution is restricted by increasing humidity (Hoogstraal 1956). *Hyalomma rufipes* is reported to infest animals in September and October in Guinea (Tomassone et al. 2004) and between June and December in Turkey (Bakirci et al. 2011).

Distribution

Hyalomma rufipes is distributed in the dryer parts of Africa. Its natural distribution covers sub-Saharan Africa and regions around the Red Sea (Apanaskevich and Horak 2008). The reports of its presence outside of these regions (Ruiz-Fons et al. 2006; Bakirci et al. 2011; Hornok and Horvath 2012) are believed to be consequence of the dissemination of immature stages by migratory birds. Apanaskevich and Horak (2008) mentioned its presence in the Republic of Macedonia, Malta, the Ukraine, Israel, Egypt and Tunisia.

Vectorial Capacity and Pathogen Burden

Hyalomma rufipes is the main vector of Crimean–Congo haemorrhagic fever virus in Africa (Turell 2007). Transovarial and transstadial transmission of *Babesia occultans* have also been demonstrated in *Hy. rufipes* (Gray and De 1981). *Rickettsia aeshlimannii*, *Ehrlichia* spp., *Coxiella burnetii* and *Borrelia burgdorferi* s.l. have been detected in nymphs collected from birds in Italy (Toma et al. 2014). It can also transmit *Rickettsia conorii* and *Anaplasma marginale* (Bakheit et al. 2012).

Z. Vatansever (✉)
Department of Parasitology, Faculty of Veterinary Medicine, Kafkas University, Kars, Turkey
e-mail: zativet@gmail.com

A. Estrada-Peña et al. (eds.), *Ticks of Europe and North Africa*,
https://doi.org/10.1007/978-3-319-63760-0_67

Fig. 142 **A** Distribution of *Hy. rufipes* in Europe and Northern Africa (10 × 10 km grid presence with black dots). **B** Countries where the species has been reported are marked in dark grey

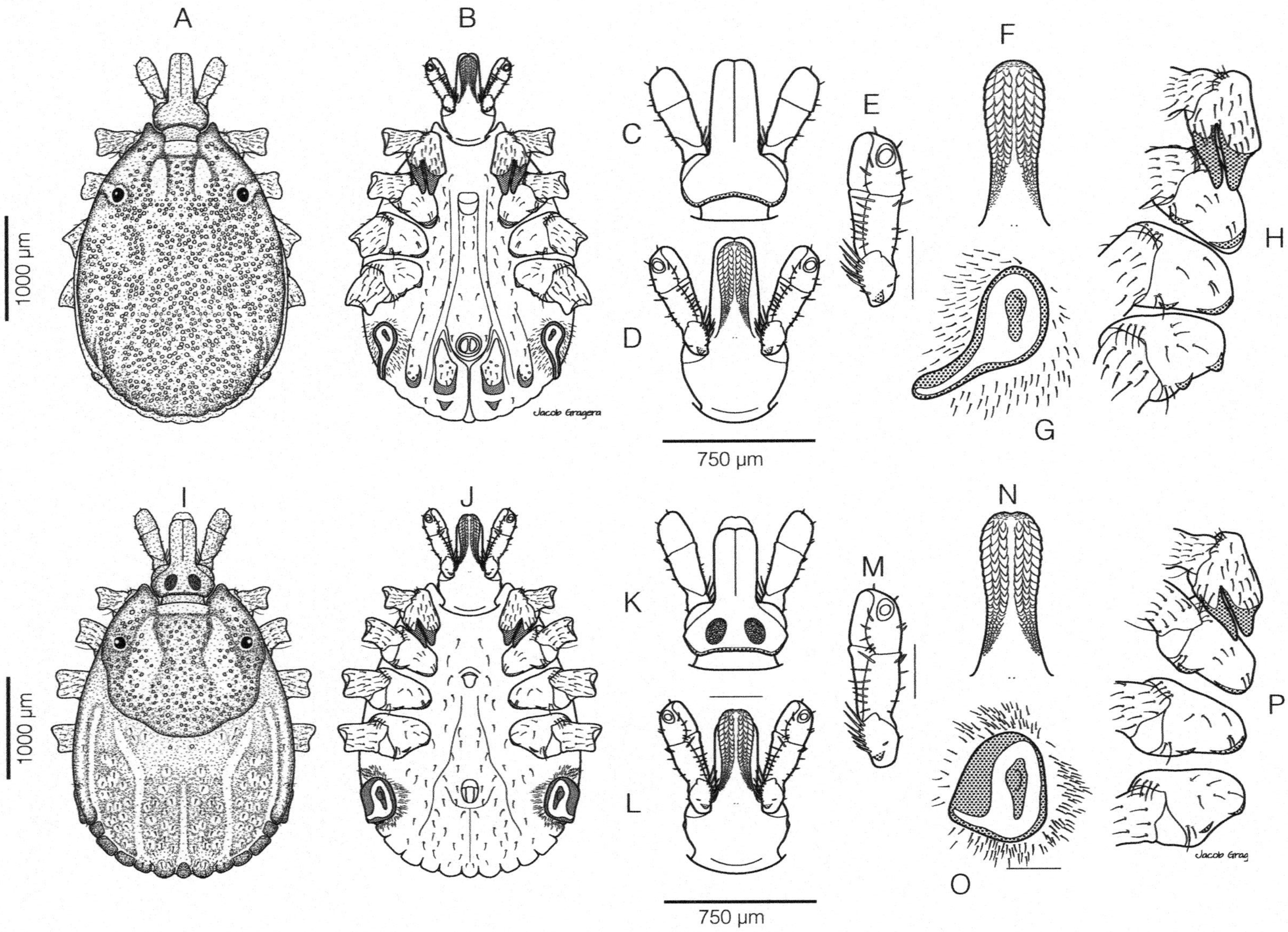

Fig. 143 **A–H** The male of *Hy. rufipes*. **A** Dorsal, **B** ventral, **C** capitulum, dorsal, **D** capitulum, ventral, **E** palpi, lateral view, **F** hypostome, **G** spiracular plate, **H** coxae and trochanters I–IV. **I–P** The female of *Hy. rufipes*. **I** Dorsal, **J** Ventral, **K** capitulum, dorsal, **L** capitulum, ventral, **M** palpi, lateral view, **N** hypostome, **O** spiracular plate, **P** coxae and trochanters I–IV. Illustrations from specimens collected in Tunisia

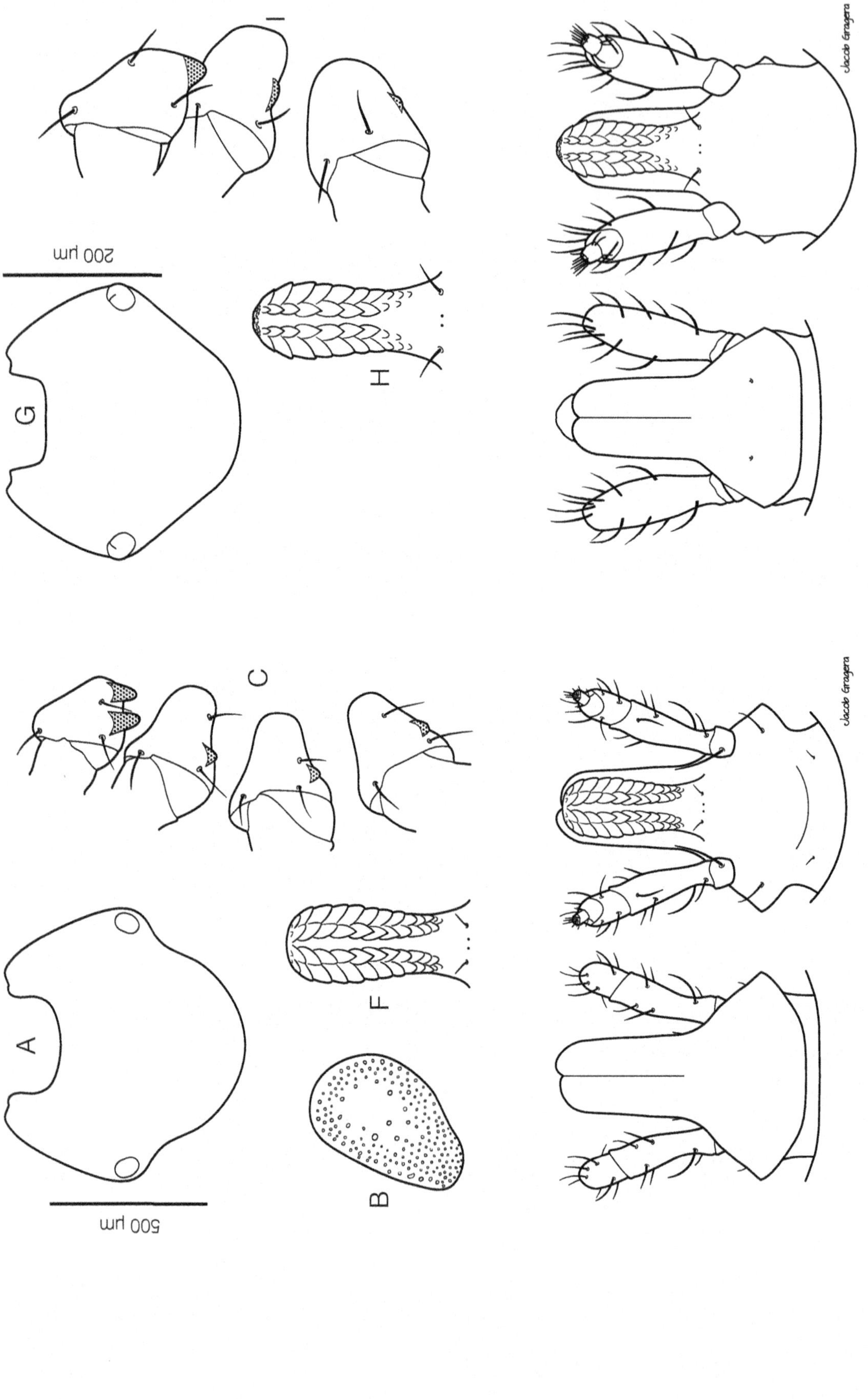

Fig. 144 **A–F** The nymph of *Hy. rufipes*. **A** Dorsal scutum, **B** spiracular plate, **C** coxae I–IV, **D** capitulum, dorsal, **E** capitulum, ventral, **F** hypostome. **G–K** The larva of *Hy. rufipes*. **G** Dorsal plate, **H** hypostome, **I** coxae I–III, **J** capitulum, dorsal, **K** capitulum, ventral. Illustrations from specimens obtained in the laboratory, from engorged females collected in Tunisia

References

Apanaskevich DA, Horak IG (2008) The genus *Hyalomma* Koch, 1844: v. Re-evaluation of the taxonomic rank of taxa comprising the *H. (Euhyalomma) marginatum* koch complex of species (Acari: Ixodidae) with redescription of all parasitic stages and notes on biology. Int J Acarol 34:13–42

Bakheit MA, Latif A, Vatansever Z, Seitzer U, Ahmed J (2012) The huge risks due to *Hyalomma* ticks. In: Mehlhorn H (ed) Arthropods as vectors of emerging diseases. Springer, Berlin, pp 167–194

Bakirci S, Sarali H, Aydin L, Latif A, Eren H, Karagenc T (2011) *Hyalomma rufipes* (Koch, 1844) infesting cattle in the West Aegean region of Turkey. Turk J Vet Anim Sci 35:359–363

Gray JS, De VAJ (1981) Studies on a bovine *Babesia* transmitted by *Hyalomma marginatum rufipes* Koch, 1844. Onderstepoort J Vet Res 48:215–223

Hoogstraal H (1956) African Ixodoidea. I. Ticks of the Sudan (with special reference to Equatoria province and with preliminary reviews of the genera *Boophilus*, *Margaropus*, and *Hyalomma*). Department of the Navy, Washington DC

Horak IG, Fourie LJ, Heyne H, Walker JB, Needham GR (2002) Ixodid ticks feeding on humans in South Africa: with notes on preferred hosts, geographic distribution, seasonal occurrence and transmission of pathogens. Exp Appl Acarol 27:113–136

Hornok S, Horvath G (2012) First report of adult *Hyalomma marginatum rufipes* (vector of Crimean-Congo haemorrhagic fever virus) on cattle under a continental climate in Hungary. Parasit Vectors 5:170

Knight MM, Norval RAI, Rechav Y (1978) The life cycle of the tick *Hyalomma marginatum rufipes* Koch (Acarina: Ixodidae) under laboratory conditions. J Parasitol 64:143–146

Ruiz-Fons F, Fernandez-de-Mera IG, Acevedo P, Hofle U, Vicente J, de la Fuente J, Gortazar C (2006) Ixodid ticks parasitizing Iberian red deer (*Cervus elaphus hispanicus*) and European wild boar (*Sus scrofa*) from Spain: geographical and temporal distribution. Vet Parasitol 140:133–142

Toma L, Mancini F, Di Luca M, Cecere JG, Bianchi R, Khoury C, Quarchioni E, Manzia F, Rezza G, Ciervo A (2014) Detection of microbial agents in ticks collected from migratory birds in central Italy. Vector Borne Zoonotic Dis 14:199–205

Tomassone L, Camicas JL, Pagani P, Diallo OT, Mannelli A, De Meneghi D (2004) Monthly dynamics of ticks (Acari: Ixodida) infesting N'Dama cattle in the Republic of Guinea. Exp Appl Acarol 32:209–218

Turell MJ (2007) Role of ticks in the transmission of Crimean-Congo hemorrhagic fever virus. In: Ergonul O, Whitehouse CA (eds) Crimean-Congo hemorrhagic fever: a global perspective. Springer, Netherlands, Dordrecht, pp 143–154

Hyalomma aegyptium (Linnaeus, 1758) (Fig. 145)

A. D. Mihalca, T. N. Petney, and M. P. Pfäffle

Life Cycle and Host Preferences

Hyalomma aegyptium is a three-host tick (Siroký et al. 2006). Under laboratory conditions, the life cycle is completed within 147 days (Siroký et al. 2011). Although tortoises (Testudinidae, mainly genus *Testudo*) are the dominant hosts for all of the developmental stages, a wide variety of other hosts are known, including cattle, deer, wild pigs, horses, camels, dogs, hedgehogs, hares, rodents, birds and lizards (Hoogstraal and Kaiser 1960; Apanaskevich 2003). Humans can be infested by adults and nymphs. It is the most common tick attaching to humans in the Istanbul area of Turkey (Vatansever et al. 2008; Bursali et al. 2013).

Ecology

Hyalomma aegyptium is the most common tick found on tortoises in the Balkan area and the Middle East, with prevalences ranging from 20 to 100% (Petney and Al-Yaman 1985; Siroký et al. 2006, 2009). Its development, engorgement and oviposition are temperature dependent, with nymphs being able to engorge only at temperatures over 20 °C and oviposition being most successful between 20 and 35 °C (Sweatman 1968; Sweatman and Gregson 1970).

Distribution

Hyalomma aegyptium has a wide distribution particularly around the Mediterranean and the Black Sea, and the Middle East reaching into Central Asia with records from as far west as Pakistan. It is co-distributed with tortoises belonging to the genus *Testudo* (Kaiser and Hoogstraal 1963; Siroký et al. 2006).

Vectorial Capacity and Pathogen Burden

The vectorial capacity has been experimentally proven for *Hemolivia mauritanica* (Sergent and Sergent 1904), *Hepatozoon kisrae* (Paperna et al. 2002), *Rickettsia aeschlimannii* (Bitam et al. 2009), *Coxiella burnetii* (Siroký et al. 2010) and *Borrelia turcica* (Kalmár et al. 2015). A number of other pathogens have been isolated from *Hy. aegyptium*, but the vectorial capacity has not been evaluated: *Anaplasma phagocytophilum*, *Ehrlichia canis* (Paştiu et al. 2012),

A. D. Mihalca
Department of Parasitology and Parasitic Diseases, University of Agricultural Sciences and Veterinary Medicine Cluj-Napoca, Cluj-Napoca, Romania
e-mail: amihalca@usamvcluj.ro

T. N. Petney (✉)
State Museum of Natural History, Erbprinzenstr. 13, 76133 Karlsruhe, Germany
e-mail: trevor.petney@kit.edu

M. P. Pfäffle
Department of Ecology and Parasitology, Institute of Zoology, Karlsruhe Institute of Technology, Karlsruhe, Germany
e-mail: miripfaeffle@web.de

A. Estrada-Peña et al. (eds.), *Ticks of Europe and North Africa*,
https://doi.org/10.1007/978-3-319-63760-0_68

Fig. 145 **A** Distribution of *Hy. aegyptium* in Europe and Northern Africa (10 × 10 km grid presence with black dots). **B** Countries where the species has been reported are marked in dark grey

Borrelia burgdorferi s.l. (Kar et al. 2011), *Rickettsia africae* (Orkun et al. 2014) and *Theileria annulata* (Ray 1950).

References

Apanaskevich DA (2003) Towards a diagnostic view of *Hyalomma (Hyalomma) aegyptium* (Acari, Ixodidae). Parazitologiia 37:47–59

Bitam I, Kernif T, Harrat Z et al (2009) First detection of *Rickettsia aeschlimannii* in *Hyalomma aegyptium* from Algeria. Clin Microbiol Infect 15(Suppl 2):253–254. https://dx.doi.org/10.1111/j.1469-0691.2008.02274.x

Bursali A, Keskin A, Tekin S (2013) Ticks (Acari: Ixodida) infesting humans in the provinces of Kelkit Valley, a Crimean-Congo Hemorrhagic Fever endemic region in Turkey. Exp Appl Acarol 59:507–515. https://dx.doi.org/10.1007/s10493-012-9608-z

Hoogstraal H, Kaiser MN (1960) Some host relationships of the tortoise tick, *Hyalomma (Hyalommasta) aegyptium* (L.) (Ixodoidea, Ixodidae) in Turkey. Ann Entomol Soc Am 53:457–458.https://dx.doi.org/10.1093/aesa/53.4.457

Kaiser MN, Hoogstraal H (1963) The *Hyalomma* ticks (Ixodoidea, Ixodidade) of Afghanistan. J Parasitol 49:130–139

Kalmár Z, Cozma V, Sprong H, Jahfari S, D'Amico G, Mărcuţan DI, Ionică AM, Magdaş C, Modrý D, Mihalca AD (2015) Transstadial transmission of *Borrelia turcica* in *Hyalomma aegyptium* ticks. PLoS One 10(2):e0115520

Kar S, Yılmazer N, Midilli K, Ergin S, Alp H, Gargılı A (2011) Presence of the zoonotic *Borrelia burgdorferi* s.l. and *Rickettsia* spp. in the ticks from wild tortoises and hedgehogs. J Marmara Univ Inst Health Sci 1:166–170

Orkun Ö, Karaer Z, Çakmak A, Nalbantoğlu S (2014) Spotted fever group rickettsiae in ticks in Turkey. Ticks Tick Borne Dis 5:213–218. https://dx.doi.org/10.1016/j.ttbdis.2012.11.018

Paperna I, Kremer-Mecabell T, Finkelman S (2002) *Hepatozoon kisrae* n. sp. infecting the lizard *Agama stellio* is transmitted by the tick *Hyalomma* cf. *aegyptium*. Parasite 9:17–27

Paştiu AI, Matei IA, Mihalca AD et al (2012) Zoonotic pathogens associated with *Hyalomma aegyptium* in endangered tortoises: evidence for host-switching behaviour in ticks? Parasit Vectors 5:301. https://dx.doi.org/10.1186/1756-3305-5-301

Petney TN, Al-Yaman F (1985) Attachment sites of the tortoise tick *Hyalomma aegyptium* in relation to tick density and physical condition of the host. J Parasitol 71:287–289

Ray HN (1950) Hereditary transmission of *Theileria annulata* infection in the tick, *Hyalomma aegyptium* Neum. Trans R Soc Trop Med Hyg 44:93–104

Sergent ED, Sergent ET (1904) Sur une hémogrégarine, parasite de *Testudo mauritanica*. CR Soc Biol 56:130–131

Siroký P, Petrzelková KJ, Kamler M et al (2006) *Hyalomma aegyptium* as dominant tick in tortoises of the genus *Testudo* in Balkan countries, with notes on its host preferences. Exp Appl Acarol 40:279–290. https://dx.doi.org/10.1007/s10493-006-9036-z

Siroký P, Mikulícek P, Jandzík D et al (2009) Co-distribution pattern of a haemogregarine *Hemolivia mauritanica* (Apicomplexa: Haemogregarinidae) and its vector *Hyalomma aegyptium* (Metastigmata: Ixodidae). J Parasitol 95:728–733. https://dx.doi.org/10.1645/GE-1842.1

Siroký P, Kubelová M, Modrý D et al (2010) Tortoise tick *Hyalomma aegyptium* as long term carrier of Q fever agent *Coxiella burnetii*—evidence from experimental infection. Parasitol Res 107:1515–1520. https://dx.doi.org/10.1007/s00436-010-2037-1

Siroký P, Erhart J, Petrželková KJ, Kamler M (2011) Life cycle of tortoise tick *Hyalomma aegyptium* under laboratory conditions. Exp Appl Acarol 54:277–284. https://dx.doi.org/10.1007/s10493-011-9442-8

Sweatman GK (1968) Temperature and humidity effects on the oviposition of *Hyalomma aegyptium* ticks of different engorgement weights. J Med Entomol 5:429–439. https://dx.doi.org/10.1093/jmedent/5.4.429

Sweatman GK, Gregson JD (1970) Feeding electrograms of *Hyalomma aegyptium* ticks at different temperatures. J Med Entomol 7:575–584. https://dx.doi.org/10.1093/jmedent/7.5.575

Vatansever Z, Gargili A, Aysul NS et al (2008) Ticks biting humans in the urban area of Istanbul. Parasitol Res 102:551–553. https://dx.doi.org/10.1007/s00436-007-0809-z

Hyalomma dromedarii Koch, 1844 (Figs. 146–148)

A. D. Mihalca, M. P. Pfäffle, and T. N. Petney

Life Cycle and Host Preferences

Hyalomma dromedarii has been recorded as a one-, two- or three-host species, although the two-host life cycle is thought to be the most common with attached larvae detaching as engorged nymphs. These nymphs may later be found unattached on the host, leading to the suggestion of a one-host life cycle (Hoogstraal et al. 1981; ElGhali and Hassan 2010a). The two-host cycle appears to be the rule under field conditions (ElGhali and Hassan 2010a). The larval/nymphal stages take 16–27 days before detaching and the females 6–9 days (ElGhali and Hassan 2010a). The main hosts are camels, although adults have also been recorded from livestock (cattle, sheep, goats, buffaloes, horses and donkeys) and from the Arabian antelope (*Oryx leucoryx*). Other hosts rarely infested by adults include hyenas (*Hyena hyena*), domestic dogs, ostriches (*Struthio camelus*), reptiles and humans (Salih et al. 2004; Apanaskevich et al. 2008; Shemshad et al. 2012). Immatures utilize the same host species as adults, but they also infest a wide range of small- and medium-sized mammal, bird and reptile hosts (Apanaskevich et al. 2008). This species is frequently (ElGhali and Hassan 2009; Nazifi et al. 2011; Fard et al. 2012; Champour et al. 2013), but not always (Zeleke and Bekele 2004; Gharbi et al. 2013) the most common tick on camels. Adults are most commonly found on the perineum and chest (Fard et al. 2012), although they can occur anywhere on the camel's body (ElGhali and Hassan 2009).

A. D. Mihalca
Department of Parasitology and Parasitic Diseases, University of Agricultural Sciences and Veterinary Medicine, Cluj-Napoca, Romania
e-mail: amihalca@usamvcluj.ro

M. P. Pfäffle (✉)
Department of Ecology and Parasitology, Institute of Zoology, Karlsruhe Institute of Technology, Karlsruhe, Germany
e-mail: miripfaeffle@web.de

T. N. Petney
State Museum of Natural History, Erbprinzenstr. 13, 76133 Karlsruhe, Germany
e-mail: trevor.petney@kit.edu

A. Estrada-Peña et al. (eds.), *Ticks of Europe and North Africa*,
https://doi.org/10.1007/978-3-319-63760-0_69

Ecology

This species is active all the year round. It occurs predominantly in desert, semi-desert and steppe habitats (Hoogstraal et al. 1981). ElGhali and Hassan (2009) found the highest intensities of infestation in the northern Sudan in September and October (both with a mean infestation/camel of over 250 ticks) and the lowest numbers in January and February (ca. 70 ticks/camel). Both engorged nymphs and females detach predominantly towards evening; respectively, at 16.00–18.00 and 10.00–20.00 (ElGhali and Hassan 2010a). Direct sun leads to high mortality rates in females and substantially reduced egg hatching (ElGhali and Hassan 2010b). Host finding appears to differ between life history stages with larvae and nymphs climbing vegetation, particularly in the afternoon, reaching maximum heights of ca. 50 cm. Adults did not climb (ElGhali and Hassan 2010c).

Distribution

The camel *Hyalomma* has a very wide distribution ranging from India throughout the Middle East and Central Asia, the Arabia Peninsula, North, Central and East Africa as well as parts of West Africa.

Vectorial Capacity and Pathogen Burden

Hyalomma dromedarii is known to harbour a variety of viral, bacterial and protozoan pathogens. The viruses include Crimean-Congo hemorrhagic fever virus, Kadam virus, Dera

Fig. 146 **A** Distribution of *Hy. dromedarii* in Europe and Northern Africa (10 × 10 km grid presence with black dots). **B** Countries where the species has been reported are marked in dark grey

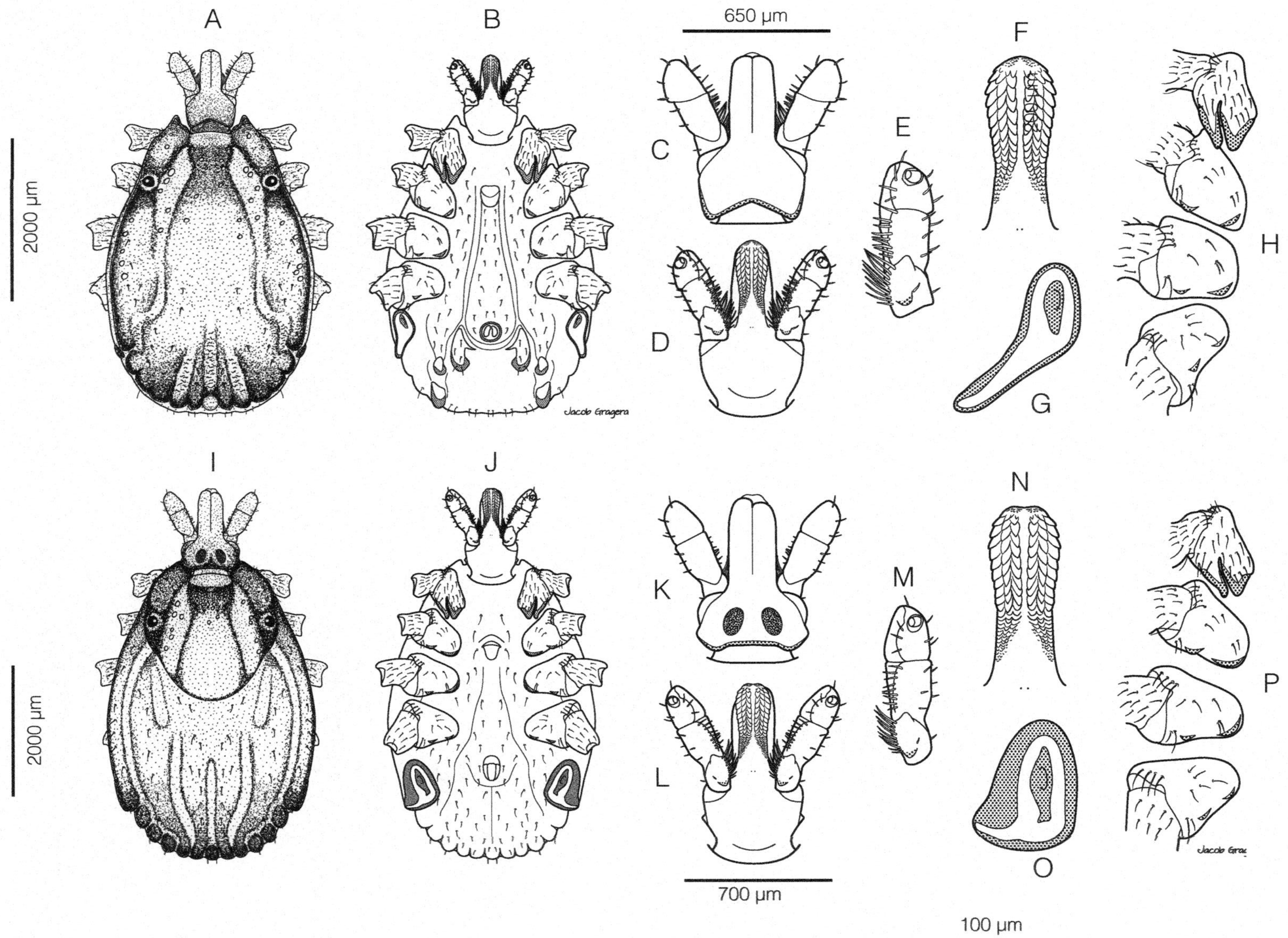

Fig. 147 **A–H** The male of *Hy. dromedarii*. **A** Dorsal, **B** ventral, **C** capitulum, dorsal, **D** capitulum, ventral, **E** palpi, lateral view, **F** hypostome, **G** spiracular plate, **H**: coxae and trochanters I–IV. **I–P** The female of *Hy. dromedarii*. **I** Dorsal, **J** ventral, **K** capitulum, dorsal, **L** capitulum, ventral, **M** palpi, lateral view, **N** hypostome, **O** spiracular plate, **P** coxae and trochanters I–IV. Illustrations from specimens collected in Tunisia

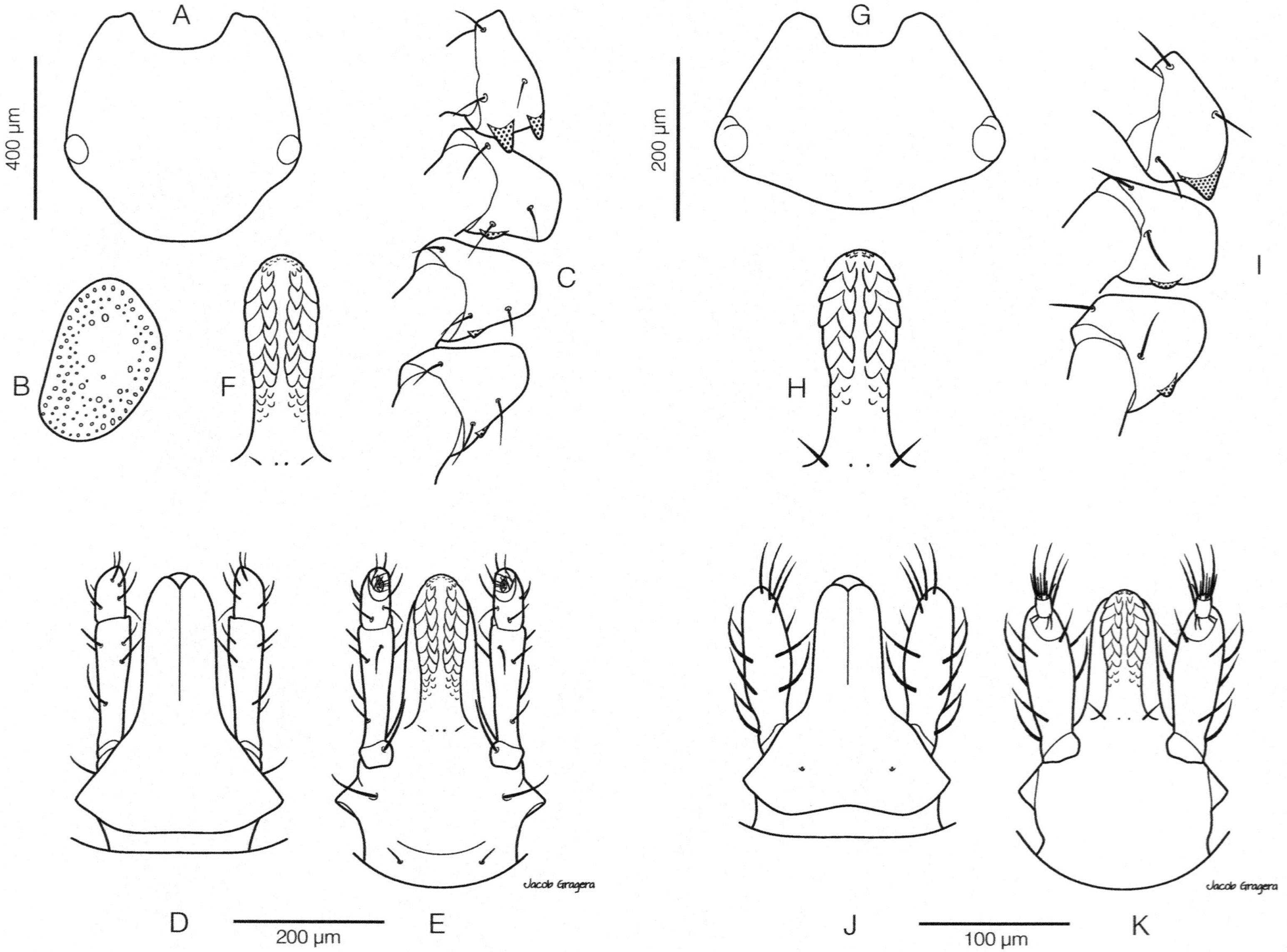

Fig. 148 **A–F** The nymph of *Hy. dromedarii*. **A** Dorsal scutum, **B** spiracular plate, **C** coxae I–IV, **D** capitulum, dorsal, **E** capitulum, ventral, **F** hypostome. **G–K** The larva of *Hy. dromedarii*. **G** Dorsal plate, **H** hypostome, **I** coxae I–III, **J** capitulum, dorsal, **K** capitulum, ventral. Illustrations from specimens obtained in the laboratory, from engorged females collected in Tunisia

Ghazi Khan, Sindbis and Dhori viruses (Hoogstraal et al. 1981; Al-Khalifa et al. 2007; Champour et al. 2016), the bacteria *Coxiella burnetti*, *Rickettsia aeschlimannii* and *R. africae* and the protozoa *Theileria camelensis* and *T. annulata* (Hoogstraal et al. 1981; Hamed et al. 2011; Abdel-Shafy et al. 2012; Demoncheaux et al. 2012; Kernif et al. 2012; Kleinerman et al. 2013).

References

Abdel-Shafy S, Allam NA, Mediannikov O, Parola P, Raoult D (2012) Molecular detection of spotted fever group rickettsiae associated with ixodid ticks in Egypt. Vector-Borne Zoonot Dis 12:346–359

Al-Khalifa MS, Diab FM, Khalil GM (2007) Man-threatening viruses isolated from ticks in Saudi Arabia. Saudi Med J 28:1864–1867

Apanaskevich DA, Schuster AL, Horak IG (2008) The genus *Hyalomma*: VII. Redescription of all parasitic stages of *H.* (*Euhyalomma*) dromedarii and *H.* (*E.*) *schulzei* (Acari: *Ixodidae*). J Med Entomol 45:817–831

Champour M, Mohammadi GR, Chinikar S, Razmi GR, Mostafavi E, Jalali T (2013) Frequency of hard-ticks and the influence of age and sex of camel on ticks' infestation rates in one-humped camel (*Camelus dromedarius*) population in the northeast of Iran. Sci Parasitol 14:89–93

Champour M, Chinikar S, Mohammadi G, Razmi G, Mostafavi E, Shah-Hosseini N, Khakifirouz S, Jalali T (2016) Crimean-Congo hemorrhagic fever in the one-humped camel (*Camelus dromedarius*) in east and northeast of Iran. J Arthrop-Borne Dis 10:168–177

Demoncheaux JP, Socolovschi C, Davoust B, Haddad S, Raoult D, Parola P (2012) First detection of *Rickettsia aeschlimannii* in *Hyalomma dromedarii* ticks from Tunisia. Ticks Tick-Borne Dis 3:398–402

Elghali A, Hassan SM (2009) Ticks (Acari: Ixodidae) infesting camels (*Camelus dromedarius*) in Northern Sudan. Onderstepoort J Vet Res 76:177–185

ElGhali A, Hassan SM (2010a) Drop-off rhythms and survival periods of *Hyalomma dromedarii* (Acari: Ixodidae) fed on camels (*Camelus dromedarius*) in the Sudan. Vet Parasitol 170:302–306

ElGhali A, Hassan SM (2010b) Life cycle of the camel tick *Hyalomma dromedarii* (Acari: Ixodidae) under field conditions in Northern Sudan. Vet Parasitol 174:305–312

ElGhali A, Hassan SM (2010c) Horizontal dispersal and climbing behaviour of the camel tick *Hyalomma dromedarii* (Acari: Ixodidae) in desert ecosystem of Northern Sudan. J Vet Med Anim Prod 1:64–75

Fard SR, Fathi S, Asl EN, Nazhad HA, Kazeroni SS (2012) Hard ticks on one-humped camel (*Camelus dromedarius*) and their seasonal population dynamics in southeast Iran. Trop Anim Health Prod 44:197–200

Gharbi M, Moussi N, Jedidi M, Mhadhbi M, Sassi L, Darghouth MA (2013) Population dynamics of ticks infesting the one-humped camel (*Camelus dromedarius*) in central Tunisia. Ticks Tick-Borne Dis 4:488–491

Hamed MI, Zaitoun AMA, El-Allawy TAA, Mourad MI (2011) Investigation of *Theileria camelensis* in camels infested by *Hyalomma dromedarii* ticks in Upper Egypt. J Adv Vet Res 1:4–7

Hoogstraal H, Wassef HY, Büttiker W (1981) Ticks (Acarina) of Saudi Arabia: fam. Argasidae. Ixodidae. Fauna Saudi Arab 3:25–110

Kernif T, Djerbouh A, Mediannikov O, Ayach B, Rolain JM, Raoult D, Parola P, Bitam I (2012) *Rickettsia africae* in *Hyalomma dromedarii* ticks from sub-Saharan Algeria. Ticks Tick-Borne Dis 3:377–379

Kleinerman G, Baneth G, Mumcuoglu KY, van Straten M, Berlin D, Apanaskevich DA, Abdeen Z, Nasereddin A, Harrus S (2013) Molecular detection of *Rickettsia africae*, *Rickettsia aeschlimannii*, and *Rickettsia sibirica mongolitimonae* in camels and *Hyalomma* spp. ticks from Israel. Vector-Borne Zoonot Dis 13:851–856

Nazifi S, Tamadon A, Behzadi MA, Haddadi S, Raayat-Jahromi AR (2011) One-humped camels (*Camelus dromedarius*) hard ticks infestation in Qeshm Island. Iran. Vet Res Forum 2:135–138

Salih DA, Hassan SM, El Hussein AM, Jongejan F (2004) Preliminary survey of ticks (Acari: Ixodidae) on cattle in northern Sudan. Onderstepoort J Vet Res 71:319–326

Shemshad K, Rafinejad J, Kamali K, Piazak N, Sedaghat MM, Shemshad M, Biglarian A, Nourolahi F, Beigi EV, Enayati AA (2012) Species diversity and geographic distribution of hard ticks (Acari: Ixodoidea: Ixodidae) infesting domestic ruminants, in Qazvin Province, Iran. Parasitol Res 110:373–380

Zeleke M, Bekele T (2004) Species of ticks on camels and their seasonal population dynamics in eastern Ethiopia. Trop Anim Health Prod 36:225–231

Hyalomma impeltatum Schulze and Schlottke, 1929 (Figs. 149–151)

T. N. Petney and M. P. Pfäffle

Life Cycle and Host Preferences

Hyalomma impeltatum can exhibit either a three-host or a two-host life cycle (Apanaskevich and Horak 2009). It occurs on a relatively wide range of mammalian hosts, with the most common hosts for adults being camels and cattle (Camelidae and Bovidae) and for immatures, hares (Leporidae) and rodents (Muridae) (Guglielmone et al. 2014). Birds are rarely infested, but this species has been recorded from adult ostriches (*Struthio camelus*) and passerine wheatear species (*Oenanthe oenanthe* and *O. isabellina*), as well as the common redstart (*Phoenicurus phoenicurus*) (Apanaskevich and Horak 2009). Both adults and immatures of *Hy. impeltatum* have been found on humans (Hoogstraal 1956; Walker 1974; Bursali et al. 2013). Recent, detailed descriptions of all life history stages are provided by (Apanaskevich and Horak 2009).

Ecology

Hyalomma impeltatum is a relatively rare species throughout much of its range and this is particularly true of the Western Palaearctic (Hoogstraal 1956; Kaiser and Hoogstraal 1963; Estrada-Peña et al. 2013). It is most common in semi-desert, steppe and savanna habitats (Hoogstraal et al. 1981). In the laboratory, the complete life cycle takes on average 108 days at 26 °C and 92–96% relative humidity (Logan et al. 1989). There is one detailed study of the population dynamics of nymphs found on the greater Egyptian gerbil (*Gerbillus pyramidum*) in Egypt which showed a single peak of abundance in September with an incidence of 13.4 nymphs/host with very low infestations from December to April (Hoogstraal and Kaiser 1958).

Distribution

This species has a wide distribution including Palaearctic North Africa, the Arabian Peninsula, the Middle East, Anatolia, reaching as far east as Afghanistan and Pakistan. It is also found in sub-Saharan Africa in a band from Senegal and Mauritania in the west across to Eritrea, Djibouti and Somalia in the east. It reaches as far south as Cameroon in Central Africa and Tanzania in the east (Apanaskevich and Horak 2009). In Turkey, it has been found in the Marmara Region which includes the European part of the country (Bursali et al. 2012).

Vectorial Capacity and Pathogen Burden

Both Crimean-Congo hemorrhagic fever (CCHF) and Wanowrie viruses have been isolated from *Hy. impeltatum* (Williams et al. 1973; Hoogstraal 1979). Dohm et al. (1996) showed efficient transmission of CCHF from one-day-old mice to *Hy. impeltatum* larvae with successful transstadial transmission through nymphs to adults. Infected nymphs were able to transmit the virus to guinea pigs. Loftis et al. (2006) found *Rickettsia aeschlimannii* in this species from cattle in Egypt and Kleinerman et al. (2013) found two *Hy. impeltatum* from a camel containing *Rickettsia africae*. This species is also a recognized vector of *Theileria annulata* (Mustafa et al. 1983) and has been associated with malignant theileriosis (*T. hirci*) in sheep in Saudi Arabia (El-Azazy et al. 2001).

T. N. Petney (✉)
State Museum of Natural History, Erbprinzenstr. 13, 76133 Karlsruhe, Germany
e-mail: trevor.petney@kit.edu

M. P. Pfäffle
Department of Ecology and Parasitology, Institute of Zoology, Karlsruhe Institute of Technology, Karlsruhe, Germany
e-mail: miripfaeffle@web.de

A. Estrada-Peña et al. (eds.), *Ticks of Europe and North Africa*,
https://doi.org/10.1007/978-3-319-63760-0_70

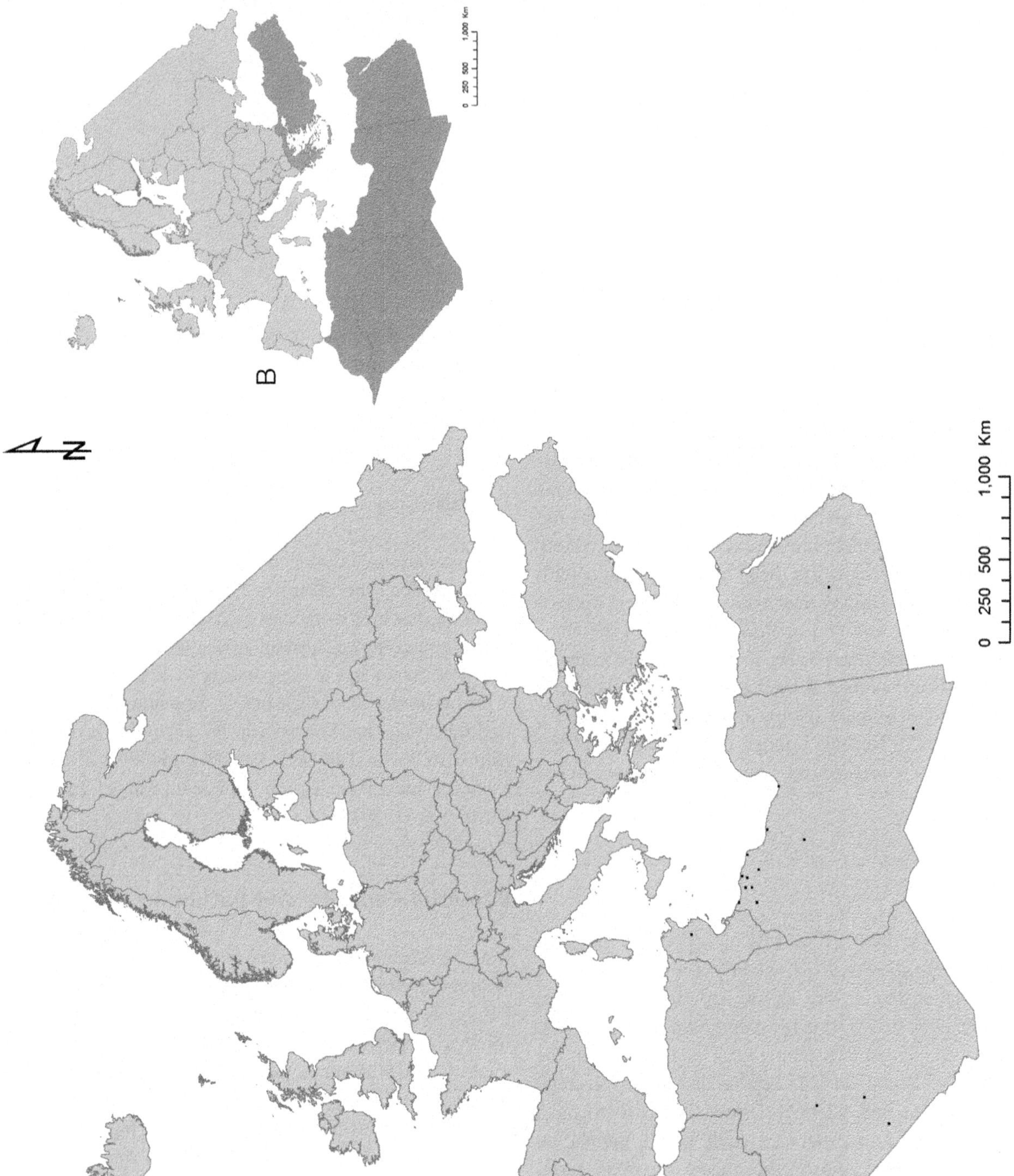

Fig. 149 **A** Distribution of *Hy. impeltatum* in Europe and Northern Africa (10 × 10 km grid presence with black dots). **B** Countries where the species has been reported are marked in dark grey

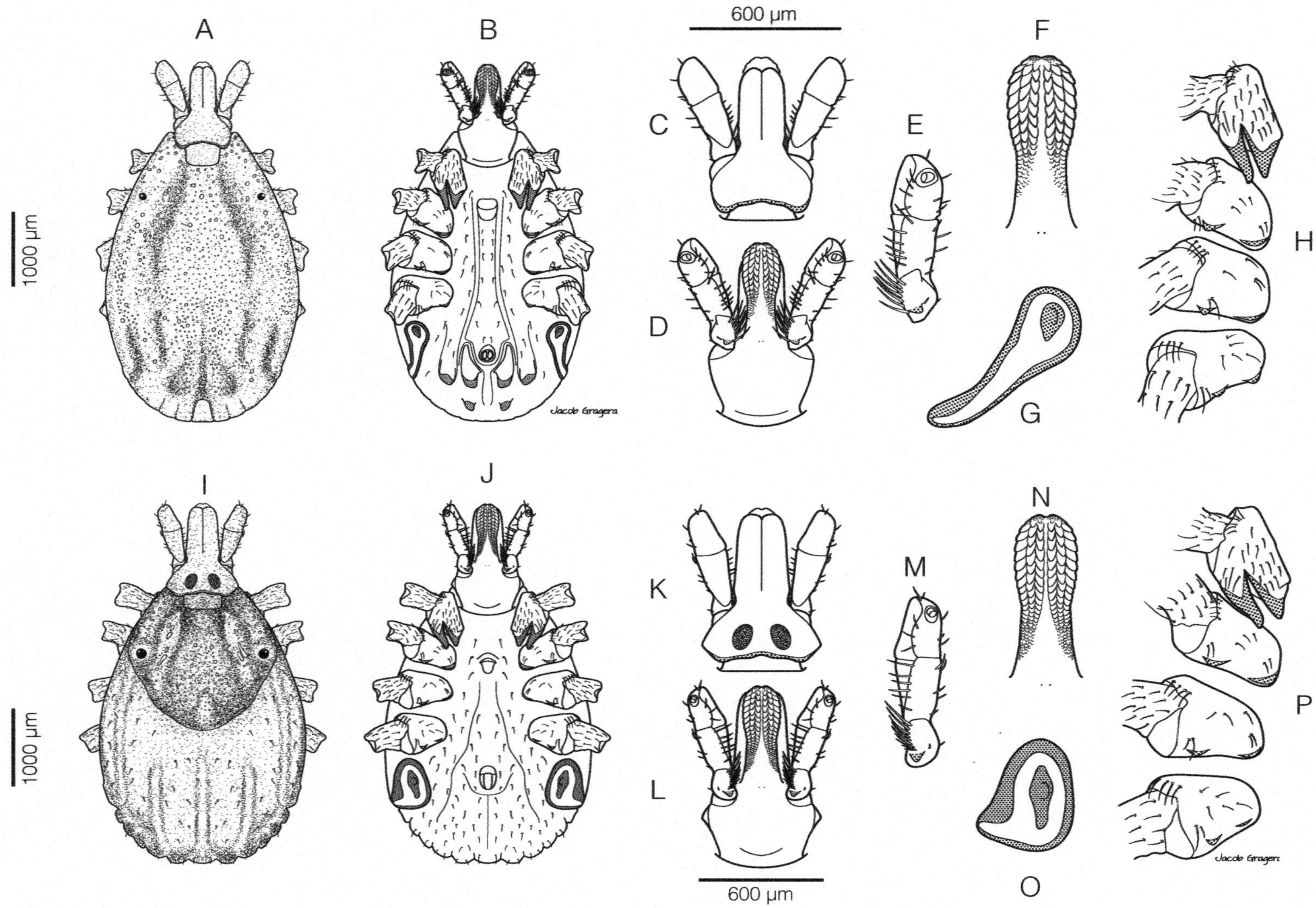

Fig. 150 **A–H** The male of *Hy. impeltatum*. **A** Dorsal, **B** ventral, **C** capitulum, dorsal, **D** capitulum, ventral, **E** palpi, lateral view, **F** hypostome, **G** spiracular plate, **H** coxae and trochanters I–IV. **I–P** The female of *Hy. impeltatum*. **I** Dorsal, **J** ventral, **K** capitulum, dorsal, **L** capitulum, ventral, **M** palpi, lateral view, **N** hypostome, **O** spiracular plate, **P** coxae and trochanters I–IV. Illustrations from specimens collected in Tunisia

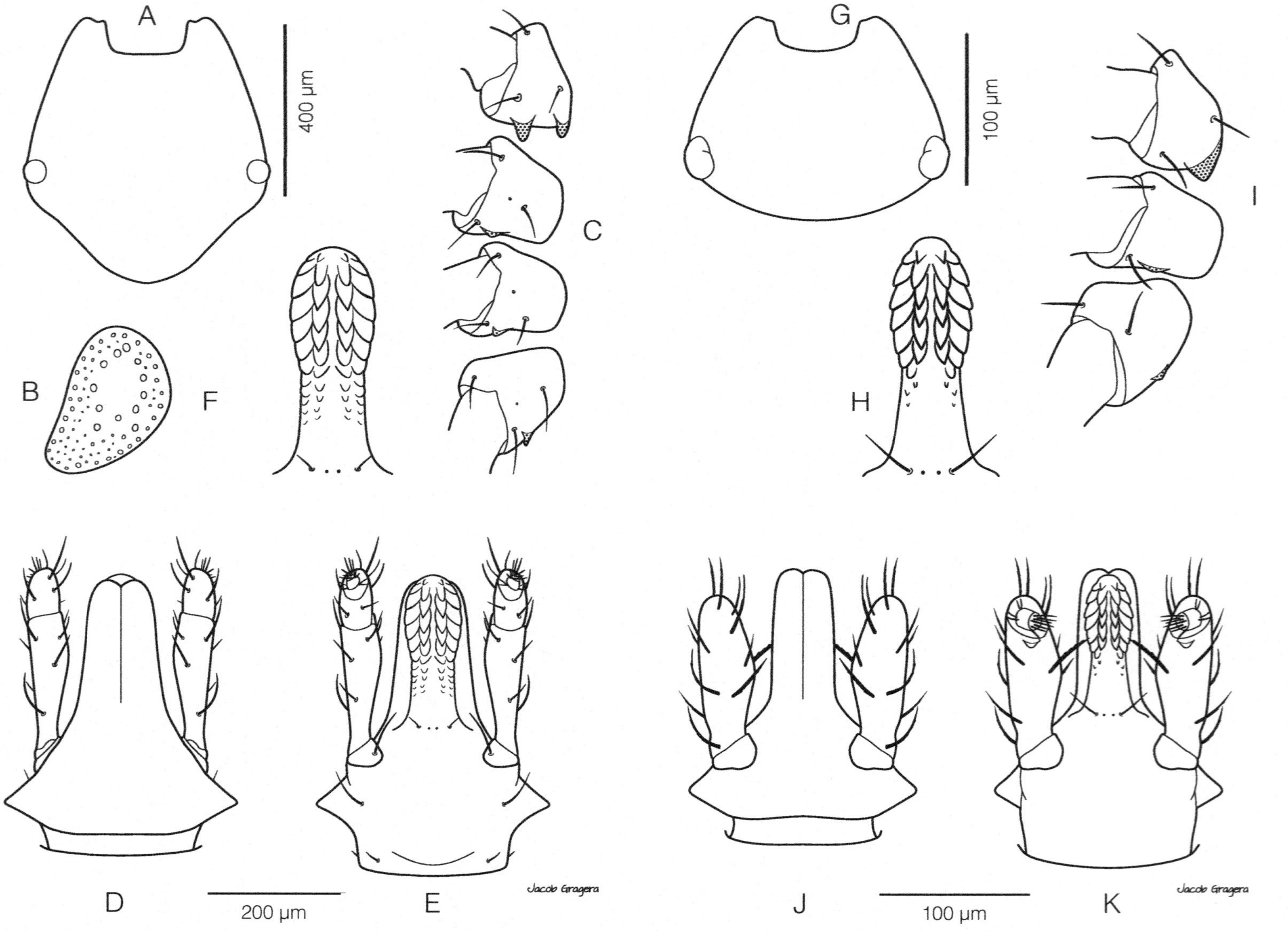

Fig. 151 **A–F** The nymph of *Hy. impeltatum*. **A** Dorsal scutum, **B** spiracular plate, **C** coxae I–IV, **D** capitulum, dorsal, **E** capitulum, ventral, **F** hypostome, **G–K** the larva of *Hy. impeltatum*. **G** Dorsal plate, **H** hypostome, **I** coxae I–III, **J** capitulum, dorsal, **K** capitulum, ventral. Illustrations from specimens obtained in the laboratory, from engorged females collected in Tunisia

References

Apanaskevich DA, Horak IG (2009) The genus *Hyalomma* Koch, 1844. IX. Redescription of all parasitic stages of *H.* (*Euhyalomma*) *impeltatum* Schulze & Schlottke, 1930 and *H.* (*E.*) *somalicum* Tonelli Rondelli, 1935 (Acari: Ixodidae). Syst Parasitol 73:199–218. https://dx.doi.org/10.1007/s11230-009-9190-x

Bursali A, Keskin A, Tekin S (2012) A review of the ticks (Acari: Ixodida) of Turkey: species diversity, hosts and geographical distribution. Exp Appl Acarol 57:91–104. https://dx.doi.org/10.1007/s10493-012-9530-4

Bursali A, Keskin A, Tekin S (2013) Ticks (Acari: Ixodida) infesting humans in the provinces of Kelkit Valley, a Crimean-Congo hemorrhagic fever endemic region in Turkey. Exp Appl Acarol 59:507–515. https://dx.doi.org/10.1007/s10493-012-9608-z

Dohm DJ, Logan TM, Linthicum KJ et al (1996) Transmission of Crimean-Congo hemorrhagic fever virus by *Hyalomma impeltatum* (Acari: Ixodidae) after experimental infection. J Med Entomol 33:848–851

El-Azazy OME, El-Metenawy TM, Wassef HY (2001) *Hyalomma impeltatum* (Acari: Ixodidae) as a potential vector of malignant theileriosis in sheep in Saudi Arabia. Vet Parasitol 99:305–309. https://dx.doi.org/10.1016/S0304-4017(01)00468-X

Estrada-Peña A, Farkas R, Jaenson TGT et al (2013) Association of environmental traits with the geographic ranges of ticks (Acari: Ixodidae) of medical and veterinary importance in the western Palaearctic. A digital data set. Exp Appl Acarol 59:351–366. https://dx.doi.org/10.1007/s10493-012-9600-7

Guglielmone AA, Robbins RG, Apanaskevich DA et al (2014) The hard ticks of the world (Acari: Ixodida: Ixodidae). Springer, Dordrecht, p 738

Hoogstraal H (1956) African Ixodoidea 1. Ticks of the Sudan. Washington, DC

Hoogstraal H (1979) Review article: the epidemiology of tick-borne Crimean-Congo hemorrhagic fever in Asia, Europe, and Africa. J Med Entomol 15:307–417. https://dx.doi.org/10.1093/jmedent/15.4.307

Hoogstraal H, Kaiser MN (1958) Observations on Egyptian *Hyalomma* Ticks (Ixodoidea, Ixodidae), 3. Infestation of greater gerbils, especially by immature *H. impeltatum* S. & S. Ann Entomol Soc Am 51:17–19. https://dx.doi.org/10.1093/aesa/51.1.17

Hoogstraal H, Wassef HY, Buttiker W (1981) Ticks (Acarina) of Saudi Arabia Fam. Argasidae. Fauna Saudi Arab 3:25–110

Kaiser MN, Hoogstraal H (1963) The *Hyalomma* ticks (Ixodoidea, Ixodidade) of Afghanistan. J Parasitol 49:130–139

Kleinerman G, Baneth G, Mumcuoglu KY et al (2013) Molecular detection of *Rickettsia africae*, *Rickettsia aeschlimannii*, and *Rickettsia sibirica mongolitimonae* in camels and *Hyalomma* spp. ticks from Israel. Vector Borne Zoonotic Dis 13:851–856. https://dx.doi.org/10.1089/vbz.2013.1330

Loftis AD, Reeves WK, Szumlas DE et al (2006) Rickettsial agents in Egyptian ticks collected from domestic animals. Exp Appl Acarol 40:67–81. https://dx.doi.org/10.1007/s10493-006-9025-2

Logan TM, Linthicum KJ, Kondig JP, Bailey CL (1989) Biology of *Hyalomma impeltatum* (Acari: Ixodidae) under laboratory conditions. J Med Entomol 26:479–483. https://dx.doi.org/10.1093/jmedent/26.5.479

Mustafa U-H, Jongejan F, Morzaria SP (1983) Note on the transmission of *Theileria annulata* by *Hyalomma* ticks in the Sudan. Vet Q 5:112–113. https://dx.doi.org/10.1080/01652176.1983.9693883

Walker JB (1974) Ixodid ticks of Kenya. Commonwealth Institute of Entomology, London

Williams RE, Hoogstraal H, Casals J et al (1973) Isolation of Wanowrie, Thogoto, and Dhori viruses from *Hyalomma* ticks infesting camels in Egypt. J Med Entomol 10:143–146. https://dx.doi.org/10.1093/jmedent/10.2.143

Hyalomma scupense Schulze, 1919 (Figs. 152–154)

Z. Vatansever

Life Cycle and Host Preferences

Hyalomma scupense is an endophilic polymorphic species with different ecological preferences and life cycles. There are two forms that prefer either one-host (also previously known as *Hy. scupense* (Pomerantzev 1950) or two-host life cycles [previously known as *Hy. detritum* (Pomerantzev 1950)] (Pomerantzev 1950; Apanaskevich et al. 2010; Guglielmone and Nava 2014). Regardless of the life cycle, all developmental stages of *Hy. scupense* prefer large ungulates, especially cattle and horses. It can rarely be observed on camels, sheep, donkeys, pigs, wild ungulates and humans (Pomerantzev 1950; Hoogstraal 1956; Apanaskevich 2004; Apanaskevich et al. 2010; Bursali et al. 2012).

Ecology

The one-host type of *Hy. scupense* is known for its winter activity. Adults of this form infest cattle from January to April. Larvae parasitize hosts in November, while nymphs have been recorded on hosts between November and March (Galuzo 1947; Pomerantzev 1950). Under field conditions in the European part of Turkey, larval activity starts in October and the maximum number of feeding and moulting nymphs is observed in December when some adults start to appear. Most of the adults stay attached but do not feed until the end of January when the last nymphs start to moult. Adult infestations have been observed between December and April, with a peak in February (Galuzo 1947; Pomerantzev 1950) (Sirri Kar, personal communication). The two-host type of *Hy. scupense* is active in summer and autumn. Adults appear in May, reach their maximum number in July and disappear until September. Larvae and nymphs appear from August and September. Most of the nymphs engorge and drop off the host in autumn, while some overwinter on hosts and drop-off in spring after feeding. Dropped nymphs move into cracks and crevices in cattle barns, where they overwinter and moult to adults in May (Galuzo 1947; Pomerantzev 1950). Both forms are nidiculous and are strongly associated with cattle barns. Females of the one-host *Hy. scupense* tend to have an affinity to cattle dung and after drop-off prefer to bury themselves there in barns (Galuzo 1947) (Sirri Kar, personal communication). Preoviposition and oviposition, as well as embryonic development, take relatively long, covering the whole summer (Pomerantzev 1950). Females of the two-host forms of *Hy. scupense* usually drop-off in the barns and rarely in pastures. Oviposition usually takes place under rocks, debris or cattle waste, and in the cracks and crevices in the barns. Fed nymphs drop-off in animal inhabited barns and their surroundings. Winter diapause takes place in cracks, wall crevices and even under dried cattle dung in the animal shelters (Galuzo 1947; Gharbi and Darghouth 2014).

Distribution

Hyalomma scupense is one of the most widely distributed *Hyalomma* species in the world. It has been reported from 42 countries ranging from Western Europe to Eastern China, including North Africa (Apanaskevich et al. 2010; Galuzo 1947; Gharbi and Darghouth 2014; Pomerantzev 1950). In the Western Palaearctic, it has been reported in Albania, Bosnia and Herzegovina, Bulgaria, Croatia, France, Greece, Italy, Macedonia, Moldova, Montenegro, Romania, Russia (southern Federal Districts), Serbia, Spain, Turkey, the Ukraine, Israel, Jordan (Galuzo 1947; Pomerantzev 1950), Algeria, Egypt, Libya, Morocco and Tunisia (Hoogstraal 1956; Gharbi and Darghouth 2014).

Z. Vatansever (✉)
Department of Parasitology, Faculty of Veterinary Medicine, Kafkas University, Kars, Turkey
e-mail: zativet@gmail.com

A. Estrada-Peña et al. (eds.), *Ticks of Europe and North Africa*,
https://doi.org/10.1007/978-3-319-63760-0_71

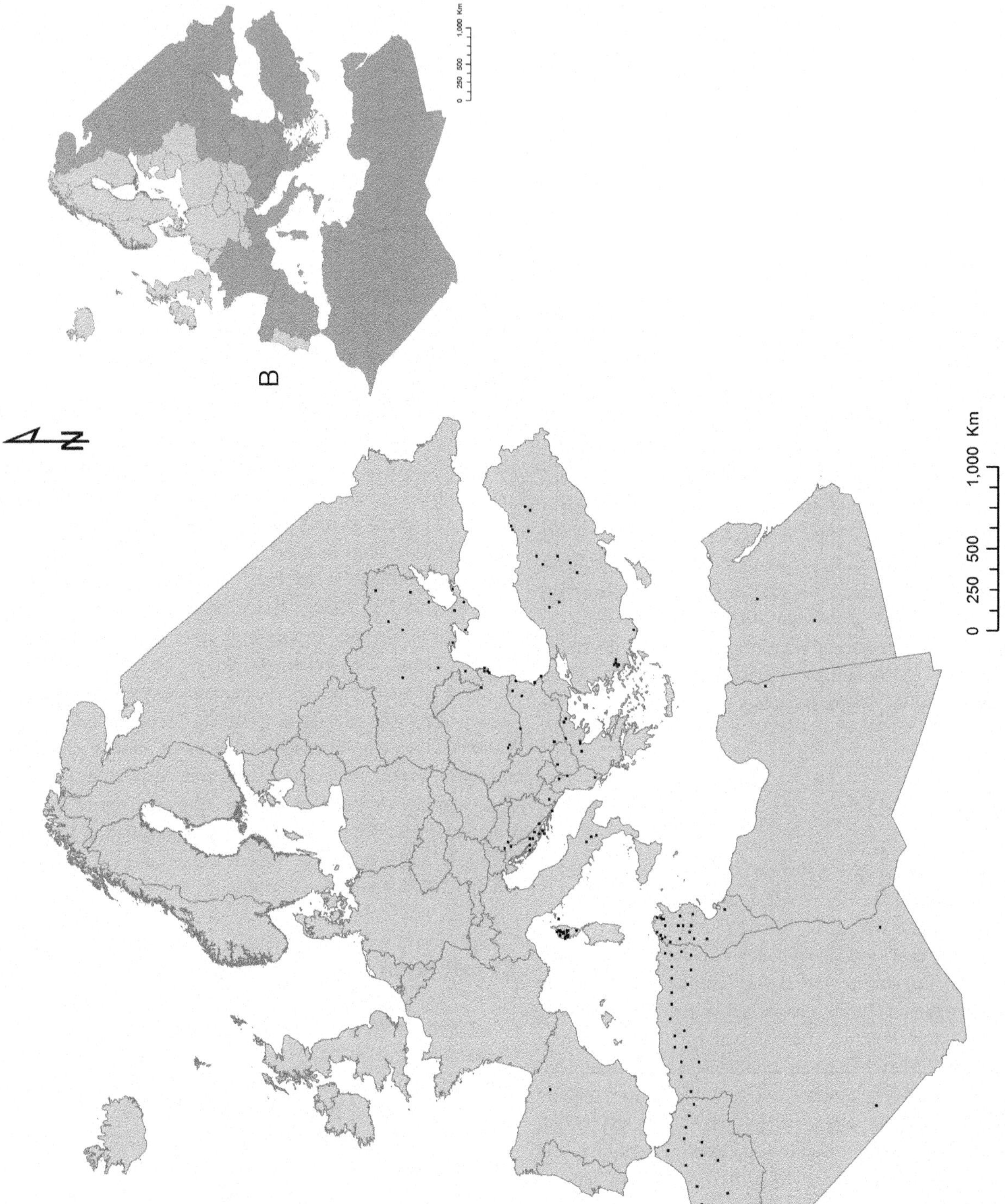

Fig. 152 **A** Distribution of *Hy. scupense* in Europe and Northern Africa (10 × 10 km grid presence with black dots). **B** Countries where the species has been reported are marked in dark grey

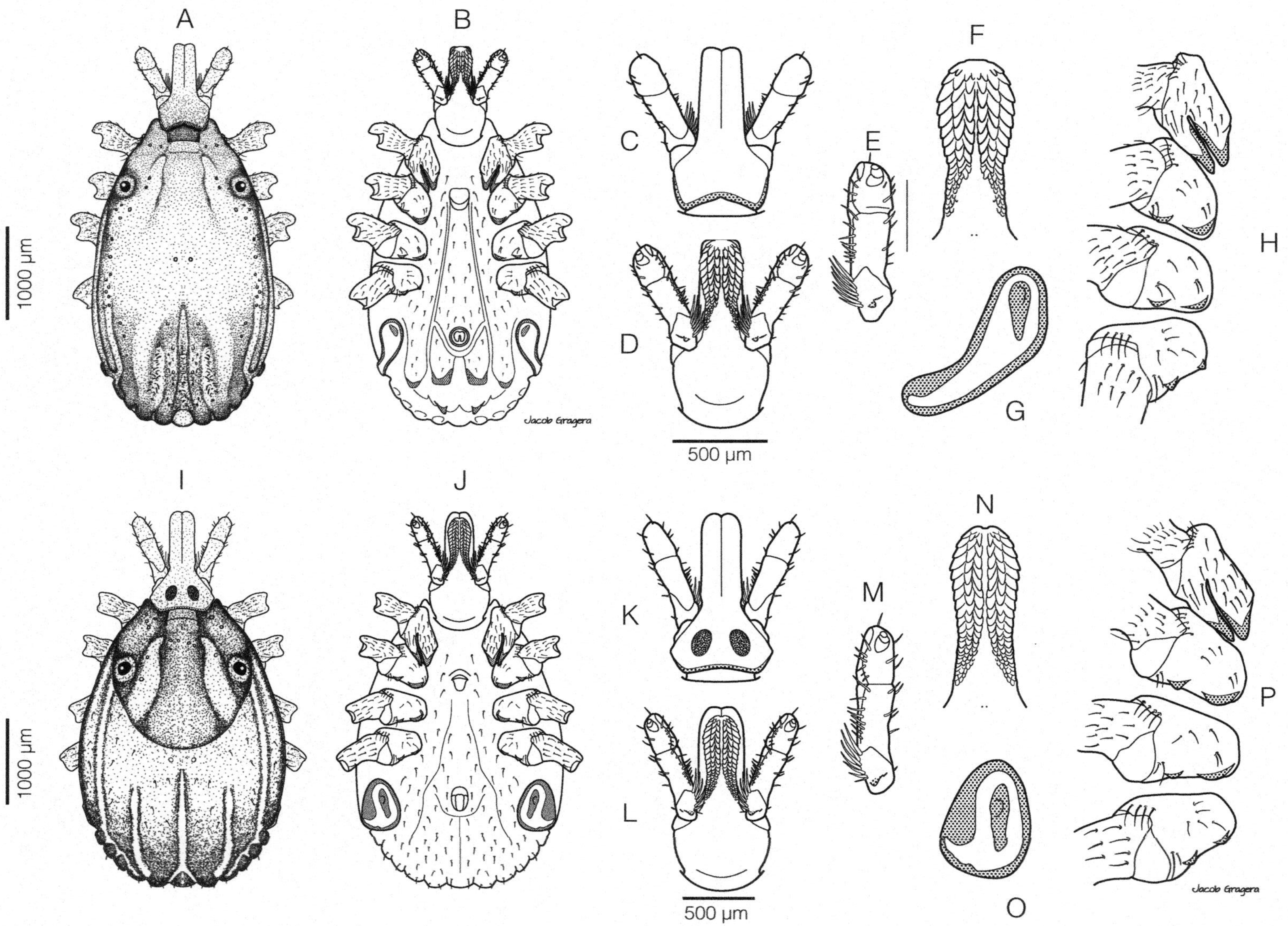

Fig. 153 **A–H** The male of *Hy. scupense*. **A** Dorsal, **B** ventral, **C** capitulum, dorsal, **D** capitulum, ventral, **E** palpi, lateral view, **F** hypostome, **G** spiracular plate, **H** coxae and trochanters I–IV. **I–P** The female of *Hy. scupense*. **I** Dorsal, **J** ventral, **K** capitulum, dorsal, **L** capitulum, ventral, **M** palpi, lateral view, **N** hypostome, **O** spiracular plate, **P** coxae and trochanters I–IV. Illustrations from specimens collected in France, Morocco, Spain and Tunisia

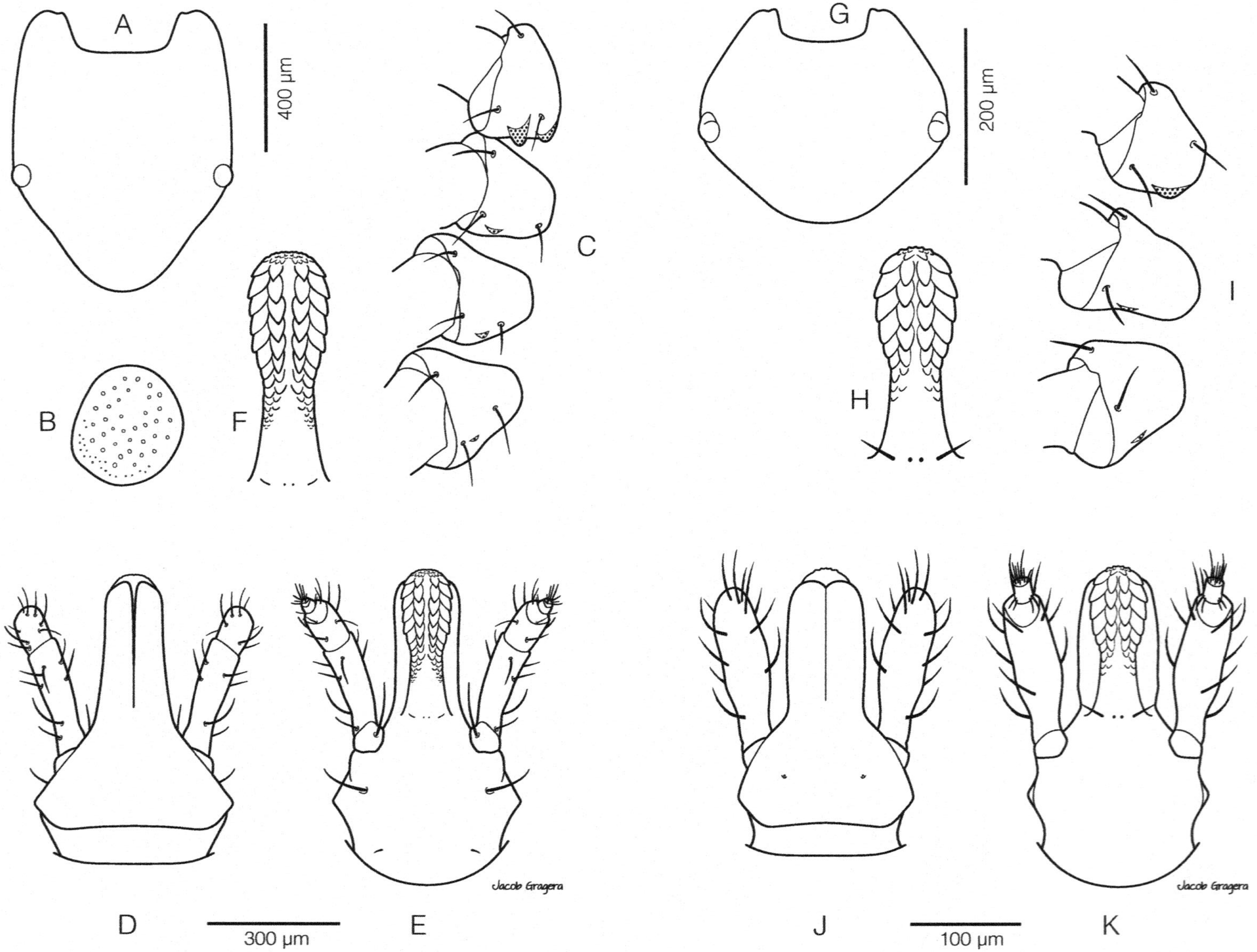

Fig. 154 **A–F** The nymph of *Hy. scupense*. **A** Dorsal scutum, **B** spiracular plate, **C** coxae I–IV, **D** capitulum, dorsal, **E** capitulum, ventral, **F** hypostome. **G–K** The larva of *Hy. scupense*. **G** Dorsal plate, **H** hypostome, **I** coxae I–III, **J** capitulum, dorsal, **K** capitulum, ventral. Illustrations from specimens obtained in the laboratory, from engorged females collected in France, Morocco, Spain and Tunisia

Vectorial Capacity and Pathogen Burden

Hyalomma scupense is one of the most important vectors of *Theileria annulata* (causing bovine tropical theileriosis) and *T. equi* (causing equine piroplasmosis) (Bakheit et al. 2012; Gharbi and Darghouth 2014). It can also transmit *Rickettsia aeshlimannii*, *Coxiella burnetii* (Q fever), Bhanja virus and is a potential vector of Crimean-Congo haemorrhagic fever virus (Bakheit et al. 2012; Hoogstraal 1979). *Hyalomma scupense* has been also historically cited to harbour *Yersinia pestis*, *Francisella tularensis* and *Brucella* spp. (Hoogstraal 1956).

References

Apanaskevich DA (2004) Host-parasite relationships of the genus *Hyalomma* Koch, 1844 (Acari, Ixodidae) and their connection with microevolutionary process. Parazitologiia 38:515–523

Apanaskevich DA, Filippova NA, Horak IG (2010) The genus *Hyalomma* Koch, 1844. x. Redescription of all parasitic stages of *H. (Euhyalomma) scupense* Schulze, 1919 (=*H. detritum* Schulze) (Acari: Ixodidae) and notes on its biology. Folia Parasitol (Praha) 57:69–78

Bakheit MA, Latif A, Vatansever Z, Seitzer U, Ahmed J (2012) The huge risks due to *Hyalomma* ticks. In: Mehlhorn H (ed) Arthropods as vectors of emerging diseases. Springer, Berlin, Heidelberg, pp 167–194

Bursali A, Keskin A, Tekin S (2012) A review of the ticks (Acari: Ixodida) of Turkey: species diversity, hosts and geographical distribution. Exp Appl Acarol 57:91–104

Galuzo IG (1947) Bloodsucking ticks of Kazakh SSR. In: Genus *Hyalomma*, vol 2 (in Russian) (Alma Ata, Inst. Zool. Akad. Nauk Kazakh SSR), p 280

Gharbi M, Darghouth MA (2014) A review of *Hyalomma scupense* (Acari, Ixodidae) in the Maghreb region: from biology to control. Parasite 21:2

Guglielmone AA, Nava S (2014) Names for Ixodidae (Acari: Ixodoidea): valid, synonyms, incertae sedis, nomina dubia, nomina nuda, lapsus, incorrect and suppressed names–with notes on confusions and misidentifications. Zootaxa 3767:1–256

Hoogstraal H (1956) African Ixodoidea. I. Ticks of the Sudan (with special reference to Equatoria province and with preliminary reviews of the genera *Boophilus*, *Margaropus*, and *Hyalomma*). Department of the Navy, Washington, D.C.

Hoogstraal H (1979) The epidemiology of tick-borne Crimean-Congo hemorrhagic fever in Asia, Europe, and Africa. J Med Entomol 15:307–417

Pomerantzev BI (1950) Ixodid ticks (Ixodidae). In: Fauna of the USSR, New series 41: Arachnoidea, vol 4(2) (in Russian). (Leningrad, Nauk), p 224

Hyalomma lusitanicum Koch, 1844 (Figs. 155–157)

M. M. Santos-Silva

Life Cycle and Host Preferences

Hyalomma lusitanicum is a three-host tick with a natural life cycle of one year. Under laboratory conditions, the life cycle is completed in 138–196 days (Ouhelli 1994). Adults usually feed on various large- and medium-size mammals such as domestic and wild ungulates, particularly, bovines (Apanaskevich et al. 2008; Estrada-Peña et al. 2013; Guglielmone et al. 2014). Infrequently, they feed on other hosts such as carnivores, insectivores, lagomorphs, galliformes, passeriformes, strigiformes, struthioniformes (Guglielmone et al. 2014). Immature stages of larvae and nymphs generally use lagomorphs to feed, while birds are exceptional hosts. Records of nymphs on birds (Perez-Eid and Cabrita 2003) and cattle (Fernández and Hueli 1984) require confirmation (Guglielmone et al. 2014). This tick has been reported from humans (Santos-Silva et al. 2011).

Ecology

Hyalomma lusitanicum is mostly adapted to mesomediterranean vegetation (Estrada-Peña et al. 2004). The typical habitats include Mediterranean forests such as Iberian sclerophyllous and semi-deciduous, woodlands, steppe and scrub (Estrada-Peña et al. 2013; Guglielmone et al. 2014). It can also be found within rock crevices or inside the burrows of lagomorphs (Estrada-Peña et al. 2004). All stages are more active during spring and summer. Immature stages are active during summer, from July to September, but larvae are reported to be active earlier (e.g. May) (Estrada-Peña et al. 2004). Adults are active all year (Caeiro 1999) with a peak between April and July (Ouhelli et al. 1985; Estrada-Peña et al. 2004). *Hyalomma lusitanicum* is a non-nidicolous tick. The host-seeking behaviour is similar to *Hy. marginatum* in which an active hunter strategy is used to find a host.

Distribution

Hyalomma lusitanicum is restricted to the western part of the Mediterranean subregion of the Palaearctic Zoogeographic Region. It has been reported in Algeria, France, Italy, Morocco, Portugal and Spain (Apanaskevich et al. 2008). There is a report on the transport of this tick species to the Nearctic region attached to ostrich imported from Portugal (Mertins and Schlater 1991). However, there is no evidence that it has become established outside of the Palaearctic area (Guglielmone et al. 2014).

Vectorial Capacity and Pathogen Burden

Experimental proof for the vectorial capacity of *Hy. lusitanicum* is available for *Theileria annulata* (Viseras et al. 1999) and *T. equi* (Zapf and Schein 1994). Other pathogens, such as *Anaplasma* spp., *Ehrlichia* spp. and *Babesia*/*Theileria* spp. (Torina et al. 2010), *Anaplasma phagocytophilum*, *Borrelia burgdorferi* s.l., *Franscisella tularensis* and co-infection with *Anaplasma phagocytophilum*/*B. burgdorferi* (Toledo et al. 2009), *Borrelia lusitaniae* (Milhano et al. 2010); *Coxiella burnetii* (Derick 1939; Santos–Silva et al. 2014) Crimean-Congo haemorrhagic fever virus (Morel 1971; Estrada-Peña 2012) have been detected by several methods. This species remains poorly understood in terms of its vectorial ability for these pathogens.

M. M. Santos-Silva (✉)
Centro de Estudos de Vectores e Doenças Infecciosas Dr. Francisco Cambournac (CEVDI), Instituto Nacional de Saúde Doutor Ricardo Jorge, Águas de Moura, Portugal
e-mail: m.santos.silva@insa.min-saude.pt

A. Estrada-Peña et al. (eds.), *Ticks of Europe and North Africa*,
https://doi.org/10.1007/978-3-319-63760-0_72

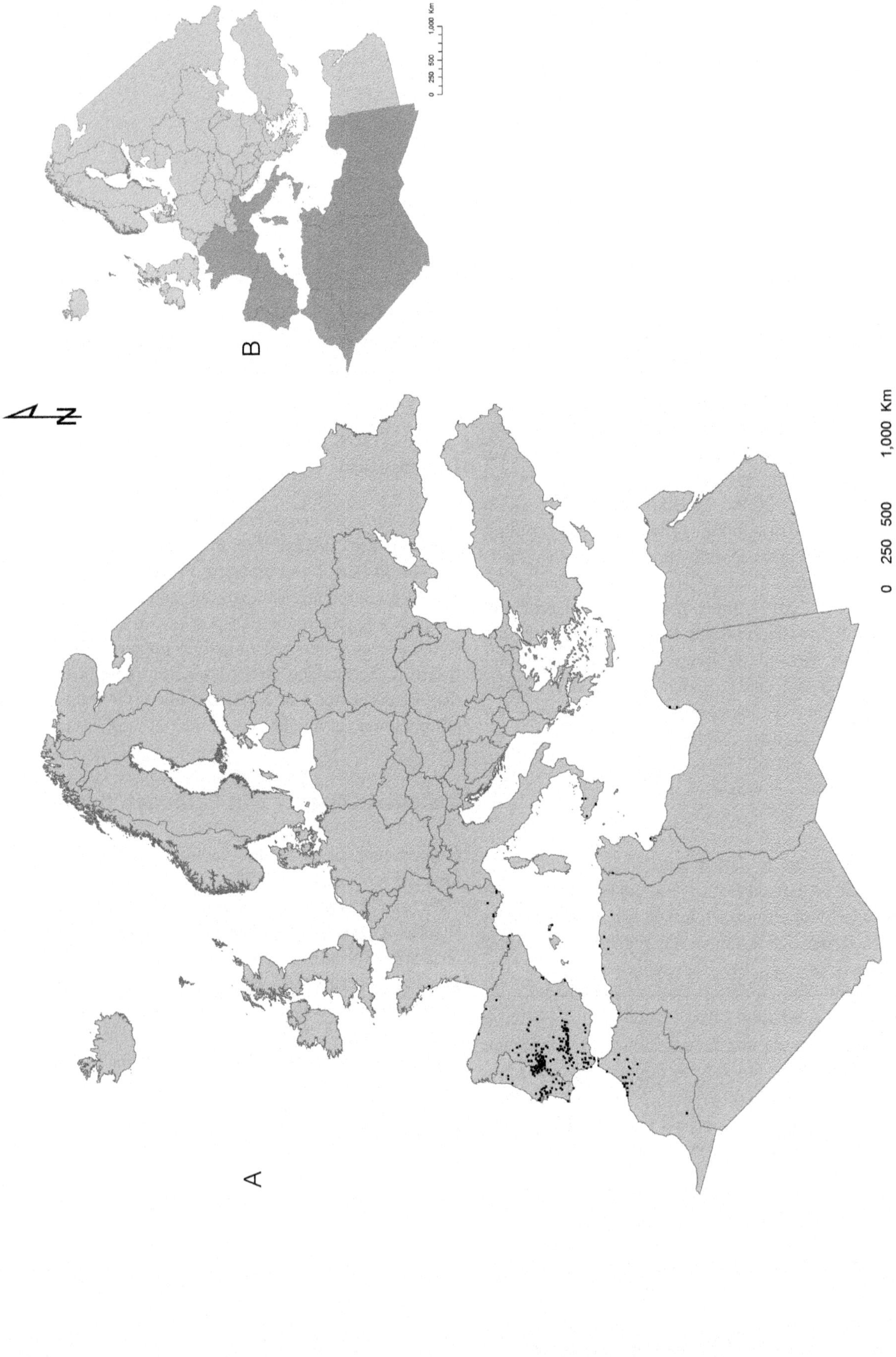

Fig. 155 **A** Distribution of *Hy. lusitanicum* in Europe and Northern Africa (10 × 10 km grid presence with black dots). **B** Countries where the species has been reported are marked in dark grey

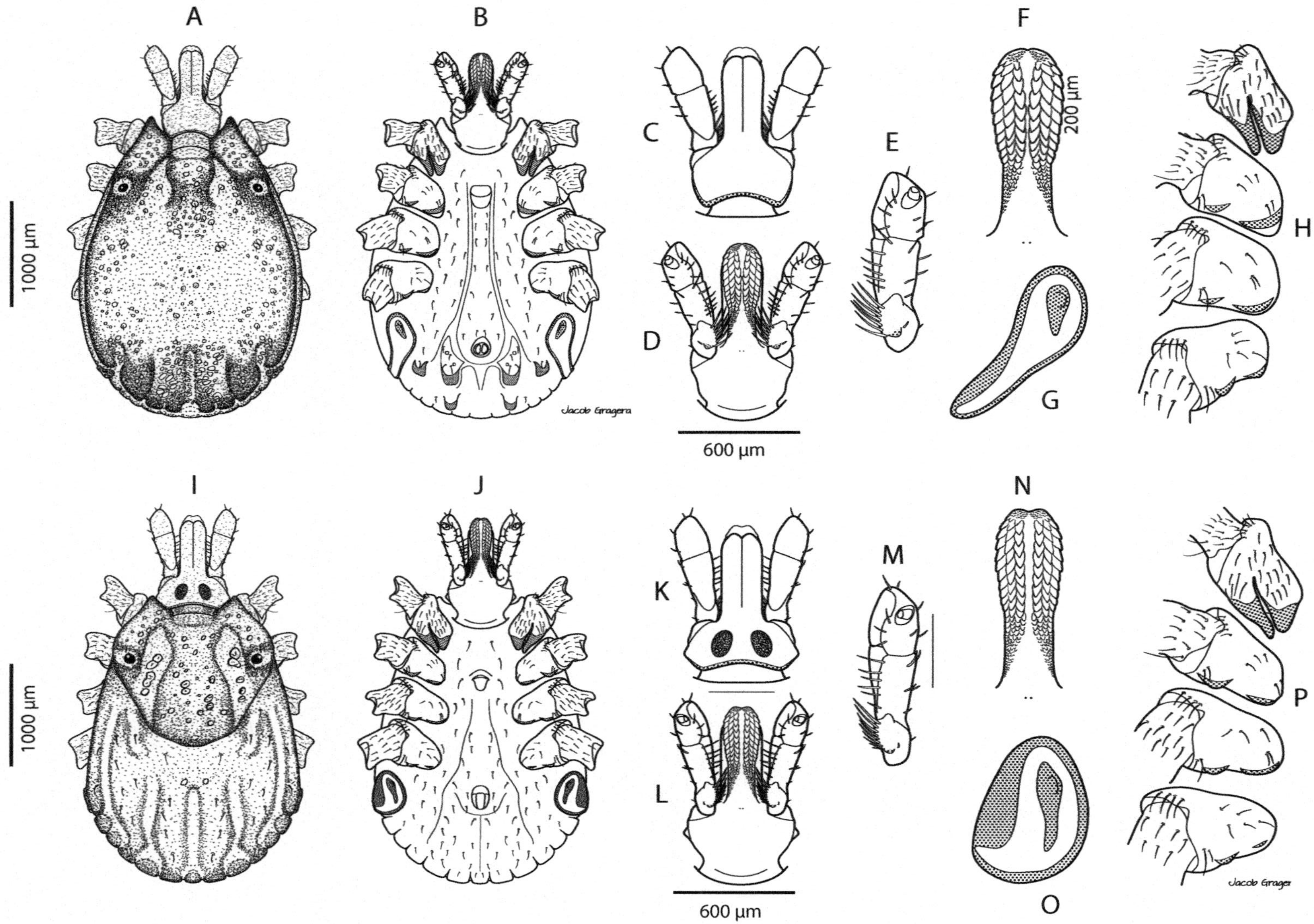

Fig. 156 **A–H** The male of *Hy. lusitanicum*. **A** Dorsal, **B** ventral, **C** capitulum, dorsal, **D** capitulum, ventral, **E** palpi, lateral view, **F** hypostome, **G** spiracular plate, **H** coxae and trochanters I–IV. **I–P** The female of *Hy. lusitanicum*. **I** Dorsal, **J** ventral, **K** capitulum, dorsal, **L** capitulum, ventral, **M** palpi, lateral view, **N** hypostome, **O** spiracular plate, **P** coxae and trochanters I–IV. Illustrations from specimens collected in Portugal and Spain

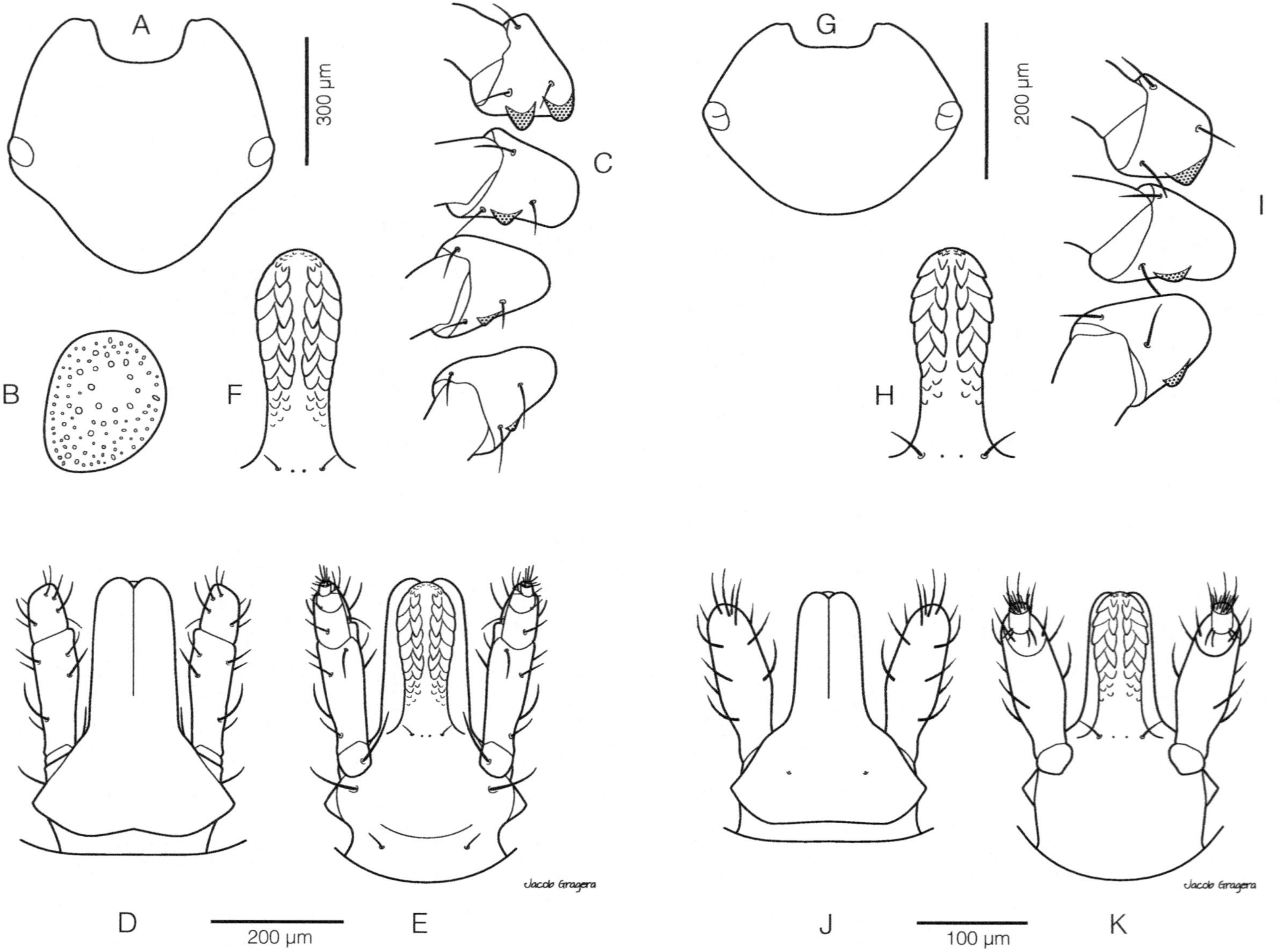

Fig. 157 **A–F** The nymph of *Hy. lusitanicum*. **A** Dorsal scutum, **B** spiracular plate, **C** coxae I–IV, **D** capitulum, dorsal, **E** capitulum, ventral, **F** hypostome. **G–K** The larva of *Hy. lusitanicum*. **G** Dorsal plate, **H** hypostome, **I** coxae I–III, **J** capitulum, dorsal, **K** capitulum, ventral. Illustrations from specimens obtained in the laboratory, from engorged females collected in Portugal and Spain

References

Apanaskevich DA, Santos-Silva MM, Horak IG (2008) The genus *Hyalomma* Koch, 1844. IV. Redescription of all parasitic stages of *H.* (*Euhyalomma*) *lusitanicum* Koch, 1844 and the adults of *H.* (*E.*) *franchinii* Tonelli Rondelli, 1932 (Acari: Ixodidae) with a first description of its immature stages. Folia Parasitol 55:61–74

Caeiro V (1999) General review of tick species present in Portugal. Parassitol 41:11–15

Derrick EH (1939) *Rickettsia burneti*: the cause of Q fever. Med J Australia 26:14

Estrada-Peña A, Farkas R, Jaenson TGT, Koenen F, Madder M, Pascucci I, Salman M, Tarrés-Call J, Jongejan F (2013) Association of environmental traits with the geographic ranges of ticks (Acari: Ixodidae) of medical and veterinary importance in the western Palaearctic. A digital data set. Exp Appl Acarol 59:351–366. https://dx.doi.org/10.1007/s10493-012-9600-7

Estrada-Peña A, Palomar AM, Santibáñez P, Sánchez N, Habela MA, Portillo A, Romero L, Oteo J (2012) Crimean-Congo hemorrhagic fever virus in ticks, southwestern Europe, 2010 [letter]. Emerg Infect Dis 18:179–180. https://dx.doi.org/10.3201/eid1801.111040

Estrada-Peña A, Bouattour A, Camicas J-L, Walker AR (2004) Ticks of domestic animals in the Mediterranean Region. Special Publication of the International Consortium on Ticks and Tick-Borne Diseases-2, Atlanta, The Netherlands Hauten

Fernández P, Hueli LE (1984) Garrapatas (Acarina, Ixodidae) parasitas del ganado bovino en el sur de España. Identificación, distribución geográfica y estacional. Rev Ibér Parasitol 44:129–138

Guglielmone AA, Robbins RG, Apanaskevich DA, Petney TN, Estrada-Peña A, Horak IG (2014) The hard ticks of the world (Acari: Ixodida: Ixodidae). Springer, Dordrecht, p 738

Mertins JW, Schlater JL (1991) Exotic ectoparasites of ostriches recently imported into the United States. J Wildl Dis 27:180–182

Milhano N, Lopes de Carvalho I, Alves AS, Arroube S, Soares J, Rodriguez P, Carolino M, Núncio MS, Piesman J, de Sousa R (2010) Coinfections of *Rickettsia slovaca* and *Rickettsia helvetica* with *Borrelia lusitaniae* in ticks collected in a Safari Park, Portugal. Ticks and Tick-borne Diseases 1 172–177

Morel PC (1971) Relations des virus d'animaux et des rickettsies avec leurs tiques vectrices. Ann Parasit Hum Comp 46:179–196

Ouhelli H, Panday VS, Benzaoui T (1985) Seasonal variation of cattle ticks in a subhumid area of Morocco. Bull Anim Hith Prod Afr 33:207–210

Ouhelli H (1994) Comparative development of Hyalomma marginatum (Koch, 1844), *H. detritum* (Schuize, 1919), *H. anatolicum excavatum* (Koch, 1844), *H. lusitanicum* (Koch, 1884) and *H. dromedarii* (Koch, 1844) under laboratory conditions. Acta Parasitol 39:153–157

Pérez-Eid C, Cabrita J (2003) La larve et la nymphe de *Hyalomma* (*Hyalomma*) *lusitanicum* Koch, 1844 (Acari: Ixodida): description morphologique, habitats, hotes. Acarologia 43:327–335

Santos-Silva MM, Beati L, Santos AS, De Sousa R, Núncio MS, Melo P, Santos-Reis M, Fonseca C, Formosinho P, Vilela C, Bacellar F (2011) The hard-tick fauna of mainland Portugal (Acari: Ixodidae): an update on geographical distribution and known associations with hosts and pathogens. Exp Appl Acarol 55:85–121. https://dx.doi.org/10.1007/s10493-011-9440-x

Santos-Silva MM, Domingues N, Melo P, Santos N, Tavares Santos P, Antunes S, Domingos A, Santos AS (2014) Surveillance for ticks and tick-borne pathogens in field and animal populations sharing Iberian lynx habitat. In: Joint 8th international ticks & tick-borne pathogens (TTP8) & Biennial Society for Tropical Veterinary Medicine (8STVM) conference, 24–29 Aug 2014, Cape Town, South Africa

Toledo A, Olmeda AS, Escudero R, Jado I, Valcárcel F, Casado-Nistal MA, Rodríguez-Vargas M, Gil H, Anda P (2009) Tick-borne zoonotic bacteria in ticks collected from central Spain. Am J Trop Med Hyg 81:67–74

Torina A, Alongi A, Scimeca S, Vicente J, Caracappa S, de la Fuente J (2010) Prevalence of tick-borne pathogens in ticks in Sicily. Transbound Emerg Dis 57:46–48. https://dx.doi.org/10.1111/j.18

Viseras J, Hueli LE, Adroher FJ, García-Fernández P (1999) Studies on the transmission of *Theileria annulata* to cattle by the tick *Hyalomma lusitanicum*. Zentralbl Veterinarmed B 46:505–509

Zapf F, Schein E (1994) New findings in the development of *Babesia* (*Theileria*) *equi* (Laveran, 1901) in the salivary glands of the vector ticks, *Hyalomma* species. Parasitol Res 80:543–548 65-1682.2010.01101.x

Hyalomma franchinii Tonelli-Rondelli, 1932

A. Estrada-Peña

Life Cycle and Host Preferences

Hyalomma franchinii is a three-host tick with one life cycle per year under natural conditions (Hoogstraal and Kaiser 1958b). The immature stages of *Hy. franchinii* parasitize lizards and, less commonly, rodents. Adult hosts are camels, cattle, sheep and donkeys. Immature stages are found on lizards all year around, having been recorded on *Acanthodactylus boskianus*, *Acanthodactylus schreiberi*, *A. scutellatus* and *Agama mutabilis* according to data summarized by Apanaskevich et al. (2008). Adults appear to be most common during the cooler months of the year. Observations in Egypt (July and August) mentioned a duration of egg development of about 25 days, and a duration of larval moult of about 14–16 days (recorded in September–October).

Ecology

Hyalomma franchinii has a Mediterranean distribution and is adapted to dry and warm environmental conditions. Its distribution area is typically desert and it has not been recorded in areas with more than 10 or 20 l of rainfall/m^2 per year. However, it can be relatively abundant near watering points. The immatures are endophilic and are commonly found in the burrows of the lizards. Some observations published by Morel (2003) state that the adults can be found at the entry point of the lizard's shelter and are exophilic.

Distribution

The known distribution of *Hy. franchinii* is restricted to the eastern part of the Mediterranean subregion of the Palaearctic zoogeographic region in Africa (Egypt, Libya, Tunisia) and Asia (Israel) (Hoogstraal 1956; Hoogstraal and Kaiser 1958a, b, 1960; Cwilich and Hadani 1962; Bouattour et al. 1999; Apanaskevich et al. 2008).

Vectorial Capacity and Pathogen Burden

The relationships between disease inducing agents and *Hy. franchinii* have not been investigated. It is common in areas where protozoa of the genus *Theileria* circulate in nature, but its role in the circulation of these agents has been never tested either under natural or laboratory conditions.

References

Apanaskevich D, Santos-Silva MM, Horak IG (2008) The genus *Hyalomma* Koch, 1844. IV. Redescription of all parasitic stages of *H. (Euhyalomma) lusitanicum* Koch, 1844 and the adults of *H. (E.) franchinii* Tonelli Rondelli, 1932 (Acari: Ixodidae) with a first description of its immature stages. Folia Parasitol 55:61–74

Bouattour A, Darghouth MA, Daoud A (1999) Distribution and ecology of ticks (Acari: Ixodidae) infesting livestock in Tunisia: an overview of eight years field collections. Parassitologia 41:5–10

Cwilich R, Hadani A (1962) The identification of the tick *Hyalomma franchinii*, Tonelli-Rondelli, 1932. Refu Vet 19:178–180

A. Estrada-Peña (✉)
Department of Animal Health, Faculty of Veterinary Medicine, Zaragoza, Spain
e-mail: aestrada@unizar.es

A. Estrada-Peña et al. (eds.), *Ticks of Europe and North Africa*,
https://doi.org/10.1007/978-3-319-63760-0_73

Hoogstraal H (1956) African Ixodoidea. I. Ticks of the Sudan (with special reference to Equatoria Province and with preliminary reviews of the genera *Boophilus*, *Margaropus*, and *Hyalomma*). Department of the Navy, Washington DC, 1101 pp

Hoogstraal H, Kaiser MN (1958a) Observations on Egyptian *Hyalomma* ticks (Ixodoidea, Ixodidae). I. Parasitism of lizards by nymphs. Ann Entomol Soc Am 51:7–12

Hoogstraal H, Kaiser MN (1958b) Observations on Egyptian *Hyalomma* ticks (Ixodoidea, Ixodidae). 4. Identity, distribution, and hosts of *H. franchinii* Tonelli-Rondelli (new combination). Systematic status of *H. tunesiacum* Sc. & Sc. and its subspecies. Ann Entomol Soc Am 51:397–400

Hoogstraal H, Kaiser MN (1960) Observations on ticks (Ixodoidea) of Libya. Ann Entomol Soc Am 53:445–457

Morel PC (2003) Les tiques d'Afrique et du Bassin méditerranéen (1965–1995). CIRAD-EMVT, 1342 pp

Hyalomma anatolicum Koch, 1844 (Figs. 158–160)

Z. Vatansever

Life Cycle and Host Preferences

Hyalomma anatolicum is a two- or three-host endophilic tick. Under natural conditions it has a monotropic life cycle, with hosts for all developmental stages being large ungulates, especially cattle, then horses, camels, sheep and goats. When feeding on its natural hosts (e.g. cattle), *Hy. anatolicum* always develops as a three-host species. Under laboratory conditions, it shows a three-host pattern when fed on gerbils and a two-host pattern when fed on rabbits (Serdyukova 1946; Hoogstraal et al. 1981; Apanaskevich 2004; Estrada-Peña et al. 2004; Ghosh and Azhahianambi 2007; Bakheit et al. 2012). Adult ticks are reported to infest humans (Jongejan and Uilenberg 2004; Karaer et al. 2011). Under field conditions, the whole life cycle is completed in one year. Preoviposition, oviposition and larval hatching take 7–9, 15–19 and 23–26 days, respectively. The feeding period lasts 2–4 days in larvae and 4–6 days in nymphs. Moulting takes 6–12 and 12–20 days in larvae and nymphs, respectively (Pomerantzev 1950). Under laboratory conditions with larvae fed on rabbits and adults on calves, *Hy. anatolicum* acts as a two-host tick. Preoviposition and oviposition take 6.4 and 21.8 days on average, respectively. Engorged nymphs drop-off after 13–18 days of feeding and moult to adults after an average of 14 days. Adults feed for an average of 10.5 days when attached to calves (Ghosh and Azhahianambi 2007).

Ecology

Hyalomma anatolicum is adapted to dry Mediterranean, semi-desert, steppe and savanna climates (Hoogstraal et al. 1981; Estrada-Peña et al. 2004). It is a dominant tick species, infesting cattle in river valleys, deltas and cultivated belts (Pomerantzev 1950; Hoogstraal and Kaiser 1959). It is well known for its endophilic character. Due to the nocturnal drop-off rhythm of nymphs and females, it is found in very large numbers in cracks and crevices in buildings housing livestock and their vicinity (Pomerantzev 1950; Kuklina 1964). It can often be found in cattle herds kept under zero grazing regimens (Estrada-Peña et al. 2004). *Hyalomma anatolicum* is active during the summer months. Adults parasitize cattle from March to September. Larvae are found between July and September, while nymphs can be found between July and October. It overwinters as unfed adults, as well as fed females and nymphs (Pomerantzev 1950).

Distribution

Hyalomma anatolicum is one of the most widely distributed tick species. Its range covers Algeria, Djiboti, Egypt, Ethiopia, Libya, Morocco, Somalia, Sudan and Tunisia in Africa and Afghanistan, Armenia, Azerbaijan, Bangladesh, India, Iran, Iraq, Israel, Jordan, Kazakhstan, Kirghizstan, Lebanon, Nepal, Oman, Pakistan, Dagestan, Saudi Arabia, Syria, Tajikistan, Turkey, Turkmenistan, the United Arab Emirates, Uzbekistan and Yemen in Asia (Pomerantzev 1950; Hoogstraal et al. 1981; Estrada-Peña et al. 2004).

Vectorial Capacity and Pathogen Burden

Hyalomma anatolicum is the main vector of *Theileria annulata*, *T. lestoquardi*, *T. equi* and *Babesia caballi* (Bakheit et al. 2012; Estrada-Peña et al. 2004). It is also accepted as a vector of Crimean-Congo haemorrhagic fever virus in Armenia, Turkmenistan, Uzbekistan, Tajikistan, Pakistan, Iran and Nigeria (Hoogstraal 1979; Telmadarraiy et al. 2015).

Z. Vatansever (✉)
Department of Parasitology, Faculty of Veterinary Medicine, Kafkas University, Kars, Turkey
e-mail: zativet@gmail.com

A. Estrada-Peña et al. (eds.), *Ticks of Europe and North Africa*,
https://doi.org/10.1007/978-3-319-63760-0_74

Fig. 158 **A** Distribution of *Hy. anatolicum* in Europe and Northern Africa (10 × 10 km grid presence with black dots). **B** Countries where the species has been reported are marked in dark grey

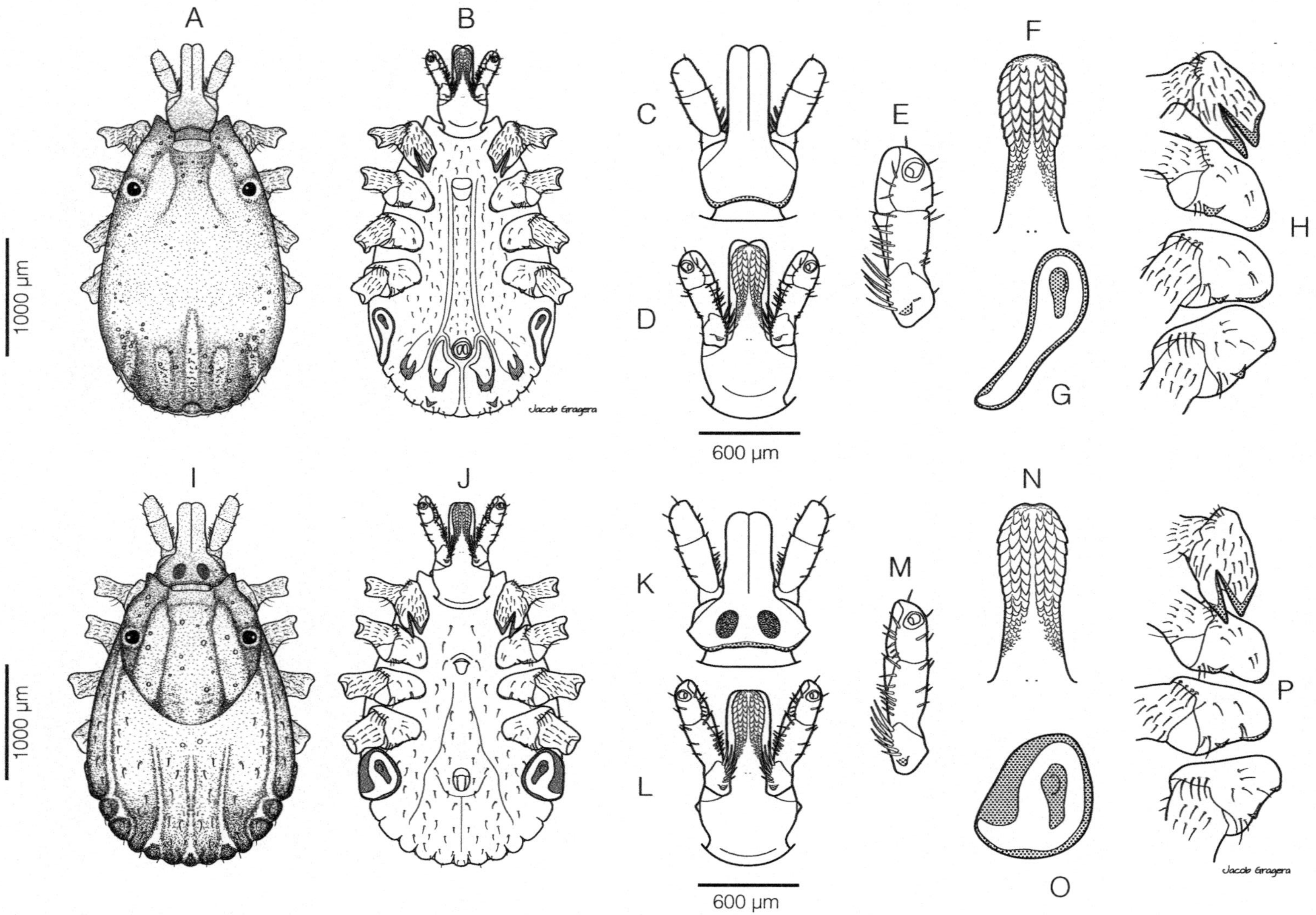

Fig. 159 **A–H** The male of *Hy. anatolicum*. **A** Dorsal, **B** ventral, **C** capitulum, dorsal, **D** capitulum, ventral, **E** palpi, lateral view, **F** hypostome, **G** spiracular plate, **H** coxae and trochanters I–IV. **I–P** The female of *Hy. anatolicum*. **I** Dorsal, **J** ventral, **K** capitulum, dorsal, **L** capitulum, ventral, **M** palpi, lateral view, **N** hypostome, **O** spiracular plate, **P** coxae and trochanters I–IV. Illustrations from specimens collected in Turkey

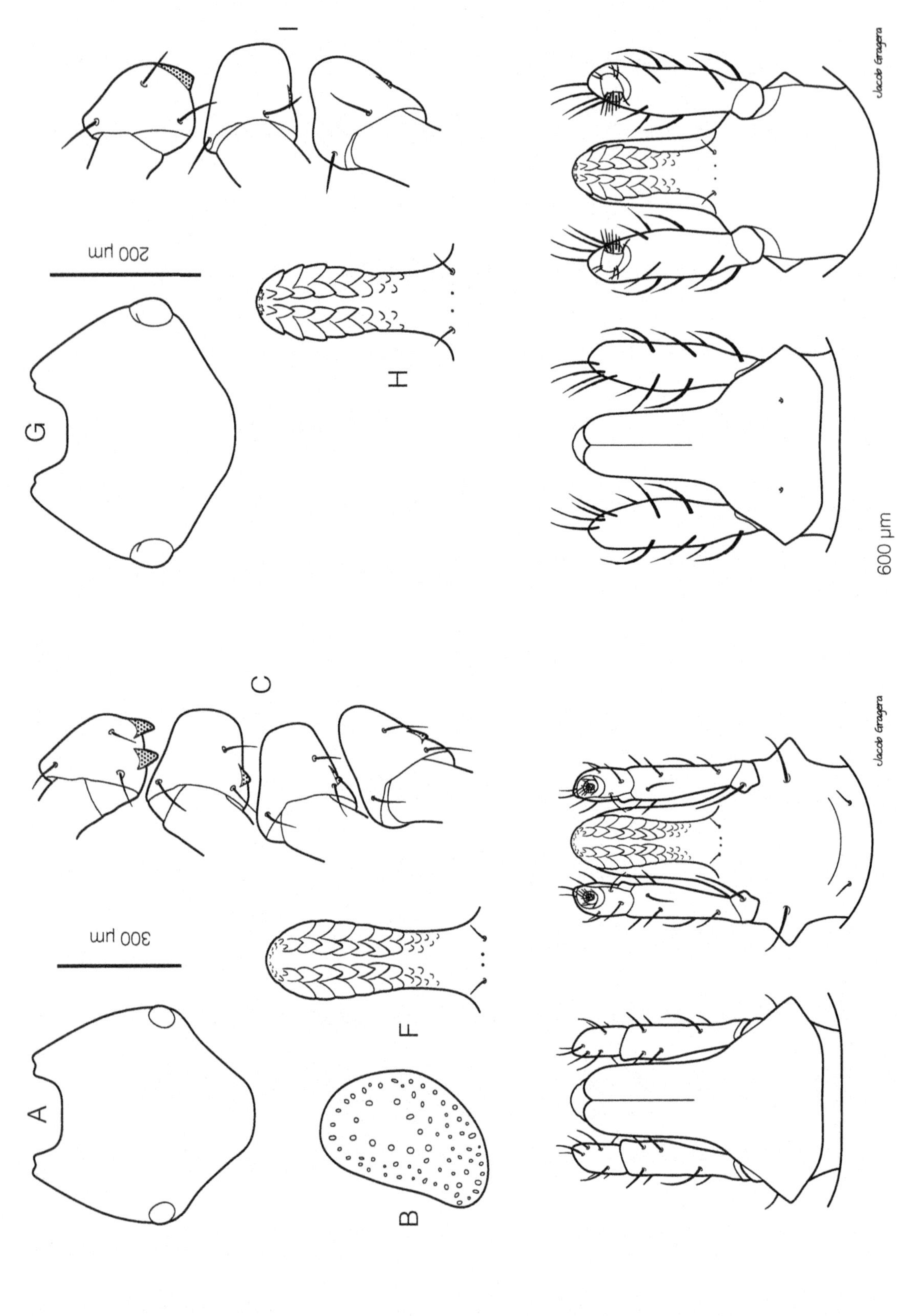

Fig. 160 **A–F** The nymph of *Hy. anatolicum*. **A** Dorsal scutum, **B** spiracular plate, **C** coxae I–IV, **D** capitulum, dorsal, **E** capitulum, ventral, **F** hypostome. **G–K** The larva of *Hy. anatolicum*. **G** Dorsal plate, **H** hypostome, **I** coxae I–III, **J** capitulum, dorsal, **K** capitulum, ventral. Illustrations from specimens obtained in the laboratory, from engorged females collected in Turkey

References

Apanaskevich DA (2004) Host-parasite relationships of the genus *Hyalomma* Koch, 1844 (Acari, Ixodidae) and their connection with microevolutionary process. Parazitologiia 38:515–523

Bakheit MA, Latif A, Vatansever Z, Seitzer U, Ahmed J (2012) The huge risks due to *Hyalomma* ticks. In: Mehlhorn H (ed) Arthropods as vectors of emerging diseases. Springer, Berlin, pp 167–194

Estrada-Peña A, Bouattour A, Camicas JL, Walker AR (2004) Ticks of domestic animals in the Mediterranean region. A guide to identification of species. University of Zaragoza (Spain)

Ghosh S, Azhahianambi P (2007) Laboratory rearing of *Theileria annulata*-free *Hyalomma anatolicum anatolicum* ticks. Exp Appl Acarol 43:137–146

Hoogstraal H (1979) The epidemiology of tick-borne Crimean-Congo hemorrhagic fever in Asia, Europe, and Africa. J Med Entomol 15:307–417

Hoogstraal H, Kaiser MN (1959) Observations on Egyptian *Hyalomma* ticks (Ixodoidea, Ixodidae). 5. Biological notes and differences in identity of *H. anatolicum* and its subspecies *anatolicum* Koch and *excavatum* Koch among Russian and other workers. Identity of *H. lusitanicum* Koch. Ann Entomol Soc Am 52:243–261

Hoogstraal H, Wassef HY, Büttiker W (1981) Ticks (Acarina) of Saudi Arabia. Fam. Argasidae, Ixodidae. Fauna of Saudi Arabia 3:25–110

Jongejan F, Uilenberg G (2004) The global importance of ticks. Parasitology 129(Suppl):S3–14

Karaer Z, Guven E, Nalbantoglu S, Kar S, Orkun O, Ekdal K, Kocak A, Akcay A (2011) Ticks on humans in Ankara, Turkey. Exp Appl Acarol 54:85–91

Kuklina TE (1964) Duration of feeding and behaviour of engorged *Hyalomma anatolicum* Koch females. Uzbek Biol Zh 8:75–78. Translated from Russian, Translation 312 (T312) US NAMRU 313, Cairo

Pomerantzev BI (1950) Ixodid ticks (Ixodidae). In: Fauna of the USSR, New series 41: Arachnoidea, vol 4(2). Leningrad, Nauk, p 224 (in Russian)

Serdyukova GV (1946) On the developmental cycle of the tick *Hyalomma anatolicum anatolicum*, Koch. Izvestiya Akademii Nauk SSSR seriya Biologicheskaya, 199–202. Translation from Russian, Translation number 236 (T236) US NAMRU No 193, Cairo

Telmadarraiy Z, Chinikar S, Vatandoost H, Faghihi F, Hosseini-Chegeni A (2015) Vectors of Crimean Congo hemorrhagic fever virus in Iran. J Arthropod Borne Dis 9:137–147

Hyalomma excavatum Koch, 1844 (Figs. 161–163)

Z. Vatansever

Life Cycle and Host Preferences

Hyalomma excavatum is a two-host or three-host exophilic tick. Adult ticks prefer mainly cattle and camels, but also frequently infest sheep, goats, horses and donkeys (Hoogstraal et al. 1981; Apanaskevich 2004; Estrada-Peña et al. 2004; Bakheit et al. 2012). Although controversial (Berdyev 1974), the immature stages are reported to prefer rodents, leporids and insectivores as hosts (Hoogstraal et al. 1981; Apanaskevich 2004; Estrada-Peña et al. 2004). Human infestations with adult ticks are also frequently reported (Karaer et al. 2011; Kar et al. 2013; Keskin et al. 2015). Under laboratory conditions, *Hy. excavatum* acts as a three-host species when larvae feed on gerbils (Ouhelli 1994; Bakirci et al. 2015), and as a two-host or three-host species when larvae feed on rabbits and calves (Hoogstraal and Kaiser 1959; Ouhelli 1994; Yukari 1993). The whole life cycle takes 113–161 days when it is fed on calves. Female ticks engorge in 8–12 days. Preoviposition, oviposition and larval hatching take 14–22, 28–38 and 48 days, respectively. The period from larval feeding to nymphal engorgement is 17–22 days. Engorged nymphs moult to adults in 18 days (Ouhelli 1994). When fed on rabbits, half of the population act as two-host and the other half as three-host ticks. The life cycle takes 80–179 days in one-host and 83–190 days in two-host populations (Yukari 1993). Under laboratory conditions of 12 °C and 85% relative humidity, flat larvae, nymphs and adults can survive up to 4, 6 and 8 months, respectively (Bakirci et al. 2015).

Ecology

The distribution, prevalence and population density *Hy. excavatum* depends on the availability of burrowing rodents and other small mammals, which its immature stages prefer as hosts (Hoogstraal 1956; Hoogstraal et al. 1981). It is common in steppe and semi-desert areas and oases, where it coexists and interbreeds to some degree with large populations of *Hy. anatolicum* (Hoogstraal 1956). In contrast to *Hy. anatolicum*, it does not occur in large river valleys and deltas, or at high altitudes (Hoogstraal et al. 1981). In Israel, the adults infest animals between April and November, while larvae are found in summer (Yeruham et al. 1996). Adults parasitize animals in all months of the year in North Africa and Turkey, being most numerous in October in Egypt (Hoogstraal 1956) and in the summer months in Turkey (Dumanli 1983).

Distribution

The distribution of *Hy. excavatum* covers Morocco, Algeria, Libya, Egypt, Sudan, Djibouti, Somalia, Albania, Greece, Cyprus, Turkey, Jordan, Lebanon, Syria, Iraq, Iran, Kazakhstan, Turkmenistan, Uzbekistan, Tajikistan and Afghanistan (Apanaskevich and Horak 2005; Hoogstraal et al. 1981).

Vectorial Capacity and Pathogen Burden

Hyalomma excavatum is well known for its ability to transmit *Theileria annulata* under laboratory conditions (Sayin et al. 2003), but since its immature stages prefer small mammals, the reports on its vectorial influence under field conditions should be interpreted carefully. *Borrelia burgdorferi* s.s., *Rickettsia africae*, *R. aeschlimannii* and *R. sibirica mongolotimonae* (Orkun et al. 2014a, b; Parola et al. 2013; Psaroulaki et al. 2005) have been detected in *Hy. excavatum* feeding on humans.

Z. Vatansever (✉)
Department of Parasitology, Faculty of Veterinary Medicine, Kafkas University, Kars, Turkey
e-mail: zativet@gmail.com

A. Estrada-Peña et al. (eds.), *Ticks of Europe and North Africa*,
https://doi.org/10.1007/978-3-319-63760-0_75

Fig. 161 **A** Distribution of *Hy. excavatum* in Europe and Northern Africa (10 × 10 km grid presence with black dots). **B** Countries where the species has been reported are marked in dark grey

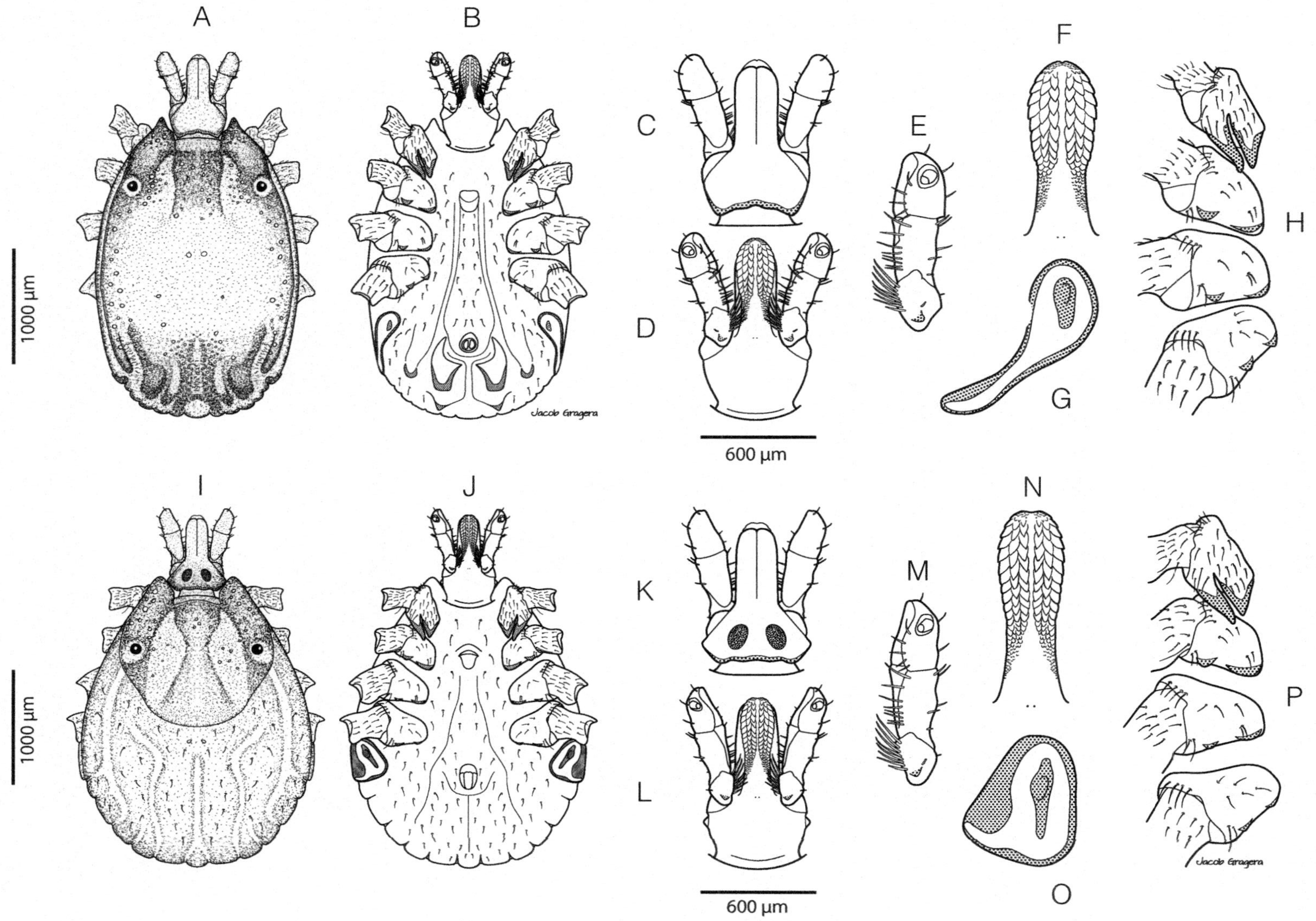

Fig. 162 **A–H** The male of *Hy. excavatum*. **A** Dorsal, **B** ventral, **C** capitulum, dorsal, **D** capitulum, ventral, **E** palpi, lateral view, **F** hypostome, **G** spiracular plate, **H** coxae and trochanters I–IV. **I–P** The female of *Hy. excavatum*. **I** Dorsal, **J** ventral, **K** capitulum, dorsal, **L** capitulum, ventral, **M** palpi, lateral view, **N** hypostome, **O** spiracular plate, **P** coxae and trochanters I–IV. Illustrations from specimens collected in Turkey

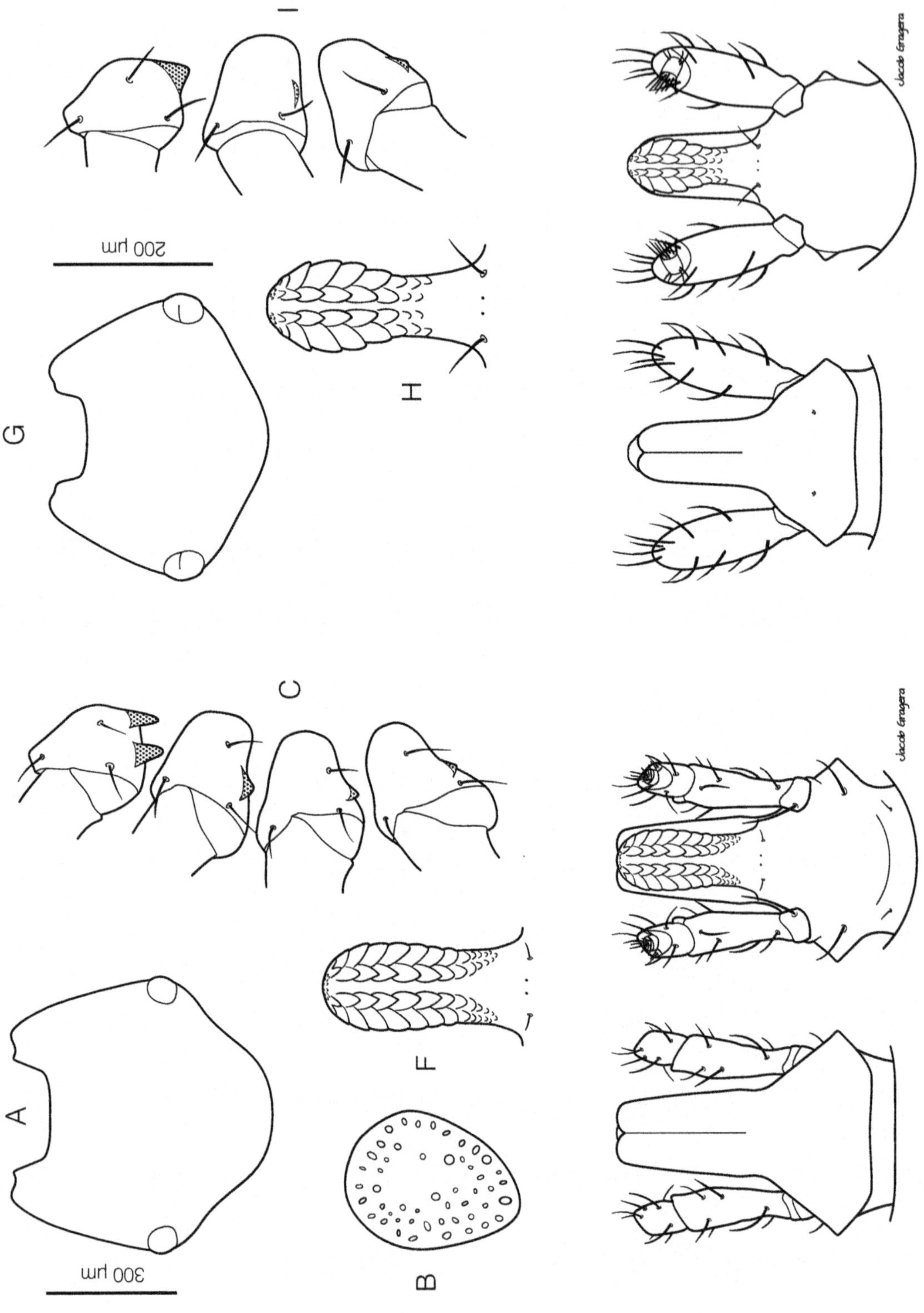

Fig. 163 **A–F** The nymph of *Hy. excavatum*. **A** Dorsal scutum, **B** spiracular plate, **C** coxae I–IV, **D** capitulum, dorsal, **E** capitulum, ventral, **F** hypostome. **G–K** The larva of *Hy. excavatum*. **G** Dorsal plate, **H** hypostome, **I** coxae I–III, **J** capitulum, dorsal, **K** capitulum, ventral. Illustrations from specimens obtained in the laboratory, from engorged females collected in Turkey

References

Apanaskevich DA (2004) Host-parasite relationships of the genus *Hyalomma* Koch, 1844 (Acari, Ixodidae) and their connection with microevolutionary process. Parazitologiia 38:515–523

Apanaskevich DA, Horak IG (2005) The genus *Hyalomma* Koch, 1844. II. Taxonomic status of *H.* (*Euhyalomma*) *anatolicum* Koch, 1844 and *H.* (*E.*) *excavatum* Koch, 1844 (Acari: Ixodidae) with redescriptions of all stages. Acarina 13:181–197

Bakheit MA, Latif A, Vatansever Z, Seitzer U, Ahmed J (2012) The Huge risks due to *Hyalomma* ticks. In: Mehlhorn H (ed) Arthropods as vectors of emerging diseases. Springer, Berlin, Heidelberg, pp 167–194

Bakirci S, Bilgic HB, Hacilarlioglu S, Unlu AH, Kose O, Aksulu A, Eren H, Karagenc T (2015) The breeding and colonization of *Hyalomma excavatum* under laboratory conditions (in Turkish). Ankara Üniversitesi Veteriner Fakültesi Dergisi 62:99–104

Berdyev A (1974) Cycle of development of *Hyalomma anatolicom excavatum* Koch. (Parasitiformes, Ixodidae) v Turkmenii. Med Parazitol (Mosk) 43:34–43

Dumanli N (1983) Observations on the bio-ecology of *Hyalomma excavatum* around Elazig (in Turkish). TUBITAK Doga Bilimleri Dergisi 7:23–31

Estrada-Peña A, Bouattour A, Camicas JL, Walker AR (2004) Ticks of domestic animals in the mediterranean region. A guide to identification of species. University of Zaragoza (Spain)

Hoogstraal H (1956) African Ixodoidea. I. Ticks of the Sudan (with special reference to Equatoria province and with preliminary reviews of the genera *Boophilus*, *Margaropus*, and *Hyalomma*). Department of the Navy, Washington D.C

Hoogstraal H, Kaiser MN (1959) Observations on Egyptian *Hyalomma* ticks (Ixodoidea, Ixodidae). 5. Biological notes and differences in identity of *H. anatolicum* and its subspecies anatolicum Koch and *excavatum* Koch among Russian and other workers. Identity of *H. lusitanicum* Koch. Ann Entomol Soc Am 52:243–261

Hoogstraal H, Wassef HY, Büttiker W (1981) Ticks (Acarina) of Saudi Arabia. Fam. Argasidae. Ixodidae. Fauna of Saudi Arabia 3:25–110

Kar S, Dervis E, Akin A, Ergonul O, Gargili A (2013) Preferences of different tick species for human hosts in Turkey. Exp Appl Acarol 61:349–355

Karaer Z, Guven E, Nalbantoglu S, Kar S, Orkun O, Ekdal K, Kocak A, Akcay A (2011) Ticks on humans in Ankara, Turkey. Exp Appl Acarol 54:85–91

Keskin A, Keskin A, Bursali A, Tekin S (2015) Ticks (Acari: Ixodida) parasitizing humans in Corum and Yozgat provinces, Turkey. Exp Appl Acarol 67:607–616

Orkun O, Karaer Z, Cakmak A, Nalbantoglu S (2014a) Identification of tick-borne pathogens in ticks feeding on humans in Turkey. PLoS Negl Trop Dis 8:e3067

Orkun O, Karaer Z, Cakmak A, Nalbantoglu S (2014b) Spotted fever group rickettsiae in ticks in Turkey. Ticks Tick Borne Dis 5:213–218

Ouhelli H (1994) Comparative development of *Hyalomma marginatum* (Koch, 1844), *H. detritum* (Schulze, 1919), *H. anatolicum excavatum* (Koch, 1844), *H. lusitanicum* (Koch, 1844) and *H. dromedarii* (Koch, 1844) under laboratory conditions. Acta Parasitol 39:153–157

Parola P, Paddock CD, Socolovschi C, Labruna MB, Mediannikov O, Kernif T, Abdad MY, Stenos J, Bitam I, Fournier PE, Raoult D (2013) Update on tick-borne rickettsioses around the world: a geographic approach. Clin Microbiol Rev 26:657–702

Psaroulaki A, Germanakis A, Gikas A, Scoulica E, Tselentis Y (2005) Simultaneous detection of "*Rickettsia mongolotimonae*" in a patient and in a tick in Greece. J Clin Microbiol 43:3558–3559

Sayin F, Karaer Z, Dincer S, Cakmak A, Inci A, Yukari BA, Eren H, Vatansever Z, Nalbantoglu S, Melrose TR (2003) A comparison of susceptibilities to infection of four species of *Hyalomma* ticks with *Theileria annulata*. Vet Parasitol 113:115–121

Yeruham I, Hadani A, Galker F, Rosen S (1996) The seasonal occurrence of ticks (Acari: Ixodidae) on sheep and in the field in the Judean area of Israel. Exp Appl Acarol 20:47–56

Yukari BA (1993) Research on the establishment and maintenance of a laboratory *Hyalomma anatolicum excavatum* colony. Ankara Üniversitesi Veteriner Fakültesi Dergisi 40:99–114

Hyalomma asiaticum Schülze and Schlottke, 1929

Z. Vatansever

Life Cycle and Host Preferences

Hyalomma asiaticum is a exophilic three-host tick. Adults feed mainly on a relatively narrow range of artiodactyl animals, especially cattle, camels and sheep. They have also been reported from humans. In contrast, larvae and nymphs have a wide host range reaching up to 50 species, mainly consisting of small mammals, mostly burrowing rodents, leporids and hedgehogs. All developmental stages can be found accidentally on other hosts (Pomerantzev 1950; Berdyev 1980; Apanaskevich 2004; Apanaskevich and Horak 2010). Females of *Hy. asiaticum* feed for about 5–8 days. Depending on the climatic conditions, the period for pre-oviposition, oviposition and larval hatching takes 4–59, 6–41 and 19–24 days, respectively. Larvae feed for 3–7 days and moult to nymphs in 6–15 days after drop-off. Nymphs feed for 4–8 days and moult to adults in 8–18 days after drop-off (Berdyev 1972; Galuzo 1947). The whole life cycle takes 103–132 days under laboratory conditions and can be as short as 90 days (from fed female to fed female) in nature (Berdyev 1972). Fed female *Hy. asiaticum* produce an average of 14,500 eggs (121 eggs per 10 mg body weight), reaching a maximum of 26,000 under experimental conditions (Berdyev and Khudainazarova 1976), which makes it one of the tick species taking the largest blood meal known.

Ecology

Hyalomma asiaticum is well adapted to the desert, semi-desert and steppe habitats of Asia, where it is the dominant tick species. Although accepted as a non-nidiculous species, all developmental stages are strongly associated with rodent burrows and their surroundings (Galuzo 1947; Pomerantzev 1950; Balashov 1998). In the climatic conditions of Turkmenistan, adults infest livestock in two waves between March and October, the first being March–May, and the second September–October (Balashov 1998; Berdyev 1980). Engorged females, which drop-off of their hosts before the end of September, oviposit in the same year, while females that drop-off after the end of September overwinter in that state and oviposit at the end of April or the beginning of May the following year (Berdyev 1972). Engorged females tend to move into rodent burrows for egg lying, as determined by their leaving typical tracks in the sand (Pomerantzev 1950; Balashov 1998). The larvae hatch in the burrows and feed on large burrowing rodents and small mammals from March to September, with a peak in June and July. Nymphs feed in the same place and group of hosts, from February to November, with a peak in August and September (Berdyev 1980; Balashov 1998). Depending on the microclimatic conditions of rodent burrows, flat adults may survive for up to 320 days, while nymphs and larvae survive for several months (Galuzo 1947; Berdyev 1972). All developmental stages of *Hy. asiaticum* can overwinter, but this is most marked in both flat and engorged adults and nymphs, as well as flat larvae. Thus, in one year both the larvae and adults of the first and second generations can be observed together (Berdyev 1972; Balashov 1998). In Turkmenistan, adults of *Hy. asiaticum* quest on the soil surface during the morning and evening when temperatures do not exceed 40–43 °C. When the soil surface reaches 60–70 °C at midday, they prefer to hide in burrows or cracks where temperatures are 30–35 °C (Balashov 1998). Host-seeking adults are well known for their 'hunter' host finding strategies and can visually recognize hosts from up to 9 m away. They move very fast and can follow the host for 10 min or more, covering a distance of up to 100 m (Romanenko 2005, 2007).

Z. Vatansever (✉)
Department of Parasitology, Faculty of Veterinary Medicine,
Kafkas University, Kars, Turkey
e-mail: zativet@gmail.com

A. Estrada-Peña et al. (eds.), *Ticks of Europe and North Africa*,
https://doi.org/10.1007/978-3-319-63760-0_76

Distribution

Hyalomma asiaticum is widely distributed in desert type habitats in Asia covering Afghanistan, Armenia, Azerbaijan, China, Georgia, Iran, Iraq, Kazakhstan, Kyrgyzstan, Mongolia, Pakistan, Russia (Dagestan), Syria, Turkey, Tajikistan, Turkmenistan and Uzbekistan (Apanaskevich and Horak 2010; Pomerantzev 1950).

Vectorial Capacity and Pathogen Burden

Hyalomma asiaticum is a possible vector of Crimean-Congo haemorrhagic fever in Uzbekistan, Xinjiang (China) and Iran (Hoogstraal 1979; Sun et al. 2009; Kayedi et al. 2015; Telmadarraiy et al. 2015). Tamdy (Lvov et al. 2014) and Wad Medani (Alkhovskii et al. 2014) viruses have also been isolated from *Hy. asiaticum*. It transmits *Coxiella burnetti* (causing Q fever) and *Rickettsia siberica* (causing Siberian tick typhus) (Berdyev 1980).

References

Alkhovskii SV, Lvov DK, Shchelkanov M, Shchetinin AM, Deriabin PG, Gitel'man AK, Aristova VA, Botikov AG (2014) [Genetic characterization of the Wad Medani virus (WMV) (*Reoviridae, Orbivirus*), isolated from the ticks *Hyalomma asiaticum* Schulze et Schlottke, 1930 (Ixodidae: Hyalomminae) in Turkmenistan, Kazakhstan, and Armenia and from the ticks *H. anatolicum* Koch, 1844 in Tajikistan]. Vop Virusol 59:25–30

Apanaskevich DA (2004) Host-parasite relationships of the genus *Hyalomma* Koch, 1844 (Acari, Ixodidae) and their connection with microevolutionary process. Parazitologiia 38:515–523

Apanaskevich DA, Horak IG (2010) The genus *Hyalomma*. XI. Redescription of all parasitic stages of *H.* (*Euhyalomma*) *asiaticum* (Acari: Ixodidae) and notes on its biology. Exp Appl Acarol 52:207–220

Balashov YS (1998) Ixodid ticks—parasites and vectors of diseases. Nauka, Sankt Petersburg, p 287 (in Russian)

Berdyev A (1972) Duration of the developmental cycle of *Hyalomma asiaticum asiaticum* Schulze et Schlottke in natural biotopes. Parazitologiia 6:30–34

Berdyev A (1980) Ecology of ixodid ticks of Turkmenistan and their importance in natural focal disease epizootiology. Ylym, Ashgabat, Turkmenistan, p 284 (in Russian)

Berdyev A, Khudainazarova SN (1976) Characteristics of acquired resistance of lambs to mature ticks *Hyalomma asiaticum asiaticum* (experimental studies). Parazitologiia 10:519–525

Galuzo IG (1947) Bloodsucking ticks of Kazakh SSR. In: Genus *Hyalomma* (Alma Ata, Inst. Zool. Akad. Nauk Kazakh SSR), vol 2, p 280 (in Russian)

Hoogstraal H (1979) The epidemiology of tick-borne Crimean-Congo hemorrhagic fever in Asia, Europe, and Africa. J Med Entomol 15:307–417

Kayedi MH, Chinikar S, Mostafavi E, Khakifirouz S, Jalali T, Hosseini-Chegeni A, Naghizadeh A, Niedrig M, Fooks AR, Shahhosseini N (2015) Crimean-Congo hemorrhagic fever virus clade IV (Asia 1) in ticks of Western Iran. J Med Entomol 52:1144–1149

Lvov DK, Alkhovskii SV, Shchelkanov M, Shchetinin AM, Aristova VA, Gitel'man AK, Deriabin PG, Botikov AG (2014) [Taxonomy of previously unclassified Tamdy virus (TAMV) (*Bunyaviridae, Nairovirus*) isolated from the *Hyalomma asiaticum asiaticum* Schulce et Schlottke, 1929 (Ixodidae, Hyalomminae) in the Middle East and transcaucasia]. Vop Virusol 59:15–22

Pomerantzev BI (1950) Ixodid ticks (Ixodidae). In: Fauna of the USSR, New series 41: Arachnoidea, vol 4(2). Leningrad, Nauk, p 224 (in Russian)

Romanenko VN (2005) Visual potentialities of the tick *Hyalomma asiaticum asiaticum* (Ixodidae). Parazitologiia 39:186–190

Romanenko VN (2007) Ecological bases of ethology of pasture ticks (Parasitiformes, Ixodidae) in finding and attacking hosts. Bull Tomsk State Univ, pp 224–228 (in Russian)

Sun S, Dai X, Aishan M, Wang X, Meng W, Feng C, Zhang F, Hang C, Hu Z, Zhang Y (2009) Epidemiology and phylogenetic analysis of crimean-congo hemorrhagic fever viruses in Xinjiang, China. J Clin Microbiol 47:2536–2543

Telmadarraiy Z, Chinikar S, Vatandoost H, Faghihi F, Hosseini-Chegeni A (2015) Vectors of Crimean-Congo hemorrhagic fever virus in Iran. J Arthropod Borne Dis 9:137–147

The manufacturer's authorised representative in the EU is Springer Nature Customer Service Centre GmbH, Europaplatz 3, 69115 Heidelberg, Germany. If you have any concerns regarding our products, please contact ProductSafety@springernature.com

Printed and bound by CPI Group (UK) Ltd, Croydon, CR0 4YY
01/10/2026
02225991-0003